Lecture Notes in Mathematics

Edited by A. Dold and B. Eckmann
Series: Mathematisches Institut der Universität Bonn
Adviser: F. Hirzebruch

570

Differential Geometrical Methods in Mathematical Physics

Proceedings of the Symposium
Held at the University of Bonn,
July 1–4, 1975

Edited by K. Bleuler und A. Reetz

Springer-Verlag
Berlin · Heidelberg · New York 1977

Editors
Konrad Bleuler
Institut für Theoretische Kernphysik
der Universität Bonn
Nussallee 14–16
5300 Bonn/BRD

Axel Reetz
Institut für Theoretische Kernphysik
der Universität Bonn
Nussallee 14–16
5300 Bonn/BRD

Library of Congress Cataloging in Publication Data

Main entry under title:

Differential geometrical methods in mathematical physics.

(Lecture notes in mathematics ; 570)
1. Geometry, Differential--Congresses.
2. Mathematical physics--Congresses. I. Bleuler, Konrad, 1912- II. Reetz, Axel, 1937- III. Bonn. Universität. IV. Series: Lecture notes in mathematics (Berlin) ; 570.
QA3.L28 no. 570 [QC20.7.D52] 510'.8s
[530.1'5'636] 77-406

AMS Subject Classifications (1970): 17A30, 53-02, 53AXX, 53BXX, 53CXX, 53C50, 55FXX, 57DXX, 57D15, 58AXX, 58F05, 70HXX, 81-02, 83CXX

ISBN 3-540-08068-6 Springer-Verlag Berlin · Heidelberg · New York
ISBN 0-387-08068-6 Springer-Verlag New York · Heidelberg · Berlin

Printed in Germany
Printing and binding: Beltz Offsetdruck, Hemsbach/Bergstr.

P R E F A C E

It is of greatest scientifical importance that physics and mathematics met again during the past years after a relatively long period of separate development. Remembering some remarks of Prof.Rolf Nevanlinna, made during one of our earlier meetings, one might state that the great progress of pure mathematics during the last 100 years, as e.g. the introduction of Riemann's geometry, the general concept of manifolds, the theory of Hilbert spaces, the abstract theory of groups together with various algebraical systems and finally topological viewpoints, has fundamentally transformed the whole of modern physics. These mathematical structures helped to clarify and to unify different empirically known but more or less isolated facts and led through such a deeper insight under certain circumstances even to the discovery of new empirical phenomena. On the other hand, experimental results and phenomenological theories gave all through the history of mathematics surprising and important impulses for building up new mathematical structures. One might thus say that there exists a kind of a deep-lying coincidence between the phenomena in the empirical world and the realm of abstract theoretical ideas. It should also be realized that several mathematical theories which, starting originally from certain domains of practical and empirical facts, may finally be, after a logical generalization, emancipated from the fortuitousness of everyday experience projected back into the world of empirical phenomena and leading thus to some new and unforeseen insight. As an example one might think on the elementary Euclidian geometry with all its developments to higher dimensions and to differential geometry with its various metrizations - definite, undefinite or symplectic forms - i.e. in all kinds of structures which by now dominate large parts of modern physics.

Our conference program is thus entirely devoted to an exchange between physics and mathematics in which differential geometric viewpoints play a predominant role. In this respect the Kostant-Souriau Theory - the so-called Geometric Quantization - represents an outstanding example for a fruitful collaboration between the two domains: Starting originally from a new differential geometric representation theory of Lie groups it leads to an intuitive mathematical interpretation of quantum mechanics which, in turn, covers, by now, an enormous domain of experimental facts. Chapter I, which is devoted to this theory, begins with a general outline of this method by D. Simms. Detailed considerations about the metalinear structure, appearing in this context, as well as a generalization of the state space by cohomology groups are then presented by R.J. Blattner and J. Sniatycki, respectively. In the next papers K. Gawedzki and E. Onofri deal with applications, whereas J. Kijowski discusses a slight modification of Kostant-Souriau's theory with respect to the interpretation of quantum states. Another new and intuitive interpretation of quantum states, based on the general principles of the theory in connection with methods from statistical mechanics, is tentatively introduced by J.M. Souriau. In contrast to these papers which deal with the conventional quantization problem B. Kostant introduces an essential mathematical enlargement of the theory which is connected with a new physical interpretation and the corresponding mathematical concepts of Graded Lie Algebras and Graded Lie groups. It is interesting to observe that the development of these new concepts was, to a large extent, induced by the related physical viewpoint, namely the fieldtheoretical notion of Supersymmetry. The second chapter is devoted entirely to this new subject and represents a typical example in which a physical idea leads to the investigation and generalization of an important mathematical structure. At first Y. Ne'eman gives a historical survey and a detailed description of the physical background, whereupon S. Sternberg presents the new mathematical viewpoints suggested by Supersymmetry. Eventually, B. Kostant's paper - a considerable enlargement of his original lecture - represents a general theory of graded structures in Differential Geometry with the special aim of constructing a graded version of Geometric Quantization.

A second domain, in which an exchange appears of great importance, is given by the physical concept of Gauge Invariance (chapter III): It is based on the differential geometric notion of Connection and plays a decisive role in recent physical theories, especially in relation to the fundamental Renormalization problem in quantum field theory. M.E. Mayer introduces his notion of a Quantized Connection Form and discusses its relation to the gauge fields, whereas W. Greub treats the well-known problem of the Magnetic Monopol (suggested originally by P.A.M. Dirac) from a differential geometric viewpoint using in an essential way some concepts from the Geometric Quantization. L.Halpern applies an extended gauge principle to Dirac's equation and P.L.Garcia deals with the interrelation between Gauge Symmetry and Symplectic Structure. This paper naturally leads to a general discussion of Symplectic Structures which - being also the starting point of Geometric Quantization and a basis for analytical mechanics - are, by now, used in a large number of physical problems (chapter IV): A.Lichnerowicz as well as P. Dedecker deal with special extensions to Symplectic Structures within the framework of classical mechanics. On the other hand, M. Moshinsky discusses the quantum mechanical representation of canonical transformations, whereas W.M. Tulczyjew gives a rigorous description to particle and field dynamics with the help of Symplectic Spaces. In the last chapter (V) we finally discuss some special points in the classical application of Differential Geometry in physics, namely General Relativity. At first, W. Szczyrba and M. Francaviglia introduce interesting relations between General Relativity and Symplectic Structure thus showing the appearence of various geometrical concepts in the same physical domain, whereupon R. Debever and R. Kerner deal with some special mathematical problems. H.J. Seifert then discusses the mathematical question of singularities of Lorentz manifolds, a problem which is intimately related to recent astrophysical investigations (search for Black Holes). The last two papers are devoted to the interrelation between Micro- and Macrocosmos: Mrs.C.DeWitt-Morette treats one of the most important and so far unsolved problems in modern physics, the quantization of the gravitational field using for this purpose the differential geometric method of path integration (according to Feynman), whereas Pascual Jordan describes his ideas (developed also by P.A.M. Dirac) concerning the variation of the "constants" of Nature within the age of the Universe.

Acknowledgements

The organizers wish to express their sincere thanks to the Volkswagenstiftung as well as to the Deutsche Forschungsgemeinschaft for their most generous financial support of the symposium.

November 1976 K. Bleuler A. Reetz

TABLE OF CONTENTS

Chapter IV. Symplectic Structures - Mechanics

Chapter V. Riemannian Spaces - General Relativity

AN OUTLINE OF GEOMETRIC QUANTISATION (d'après KOSTANT)

D. J. Simms
School of Mathematics
Trinity College, Dublin.

INTRODUCTION

Geometric quantisation is a technique which has developed from two sources; (i) the study of unitary irreducible representations of Lie groups (work of Borel, Weil, Bott, Kirillov, Kostant, Auslander, Dixmier), and (ii) analysis of the procedure used to obtain the quantum mechanical description of a given physical system (van Hove, Segal, Souriau). It is based on the notion of a symplectic manifold (M,ω) where M denotes a real C^∞ manifold and ω a non-degenerate, real, closed differential 2-form (symplectic form) on M. The following are examples of symplectic manifolds

(i) cotangent bundles

(ii) Kähler manifolds

(iii) orbits under the coadjoint action of a Lie group on the dual of its Lie algebra

Each of these examples carries a natural symplectic form. The momentum phase space of a classical mechanical system is represented by a symplectic manifold. For systems arising from a configuration space X, the phase space is the cotangent bundle of X. We may however consider more general systems, as suggested by Souriau.

If ϕ is a smooth function on a symplectic manifold M, there is a unique vector field ξ_ϕ on M whose contraction with the symplectic form ω is equal to the differential $d\phi$. The local flow generated by ξ_ϕ leaves ω invariant. In the case when M is the phase space of a classical system and ϕ is the energy function (Hamiltonian), the vector field ξ_ϕ is the one which generates the classical motion. Given two smooth functions ϕ and ψ on M, the contraction of ω with ξ_ϕ and ξ_ψ is a smooth function on M called the Poisson bracket of ϕ and ψ. In this way $C^\infty(M)$ becomes a Lie algebra, and the map $\phi \to \xi_\phi$ is a representation of this algebra by vector fields on M which preserve ω.

It is a basic aim of geometric quantisation to study other, and related, representations of the Lie algebra $C^\infty(M)$, or of suitable subalgebras. In particular, the theory seeks to construct Hilbert spaces, and to assign self-adjoint operators on these spaces to some class of elements of $C^\infty(M)$. In the applications to mechanics, the symplectic manifold M represents the classical phase space, while the relevant

Hilbert space H represents the quantum mechanical phase space. Elements of $C^{\infty}(M)$ represent classical observables, while self-adjoint operators on H represent the corresponding quantum mechanical observables.

METAPLECTIC STRUCTURE

The process of geometric quantisation is divided into two stages: *prequantisation* and *polarisation*.

The first stage, prequantisation, requires (i) a choice of metaplectic structure for (M,ω) and (ii) a choice of hermitian line bundle L over M, with connection, having ω as curvature form. To explain the notion of metaplectic structure we note that, associated with ω, we have the symplectic frame bundle E consisting of all frames of M relative to which ω has matrix $\begin{pmatrix} 0 & I \\ -I & 0 \end{pmatrix}$. E is a principal $Sp(2n, R)$-bundle where $2n$ is the dimension of M, and $Sp(2n, R)$ is the subgroup of $GL(2n, R)$ which leaves invariant the scalar product ω_o on R^{2n} whose matrix is $\begin{pmatrix} 0 & I \\ -I & 0 \end{pmatrix}$. The metaplectic group $Mp(2n, R)$ is the unique connected double cover of the symplectic group $Sp(2n, R)$. Choice of a metaplectic structure for ω amounts to choosing a principal $Mp(2n, R)$-bundle $\tilde{E}$ which covers E twice. This can be done if and only if a certain cohomology class in $H^2(M, Z_2)$ vanishes. When one metaplectic structure is fixed, the set of all inequivalent metaplectic structures is parametrised by $H^1(M, Z_2)$. We call $\tilde{E}$ a metaplectic frame bundle for M.

As regards (ii) we remark that such a line bundle L with connection exists if and only if ω represents an integral de Rham cohomology class. When one choice of L is fixed, the set of all inequivalent choices is parametrised by the character group of the fundamental group of M.

SYMPLECTIC SPINORS

Consider the symplectic vector space R^{2n} with the symplectic form ω_o. Denote by σ the restriction to R^{2n} of the representation of the Heisenberg algebra $R^{2n} \oplus R$ on the space S' of tempered distributions on R^n, which is given by the Schrödinger quantisation prescription. Explicitly, $\sigma(x_j) = 2\pi i t_j$ and $\sigma(y_j) = \frac{\partial}{\partial t_j}$ where $x_1,\dots,x_n$, $y_1,\dots,y_n$ is the standard basis of R^{2n} and $t_1,\dots,t_n$ are standard coordinates on R^n. The map $R^{2n} \times S' \to S'$ given by σ commutes with the action of the metaplectic group, where $Mp(2n, R)$ acts on R^{2n} via the symplectic group and acts on S' via the van Hove - Shale - Segal - Weil representation. We can therefore define a Dirac operator:

$$S' \otimes \Lambda(R^{2n})^*_C \xrightarrow{\kappa} S' \otimes \Lambda(R^{2n})^*_C$$

which commutes with the action of the metaplectic group. Here $\Lambda(R^{2n})^*_C$ denotes the exterior algebra over the dual of $(R^{2n})_C$ and κ is given by

$$\kappa(\phi \otimes \eta) = \sum_{i=1}^{2n} \left(\sigma(e_i)\phi\right) \otimes (e^i \wedge \eta)$$

where $e_1,\dots,e_{2n}$ is any basis for R^{2n} and $e^1,\dots e^{2n}$ is the dual basis. We denote by B_o the set of elements of $S' \otimes \Lambda^n(R^{2n})^*_C$ which lie in the kernel of K.

If V is any real 2n-dimensional vector space with non-singular skew-symmetric scalar product η then we can extend η to the complexification V_C by complex bilinearity. A complex vector subspace W of V_C is said to be Lagrangian if (i) W is n-dimensional (ii) η vanishes identically on W. A Lagrangian subspace W is said to be Kähler if $W \cap \bar{W} = \{0\}$, is positive definite Kähler if the hermitian form $X \to i\eta(X,\bar{X})$ is positive definite on W, and is said to be real if $\bar{W} = W$. The space of all positive definite Kählerian subspaces of V_C is isomorphic to the Siegel upper half plane, and the space of all real Lagrangian subspaces is the Shilov boundary.

Each element of $(R^{2n})_C$ defines via σ an operator on S' which can be extended to $S' \otimes \Lambda(R^{2n})^*_C$ by trivial action on the second factor. For each Lagrangian subspace W of $(R^{2n})_C$ we denote by $N_o^{\frac{1}{2}}(W)$ the subspace of B_o annihilated by all members of the family of operators $\sigma(W)$. If W is positive definite Kähler then $N_o^{\frac{1}{2}}(W)$ is one-dimensional and is contained in the space $S \otimes \Lambda^n(R^{2n})_C$ of complex n-forms on R^{2n} with coefficients in the Schwartz space S of rapidly decreasing functions on R^n. We may call the elements of $N_o^{\frac{1}{2}}(W)$ the pure symplectic spinors associated with W.

For any two positive definite Kähler subspaces W, U we define a hermitian pairing

$$N_o^{\frac{1}{2}}(W) \times N_o^{\frac{1}{2}}(U) \to \Lambda^{2n}(R^{2n})^*_C$$

by

$$(\phi \otimes \alpha,\ \psi \otimes \beta) \to (-1)^{\frac{1}{2}n(n-1)} < \phi\ ,\ \psi > \alpha \wedge \bar{\beta}$$

where $<\phi,\psi>$ denotes the inner product of ϕ and ψ in the Hilbert space $L^2(R^n)$. If W, U are real Lagrangian subspaces such that $W \cap U = \{0\}$, then we can define a hermitian pairing

$$N_o^{\frac{1}{2}}(W) \times N_o^{\frac{1}{2}}(U) \to \Lambda^{2n}(R^{2n})^*_C$$

such that

$$(\phi \otimes \alpha,\ \psi \otimes \beta) \to \lim_{k\to\infty} (-1)^{\frac{1}{2}n(n-1)} <\phi_k,\ \psi_k> \alpha \wedge \bar{\beta}$$

where $\phi_k \otimes \alpha \in N_o^{\frac{1}{2}}(W_k)$ and $\psi_k \otimes \beta \in N_o^{\frac{1}{2}}(U_k)$ are such that the sequences $\{\phi_k\}$ and $\{\psi_k\}$ converge in S' to ϕ and ψ respectively and $\{W_k\}$, $\{U_k\}$ are any sequences of positive definite Kähler subspaces such that $W = \lim_k W_k$ and $U = \lim_k U_k$ in the Grassmannian topology.

The bundle with fibre R^{2n} associated to the principal $Mp(2n, R)$-bundle $\tilde{E}$ is the tangent bundle $T(M)$. The associated bundle with fibre $(R^{2n})_C$ is the complexified tangent bundle $T(M)_C$. We denote by B the associated bundle with fibre B_o. For each $x \in M$, the tangent space $T_x(M)$ is a 2n-dimensional real symplectic space with respect to the bilinear form ω_x. Let W_x and U_x be

Lagrangian subspaces of $T_x(M)$. Then a choice of real symplectic frame at x defines isomorphisms of $T_x(M)$, $T_x(M)_C$, B_x, W_x, U_x with R^{2n}, $(R^{2n})_C$, B_o, W_o, U_o respectively where W_o and U_o are Lagrangian subspaces of $(R^{2n})_C$. The vector subspace $N_o^{\frac{1}{2}}(W_o)$ of B_o annihilated by the operators $\sigma(W_o)$ therefore corresponds to a vector subspace $N_x^{\frac{1}{2}}(W_x)$ (say) of B_x. The subspace $N_x^{\frac{1}{2}}(W_x)$ depends only on W_x and not on the choice of symplectic frame at x.

If W_x and U_x are real Lagrangian subspaces, then the hermitian pairing of $N_o^{\frac{1}{2}}(W_o) \times N_o^{\frac{1}{2}}(U_o)$ into $\Lambda^{2n}(R^{2n})_C^*$ defines a hermitian pairing of $N_x^{\frac{1}{2}}(W_x) \times N_x^{\frac{1}{2}}(U_x)$ into $\Lambda^{2n} T_x(M)_C^*$.

PREQUANTISATION

Let E be the symplectic frame bundle and $\tilde{E}$ a metaplectic frame bundle for a symplectic manifold (M, ω). Let L be a hermitian line bundle with connection form α having ω as curvature form. Let B be the bundle associated to $\tilde{E}$ with fibre the kernel B_o of the Dirac operator in $S' \otimes \Lambda^r(R^{2n})_C^*$. Let $\Gamma(A)$ denote the space of smooth sections of a vector bundle A over M.

We have a natural representation of the Lie algebra $C^\infty(M)$ on $\Gamma(L)$ and on $\Gamma(B)$ and hence on $\Gamma(L \otimes B)$. This is called <u>prequantisation</u> and is obtained as follows. Let $\phi \in C^\infty(M)$ and let ξ_ϕ be the vector field generated by ϕ. The local flow of ξ_ϕ leaves ω invariant and therefore generates a local flow on E and hence on $\tilde{E}$. Any element of $\Gamma(B)$ can be considered as a function on the principal bundle $\tilde{E}$ and its Lie derivative along the local flow is again a section of B. This gives the action of $C^\infty(M)$ on $\Gamma(B)$.

To obtain the action of $C^\infty(M)$ on $\Gamma(L)$ we first note that we have a natural action on $\Gamma(L)$ of the Lie algebra aut (L^*) of all vector fields on the principal bundle L^* which are C^*-invariant and preserve both the hermitian structure and the connection form α. This is because each element of $\Gamma(L)$ can be considered as a function on the principal bundle L^*. Now the Lie algebras aut(L^*) and $C^\infty(M)$ are isomorphic under the map which sends η to $\langle\eta, \alpha\rangle$, which is a function on L^* constant on the fibres and therefore corresponds to an element of $C^\infty(M)$. The operator on $\Gamma(L)$ assigned to a smooth function ϕ in this way turns out to be

$$\nu(\phi) = \nabla_{\xi_\phi} + 2\pi i\phi$$

where ∇ denotes covariant differentiation with respect to the connection.

There is a more conceptual way of obtaining the prequantisation of ϕ as an operator on $\Gamma(L \otimes B)$ in the case when ξ_ϕ is a complete vector field and hence generates a 1-parameter family of automorphisms $t \to \sigma_t$ of (M, ω). We lift σ_t to automorphisms of the principal bundles L^* and $\tilde{E}$ and hence get a family of automorphisms of $\Gamma(L \otimes B)$. The derivative at $t = 0$ is the prequantisation operator on $\Gamma(L \otimes B)$.

POLARISATION

By a polarisation of a symplectic manifold (M, ω) we mean an involutive sub-bundle F of the complexified tangent bundle to M, such that the fibre F_x is a Lagrangian subspace of $T_x(M)_C$ for each $x \in M$. We say that F is a real polarisation if each F_x is real Lagrangian and that F_x is Kähler if each F_x is Kähler.

A choice of polarisation is, in contrast to the choices involved in pre-quantisation, highly non-unique even when M is simply connected. For example, if M is 2-dimensional then every non-vanishing real or complex vector field on M will generate a polarisation. Standard examples of polarisation are:

(i) on a cotangent bundle, the tangent spaces to the fibres generate a real polarisation

(ii) on a Kähler manifold, the anti-holomorphic tangent vectors generate a Kähler polarisation.

From now on we shall suppose that F is a real polarisation. The results in the case of a general complex polarisation are less complete although many of the definitions extend in a straight-forward way. Let $N^{\frac{1}{2}} = N^{\frac{1}{2}}(F)$ denote the one-dimensional sub-bundle of B whose fibre over x is $N_x^{\frac{1}{2}}(F_x)$. Let E^F denote the sub-bundle of the symplectic frame bundle E consisting of all frames of the form $(u_1,\ldots,u_n, v_1,\ldots,v_n)$ with $u_1,\ldots,u_n \in F$. Let $\tilde{E}^F$ denote the corresponding sub-bundle of $\tilde{E}$. Let P denote the subgroup of $Sp(2n, R)$ which stabilises the subspace $R^n = R^n \times \{0\}$ of R^{2n} and let $\tilde{P}$ be the corresponding subgroup of $Mp(2n, R)$. Then $\tilde{E}^F$ is a principal bundle with group $\tilde{P}$. The bundle with fibre S' associated to $\tilde{E}$ is isomorphic to the bundle with fibre S' associated to $\tilde{E}^F$, and this in turn contains the line bundle $N^{\frac{1}{2}}(F)$ as the bundle associated to $\tilde{E}^F$ with fibre $N_0^{\frac{1}{2}}(R^n)$.

Let N be the simply connected subgroup of $Sp(2n, R)$ which leaves the subspace R^n point-wise fixed and let $\tilde{N}$ denote the corresponding connected subgroup of $Mp(2n, R)$. Then the quotient P/N is isomorphic to the general linear group $GL(n, R)$ and the quotient E^F/N is isomorphic to the real frame bundle of F. Moreover $\tilde{P}/\tilde{N}$ is isomorphic to the unique connected double cover $ML(n, R)$ of $GL(n, R)$, which is called the metalinear group, and $\tilde{E}^F/\tilde{N}$ is a principal $ML(n, R)$-bundle which double covers the real frame bundle E^F/N. The bundle $\tilde{E}^F/\tilde{N}$ is called a metalinear frame bundle of F and it represents a metalinear structure for the polarisation F.

Now N acts trivially on $N_0^{\frac{1}{2}}(R^n)$ and hence $\tilde{P}/\tilde{N} = ML(n, R)$ acts on $N_0^{\frac{1}{2}}(R)$. Therefore the line bundle $N^{\frac{1}{2}}(F)$, which is the bundle associated with $\tilde{E}^F$ having fibre $N_0^{\frac{1}{2}}(R)$, may also be regarded as the bundle associated with the metalinear frame bundle $\tilde{E}^F/\tilde{N}$ having $N_0^{\frac{1}{2}}(R)$ as fibre. The restriction of E^F/N to any leaf of the foliation associated with F is the linear frame bundle of that leaf.

Therefore the restriction of $\tilde{E}^F/\tilde{N}$ to any leaf is a metalinear frame bundle for that leaf, and the restriction of $N^{\frac{1}{2}}(F)$ to the leaf is an associated line bundle. Let $\pi : ML(n, R) \to GL(n, R)$ denote the double cover, then the action of $g \in ML(n, R)$ on the tensor product $N_o^{\frac{1}{2}}(R^n) \otimes N_o^{\frac{1}{2}}(R^n)$ is multiplication by $\det \pi(g)$. Therefore the line bundle $N^{\frac{1}{2}}(F) \otimes N^{\frac{1}{2}}(F)$ is isomorphic to the exterior nth power $\Lambda^n F$ of the vector bundle F.

Denote by M/F the quotient of M by the foliation and assume that the projection $M \to M/F$ is a smooth submersion onto a Hausdorff manifold. If q is the leaf through $x \in M$ then $T_q(M/F)$ is canonically isomorphic to $T_x(M)/D_x$ where $D_x = F_x \cap T_x(M)$. Now D_x is isomorphic via ω_x to the dual of $T_x(M)/D_x$. Thus for all x in the leaf q, the tangent space D_x to the leaf at x is canonically isomorphic to the dual of the fixed vector space $T_q(M/F)$. This determines a flat affine connection on the leaf. A section s of the line bundle $N^{\frac{1}{2}}(F) \otimes N^{\frac{1}{2}}(F)$ will have vanishing covariant derivative along each leaf if and only if s corresponds to a complex n-form on M/F under the isomorphisms

$$N_x^{\frac{1}{2}}(F) \otimes N_x^{\frac{1}{2}}(F) \approx \Lambda^n F_x \approx \Lambda^n T_q(M/F)_C.$$

For each $\xi \in \Gamma(F)$ we have a covariant derivative ∇_ξ on $\Gamma(L)$ and, using the affine connections on the leaves, we have a covariant derivative ∇_ξ on $\Gamma(N^{\frac{1}{2}}(F))$ and hence also on $\Gamma(L \otimes N^{\frac{1}{2}}(F))$. For each open set V in M we denote by $S_F(V)$ the set of all sections of $L \otimes N^{\frac{1}{2}}(F)$ over V whose covariant derivative vanishes along all sections of F over V. The corresponding sheaf S_F is called the sheaf of germs of <u>L-valued half-forms normal to F and covariant constant along F</u>. We have then, for each integer $r \geq 0$, the r-dimensional cohomology space $H^r(M,S_F)$ of M with coefficients in the sheaf S_F. In particular $H^0(M,S_F)$ is the space of sections of $L \otimes N^{\frac{1}{2}}(F)$ over M which are covariant constant along F.

We use the hermitian structure on L to define a hermitian product $(.,.)$ on $\Gamma(L \otimes N^{\frac{1}{2}}(F))$ with values in $\Gamma(N^{\frac{1}{2}}(F) \otimes N^{\frac{1}{2}}(F))$. If $s, t \in H^0(M,S_F)$ then the section (s,t) of $N^{\frac{1}{2}}(F) \otimes N^{\frac{1}{2}}(F)$ will have vanishing covariant derivative along each leaf and hence may be considered as a complex n-form on M/F. We denote by $\langle s,t \rangle$ the integral of (s,t) over M/F. We take the pre-Hilbert space of all elements s of $H^0(M,S_F)$ for which $\langle s,s \rangle$ is finite and complete it to get a Hilbert space H_F.

Let F and G be real polarisations of M which are transverse in the sense that $F_x \cap H_x = 0$ for all $x \in M$. Then for each $x \in M$ we have a hermitian pairing of $N_x^{\frac{1}{2}}(F_x) \times N_x^{\frac{1}{2}}(G_x)$ into $\Lambda^{2n} T_x(M)^*_C$ and hence a hermitian pairing of $s \in \Gamma(M, L \otimes N^{\frac{1}{2}}(F))$ and $t \in \Gamma(M, L \otimes N^{\frac{1}{2}}(G))$ to $(s,t) \in \Gamma(M, \Lambda^{2n} T^*(M)_C)$. We denote by $\langle s,t \rangle$ the integral over M of the complex 2n-form (s,t). The pairing $\langle .,. \rangle$ is called the BKS pairing (Blattner-Kostant-Sternberg) between sections of $L \otimes N^{\frac{1}{2}}(F)$ and $L \otimes N^{\frac{1}{2}}(G)$.

QUANTISATION

Fix a real polarisation F of a symplectic manifold (M, ω). Let $\phi \in C^\infty(M)$ be such that ξ_ϕ is a complete vector field and let $t \to \sigma_t$ be the 1-parameter group of automorphisms of (M, ω) generated by ξ_ϕ. Lift σ_t to a 1-parameter family of automorphisms of the bundle $L \otimes B$, also denoted by σ_t as in prequantisation. Then σ_t maps $L \otimes N^{\frac{1}{2}}(F)$ onto $L \otimes N^{\frac{1}{2}}(F^t)$ and hence gives a homomorphism of S_F into S_{F^t}, where F^t is the transform of F under σ_t. We thus get, for each $r \geq 0$, a family of operators σ_t from $H^r(M, S_F)$ to $H^r(M, S_{F^t})$.

Now let $\psi \in H^O(M, S_F)$ and suppose that for all sufficiently small t there is a unique $U_t\psi \in H^O(M, S_F)$ such that

$$<U_t\psi, \zeta> = <\sigma_t\psi, \zeta>$$

for all $\zeta \in H^O(M, S_F)$ for which the BKS pairing on the right and the pairing in $H^O(M, S_F)$ on the left are finite. The derivative at $t = 0$ of the family of operators

$$U_t : H^O(M, S_F) \to H^O(M, S_F)$$

is sometimes taken as a quantisation of ϕ.

Other proposals for the quantisation of ϕ involve representing it by operators on

(i) the other cohomology spaces $H^r(M, S_F)$, $r \geq 0$

(ii) distribution spaces associated with F.

To do this we denote by $S_r(V)$ the space of smooth sections over an open set V of the vector bundle $L \otimes N^{\frac{1}{2}}(F) \otimes \Lambda^r F^*$ where $\Lambda^r F^*$ denotes the rth exterior power of the dual of F. We define $d : S_r(V) \to S_{r+1}(V)$ by

$$(d\beta)(\xi_1,\ldots,\xi_{r+1}) = \sum_{i=1}^{r+1} (-1)^{i+1} \nabla_{\xi_i}\left[\beta(\xi_1,\ldots,\hat{\xi}_i,\ldots,\xi_{r+1})\right]$$

$$+ \sum_{i<j} (-1)^{i+j}\beta([\xi_i,\xi_j], \xi_1,\ldots,\hat{\xi}_i,\ldots,\hat{\xi}_j,\ldots,\xi_{r+1})$$

for all sections $\xi_1,\ldots,\xi_{r+1}$ of F over V. Let S_r denote the corresponding sheaf, then

$$0 \to S_F \to S_o \to S_1 \to \ldots \to S_n \to 0$$

is a fine resolution of S_F. Consider the corresponding sequence of smooth sections

$$\Gamma(S_F) \to \Gamma(S_o) \to \Gamma(S_1) \to \ldots \to \Gamma(S_n)$$

endowed with the C^∞ topology in each case. We then have the dual sequence of distributions

$$\Gamma(S_F)' \leftarrow \Gamma(S_o)' \leftarrow \Gamma(S_1) \leftarrow \ldots \leftarrow \Gamma(S_n)'.$$

Since the elements in the first sequence are $L \otimes N^{\frac{1}{2}}(F)$-valued differential forms based on F, we may regard the elements of the second sequence as types of currents based on F.

If $\phi \in C^{\infty}(M)$ is such that σ_t leaves F invariant, then σ_t acts on each of these sequences and on each of the associated cohomology spaces. The derivative at $t = 0$ yields operators which represent the quantised action of ϕ in each case.

A BASIC EXAMPLE

Let $M = R^2$ with coordinate functions q,p and symplectic form $\omega = dp \wedge dq$. Let L be a hermitian line bundle with connection having ω as curvature form. The symplectic frame bundle is $\tilde{E} = M \times Sp(2, R)$ and the unique metaplectic frame bundle is $E = M \times Mp(2, R)$. Then $Mp(2, R)$ acts on $S \subset L^2(R) \subset S'$ where S is the Schwartz space and S' the space of tempered distributions. If e_1, e_2 is the usual basis of R^2 and e^1, e^2 the dual basis then $\sigma : R^2 \to End(S')$ is given by $\sigma(e_1) = 2\pi it$, $\sigma(e_2) = \frac{d}{dt}$ where t is the usual coordinate on R.

The Dirac operator κ from $S' \otimes (R^2)^*_C$ to $S' \otimes \Lambda^2(R^2)^*_C$ is given by

$$\kappa\left[\phi \otimes (\alpha e^1 + \beta e^2)\right] = \left[2\pi i\beta t\phi - \alpha\frac{d\phi}{dt}\right] \otimes (e^1 \wedge e^2).$$

We find that

$$N_o^{\frac{1}{2}}\left[C \langle\lambda e_1 + \mu e_2\rangle\right] = \begin{cases} Ce^{-\pi i\frac{\lambda}{\mu}t^2} \otimes (\mu e^1 - \lambda e^2) & \text{if } \mu \neq 0 \\ C\,\delta_o(t) \otimes e^2 & \text{if } \mu = 0 \end{cases}$$

where δ_o is the Dirac measure at the origin.

Now put $q = r\cos\theta$, $p = r\sin\theta$ and let F be the polarisation of $M_o = M - \{0\}$ given by $F_{(r,\theta)} = C\frac{\partial}{\partial\theta} = C\left(-p\frac{\partial}{\partial q} + q\frac{\partial}{\partial p}\right)$. The metaplectic frame bundle restricts to $M_o \times Mp(2, R)$ and the associated bundle with fibre B_o = kernel of κ in $S' \otimes (R^2)^*_C$ is a product $M_o \times B_o$. The line bundle $N^{\frac{1}{2}} = N^{\frac{1}{2}}(F)$ is the sub-bundle of $M_o \times B_o$ with fibre

$$N^{\frac{1}{2}}_{(r,\theta)}(F) = \begin{cases} Ce^{\pi i\frac{p}{q}t^2} \otimes (q\,dq + p\,dp) = Ce^{(\pi i\tan\theta)t^2} \otimes dr \\ C\,\delta_o(t) \otimes dp = C\,\delta_o(t) \otimes dr \end{cases}$$

according as $q \neq 0$ or $q = 0$.

We use the vector field $\frac{\partial}{\partial\theta}$ to identify F^* with the product bundle $M_o \times C$, so that the map $\Gamma(L \otimes N^{\frac{1}{2}}) \xrightarrow{d} \Gamma(L \otimes N^{\frac{1}{2}} \otimes F^*)$ may be identified with the map $D : \Gamma(L \otimes N^{\frac{1}{2}}) \to \Gamma(L \otimes N^{\frac{1}{2}})$ given by covariant differentiation along $\frac{\partial}{\partial\theta}$. We have the exact sequence

$$0 \to H^0(M_o, S_F) \to \Gamma(L \otimes N^{\frac{1}{2}}) \xrightarrow{D} \Gamma(L \otimes N^{\frac{1}{2}}) \to H^1(M_o, S_F) \to 0$$

and its dual

$$0 \leftarrow H^0(M_o, S_F)' \leftarrow \Gamma(L \otimes N^{\frac{1}{2}})' \xleftarrow{D'} \Gamma(L \otimes M^{\frac{1}{2}})' \leftarrow H^1(M_o, S_F)' \leftarrow 0$$

The kernel of D is the cohomology space $H^0(M_o, S_F)$ and consists of all C^{∞} solutions of the equation

$$\nabla_{\frac{\partial}{\partial\theta}} \psi = 0 ,$$

while the kernel of D' is isomorphic to $H^1(M_o, S_F)'$ and consists of all distributional solutions (of compact support) of this equation. The cokernel of D is isomorphic to $H^1(M_o, S_F)$. The spectrum of any operator on $H^1(M_o, S_F)$ is the same as the spectrum of its transpose on $H^1(M_o, S_F)'$. It follows that if F is invariant under ξ_ϕ for some $\phi \in C^\infty(M)$, then the associated operator on $H^1(M_o, S_F)$ has the same spectrum as the associated operator on the space of distributional sections of $L \otimes N^{\frac{1}{2}}$ which are of compact support and covariant constant along F.

We now determine the cohomology spaces. The frame $\frac{\partial}{\partial\theta}$ is covariant constant along each leaf of F, and the corresponding symplectic frame $\{\frac{\partial}{\partial\theta}, \frac{\partial}{\partial\rho}\}$, where $\rho = \frac{1}{2} r^2$, describes a generator of the fundamental group in $Sp(2, R)$ around each leaf and therefore corresponds to -1 in the representation of $Mp(2, R)$ on S'. It follows that a section of $N^{\frac{1}{2}}$ which is covariant constant along F will be multiplied by -1 on going around a leaf of F.

Let s_1 and s_2 be sections of $N^{\frac{1}{2}}$ over the open sets $0 < \theta < 2\pi$ and $-\pi < \theta < \pi$ respectively in M_o and which are covariant constant along F and such that $s_1 = s_2$ in $0 < \theta < \pi$. Then $s_1 = -s_2$ in $\pi < \theta < 2\pi$.

Since ω is exact and M is simply connected, L is equivalent to the product bundle $M \times C$ and we can identify $\Gamma(L)$ with $C^\infty(M_o)$. We can also identify $\Gamma(L \otimes N^{\frac{1}{2}})$ with $C^\infty(M_o)$ by associating with each $\psi \in C^\infty(M_o)$ the section of $L \otimes N^{\frac{1}{2}}$ which equals $e^{\frac{1}{2}i\theta}\psi\, s_1$ for $0 < \theta < 2\pi$ and equals $e^{\frac{1}{2}i\theta}\psi\, s_2$ for $-\pi < \theta < \pi$.

Since $\omega = d\alpha$ on M_o with $\alpha = -\frac{1}{2}r^2 d\theta$, the covariant derivative along ξ acting on $\Gamma(L) = C^\infty(M_o)$ can be taken as $\nabla_\xi = \xi + 2\pi i < \alpha,\xi>$. Thus

$$D(e^{\frac{1}{2}i\theta}\psi\, s_1) = \nabla_{\frac{\partial}{\partial\theta}}\left[e^{\frac{1}{2}i\theta}\psi\, s_1\right]$$

$$= \left(\frac{\partial}{\partial\theta} - \pi i r^2\right)\left[e^{\frac{1}{2}i\theta}\psi\right]s_1$$

$$= e^{\frac{1}{2}i\theta}\left[\frac{\partial\psi}{\partial\theta} + i(\tfrac{1}{2} - \pi r^2)\psi\right]s_1 .$$

The operator D on $\Gamma(L \otimes N^{\frac{1}{2}})$ therefore corresponds to the operator D on $C^\infty(M_o)$ given by

$$D\psi = \frac{\partial\psi}{\partial\theta} + i(\tfrac{1}{2} - \pi r^2)\psi.$$

Now consider the exact sequence

$$0 \to H^0(M_o, S_F) \to C^\infty(M_o) \overset{D}{\to} C^\infty(M_o) \to H^1(M_o, S_F) \to 0.$$

The only C^∞ solution of $D\psi = 0$ is $\psi = 0$. Therefore $H^0(M_o, S_F) = \{0\}$. Also, if $D\psi = \phi$ then

$$\psi(r\cos\theta, r\sin\theta) = e^{-i(\frac{1}{2}-\pi r^2)\theta}\left[\int_0^\theta e^{i(\frac{1}{2}-\pi r^2)t}\,\phi(r\cos t, r\sin t)dt + \psi(1,0)\right].$$

By periodicity we have

$$\int_0^{2\pi} e^{i(\frac{1}{2}-\pi r^2)t}\phi(r\cos t, r\sin t)dt = \left[e^{2\pi i(\frac{1}{2}-\pi r^2)} - 1\right]\psi(1, 0).$$

For each $\phi \in C^\infty(M_o)$ we write

$$\phi_N = \int_0^{2\pi} e^{-iNt}\,\phi\left(\left(\frac{2N+1}{2\pi}\right)^{\frac{1}{2}}\cos t, \left(\frac{2N+1}{2\pi}\right)^{\frac{1}{2}}\sin t\right)dt$$

and conclude that if $D\psi = \phi$ then $\phi_N = 0$ for all integers $N \geq 0$. Conversely, if $\phi_N = 0$ for all integers $N \geq 0$ then $\phi = D\psi$ for some $\psi \in C^\infty(M_o)$.

Hence the map

$$\phi \to (\phi_o, \phi_1, \phi_2, \ldots, \phi_N, \ldots)$$

induces a bijection of $H^1(M_o, S_F)$ onto the space of all complex sequences. We note that the cohomology class of the function $e^{i\lambda\theta}$ is non-zero only if $\lambda \geq 0$, in which case it is represented by the sequence

$$\{2\pi\delta_{N,\lambda}\} = (0, 0, \ldots, 0, 2\pi, 0, \ldots, 0).$$

We now consider the quantisation of the function $\phi = \frac{1}{2}(p^2 + q^2) = \frac{1}{2} r^2$ as an operator on $H^1(M_o, S_F)$. We have $\xi_\phi = \frac{\partial}{\partial\theta}$ and therefore prequantisation on $\Gamma(L)$ is given by

$$\nabla_{\xi_\phi} + 2\pi i\phi = \frac{\partial}{\partial\theta} - \pi i r^2 + \pi i r^2 = \frac{\partial}{\partial\theta}.$$

Prequantisation on $\Gamma(N^{\frac{1}{2}})$ is given by covariant differentiation along $\frac{\partial}{\partial\theta}$. Therefore the prequantisation operator $\nu(\phi)$ on $\Gamma(L \otimes N^{\frac{1}{2}})$ is given by

$$\nu(\phi)\left[e^{\frac{1}{2}i\theta}\psi\, s_1\right] = \frac{\partial}{\partial\theta}\left[e^{\frac{1}{2}i\theta}\psi\right]s_1 = e^{\frac{1}{2}i\theta}\left[\frac{\partial\psi}{\partial\theta} + \tfrac{1}{2}\, i\psi\right]s_1 .$$

Therefore prequantisation of ϕ on $\Gamma(L \otimes N^{\frac{1}{2}}) = C^\infty(M_o)$ is given by $\frac{\partial}{\partial\theta} + \frac{1}{2} i$. Now $\left(\frac{\partial}{\partial\theta} + \frac{1}{2} i\right) e^{i\lambda\theta} = i(\lambda + \frac{1}{2})e^{i\lambda\theta}$. Therefore the spectrum on $H^1(M_o, S_F)$ of the operator $\frac{1}{i}\nu(\phi)$ is

$$\{\lambda + \tfrac{1}{2} \mid \lambda \text{ an integer } \geq 0\}.$$

<u>LITERATURE</u> A very thorough account of line bundles with connection, and prequantisation, is given by Kostant in Springer Lecture Notes <u>170</u>. In the A.M.S. Proceedings of Symposia in Pure Mathematics XXVI Blattner uses polarisation and half-forms to obtain quantisation. In the 1973 Rome conference, Symposia Mathematica XIV, Kostant introduced symplectic spinors. Some applications and further developments are given in the proceedings of the Rome conference and of the 1974 C.N.R.S. conference in Aix.

THE METALINEAR GEOMETRY OF NON-REAL POLARIZATIONS

Robert J. Blattner

1. Introduction

This paper is a continuation of a program begun in previous papers ([3], [4]) the aim of which has been to lay the foundations of metaplectic and metalinear geometry in way to be of use in the geometric quantization of Kostant and Souriau. Whereas the previous papers dealt exclusively with real polarizations, we treat here the positive complex case by making use of the fact that the space of positive lagrangian subspaces of C^{2n} is isomorphic to the closed Siegel unit ball. The pairing of half-forms normal to such subspaces is phrased in terms of a formal kernel which generalizes the kernels of the unitary transforms introduced by Satake [11] in his treatment of Fock representations parametrized by the Siegel upper half plane.

Our paper is divided as follows: Section 2 constructs the space of frames normal to positive lagrangian subspaces of C^{2n}. Section 3 uses this space of frames to define half-forms in C^{2n} and their pairing. Sections 2 and 3 are purely local. The extension of this theory to the global situation on metaplectic manifolds carrying a Kostant line bundle is accomplished in Section 4. Throughout, we use $\hbar = h/2\pi$, h = Planck's constant, as a parameter in order to facilitate application to physical situations.

We would like to express our thanks to Bertram Kostant, Shlomo Sternberg, Victor Guillemin, Eduardo Cattani, Linda Rothschild, and Joseph Wolf for conversations bearing on this paper, and to the Rice University Mathematics Department for its kind hospitality while a preliminary sketch of this paper

---

This work was partially supported by NSF grants GP-43376 and MPS 75-17621.

was being written.

2. Positive lagrangian frames over $\mathbb{C}^{2n}$.

We let $\mathcal{V}$ denote the category whose objects are finite dimensional vector spaces over $K = \mathbb{C}$ or $\mathbb{R}$ and whose morphisms are K-linear isomorphisms between objects of $\mathcal{V}$. We next define a category $\mathfrak{F}(\mathcal{V})$ whose objects are the sets of all frames (ordered bases) of objects of $\mathcal{V}$. Let $(v_1,\ldots,v_n)$ be a frame of some vector space E. Then $GL(n,K)$ operates on $(v_1,\ldots,v_n)$ on the right according to the rule

$$(2.1) \qquad (v_1,\ldots,v_n)C = (v_1',\ldots,v_n'), \quad \text{where}$$

$$v_i' = \sum_j{}_1^n c_{ji} v_j \quad \text{and} \quad C = (c_{ji}) \in GL(n,K).$$

$(v_1',\ldots,v_n')$ is again a frame of E. Let $\mathfrak{F}(E)$ denote the set of all frames of E. Then (2.1) makes $\mathfrak{F}(E)$ a right principal homogeneous space for $GL(n,K)$. A morphism of $\mathfrak{F}(\mathcal{V})$ will be a bijection $\tau : \mathfrak{F}(E_1) \to \mathfrak{F}(E_2)$ such that $\dim E_1 = \dim E_2 = n$, say, and τ commutes with the right action of $GL(n,K)$. Let $\sigma : E_1 \to E_2$ be a morphism of $\mathcal{V}$. Define $\mathfrak{F}(\sigma) : \mathfrak{F}(E_1) \to \mathfrak{F}(E_2)$ by $\mathfrak{F}(\sigma)(v_1,\ldots,v_n) = (\sigma v_1,\ldots,\sigma v_n)$ for $(v_1,\ldots,v_n) \in \mathfrak{F}(E_1)$. Then $\mathfrak{F}(\sigma)$ is a morphism of $\mathfrak{F}(\mathcal{V})$, and $\mathfrak{F}$ is a bijective functor from $\mathcal{V}$ onto $\mathfrak{F}(\mathcal{V})$.

In the foregoing we make the convention that $\mathfrak{F}(\{0\}) = \{\emptyset\}$ and that $GL(0,K) = \{1\}$.

Let $(v_1,\ldots,v_n) \in \mathfrak{F}(E)$. We shall let $\delta(v_1,\ldots,v_n) = (w_1,\ldots,w_n) \in \mathfrak{F}(E^*)$ be the dual frame; i.e., $\langle w_i,v_j\rangle = \delta_{ij}$. δ is a bijection of $\mathfrak{F}(E) \to \mathfrak{F}(E^*)$ but is not a morphism in $\mathfrak{F}(\mathcal{V})$. Indeed

$$(2.2) \qquad \delta[(v_1,\ldots,v_n)C] = [\delta(v_1,\ldots,v_n)]^t C^{-1}, \quad C \in GL(n,K).$$

In the sequel we shall have occasion to apply $\mathfrak{F}$ to the category of finite dimensional vector bundles over a manifold M. If E is such a bundle, then $\mathfrak{F}(E)$ is the fibre bundle over M whose fibre $\mathfrak{F}(E)_m$ over $m \in M$ is just $\mathfrak{F}(E_m)$ and whose manifold structure is the obvious one. If fibre-dim $E = n$, then $\mathfrak{F}(E)$ is a right principal $GL(n,K)$-bundle. A vector bundle isomorphism σ gives rise to a principle $GL(n,K)$-bundle isomorphism $\mathfrak{F}(\sigma)$ in the obvious way. We also have the map $\delta : \mathfrak{F}(E) \to \mathfrak{F}(E^*)$. δ is a fibre bundle map but, according to (2.2), not a principal fibre bundle map.

Now let E be a finite dimensional vector space over K and let Ω be a bilinear form on E. We define a linear map $\Omega_* : E \to E^*$ by

$$\text{(2.3)} \qquad \langle \Omega_* v, w\rangle = \Omega(v,w) \quad \text{for} \quad v,w \in E.$$

If V is a subspace of E, we define $V^\perp$ by

$$\text{(2.4)} \qquad V^\perp = \{w \in E : \Omega(v,w) = 0 \quad \text{for} \quad v \in V\}.$$

Ω is <u>non-degenerate</u> if Ω_* is bijective, that is, if $E^\perp = \{0\}$. If Ω is non-degenerate and if V is a subspace of E, then Ω_* induces an isomorphism $\Omega_*^V : V \to (E/V^\perp)^*$. Thus we have a bijection $\Omega_\# : \mathfrak{F}(V) \to \mathfrak{F}(E/V^\perp)$ defined by $\Omega_\# = \delta^{-1} \circ \Omega_*^V$ on $\mathfrak{F}(V)$.

Let Ω be non-degenerate and let $g \in GL(E)$ preserve Ω. Let V be a subspace of E. Then $(gV)^\perp = gV^\perp$. We have a map $\mathfrak{F}(g) : \mathfrak{F}(V) \to \mathfrak{F}(gV)$ and an induced map $\mathfrak{F}(g) : \mathfrak{F}(E/V^\perp) \to \mathfrak{F}(E/gV^\perp)$.

<u>Lemma</u> 2.5: $\Omega_\# \circ \mathfrak{F}(g) = \mathfrak{F}(g) \circ \Omega_\#$

Proof: Let $(v_1,\dots,v_n) \in \mathfrak{F}(V)$ and let $(\hat{w}_1,\dots,\hat{w}_n) = \Omega_\#(v_1,\dots,v_n) \in \mathfrak{F}(E/V^\perp)$. Choose $w_i \in \hat{w}_i$, $i = 1,\dots,n$. Then $\Omega(v_i,w_j) = \delta_{ij}$. Now $\mathfrak{F}(g)(v_1,\dots,v_n) = (gv_1,\dots,gv_n)$ and $\mathfrak{F}(g)(\hat{w}_1,\dots,\hat{w}_n) = (\widehat{gw}_1,\dots,\widehat{gw}_n)$. Since $\Omega(gv_i,gw_j) = \Omega(v_i,w_j) = \delta_{ij}$, we have $\Omega_\#(gv_1,\dots,gv_n) = (\widehat{gw}_1,\dots,\widehat{gw}_n)$, and the lemma is proved.

(E,Ω) is a <u>symplectic vector space</u> if Ω is skew symmetric and non-degenerate. A subspace V of E is isotropic (with respect to Ω) if Ω vanishes on V, that is, if $V \subseteq V^\perp$ and is <u>lagrangian</u> if it is maximal isotropic. Thus a subspace V of a symplectic vector space E is lagrangian if and only if $V = V^{\perp}$. Moreover $\dim V = \frac{1}{2} \dim E$.

Now let (E,Ω) be a symplectic vector space over C and let J be a conjugation on E such that $\Omega(Jv,Jw) = \overline{\Omega(v,w)}$ for $v,w \in E$. The triple (E,Ω,J) will be called a symplectic vector space with conjugation.

Definition 2.6: A lagrangian subspace V of E is called positive (with respect to Ω and J) if the restriction to V of the hermitian form $\langle\cdot,\cdot\rangle$ on E defined by

$$\langle v,w\rangle = \frac{i}{2\pi\hbar}\,\Omega(v,Jw)$$

is positive semi-definite.

$\mathcal{L}_+(E,\Omega,J)$ will denote the manifold of all positive lagrangian subspaces V of E, while $\mathcal{L}_+^\sharp(E,\Omega,J)$ [resp. $\mathcal{L}_+^\flat(E,\Omega,J)$] will denote the vector bundle over $\mathcal{L}_+(E,\Omega,J)$ whose fibre over V is just V [resp. E/V]. We denote the bundle projections by ρ. We have the bijection $\Omega_\sharp : \mathfrak{F}(\mathcal{L}_+^\sharp(E,\Omega,J))$ onto $\mathfrak{F}(\mathcal{L}_+^\flat(E,\Omega,J))$.

Let $Sp(E,\Omega)$ consist of all $g \in GL(E)$ which preserve Ω, and let $Sp(E,\Omega,J) = \{g \in Sp(E,\Omega) : J \circ g = g \circ J\}$. If $V \in \mathcal{L}_+(E,\Omega,J)$ and $g \in Sp(E,\Omega,J)$, so does gV. In this way $\mathcal{L}_+(E,\Omega,J)$ becomes a left $Sp(E,\Omega,J)$ space. Moreover, $\mathfrak{F}(g) : \mathfrak{F}(V) \to \mathfrak{F}(gV)$ defines a left action of $Sp(E,\Omega,J)$ on $\mathfrak{F}(\mathcal{L}_+^\sharp(E,\Omega,J))$ and the induced action $\mathfrak{F}(g) : \mathfrak{F}(E/V) \to \mathfrak{F}(E/gV)$ defines a left action of $Sp(E,\Omega,J)$ on $\mathfrak{F}(\mathcal{L}_+^\flat(E,\Omega,J))$. These actions commute with projection ρ on $\mathcal{L}(E,\Omega,J)$ and, according to (2.5), with $\Omega_\sharp$.

We shall write members of C^{2n} as column vectors $\begin{bmatrix} v \\ w \end{bmatrix}$, where v and w are column vectors in C^n. C^{2n} has a natural symplectic structure given by

$$\Omega\left(\begin{bmatrix} v_1 \\ w_1 \end{bmatrix}, \begin{bmatrix} v_2 \\ w_2 \end{bmatrix}\right) = {}^t w_2 v_1 - {}^t v_2 w_1,$$

1×1 matrices being identified with complex numbers; it also has a natural conjugation: $J = {}^-$. We shall write $\mathcal{L}_n$ for $\mathcal{L}_+(\mathbb{C}^{2n},\Omega,{}^-)$, $\mathfrak{F}_n^\#$ for $\mathfrak{F}(\mathcal{L}_+^\#(\mathbb{C}^{2n},\Omega,{}^-))$, and $\mathfrak{F}_n^b$ for $\mathfrak{F}(\mathcal{L}_+^b(\mathbb{C}^{2n},\Omega,{}^-))$. Thus $\mathfrak{F}_n^\#$ and $\mathfrak{F}_n^b$ are right principal $GL(n,\mathbb{C})$-bundles over $\mathcal{L}_n$ with bundle projection ρ. Each frame in $\mathfrak{F}_n^\#$ can be represented by a $2n \times n$ complex matrix, the ordered n-tuple of columns of the matrix being the frame. We will write such matrices in block form

$$\begin{bmatrix} A \\ B \end{bmatrix},$$

A and B being $n \times n$ complex matrices. Our first task is to obtain a good parametrization of $\mathfrak{F}_n^\#$ and hence of $\mathfrak{F}_n^b$.

Now let $\bar{D}_n = \{W \in \mathbb{C}^{n\times n} : W = {}^tW, \|W\| \leq 1\}$, where $\|\cdot\|$ is the norm of W regarded as an operator on $\mathbb{C}^n$ with respect to the usual hermitian structure. $\bar{D}_n$ is the <u>closed Siegel unit ball</u>. Set $\mathcal{P}_n = \bar{D}_n \times GL(n,\mathbb{C})$. $\mathcal{P}_n$ is the trivial right principal $GL(n,\mathbb{C})$-bundle over $\bar{D}_n$. We map $\mathcal{P}_n$ into $2n \times n$ matrices as follows:

$$(2.8) \qquad \varphi^\#(W,C) = \begin{bmatrix} \frac{1}{2}(W + I) \\ \frac{1}{2i}(W - I) \end{bmatrix} {}^tC^{-1}.$$

We set $\varphi^b = \Omega_\# \circ \varphi^\#$.

<u>Theorem</u> 2.9: $\varphi^\#$ is a bijection of $\mathcal{P}_n$ onto $\mathfrak{F}_n^\#$. φ^b is a right principal $GL(n,\mathbb{C})$-bundle isomorphism of $\mathcal{P}_n$ onto $\mathfrak{F}_n^b$.

Proof: In view of (2.8), (2.1), (2.2) and the definition of $\Omega_\#$ it suffices to prove the first assertion. So let $\begin{bmatrix} A \\ B \end{bmatrix}$ be a $2n \times n$ matrix and let V be the subspace of $\mathbb{C}^{2n}$ spanned by its columns. Then it follows easily from (2.7) that V is lagrangian if and only if

$$(1) \qquad \operatorname{rank} \begin{bmatrix} A \\ B \end{bmatrix} = n, \text{ and}$$

$$(2) \qquad {}^tAB = {}^tBA.$$

Moreover, if (1) and (2) hold, then (2.6) says that V is positive if and only if

(3) $$i\{{}^t\bar{A}B - {}^t\bar{B}A\} \text{ is positive semi-definite.}$$

Note that (2) says that ${}^tAB$ is symmetric, while (3) says that if we write $A^*B = H + iK$ with H and K self-adjoint, then $K \geq 0$.

Now denote the right hand side of (2.8) by $\begin{bmatrix} A \\ B \end{bmatrix}$. Then ${}^tAB = \frac{1}{4i} C^{-1}({}^tW + I)(W - I){}^tC^{-1}$, which is symmetric since W is. Again $i(A^*B - B^*A) = i\{-\frac{1}{4i}(W^* - I)(W + I) - \frac{1}{4i}(W^* + I)(W - I)\} = \frac{1}{2}(I - W^*W)$, which is positive semi-definite since $\|W\| \leq 1$. Finally, $A - iB = {}^tC^{-1}$, which has trivial kernel. Therefore so does $\begin{bmatrix} A \\ B \end{bmatrix}$. This shows that $\varphi^\#$ maps $\mathcal{P}_n$ into $\mathfrak{F}_n^\#$.

Now let $\begin{bmatrix} A \\ B \end{bmatrix}$ be any matrix satisfying (1 - 3). Then $(A - iB)^*(A - iB) = A^*A + B^*B + 2K \geq A^*A + B^*B = \begin{bmatrix} A \\ B \end{bmatrix}^* \begin{bmatrix} A \\ B \end{bmatrix}$, where $A^*B = H + iK$ as above. Since $\operatorname{Ker}\begin{bmatrix} A \\ B \end{bmatrix} = \{0\}$, $A - iB$ is non-singular. Thus we can set $C = {}^t(A - iB)^{-1}$ and $W = (A + iB)(A - iB)^{-1}$. If we can show that $W \in \bar{D}_n$, then we will have $(W,C) \in \mathcal{P}_n$ and $\varphi^\#(W,C) = \begin{bmatrix} A \\ B \end{bmatrix}$. Moreover (2.8) will show that (W,C) is the <u>unique</u> preimage of $\begin{bmatrix} A \\ B \end{bmatrix}$ in $\mathcal{P}_n$ under $\varphi^\#$, and our theorem will be proved.

Now ${}^t(A + iB)(A - iB) = {}^t(A - iB)(A + iB)$ because ${}^tAB$ is symmetric. Therefore ${}^tW = W$. And finally, $W^*W = (A - iB)^{-1*}(A + iB)^*(A + iB)(A - iB)^{-1} = (A - iB)^{-1*}\{A^*A + B^*B - 2K\}(A - iB)^{-1} \leq (A - iB)^{-1*}\{A^*A + B^*B + 2K\}(A - iB)^{-1} = I$, so that $\|W\| \leq I$. Therefore $W \in \bar{D}_n$.

<u>Corollary</u> 2.10: φ^b induces a diffeomorphism $\check{\varphi}$ of $\bar{D}_n$ onto $\mathcal{L}_n$.

As usual we write $Sp(n,\mathbb{C})$ for $Sp(\mathbb{C}^{2n},\Omega)$ and $Sp(n,\mathbb{R})$ for $Sp(\mathbb{C}^{2n},\Omega,^-)$. These groups consist complex (resp. real) $2n \times 2n$ matrices which when written in block form $\begin{bmatrix} T_1 & T_2 \\ T_3 & T_4 \end{bmatrix}$ with $T_i \in K^{n\times n}$, $K = \mathbb{C}$ or $\mathbb{R}$, satisfy

(2.11) $${}^tT_3T_1 \text{ and } {}^tT_4T_2 \text{ are symmetric, and}$$

$$(2.12) \qquad {}^tT_4T_1 - {}^tT_2T_3 = I.$$

The left action of $Sp(n, \mathbb{R})$ on $\mathfrak{F}_n^{\#}$ is just given by matrix multiplication. Using $\varphi^{\#}$, we transfer this action to $\mathcal{P}_n$. Using this action of $Sp(n, \mathbb{R})$ on $\mathcal{P}_n$ and the usual action of $Sp(n, \mathbb{R})$ on $\mathfrak{F}_n^{\flat}$, $\varphi^{\flat}$ is an $Sp(n, \mathbb{R})$-equivariant map, according to (2.5).

<u>Proposition</u> 2.13: Let $g = \begin{bmatrix} T_1 & T_2 \\ T_3 & T_4 \end{bmatrix} \in Sp(n, \mathbb{R})$ and let $(W,C) \in \mathcal{P}_n$. Then $g(W,C) = (gW, \alpha(g,W)C)$, where $gW = \{[(T_1 + T_4) + i(T_3 - T_2)]W + [(T_1 - T_4) + i(T_3 + T_2)]\} \cdot \{[(T_1 - T_4) - i(T_3 + T_2)]W + [(T_1 + T_4) - i(T_3 - T_2)]\}^{-1}$ and $\alpha(g,W) = {}^t\{\tfrac{1}{2}[(T_1 - T_4) - i(T_3 + T_2)]W + \tfrac{1}{2}[(T_1 + T_4) - i(T_3 - T_2)]\}^{-1}$.

Proof: Since the left action of $Sp(n, \mathbb{R})$ commutes with the right action of $GL(n, \mathbb{C})$ on $\mathfrak{F}_n^{\flat}$ and hence on $\mathcal{P}_n$, it suffices to check these formulae in the case $C = I$. The messy but routine calculations are left to the reader.

<u>Remark</u> 2.14: Since we have a group action of $Sp(n, \mathbb{R})$ on $\mathcal{P}_n$, the formula of (2.13) defines a group action of $Sp(n, \mathbb{R})$ on $\overline{D}_n$ and α must satisfy the cocycle condition

$$\alpha(g_1g_2, W) = \alpha(g_1, g_2W)\alpha(g_2, W).$$

Moreover we must have $\alpha(1,W) = I$ for all $W \in \overline{D}_n$.

Theorem 2.9 and Proposition 2.13 show that $\varphi^{\flat} : \mathcal{P}_n \to \mathfrak{F}_n^{\flat}$ is a good parametrization of $\mathfrak{F}_n^{\flat}$ insofar as the actions of $GL(n, \mathbb{C})$ and $Sp(n, \mathbb{R})$ are concerned. Our next task is to use this parametrization to obtain the double coverings of frame bundles needed, as in [9] and [3], to do metalinear geometry. Now $\mathfrak{F}_n^{\flat}$ is the bundle (with base $\mathcal{L}_n$) of linear frames normal to positive lagrangian subspaces of $\mathbb{C}^{2n}$. Let $ML(n, \mathbb{C})$, the $n \times n$ complex <u>metalinear</u> group, be the complex analytic double covering of $GL(n, \mathbb{C})$ with covering projection p. We must construct a double covering $\widetilde{\mathfrak{F}}_n^{\flat}$ of $\mathfrak{F}_n^{\flat}$ with covering projection p such that $\widetilde{\mathfrak{F}}_n^{\flat}$ is a right principal $ML(n, \mathbb{C})$-bundle over $\mathcal{L}_n$ and such that the diagram

$$(2.15)\qquad \begin{array}{ccc} \tilde{\mathfrak{F}}_n^{\flat} \times ML(n,\mathbb{C}) & \longrightarrow & \tilde{\mathfrak{F}}_n^{\flat} \\ \downarrow{\scriptstyle p \times p} & & \downarrow{\scriptstyle p} \\ \mathfrak{F}_n^{\flat} \times GL(n,\mathbb{C}) & \longrightarrow & \mathfrak{F}_n^{\flat} \end{array}$$

commutes, where the horizontal arrows are the right group actions. This is now easy.

Definition 2.16: $\tilde{\mathfrak{F}}_n^{\flat} = \bar{D}_n \times ML(n,\mathbb{C})$, the trivial right principal $ML(n,\mathbb{C})$ bundle over $\bar{D}_n$ with bundle projection ρ_1. Then $\rho = \check{\phi}^{-1} \circ \rho_1 : \tilde{\mathfrak{F}}_n^{\flat} \to \mathcal{L}_n$ makes $\tilde{\mathfrak{F}}_n^{\flat}$ into a right principal $ML(n,\mathbb{C})$-bundle over $\mathcal{L}_n$. The covering projection of $\tilde{\mathfrak{F}}_n^{\flat}$ onto $\mathfrak{F}_n^{\flat}$ is just $\varphi^{\flat} \circ (\mathrm{id} \times p)$. We call it p also. $\tilde{\mathfrak{F}}_n^{\flat}$ is the bundle of <u>metalinear frames normal to positive lagrangian subspaces of</u> $\mathbb{C}^{2n}$ or, for short, the bundle of <u>positive lagrangian frames over</u> $\mathbb{C}^{2n}$.

Let $Mp(n,\mathbb{R})$, the metaplectic group of rank n, is the double covering of $Sp(n,\mathbb{R})$. We will also let p denote the covering projection of $Mp(n,\mathbb{R})$ onto $Sp(n,\mathbb{R})$.

Theorem 2.17: There is a unique smooth left group action of $Mp(n,\mathbb{R})$ on $\tilde{\mathfrak{F}}_n^{\flat}$ such that the diagram

$$\begin{array}{ccc} Mp(n,\mathbb{R}) \times \tilde{\mathfrak{F}}_n^{\flat} & \longrightarrow & \tilde{\mathfrak{F}}_n^{\flat} \\ \downarrow{\scriptstyle p \times p} & & \downarrow{\scriptstyle p} \\ Sp(n,\mathbb{R}) \times \mathfrak{F}_n^{\flat} & \longrightarrow & \mathfrak{F}_n^{\flat} \end{array}$$

commutes, where the horizontal arrows are the left group actions.

Proof: The uniqueness is immediate from continuity considerations. Moreover, (2.18) implies that if the action exists we must have that

$g(W,c) = ((pg)W, \tilde{\alpha}(g,W)c)$ for all $g \in Mp(n,\mathbb{R})$ and $(W,c) \in \tilde{\mathfrak{F}}^b_n$, where $\tilde{\alpha}$ is a smooth $ML(n,\mathbb{C})$-valued $Mp(n,\mathbb{R})$-cocycle on $\overline{D}_n$ such that $\tilde{\alpha}(1,W) = 1$ for all $W \in \overline{D}_n$ and such that the diagram

$$\begin{array}{ccc} Mp(n,\mathbb{R}) \times \overline{D}_n & \xrightarrow{\tilde{\alpha}} & ML(n,\mathbb{C}) \\ \big\downarrow p \times id & & \big\downarrow p \\ Sp(n,\mathbb{R}) \times \overline{D}_n & \xrightarrow{\alpha} & GL(n,\mathbb{C}) \end{array} \tag{2.19}$$

commutes. Conversely, the existence of such a cocycle would imply the existence of the desired action.

As usual, we identify $U(n)$ as a subgroup of $Sp(n,\mathbb{R})$ by means of

$$A + iB \longmapsto \begin{bmatrix} A & -B \\ B & A \end{bmatrix}. \tag{2.20}$$

$U(n)$ is a maximal compact subgroup of $GL(n,\mathbb{C})$ and of $Sp(n,\mathbb{R})$. We let $MU(n)$ denote the double cover of $U(n)$. $MU(n)$ is a maximal compact subgroup of $ML(n,\mathbb{C})$ and of $Mp(n,\mathbb{R})$.

Let $f = \alpha \circ (p \times id) : Mp(n,\mathbb{R}) \times \overline{D}_n \to GL(n,\mathbb{C})$. Then $f(1,0) = 1$. Moreover, any loop in $Mp(n,\mathbb{R}) \times \overline{D}_n$ based at $(1,0)$ is homotopic to a loop in $MU(n) \times \{0\}$ based at $(1,0)$. So let γ be a loop in $MU(n)$ based at 1 and let ε be the constant loop at 0 in $\overline{D}_n$. The explicit formula for α in (2.13) implies that $f \circ (\gamma \times \varepsilon) = p \circ \gamma$. But this says that $f_*\pi_1(Mp(n,\mathbb{R}) \times \overline{D}_n, (1,0)) = p_*\pi_1(ML(n,\mathbb{C}),1)$. By the Covering Space Lifting Theorem ([12], Theorem 5, p. 76), there is a unique smooth map $\tilde{\alpha}$ completing (2.19) such that $\tilde{\alpha}(1,0) = 1$. It is an easy consequence of continuity that $\tilde{\alpha}$ is a cocycle such that $\tilde{\alpha}(1,W) = 1$ for all $W \in \overline{D}_n$. Q.E.D.

3. Half-forms and their pairing

Let E be a vector space of dimension n over $K = \mathbb{C}$ or $\mathbb{R}$. A density of order α is a map γ from $\mathfrak{F}(E)$ to $\mathbb{C}$ which transforms under coordinate changes according to $|\mathrm{Det}|^{\alpha}$ (cf. [3], p. 150). An equivalent, but more convenient from our viewpoint, definition is the following:

<u>Definition</u> 3.1: The space $\mathcal{D}^{\alpha}(E)$ of <u>densities of order</u> α (α-<u>densities</u>) <u>on</u> E is $\mathfrak{F}(E) \times_{GL(n,K)} \mathbb{C}$, where the left action of $C \in GL(n,K)$ on $\mathbb{C}$ is left multiplication by $|\mathrm{Det}\ C|^{-\alpha}$. If $\gamma \in \mathcal{D}^{\alpha}(E)$, if $\underline{e} \in \mathfrak{F}(E)$, and if $\gamma = \underline{e} \times_{GL(n,K)} \lambda$, $\lambda \in \mathbb{C}$, then λ is the <u>value</u> of γ on $\underline{e}$, and we write $\lambda = \gamma(\underline{e})$.

Clearly, if $C \in GL(n,K)$, then

$$(3.2) \qquad \gamma(\underline{e}\ C) = |\mathrm{Det}\ C|^{\alpha}\ \gamma(\underline{e}).$$

Moreover, since $\mathfrak{F}(E)$ is a right principal homogenous space for $GL(n,\mathbb{C})$, $\gamma(\underline{e})$ is defined for all $\underline{e} \in \mathfrak{F}(E)$. Given $\lambda \in \mathbb{C}$ and $\underline{e} \in \mathfrak{F}(E)$, there is a unique $\gamma \in \mathcal{D}^{\alpha}(E)$ such that $\gamma(\underline{e}) = \lambda$. Clearly, $\mathcal{D}^{\alpha}(E)$ is a one dimensional complex vector space. Thus (3.1) is a good definition of the concept of α-density.

Let $0 \to E_1 \to E_2 \to E_3 \to 0$ be an exact sequence of finite dimensional vector spaces. We shall define an isomorphism

$$(3.3) \qquad \mu : \mathcal{D}^{\alpha}(E_1) \otimes \mathcal{D}^{\alpha}(E_3) \to \mathcal{D}^{\alpha}(E_2),$$

as follows: Let $\gamma_1 = (v_1,\ldots,v_n) \times_{GL(n,K)} \lambda_1 \in \mathcal{D}^{\alpha}(E_1)$ and $\gamma_2 = (\hat{w}_1,\ldots,\hat{w}_m) \times_{GL(m,K)} \lambda_2 \in \mathcal{D}^{\alpha}(E_3)$. Choose vectors $w_j \in E_2$ mapping onto $\hat{w}_j \in E_3$. Then $(v_1,\ldots,v_n, w_1,\ldots,w_m) \in \mathfrak{F}(E_2)$ and we set

$$(3.4) \qquad \mu(\gamma_1 \otimes \gamma_2) = (v_1,\ldots,v_n, w_1,\ldots,w_m) \times_{GL(n+m,K)} \lambda_1\lambda_2.$$

<u>Lemma</u> 3.5: μ is a well defined isomorphism.

Proof: The only part needing proof is to show that μ is well defined. Let

$(v_1',\dots,v_n') \in \mathfrak{F}(E_1)$ and $(\hat{w}_1',\dots,\hat{w}_m') \in \mathfrak{F}(E_3)$. Then $(v_1,\dots,v_n) = (v_1',\dots,v_n')C_1$ and $(\hat{w}_1,\dots,\hat{w}_m) = (\hat{w}_1',\dots,\hat{w}_m')C_2$ for unique $C_1 \in GL(n,K)$ and $C_2 \in GL(m,K)$. Choose w_j' mapping into $\hat{w}_j'$. Then $(v_1,\dots,v_n, w_1,\dots,w_m) =$

$(v_1',\dots,v_n', w_1',\dots,w_m')\begin{bmatrix} C_1 & C_3 \\ 0 & C_2 \end{bmatrix}$ for some $n \times m$ complex matrix C_3. Now $\gamma_1 = (v_1',\dots,v_n') \times_{GL(n,K)} |\mathrm{Det}\, C_1|^{\alpha} \lambda_1$ and $\gamma_2 = (\hat{w}_1',\dots,\hat{w}_m') \times_{GL(m,K)} |\mathrm{Det}\, C_2|^{\alpha} \lambda_2$. Therefore $\mu(\gamma_1 \otimes \gamma_2)$, calculated using the primed frames, is

$$(v_1',\dots,v_n', w_1',\dots,w_m') \times_{GL(n+m,K)} |\mathrm{Det}\, C_1|^{\alpha}|\mathrm{Det}\, C_2|^{\alpha} \lambda_1\lambda_2 =$$

$$(v_1',\dots,v_n', w_1',\dots,w_m') \times_{GL(n+m,K)} \left|\mathrm{Det} \begin{bmatrix} C_1 & C_3 \\ 0 & C_2 \end{bmatrix}\right|^{\alpha} \lambda_1\lambda_2 =$$

$$(v_1,\dots,v_n, w_1,\dots,w_m) \times_{GL(n+m,K)} \lambda_1\lambda_2,$$ as desired.

Now $\mathfrak{D}^{\alpha}$ can be regarded as a functor on the category $\mathcal{V}$ in the obvious way: Let $\sigma : E_1 \to E_2$ be an isomorphism of n-dimensional vector spaces. Then $\mathfrak{F}(\sigma) : \mathfrak{F}(E_1) \to \mathfrak{F}(E_2)$ is right $GL(n,K)$-equivariant. Hence we get a morphism $\mathfrak{D}^{\alpha}(\sigma) : \mathfrak{D}^{\alpha}(E_1) \to \mathfrak{D}^{\alpha}(E_2)$ as follows: If $\underline{e}_1 \in \mathfrak{F}(E_1)$, $\lambda \in \mathbb{C}$, then

(3.6) $$\mathfrak{D}^{\alpha}(\sigma) : \underline{e}_1 \times_{GL(n,K)} \lambda \mapsto (\mathfrak{F}(\sigma)\underline{e}_1) \times_{GL(n,K)} \lambda,$$

which is well defined, linear, and a bijection. As in the case of the functors of Section 2, we can apply $\mathfrak{D}^{\alpha}$ to vector bundles.

For any vector space E of dimension n, let $G_k(E)$ denote the Grassman manifold of k dimensional linear subspaces of E. We let $G_k^{\#}(E)$ [resp. $G_k^{\flat}(E)$] denote the vector bundle over $G_k(E)$ with bundle projection ρ whose fibre over $V \in G_k(E)$ is just V [resp. E/V]. If Ω is a non-degenerate bilinear form on E, then $V \mapsto V^{\perp}$ establishes a diffeomorphism of $G_k(E)$ onto $G_{n-k}(E)$. Moreover, the map $\Omega_{\#} : \mathfrak{F}(V) \to \mathfrak{F}(E/V^{\perp})$ gives rise to a map from $\mathfrak{D}^{\alpha}(V) \to \mathfrak{D}^{-\alpha}(E/V^{\perp})$ in view of (2.2). We call this map $\Omega_{\#}$ also and we have that the following diagram

$$(3.7)\qquad \begin{array}{ccc} \mathcal{D}^{\alpha}(\mathcal{G}_k^{\#}(E)) & \xrightarrow{\Omega_{\#}} & \mathcal{D}^{-\alpha}(\mathcal{G}_{n-k}^{\flat}(E)) \\ \downarrow \rho & & \downarrow \rho \\ \mathcal{G}_k(E) & \xrightarrow{\perp} & \mathcal{G}_{n-k}(E) \end{array}$$

commutes.

Let (E,Ω) be a symplectic vector space of dimension $2n$ over $K = \mathbb{C}$ or $\mathbb{R}$.

<u>Definition</u> 3.8: A frame $(v_1,\ldots,v_n, w_1,\ldots,w_n) \in \mathcal{F}(E)$ is called <u>symplectic</u> if $\Omega(v_i,v_j) = \Omega(w_i,w_j) = 0$ and $\Omega(v_i,w_j) = \delta_{ij}$ for $i,j = 1,\ldots,n$. The set of all symplectic frames of (E,Ω) will be denoted by $\mathcal{B}(E,\Omega)$.

Thus $\mathcal{B}(E,\Omega) \subseteq \mathcal{F}(E)$. Now $Sp(n,K)$ is a subgroup of $GL(2n,K)$ and so operates on $\mathcal{F}(E)$ on the right. The following lemma is an easy exercise.

<u>Lemma</u> 3.9: $\mathcal{B}(E,\Omega)$ is an $Sp(n,K)$ - orbit in $\mathcal{F}(E)$ so that $\mathcal{B}(E,\Omega)$ is a right principal $Sp(n,K)$ - space.

Now $g \in Sp(n,K)$ implies $\operatorname{Det} g = 1$. Therefore every symplectic (E,Ω) has a canonical α-density given by $\underline{e} \times_{GL(2n,K)} 1$, where $\underline{e}$ is any frame in $\mathcal{B}(E,\Omega)$. We will denote this α-density by $|\Omega^n|^{\alpha}$. It will be called the <u>Liouville</u> α-density on (E,Ω).

Now let us return to $(\mathbb{C}^{2n},\Omega,-)$. We shall write $\mathcal{G}_{n,k}$ for $\mathcal{G}_k(\mathbb{C}^n)$, $\mathcal{G}_{n,k}^{\#}$ for $\mathcal{G}_k^{\#}(\mathbb{C}^n)$, $\mathcal{G}_{n,k}^{\flat}$ for $\mathcal{G}_k^{\flat}(\mathbb{C}^n)$, $\mathcal{D}_{n,k}^{\alpha,\#}$ for $\mathcal{D}^{\alpha}(\mathcal{G}_{n,k}^{\#})$, and $\mathcal{D}_{n,k}^{\alpha,\flat}$ for $\mathcal{D}^{\alpha}(\mathcal{G}_{n,k}^{\flat})$. We wish to define the bundle of half-forms normal to positive lagrangian subspaces of $\mathbb{C}^{2n}$ in a way analogous to the way in which $\mathcal{D}_{n,k}^{\alpha,\flat}$ is defined. This is done as follows:

<u>Definition</u> 3.10: The bundle $\mathcal{E}_n^{1/2}$ of <u>half-forms normal to positive lagrangian subspaces of</u> $\mathbb{C}^{2n}$ is just $\tilde{\mathcal{F}}_n^{\flat} \times_{ML(n,\mathbb{C})} \mathbb{C}$, where the left action of $c \in ML(n,\mathbb{C})$ on $\mathbb{C}$ is left multiplication by $\chi(c)^{-1}$, where χ is the unique holomorphic character on $ML(n,\mathbb{C})$ such that $\chi^2 = \operatorname{Det} \circ p$. The bundle

projection ρ of $\mathcal{E}_n^{1/2}$ onto $\mathcal{L}_n$ is just the map induced by $\rho : \tilde{\mathfrak{F}}_n^b \to \mathcal{L}_n$. If $\beta \in \mathcal{E}_n^{1/2}$, if $\underset{\sim}{\tilde{e}} \in \tilde{\mathfrak{F}}_n^b$ is such that $\rho\underset{\sim}{\tilde{e}} = \rho\beta$, and if $\beta = \underset{\sim}{\tilde{e}} \times_{ML(n,C)} \lambda$, $\lambda \in C$, then λ is the value of β on $\underset{\sim}{\tilde{e}}$, and we write $\lambda = \beta(\underset{\sim}{\tilde{e}})$.

The problem to be solved in this section is that of extending the pairing of half-forms as discussed in [3] and [4] to cover the case of positive complex lagrangian subspaces. Here we will deal exclusively with the theory in C^{2n} as a model of what happens at a point on a metaplectic manifold. The global situation on metaplectic manifolds is discussed in the next section.

Now the philosophy of [4] tells us that if $\beta_1, \beta_2 \in \mathcal{E}_n^{1/2}$, then we must construct a density $\langle\beta_1,\beta_2\rangle$ of order 1, depending linearly on β_1 and conjugate linearly on β_2, and normal to some subspace of C^{2n}. The first question is: which subspace? Let $V_i = \rho\,\beta_i$ and suppose that $\overline{V}_i = V_i$, $i = 1,2$. We are then in the situation in which V_i is the complexification of a real lagrangian subspace of $\mathbb{R}^{2n}$, which is the situation dealt with in [4]. In this case, therefore, we want $\langle\beta_1,\beta_2\rangle \in \mathfrak{D}^1(C^{2n}/(V_1 \cap V_2))$. On the other hand, suppose that $V_1 = V_2 = V$, say. Then V gives rise to a partially complex structure on $\mathbb{R}^{2n}$ and we would expect from the work of Gel'fand and Graev [7], Dixmier [6], and Auslander and Kostant [1] that $\langle\beta_1,\beta_2\rangle \in \mathfrak{D}^1(C^{2n}/(V \cap \overline{V}))$. There are indications in ([2], Theorem 3) that the natural generalization of this would be to try to define $\langle\beta_1,\beta_2\rangle$ in $\mathfrak{D}^1(C^{2n}/(V_1 \cap \overline{V}_2))$. We shall do just that.

An additional condition we would like to have fulfilled is that $\langle\beta_1,\beta_2\rangle = \overline{\langle\beta_2,\beta_1\rangle}$. For this even to make sense, we need

<u>Lemma</u> 3.11: Let (E,Ω,J) be as in Section 2. Let $V_1,V_2 \in \mathcal{L}_+(E,\Omega,J)$. Then $V_1 \cap JV_2 = V_2 \cap JV_1$.

Proof: If $V \in \mathcal{L}_+(E,\Omega,J)$, then $\langle\cdot,\cdot\rangle$, as defined in (2.6), is positive semi-definite on V. The Schwarz inequality then implies that $\{v \in V : \langle v,v\rangle = 0\} = \{v \in V : \langle v,w\rangle = 0 \text{ for all } w \in V\} = V \cap (JV)^{\perp}$, and this is just $V \cap JV$ because JV is maximally isotropic. Now $\langle\cdot,\cdot\rangle$ is positive semi-definite on V_1 and negative semi-definite on JV_2 and hence vanishes identically on $V_1 \cap JV_2$. Therefore, by the first part of this proof, $V_1 \cap JV_2 \subseteq V_1 \cap JV_1$. Similarly $V_2 \cap JV_1 \subseteq V_2 \cap JV_2$ so that $V_1 \cap JV_2 \subseteq V_2 \cap JV_2$ also. Therefore $V_1 \cap JV_2 = (V_1 \cap JV_1) \cap (V_2 \cap JV_2)$, and our lemma follows from symmetry.

In defining $\langle\beta_1,\beta_2\rangle$, it will be convenient to make use of an isomorphism derived from (3.5) and (3.7). Let (E,Ω) be a symplectic vector space over $\mathbb{C}$ or $\mathbb{R}$. Let V be an isotropic subspace of E. Then Ω induces a symplectic bilinear form Ω_V on $V^\perp/V$. The exact sequence $0 \to V^\perp/V \to E/V \to E/V^\perp \to 0$ gives an isomorphism $\mathcal{S}^\alpha(E/V^\perp) \to \mathcal{S}^\alpha(E/V)$ according to $\gamma \to \mu(|\Omega_V|^\alpha \otimes \gamma)$, where $\dim V = n - k$. Composing with $\Omega_\#$ we get the isomorphism

$$(3.12) \qquad \Omega^\circ : \mathcal{S}^{-\alpha}(V) \to \mathcal{S}^\alpha(E/V)$$

$$\gamma \mapsto \mu(|\Omega_V|^\alpha \otimes \Omega_\#\gamma).$$

Thus we can define $\langle\beta_1,\beta_2\rangle$ as a member of $\Omega^\circ\mathcal{S}^{-1}(V_1 \cap \overline{V}_2)$, say $\langle\beta_1,\beta_2\rangle = \Omega^\circ\langle\beta_1,\beta_2\rangle_\circ$.

Before giving the formal definition of $\langle\beta_1,\beta_2\rangle_\circ$, we shall sketch the basic idea behind it in the transverse case: $V_1 \cap \overline{V}_2 = \{0\}$. Since $\mathfrak{F}(V_1 \cap \overline{V}_2) = \{\emptyset\}$, $\langle\beta_1,\beta_2\rangle_\circ$ will just be a complex number. Now Ω is a non-degenerate bilinear form on $V_1 \times \overline{V}_2$ so that $\langle\cdot,\cdot\rangle$ (2.6) is a non-degenerate sesquilinear form on $V_1 \times V_2$. Thus $(v_1,\ldots,v_n) \in \mathfrak{F}(V_1)$ and $(w_1,\ldots,w_n) \in \mathfrak{F}(V_2)$ will be dual under $\langle\cdot,\cdot\rangle$ if $\langle v_i,w_j\rangle = \delta_{ij}$ for $i,j = 1,\ldots,n$. Such dual pairs exist. Set $\varphi^\#(W_1,C_1) = (v_1,\ldots,v_n)$ and $\varphi^\#(W_2,C_2) = (w_1,\ldots,w_n)$. Then the condition for being a dual pair is just

$$(3.13) \qquad \frac{i}{2\pi\hbar}\, \varphi^\#(W_2,C_2)^* \, J \, \varphi(W_1,C_1) = I,$$

where $J = \begin{bmatrix} 0 & -I \\ I & 0 \end{bmatrix}$ is the matrix of the bilinear form Ω. When simplified, (3.13) becomes

$$\overline{C}_2^{-1}\{\tfrac{i}{4\pi\hbar}(I - \overline{W}_2W_1)\}^t C_1^{-1} = I. \tag{3.14}$$

The idea is to replace $C_j \in GL(n,\mathbb{C})$ by $c_j \in ML(n,\mathbb{C})$ and to interpret (3.14) as an equation in $ML(n,\mathbb{C})$. We then set $\langle\beta_1,\beta_2\rangle_\circ = \beta_1(W_1,c_1)\overline{\beta_2(W_2,c_2)}$ and show that this is independent of the choice of "dual" metalinear frames. It will be somewhat more convenient to define $\langle\beta_1,\beta_2\rangle_\circ$ as $\chi(\overline{c}_2^{-1}\{\frac{1}{4\pi\hbar}(I - \overline{W}_2W_1)\}^t c_1^{-1})\ \beta_1(W_1,c_1)\overline{\beta_2(W_2,c_2)}$ for arbitrary metalinear frames normal to V_1 and V_2, respectively. This is the formulation we shall generalize to the non-transverse case.

So let $(z_{k+1},\ldots,z_n) \in \mathfrak{F}(V_1 \cap \overline{V}_2)$. As usual, we regard this as a $2n \times (n - k)$ complex matrix Z. Choose $(W_j,c_j) \in \tilde{\mathfrak{F}}_n^\flat$ such that $\rho(W_j,c_j) = V_j$. Now (3.11) implies that $V_1 \cap \overline{V}_2 \subseteq V_1 \cap V_2$. And $\varphi^\#(W_j,I) \in \mathfrak{F}(V_j)$. Hence the columns of Z are linear combinations of the columns of $\varphi^\#(W_j,I)$, and we write

$$Z = \varphi^\#(W_j,I)\hat{Z}_j = \begin{bmatrix} \frac{1}{2}(W_j + I) \\ \frac{1}{2i}(W_j - I) \end{bmatrix}\hat{Z}_j, \tag{3.15}$$

where $\hat{Z}_j$ is a complex $n \times (n - k)$ matrix. But the equations $\frac{1}{2}(W_1 + I)\hat{Z}_1 = \frac{1}{2}(W_2 + I)\hat{Z}_2$ and $\frac{1}{2i}(W_1 - I)\hat{Z}_1 = \frac{1}{2i}(W_2 - I)\hat{Z}_2$ show that $\hat{Z}_1 = \hat{Z}_2$. We therefore drop the subscript and simply write $\hat{Z}$. $\hat{Z}$ has rank $n - k$.

<u>Lemma</u> 3.16: The matrices $I - \overline{W}_2W_1$ and $I - \overline{W}_1W_2$ have the same null space, and this is equal to the range of $\hat{Z}$.

Proof: $\overline{\varphi^\#(W_j,I)} \in \mathfrak{F}(\overline{V}_j)$. Let $w \in \mathbb{C}^{2n}$. Then $w \in V_1 \cap \overline{V}_2$ if and only if there exist column vectors $v_1,v_2 \in \mathbb{C}^n$ such that $w = \varphi^\#(W_1,I)v_1 = \overline{\varphi^\#(W_2,I)}v_2$.

But this says that $\frac{1}{2}(W_1 + I)v_1 = \frac{1}{2}(\overline{W}_2 + I)v_2$ and $\frac{1}{2i}(W_1 - I)v_1 = -\frac{1}{2i}(\overline{W}_2 - I)v_2$. Therefore $W_1v_1 = v_2$ and $\overline{W}_2v_2 = v_1$, and we have that v_1 is in the null space $n(I - \overline{W}_2W_1)$ of $I - \overline{W}_2W_1$. Conversely, if v_1 is in that null space, we may set $v_2 = W_1v_1$ and obtain $\varphi^{\#}(W_1,I)v_1 = \overline{\varphi^{\#}(W_2,I)v_2} = w$, say, in $V_1 \cap \overline{V}_2$. This implies that the range $\mathcal{R}(\hat{Z})$ of $\hat{Z}$ precisely equals $n(I - \overline{W}_2W_1)$. Reversing the roles of V_1 and V_2, we obtain the lemma.

Now $n(I - \overline{W}_2W_1) = n(I - \overline{W}_1W_2) = \mathcal{R}((I - \overline{W}_1W_2)^*)^{\perp} = \mathcal{R}(I - \overline{W}_2W_1)^{\perp}$ because of (3.16) and the fact that ${}^tW_j = W_j$. Here orthogonal complement is taken with respect to the usual hermitian inner product $(\cdot,\cdot)$ on $\mathbb{C}^n$. Let T_Z be the unique $n \times n$ complex matrix such that

(a) $n(T_Z) = \mathcal{R}(I - \overline{W}_2W_1)$

(3.17) (b) $\mathcal{R}(T_Z) = n(I - \overline{W}_2W_1)$

(c) $\hat{Z}^*T_Z\hat{Z} = I_{n-k}$, the identity of $GL(n - k, \mathbb{C})$.

Let $\Lambda_n = \{S \in GL(n,\mathbb{C}) : Re(Sv,v) \geq 0 \text{ for } v \in \mathbb{C}^n\}$.

Lemma 3.18: $\frac{1}{4\pi\hbar}(I - \overline{W}_2W_1) + T_Z \in \Lambda_n$.

Proof: By construction, $T_Z \geq 0$ and $\frac{1}{4\pi\hbar}(I - \overline{W}_2W_1) + T_Z \in GL(n,\mathbb{C})$. Therefore it suffices to show that $Re((I - \overline{W}_2W_1)v,v) \geq 0$ for all $v \in \mathbb{C}^n$. But this is an immediate consequence of the fact that $\|\overline{W}_2W_1\| \leq 1$.

Clearly Λ_n is a contractible closed subset of $GL(n,\mathbb{C})$ containing I. Therefore $p^{-1}\Lambda_n$ has two components, each diffeomorphic to Λ_n under p, where p is the projection of $ML(n,\mathbb{C})$ onto $GL(n,\mathbb{C})$. One of these components contains 1, and that component we henceforth denote by $\tilde{\Lambda}_n$ and identify (via p) with Λ_n. Clearly also the anti-holomorphic isomorphism $^-$ and the holomorphic anti-isomorphism ^{t} on $GL(n,\mathbb{C})$ lift uniquely to $ML(n,\mathbb{C})$. Using these facts and (3.18) we make the following

Definition 3.19: Let $\beta_1, \beta_2 \in \mathcal{E}_n^{1/2}$. Set $V_j = \rho\,\beta_j \in \mathcal{L}_n$ and $n - k = \dim(V_1 \cap \overline{V}_2)$. Then $\langle \beta_1, \beta_2 \rangle_\circ \in \wp^{-1}(V_1 \cap \overline{V}_2)$ is defined as
$Z \times_{GL(n-k,\mathbb{C})} \chi(\overline{c}_2^{-1}\{\frac{1}{4\pi h}(I - \overline{W}_2 W_1) + T_Z\}^t c_1^{-1})\ \beta_1(W_1, c_1)\overline{\beta_2(W_2, c_2)}$, where $Z \in \mathfrak{F}(V_1 \cap \overline{V}_2)$ and $(W_j, c_j) \in \tilde{\mathfrak{F}}_n^b$ are arbitrary subject to $\rho(W_j, c_j) = V_j$, and where $\frac{1}{4\pi h}(I - \overline{W}_2 W_1) + T_Z$ is regarded as a member of $\tilde{\Lambda}_n$.

We must check that this definition makes sense. Now $\check{\varphi}(W_j) = V_j$ according to (2.16), so that W_1 and W_2 are fixed. Moreover (3.10) and the facts that $\chi({}^t c) = \chi(c)$ and $\chi(\overline{c}) = \overline{\chi(c)}$ imply that $\langle \beta_1, \beta_2 \rangle_\circ$ is independent of the choice of c_1 and c_2. We are left with showing that (3.19) is independent of the choice of Z. That will follow from the next theorem, which gives an interpretation of $\langle \beta_1, \beta_2 \rangle_\circ$ in the spirit of the paragraph containing (3.13) and (3.14).

In order to formulate the theorem, we introduce a subset $\Lambda_{n,k}$ of $GL(n,\mathbb{C})$ consisting of block matrices $\begin{bmatrix} S_1 & 0 \\ S_2 & I_{n-k} \end{bmatrix}$, where $S_1 \in \Lambda_k$ and S_2 is an arbitrary $(n-k) \times n$ matrix. The subgroup of $\Lambda_{n,k}$ consisting of those matrices for which $S_1 = I$ will be denoted by $G_{n,k}$. Clearly $G_{n,k}\,\Lambda_{n,k}\,G_{n,k} = \Lambda_{n,k}$. Now $\Lambda_{n,k}$ is contractible and contains I. Letting $\tilde{\Lambda}_{n,k}$ denote the component of $p^{-1}\Lambda_{n,k}$ in $ML(n,\mathbb{C})$ containing 1, we identify $\tilde{\Lambda}_{n,k}$ and $\Lambda_{n,k}$ via p.

Theorem 3.20: Let $Z = (z_{k+1}, \ldots, z_n) \in \mathfrak{F}(V_1 \cap \overline{V}_2)$ be given. There exist frames $\underline{e}_1 = (v_1, \ldots, v_k, z_{k+1}, \ldots, z_n) \in \mathfrak{F}(V_1)$ and $\underline{e}_2 = (w_1, \ldots, w_k, z_{k+1}, \ldots, z_n) \in \mathfrak{F}(V_2)$ such that $\langle v_i, w_j \rangle = \delta_{ij}$ for $i,j = 1,\ldots,k$. For any such $\underline{e}_1$ and $\underline{e}_2$ set $(W_j, C_j) = \varphi^{\#-1}\,\underline{e}_j$. Then ${}^tC_1\,{}^tC_2^{-1} \in \Lambda_{n,k}$. Choose $c_1, c_2 \in ML(n,\mathbb{C})$ such that $pc_j = C_j$ and ${}^tc_1\,{}^tc_2^{-1} \in \tilde{\Lambda}_{n,k}$. Then $\langle \beta_1, \beta_2 \rangle_\circ = Z \times_{GL(n-k,\mathbb{C})} \beta_1(W_1, c_1)\overline{\beta_2(W_2, c_2)}$.

Proof: The restriction of $\langle \cdot, \cdot \rangle$ (2.6) to $V_1 \times V_2$ vanishes on $(V_1 \cap \overline{V}_2) \times V_2$ and on $V_1 \times (V_2 \cap \overline{V}_1)$ and induces a non-degenerate sesquilinear form on $(V_1/(V_1 \cap \overline{V}_2)) \times (V_2/(V_2 \cap \overline{V}_1))$. Therefore frames of the type of $\underline{e}_1$ and $\underline{e}_2$ exist. Now an easy computation shows that

$$\langle \varphi^{\#}(W_1,I)u_1, \varphi^{\#}(W_2,I)u_2\rangle =$$

(3.21)

$$\frac{i}{2\pi\hbar} u_2^* \varphi^{\#}(W_2,I)^* J \varphi^{\#}(W_1,I)u_1 = \frac{1}{4\pi\hbar} u_2^*(I - \overline{W}_2 W_1)u_1.$$

Applying (3.21) to the columns of ${}^tC_1^{-1}$ and ${}^tC_2^{-1}$ we get

$$\frac{1}{4\pi\hbar} \overline{C}_2^{-1}(I - \overline{W}_2 W_1)^t C_1^{-1} = \begin{bmatrix} I_k & 0 \\ 0 & 0 \end{bmatrix}, \tag{3.22}$$

where the rows and columns of the block matrix are split according to $n = k + (n - k)$. Clearly (3.22) remains true if ${}^tC_j^{-1}$ is replaced by ${}^tC_j^{-1} B_j$, where $B_j \in G_{n,k}$.

Choose B_j so that the first k columns of ${}^tC_j^{-1} B_j$ span $\mathcal{R}(I - \overline{W}_2 W_1)$. Then we will have

$$B_2^* \overline{C}_2^{-1} T_Z {}^tC_1^{-1} B_1 = \begin{bmatrix} 0 & 0 \\ 0 & I_{n-k} \end{bmatrix}, \tag{3.23}$$

because of (3.17). From (3.22), (3.23), and (3.18) we get

$$\overline{C}_2 B_2^{*-1} B_1^{-1} {}^tC_1 = \frac{1}{4\pi\hbar}(I - \overline{W}_2 W_1) + T_Z \in \Lambda_n. \tag{3.24}$$

Now $S^* \Lambda_n S = \Lambda_n$ for all $S \in GL(n,C)$. Applying this to (3.24) with $S = {}^tC_2^{-1} B_2$, we get

$$B_1^{-1} {}^tC_1 {}^tC_2^{-1} B_2 = ({}^tC_2^{-1} B_2)^* \{\frac{1}{4\pi\hbar}(I - \overline{W}_2 W_1) + T_Z\}({}^tC_2^{-1} B_2) \in \Lambda_n. \tag{3.25}$$

Now the last $n - k$ columns of ${}^tC_j^{-1}$ comprise $\hat{Z}$, $j = 1,2$. Therefore ${}^tC_1 {}^tC_2^{-1}$ is of the form $\begin{bmatrix} S_1 & 0 \\ S_2 & I \end{bmatrix}$, with blocks split according to $n = k + (n - k)$, and hence so is $B_1^{-1} {}^tC_1 {}^tC_2^{-1} B_2$. Together with (3.25), this says that $B_1^{-1} {}^tC_1 {}^tC_2^{-1} B_2 \in \Lambda_{n,k}$. Therefore so does ${}^tC_1 {}^tC_2^{-1}$.

Now choose c_1 and c_2 as in the statement of the theorem. Let

$\tilde{G}_{n,k} = (p^{-1}G_{n,k}) \cap \tilde{\Lambda}_{n,k}$, a connected subgroup of $ML(n,C)$. Continuity of multiplication and the connectedness of $\tilde{G}_{n,k}$ and $\tilde{\Lambda}_{n,k}$ imply that $\tilde{G}_{n,k}\,\tilde{\Lambda}_{n,k}\,\tilde{G}_{n,k} = \tilde{\Lambda}_{n,k}$. Choosing $b_j \in \tilde{G}_{n,k}$ so that $p\,b_j = B_j$, we get that $b_1^{-1}\,{}^tc_1\,{}^tc_2^{-1}\,b_2 \in \tilde{\Lambda}_{n,k}$. Again, continuity of multiplication and of $*$ and the connectedness of $ML(n,C)$ and $\tilde{\Lambda}_n$ imply that $s^*\tilde{\Lambda}_n s = \tilde{\Lambda}_n$ for all $s \in ML(n,C)$. Therefore $({}^tc_2^{-1}\,b_2)^*\{\frac{1}{4\pi\hbar}(I - \overline{W}_2W_1) + T_Z\}({}^tc_2^{-1}\,b_2) \in \tilde{\Lambda}_n$. Moreover, by (3.25), this and $b_1^{-1}\,{}^tc_1\,{}^tc_2^{-1}\,b_2$ have the same image in $GL(n,C)$ under p. That they are in fact equal follows from the fact that $p(\tilde{\Lambda}_n \cap \tilde{\Lambda}_{n,k}) \supseteq \Lambda_n \cap \Lambda_{n,k}$, which is true because $\Lambda_n \cap \Lambda_{n,k}$ is connected and contains I.

We have shown that

$$(3.26) \qquad b_1^{-1}\,{}^tc_1\,{}^tc_2^{-1}\,b_2 = ({}^tc_2^{-1}\,b_2)^*\{\tfrac{1}{4\pi\hbar}(I - \overline{W}_2W_1) + T_Z\}({}^tc_2^{-1}\,b_2).$$

Now $\operatorname{Det} G_{n,k} = \{1\}$ so that $\chi^2(\tilde{G}_{n,k}) = \{1\}$. Since $\tilde{G}_{n,k}$ is a connected group, $\chi(\tilde{G}_{n,k}) = \{1\}$. Applying χ to (3.26), we get

$$\chi({}^tc_1) = \chi(\overline{c}_2^{-1})\chi(\tfrac{1}{4\pi\hbar}(I - \overline{W}_2W_1) + T_Z).$$

Applying this to (3.19) completes the proof of the theorem.

<u>Corollary</u> 3.27: $\langle\beta_1,\beta_2\rangle_o$ is independent of the choice of Z.

Proof: Let $Z,\underline{e}_1,\underline{e}_2,W_1,W_2,C_1,C_2,c_1,c_2$ be as in the theorem. Let $C \in GL(n-k,C)$, set $ZC = Z' = (z'_{k+1},\ldots,z'_n)$, $\underline{e}'_1 = (v_1,\ldots,v_k, z'_{k+1},\ldots,z'_n)$, and $\underline{e}'_2 = (w_1,\ldots,w_k, z'_{k+1},\ldots,z'_n)$. Set $(W_j,C'_j) = \varphi^{\#-1}\,\underline{e}'_j$ and choose $c'_1,c'_2 \in ML(n,C)$ such that $p\,c'_j = C'_j$ and ${}^tc'_1\,{}^tc'^{-1}_2 \in \tilde{\Lambda}_{n,k}$. Clearly C'_j depends continuously on C. p is a diffeomorphism of $\tilde{\Lambda}_{n,k}$ onto $\Lambda_{n,k}$. Since ${}^tC'_1\,{}^tC'^{-1}_2$ is continuous in C, so is ${}^tc'_1\,{}^tc'^{-1}_2$, and hence so is $c'^{-1}_2\,c'_1$. Now the map $(s,t) \to s\,t\,s^*$ is continuous on $ML(n,C) \times ML(n,C)$. Let $\{1,y\} = p^{-1}\{I\}$, $p : ML(n,C) \to GL(n,C)$. Since $y^* = y$ and $y^2 = 1$, we have that $s\,t\,s^*$ depends only on (ps,t), continuously. Therefore

$c_1' c_2'^* = c_2'(c_2'^{-1} c_1')c_2'^*$ is continuous in C since $(p\ c_2', c_2'^{-1} c_1')$ is.

With notation as in (3.15-16), $\widehat{ZC} = \hat{Z}C$. Moreover the first k columns of ${}^tC_j'^{-1}$ and of ${}^tC_j^{-1}$ are the same, whereas the last $n - k$ columns of ${}^tC_j'^{-1}$ comprise $\hat{Z}C$. It follows that $\operatorname{Det} C_j' = (\operatorname{Det} C_j)(\operatorname{Det} C)^{-1}$. Therefore $\chi^2(c_1' c_2'^*) = \chi^2(c_1 c_2^*)|\operatorname{Det} C|^{-2}$. By continuity and the connectedness of $GL(n - k, \mathbb{C})$ we get

$$\chi(c_1' c_2'^*) = \chi(c_1 c_2^*)|\operatorname{Det} C|^{-1}. \tag{3.28}$$

Therefore $\beta_1(W_1,c_1')\overline{\beta_2(W_2,c_2')} = \beta_1(W_1,c_1(c_1^{-1} c_1'))\overline{\beta_2(W_2,c_2(c_2^{-1} c_2'))} = \chi(c_1^{-1} c_1')\chi(c_2^{-1} c_2')\ \beta_1(W_1,c_1)\overline{\beta_2(W_2,c_2)} = |\operatorname{Det} C|^{-1}\ \beta_1(W_1,c_1)\overline{\beta_2(W_2,c_2)}$ by (3.28) and the fact that $\chi(c^*) = \overline{\chi(c)}$, $c \in ML(n,\mathbb{C})$. The corollary now follows from the theorem together with (3.1).

We finish this section by showing that our theory is invariant under the left action of $Mp(n, \mathbb{R})$, a fact needed in formulating the theory on metaplectic manifolds in the next section. We have constructed a pairing map $\langle\cdot,\cdot\rangle$ making the following diagram commute:

$$\begin{array}{ccc} \mathcal{E}_n^{1/2} \times \mathcal{E}_n^{1/2} & \xrightarrow{\langle\cdot,\cdot\rangle} & \bigcup_{k=0}^{n} \mathcal{D}^{1,\flat}_{2n,n-k} \\ \Big\downarrow \rho \times \rho & & \Big\downarrow \rho \\ \mathcal{L}_n \times \mathcal{L}_n & \xrightarrow{\cdot\cap\bar{\cdot}} & \bigcup_{k=0}^{n} \mathcal{G}_{2n,2-k} \end{array} \tag{3.29}$$

Each object in the diagram has a left action by $Mp(n, \mathbb{R})$, or by $Sp(n, \mathbb{R})$ and hence by $Mp(n, \mathbb{R})$ via ρ.

Theorem 3.30: Diagram (3.20) is $Mp(n, \mathbb{R})$-equivariant.

Proof: Everything is obvious except for the equivariance of $\langle\cdot,\cdot\rangle$. We first show that $\langle g\ \beta_1, g\ \beta_2\rangle_\circ(g\ Z) = \langle\beta_1,\beta_2\rangle_\circ(Z)$ for all $\beta_1,\beta_2 \in \mathcal{E}_n^{1/2}$, all $Z \in \mathcal{K}(\rho\ \beta_1) \cap \overline{(\rho\ \beta_2)})$, and all $g \in Mp(n, \mathbb{R})$. Given Z, choose $\underline{e}_j$, define (W_j,C_j), and choose c_j as in Theorem 3.20. Then $g\,\underline{e}_j$ and $g(W_j,C_j)$ are good data for $g\,Z$, as follows from $\langle g\,v_i,\, g\,w_j\rangle = \langle v_i,w_j\rangle = \delta_{ij}$. Now $g(W_j,C_j) = (g\,W_j,\, \alpha(g,W_j)C_j)$ by (2.13) and $g(W_j,c_j) = (g\,W_j,\, \tilde{\alpha}(g,W_j)c_j)$ by the proof of (2.17). But $g \to \tilde{\alpha}(g,W_j)c_j$ is continuous and hence so is $g \mapsto {}^t(\tilde{\alpha}(g,W_1)c_1)^t(\tilde{\alpha}(g,W_2)c_2)^{-1}$. Since this maps under p to ${}^t(\alpha(g,W_1)C_1)^t(\alpha(g,W_2)C_2)^{-1} \in \Lambda_{n,k}$, we conclude by continuity and connectedness that ${}^t(\tilde{\alpha}(g,W_1)c_1)^t(\tilde{\alpha}(g,W_2)c_2)^{-1} \in \tilde{\Lambda}_{n,k}$ for all $g \in Mp(n,\mathbb{R})$. Therefore $\langle g\,\beta_1,\, g\,\beta_2\rangle_\circ(g\,Z) = (g\,\beta_1)(g\,W_1,\, \tilde{\alpha}(g,W_1)c_1)\overline{(g\,\beta_2)(g\,W_2,\, \tilde{\alpha}(g,W_2)c_2)} = (g\,\beta_1)(g(W_1,c_1))\overline{(g\,\beta_2)(g(W_2,c_2))}$. But $(g\,\beta_j)(g(W_j,c_j)) = \beta_j(W_j,c_j)$, so that $\langle g\,\beta_1,\, g\,\beta_2\rangle_\circ(g\,Z) = \langle\beta_1,\beta_2\rangle_\circ(Z)$. We have proved that $\langle g\,\beta_1,\, g\,\beta_2\rangle_\circ = g\langle\beta_1,\beta_2\rangle_\circ$.

To finish the proof, let V be an isotropic subspace of $\mathbb{C}^{2n}$. Observe that the diagram

$$(3.31)\qquad \begin{array}{ccccccccc} & & V^\perp/V & \longrightarrow & \mathbb{C}^{2n}/V & \longrightarrow & \mathbb{C}^{2n}/V^\perp & & \\ & \nearrow & \downarrow g & & \downarrow g & & \downarrow g & \searrow & \\ 0 & & & & & & & & 0 \\ & \searrow & & & & & & \nearrow & \\ & & (g\,V)^\perp/g\,V & \longrightarrow & \mathbb{C}^{2n}/g\,V & \longrightarrow & \mathbb{C}^{2n}/(g\,V)^\perp & & \end{array}$$

is exact and commutative and that the vertical induced maps are isomorphisms. The first vertical map is a symplectomorphism. Therefore $g|\Omega_V|^\alpha = |\Omega_{gV}|^\alpha$. Moreover, if $\gamma \in \mathcal{D}^{-\alpha}(V)$, then $\Omega_\#(g\,\gamma) = g(\Omega_\#\gamma)$, according to (2.5), since g preserves Ω. It is an easy matter to check, using (3.4) and (3.31), that $\mu(g\,\gamma_1 \otimes g\,\gamma_2) = g\,\mu(\gamma_1 \otimes \gamma_2)$ for $\gamma_1 \in \mathcal{D}^\alpha(V^\perp/V)$ and $\gamma_2 \in \mathcal{D}^\alpha(\mathbb{C}^{2n}/V^\perp)$. Therefore Ω°, as defined in (3.12), is $Mp(n,\mathbb{R})$-equivariant. Therefore $\langle g\,\beta_1,\, g\,\beta_2\rangle = \Omega^\circ g\langle\beta_1,\beta_2\rangle_\circ = g\langle\beta_1,\beta_2\rangle$, as desired.

<u>Remark</u> 3.32: Let $\beta_1,\beta_2,\underline{e}_1$, and $\underline{e}_2$ be as in Theorem (3.20). Let $u_{k+1},\dots,u_n$ be any vectors in $\mathbb{C}^{2n}$ such that

$$(3.33)\qquad \Omega(z_i,u_j) = \delta_{ij} \quad \text{for } i,j = k+1,\dots,n.$$

Set

$$(3.34) \qquad y_j = \begin{cases} v_j & \text{for } j = 1,\dots,k \\ \frac{i}{2\pi\hbar}\,\overline{w}_{j-k} & \text{for } j = k+1,\dots,2k \\ u_{j-k} & \text{for } j = 2k+1,\dots,n+k. \end{cases}$$

Let $\hat{y}_j$ be the image of y_j in $\mathbb{C}^{2n}/(V_1 \cap \overline{V}_2)$. Then $(\hat{y}_1,\dots,\hat{y}_{n+k})$ is a frame of $\mathbb{C}^{2n}/(V_1 \cap \overline{V}_2)$. Moreover

$$(3.35) \qquad \langle \beta_1,\beta_2\rangle(\hat{y}_1,\dots,\hat{y}_{n+k}) = \beta_1(W_1,c_1)\overline{\beta_2(W_2,c_2)},$$

where (W_j,c_j) is as in (3.20). This is easily checked using (3.20) and the definitions of μ, $\Omega_{\#}$ and $|\Omega_{V_1\cap\overline{V}_2}|$. Finally, (3.35) and (3.20) show that $\langle\beta_2,\beta_1\rangle = \overline{\langle\beta_1,\beta_2\rangle}$, because $({}^tc_1\,{}^tc_2^{-1})^{-1} = {}^tc_2\,{}^tc_1^{-1}$ and $\tilde{\Lambda}_{n,k}^{-1} = \tilde{\Lambda}_{n,k}$, and that $\langle\beta_1,\beta_1\rangle \geq 0$, because one can choose $v_j = w_j$ for $j = 1,\dots,k$ in that case.

4. Pairing on metaplectic manifolds

Let (X,ω) be a symplectic manifold of dimension $2n$. The bundle projection for every bundle over X will be denoted by π. (TX,ω) is a bundle of symplectic vector spaces over X and one may form, following (3.8), the bundle $\mathcal{B}(TX,\omega)$ of symplectic frames over (X,ω). $\mathcal{B} = \mathcal{B}(TX,\omega)$ is a right principal $Sp(n,\mathbb{R})$ - bundle. As in [9] and [3], we say that (X,ω) is given a metaplectic structure if it is provided with a right principal $Mp(n,\mathbb{R})$ - bundle $\tilde{\mathcal{B}}$ over X and a map $p : \tilde{\mathcal{B}} \to \mathcal{B}$ such that the diagram

(4.1)

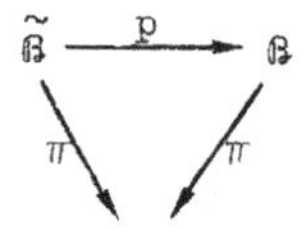

commutes and such that p is $Mp(n,\mathbb{R})$-equivariant. In this case and in all other cases, a left $Sp(n,\mathbb{R})$-space may also be regarded as a left $Mp(n,\mathbb{R})$-space via the map $p : Mp(n,\mathbb{R}) \to Sp(n,\mathbb{R})$. Clearly $p : \tilde{\mathcal{B}} \to \mathcal{B}$ is a two-fold covering projection.

Consider the bundle $\tilde{\mathcal{B}} \times_{Mp(n,\mathbb{R})} \mathbb{C}^{2n}$. Since the left action of $Mp(n,\mathbb{R})$ on $\mathbb{C}^{2n}$ preserves Ω and $-$, $\tilde{\mathcal{B}} \times_{Mp(n,\mathbb{R})} \mathbb{C}^{2n}$ becomes a bundle over X of symplectic vector spaces with conjugation. Let $(TX)_{\mathbb{C}}$ be the complexification of TX. Then $((TX)_{\mathbb{C}}, \omega, -)$ is also a bundle over X of symplectic vector spaces with conjugation. It is easy to see that the map of $\tilde{\mathcal{B}} \times_{Mp(n,\mathbb{R})} \mathbb{C}^{2n}$ into $(TX)_{\mathbb{C}}$ given by

$$(4.2) \qquad \tilde{b} \times_{Mp(n,\mathbb{R})} \begin{bmatrix} s \\ t \end{bmatrix} \mapsto \sum_{j=1}^{n} (s_j v_j + t_j w_j),$$

where $p\,\tilde{b} = (v_1,\dots,v_n, w_1,\dots,w_n)$, $s = \begin{bmatrix} s_1 \\ \vdots \\ s_n \end{bmatrix}$, and $t = \begin{bmatrix} t_1 \\ \vdots \\ t_n \end{bmatrix}$, establishes an isomorphism of bundles of symplectic vector spaces with conjugation. Using the functors of Sections 2 and 3, (4.2) immediately induces canonical isomorphisms of $\tilde{\mathcal{B}} \times_{Mp(n,\mathbb{R})} \mathcal{L}_n$ with $\mathcal{L}_+((TX)_{\mathbb{C}}, \omega, -)$, $\tilde{\mathcal{B}} \times_{Mp(n,\mathbb{R})} \mathcal{G}_{2n,m}$ with $\mathcal{G}_m((TX)_{\mathbb{C}})$, $\tilde{\mathcal{B}} \times_{Mp(n,\mathbb{R})} \mathcal{F}_n^{\flat}$ with $\mathcal{F}(\mathcal{L}_+^{\flat}((TX)_{\mathbb{C}}, \omega, -))$, $\tilde{\mathcal{B}} \times_{Mp(n,\mathbb{R})} \mathcal{F}(\mathcal{G}_{2n,m}^{\flat})$ with $\mathcal{F}(\mathcal{G}_m^{\flat}((TX)_{\mathbb{C}}))$, and $\tilde{\mathcal{B}} \times_{Mp(n,\mathbb{R})} \mathcal{D}_{2n,m}^{1,\flat}$ with $\mathcal{D}^1(\mathcal{G}_m^{\flat}((TX)_{\mathbb{C}}))$. We will denote these bundles by $\mathcal{L}_+(X)$, $\mathcal{G}_m(X)$, $\mathcal{F}^{\flat}(X)$, $\mathcal{G}_m^{\flat}(X)$, $\mathcal{F}(\mathcal{G}_m^{\flat}(X))$, and $\mathcal{D}_m^{1,\flat}(X)$, respectively. The maps $\rho : \mathcal{F}_n^{\flat} \to \mathcal{L}_n$, $\rho : \mathcal{G}_{2n,m}^{\flat} \to \mathcal{G}_{2n,m}$, $\rho : \mathcal{F}(\mathcal{G}_{2n,m}^{\flat}) \to \mathcal{G}_{2n,m}$, and $\rho : \mathcal{D}_{2n,m}^{1,\flat} \to \mathcal{G}_{2n,m}$ are left $Mp(n,\mathbb{R})$-equivariant and hence induce the canonical maps $\rho : \mathcal{F}^{\flat}(X) \to \mathcal{L}_+(X)$, $\rho : \mathcal{G}_m^{\flat}(X) \to \mathcal{G}_m(X)$, $\rho : \mathcal{F}(\mathcal{G}_m^{\flat}(X)) \to \mathcal{G}_m(X)$, and $\rho : \mathcal{D}_m^{1,\flat}(X) \to \mathcal{G}_m(X)$, respectively.

With the foregoing in mind we make

<u>Definition</u> 4.3: $\tilde{\mathcal{B}} \times_{Mp(n,\mathbb{R})} \tilde{\mathcal{F}}_n^{\flat}$ is the <u>bundle of positive lagrangian frames on</u> X and is denoted by $\tilde{\mathcal{F}}^{\flat}(X)$. $\tilde{\mathcal{B}} \times_{Mp(n,\mathbb{R})} \mathcal{E}_n^{1/2}$ is the <u>bundle of half-forms on</u> X and is denoted by $\mathcal{E}^{1/2}(X)$. The maps $p : \tilde{\mathcal{F}}^{\flat}(X) \to \mathcal{F}^{\flat}(X)$,

$\rho : \tilde{\mathfrak{F}}^\flat(X) \to \mathcal{L}_+(X)$, and $\rho : \mathcal{E}^{1/2}(X) \to \mathcal{L}_+(X)$ are those induced by $p : \tilde{\mathfrak{F}}_n^\flat \to \mathfrak{F}_n^\flat$, $\rho : \tilde{\mathfrak{F}}_n^\flat \to \mathcal{L}_n$, and $\rho : \mathcal{E}_n^{1/2} \to \mathcal{L}_n$, respectively. p is a double covering and is right $ML(n,\mathbb{C})$-equivariant.

Following Kostant [8], as modified in [4], suppose that we have a hermitian line bundle L over X with bundle projection π and a connection ∇ on L preserving the hermitian structure on L. Associated with ∇ is a unique 1-form α on the complement $L^\times$ of the zero section in L invariant under multiplication of $L^\times$ by non-zero complex numbers, such that $\nabla_\xi s = i\langle s^*\alpha, \xi\rangle s$ for any smooth local section s of $L^\times$ and any tangent vector field ξ defined on Dom s, and such that α pulled back to any fibre of $L^\times$ is just $\frac{1}{i}\frac{dz}{z}$. Our fundamental assumption is

$$d\alpha = -\frac{1}{\hbar}\pi^*\omega, \tag{4.4}$$

which requires $h^{-1}\omega$ to define an integral de Rham class.

$\mathcal{E}^{1/2}(X,L) = \mathcal{E}^{1/2}(X) \otimes L$ is the <u>bundle of L-valued half-forms on</u> X. Since the map which assigns to any pair $(\beta,\tilde{\underline{e}}) \in \mathcal{E}_n^{1/2} \times \tilde{\mathfrak{F}}_n^\flat$ such that $\rho\beta = \rho\tilde{\underline{e}}$ the number $\beta(\tilde{\underline{e}})$ is left $Mp(n,\mathbb{R})$-equivariant, we have an evaluation map sending $(\beta,\tilde{\underline{e}}) \in \mathcal{E}^{1/2}(X,L) \times \tilde{\mathfrak{F}}^\flat(X)$ such that $\rho\beta = \rho\tilde{\underline{e}}$ into $\beta(\tilde{\underline{e}}) \in L$, and $\pi(\beta(\tilde{\underline{e}})) = \pi\beta = \pi\tilde{\underline{e}}$. Similarly, we have evaluation map sending $(\gamma,\underline{e}) \in \mathcal{B}_m^{1,\flat}(X) \times \mathfrak{F}(G_m^\flat(X))$ such that $\rho\gamma = \gamma\underline{e}$ into $\gamma(\underline{e}) \in \mathbb{C}$. Finally, Theorem 3.30 tells us that $\langle\cdot,\cdot\rangle$ on $\mathcal{E}_n^{1/2} \times \mathcal{E}_n^{1/2}$ induces a sesqui-linear pairing $\mathcal{E}^{1/2}(X) \hat{\times} \mathcal{E}^{1/2}(X) \to \bigcup_{k}{}_0^n \mathcal{B}_{n-k}^{1,\flat}(X)$, where $\hat{\times}$ has the following meaning: if E_1 and E_2 are bundles over X, then $E_1 \hat{\times} E_2$ is the bundle over X consisting of $\{(e_1,e_2) \in E_1 \times E_2 : \pi e_1 = \pi e_2\}$. We extend this pairing to $\mathcal{E}^{1/2}(X,L)$ by defining

$$\langle\beta_1 \otimes \ell_1, \beta_2 \otimes \ell_2\rangle = (\ell_1,\ell_2)\langle\beta_1,\beta_2\rangle \tag{4.5}$$

for $\pi(\beta_1 \otimes \ell_1) = \pi(\beta_2 \otimes \ell_2)$, where $(\cdot,\cdot)$ is the hermitian structure on L. $\rho : \mathcal{E}^{1/2}(X) \to \mathcal{L}_+(X)$ induces a map $\rho : \mathcal{E}^{1/2}(X,L) \to \mathcal{L}_+(X)$. The diagram

$$\begin{array}{ccc} \mathcal{E}^{1/2}(X,L) \hat{\times} \mathcal{E}^{1/2}(X,L) & \xrightarrow{\langle\cdot,\cdot\rangle} & \bigcup_{k}^{n} {}_0\, \mathcal{D}^{1,\flat}_{n-k}(X) \\ \downarrow \rho\times\rho & & \downarrow \rho \\ \mathcal{L}_+(X) \hat{\times} \mathcal{L}_+(X) & \xrightarrow{\cdot\cap\cdot} & \bigcup_{k}^{n} {}_0\, \mathcal{G}_{n-k}(X) \end{array}$$

commutes.

According to the philosophy of [9] and [3], geometric quantization should be formulated in terms of L-valued half-forms normal to a positive polarization and covariant constant along the polarization. Recall that a positive polarization is a section F of $\mathcal{L}_+(X)$ which, regarded as a subbundle of $(TX)_C$, is involutive and such that $F \cap \overline{F}$ has constant fibre dimension. We do not, however, require at this stage that $F + \overline{F}$ be involutive. An L-valued half-form β on X is a smooth section of $\mathcal{E}^{1/2}(X,L)$. β is normal to the positive polarization F if $\rho \circ \beta = F$. Similarly, a density γ on X is a smooth section of $\mathcal{D}^{1,\flat}_m(X)$, and if F is a smooth involutive section of $\mathcal{G}_m(X)$ such that $\rho \circ \gamma = F$ we will say that γ is normal to F.

To define covariant constancy, we introduces, as in [4], a way of differentiating a half-form or density on X with respect to a tangent vector to which the half-form or density is normal. In the following two lemmas we make use of the fact that the canonical isomorphisms given earlier in this section show that $\mathcal{E}^{1/2}(X,L)$ is canonically isomorphic to $\tilde{\mathcal{F}}^{\flat}(X) \times_{ML(n,C)} L$ and that $\mathcal{D}^{1,\flat}_m(X)$ is canonically isomorphic to $\mathcal{F}(\mathcal{G}^{\flat}_m(X)) \times_{GL(2n-m,C)} C$, where $c \in ML(n,C)$ acts on L via multiplication by $\chi(c)^{-1}$ and $C \in GL(2n - m, C)$ acts on C via multiplication by $(\mathrm{Det}\ C)^{-1}$. We shall use $C^\infty_U E$ to denote the space of smooth sections of a bundle E over the open set U.

Lemma 4.7: Let $\beta \in C^\infty_X\, \mathcal{E}^{1/2}(X,L)$ and set $F = \rho \circ \beta \in C^\infty_X\, \mathcal{L}_+(X)$. Suppose $\xi \in C^\infty_X(TX)_C$ and satisfies $[\xi, C^\infty_X F] \subseteq C^\infty_X F$. Then there exists a unique L-valued half-form $\nabla_\xi \beta \in C^\infty_X\, \mathcal{E}^{1/2}(X,L)$ such that $F = \rho \circ \nabla_\xi \beta$ and such that

$$(4.8) \qquad (\nabla_\xi \beta) \circ \underset{\sim}{\tilde{e}} = \nabla_\xi(\beta \circ \underset{\sim}{\tilde{e}}) - \tfrac{1}{2}(\mathrm{Tr}\, A)(\beta \circ \underset{\sim}{\tilde{e}})$$

for every $\underline{\tilde{e}} \in C_U^\infty \tilde{\mathfrak{F}}^{b}(X)$ with $\rho \circ \underline{\tilde{e}} = F$, U open in X, where A is the smooth $n \times n$ matrix valued function on U defined by

$$(4.9) \qquad (\mathrm{Ad}\, \xi)\underline{e} = \underline{e}\, A, \quad \text{where } \underline{e} = p \circ \underline{\tilde{e}} \quad C_U^\infty \mathfrak{F}^{b}(X).$$

(The meaning of $(\mathrm{Ad}\,\xi)\underline{e}$ is the following: $\mathrm{Ad}\,\xi$ acts on $C_X^\infty(TX)_{\mathbb{C}}$ as $[\xi, \cdot]$. Since $(\mathrm{Ad}\,\xi)C_X^\infty F \subseteq C_X^\infty F$, we have an induced action of $\mathrm{Ad}\,\xi$ on $C_X^\infty[(TX)_{\mathbb{C}}/F]$. $\underline{e}$ is an n-tuple of elements of $C_U^\infty[(TX)_{\mathbb{C}}/F]$. $\mathrm{Ad}\,\xi$ acts on $\underline{e}$ by acting on each component of $\underline{e}$.)

Proof: Let $\ell(\underline{\tilde{e}}) \in C_U^\infty L$ denote the right hand side of (4.8). Since $\mathcal{E}^{1/2}(X,L) \cong \tilde{\mathfrak{F}}^{b}(X) \times_{ML(n,\mathbb{C})} L$, it suffices to show that $\ell(\underline{\tilde{e}}\, c) = (\chi \circ c)\, \ell(\underline{\tilde{e}})$ for any $c \in C^\infty ML(n,\mathbb{C})$. Now

$$(4.10) \qquad \nabla_\xi(\beta \circ (\underline{\tilde{e}}\, c)) = \xi(\chi \circ c)(\beta \circ \underline{\tilde{e}}) + (\chi \circ c)\, \nabla_\xi(\beta \circ \underline{\tilde{e}})$$

whereas, letting $C = p \circ c \in C^\infty GL(n,\mathbb{C})$,

$$(4.11) \qquad (\mathrm{Ad}\,\xi)(\underline{e}\, C) = (\underline{e}\, C)[C^{-1}AC + C^{-1}(\xi C)].$$

Moreover, since $(\chi \circ c)^2 = \mathrm{Det}\, C$,

$$(4.12) \qquad \xi(\chi \circ c) = \tfrac{1}{2}(\chi \circ c)\, \mathrm{Tr}[C^{-1}(\xi C)].$$

Therefore, from (4.10), (4.11), and (4.12), $\ell(\underline{\tilde{e}}\, c) = \{\tfrac{1}{2}(\chi \circ c)\, \mathrm{Tr}[C^{-1}(\xi C)](\beta \circ \underline{\tilde{e}}) + (\chi \circ c)\, \nabla_\xi(\beta \circ \underline{\tilde{e}})\} - \tfrac{1}{2}(\mathrm{Tr}[C^{-1}AC + C^{-1}(\xi C)])(\chi \circ c)(\beta \circ \underline{\tilde{e}}) = (\chi \circ c)\ell(\underline{\tilde{e}})$, as desired.

In a similar way, we have

<u>Lemma</u> 4.13: Let $\gamma \in C_X \mathcal{D}_m^{1, b}(X)$ and set $F = \rho \circ \gamma \in C_X^\infty \mathcal{G}_m(X)$. Suppose $\xi \in C_X^\infty TX$ and satisfies $[\xi, C_X^\infty F] \subseteq C_X^\infty F$. Then there exists a unique density $\nabla_\xi \gamma \in C_X^\infty \mathcal{D}_m^{1, b}(X)$ such that $F = \rho \circ \nabla_\xi \gamma$ and such that

$$(4.14) \qquad (\nabla_\xi \gamma) \circ \underline{e} = \xi(\gamma \circ \underline{e}) - (\mathrm{Re\,Tr}\, A)(\gamma \circ \underline{e})$$

for every $\underline{e} \in C_U^\infty \mathcal{F}(\mathcal{G}_m^b(X))$ with $\rho \circ \underline{e} = F$, U open in X, where A is the smooth $(2n - m) \times (2n - m)$ matrix valued function on U defined by

$$(4.15) \qquad (\mathrm{Ad}\,\xi)\underline{e} = \underline{e}\, A.$$

Proof: The proof is exactly as in (4.7) except that we must show that the right hand side of (4.14) is multiplied by $|\mathrm{Det}\, C|$ when $\underline{e}$ is replaced by $\underline{e}\, C$, $C \in C_U^\infty GL(2n - m, \mathbb{C})$. Also (4.12) must be replaced by

$$(4.16) \qquad \xi\, |\mathrm{Det}\, C| = |\mathrm{Det}\, C|\ \mathrm{Re\,Tr}[C^{-1}(\xi C)].$$

Here the fact that ξ is real is crucial.

<u>Definition</u> 4.17: Let β, F (resp. γ, F) be as in (4.7) (resp. (4.13)). Suppose $[C_X^\infty F, C_X^\infty F] \subseteq C_X^\infty F$ (and also $F = \overline{F}$ in (4.13)). Then β (resp. γ) is <u>covariant constant along</u> F if $\nabla_\xi \beta$ (resp. $\nabla_\xi \gamma$) $= 0$ for all $\xi \in C_X^\infty F$ (and also ξ real in (4.13)).

Let $\beta_i \in C_X^\infty\, \mathcal{C}^{1/2}(X,L)$ and set $F_i = \rho \circ \beta_i$. Assume that $[C_X^\infty F_i, C_X^\infty F_i] \subseteq C_X^\infty F_i$ and that $F_1 \cap \overline{F}_2$ has constant fibre dimension. Let ξ be a real vector field in $C_X^\infty(F_1 \cap \overline{F}_2)$. The main result of this section relates $\nabla_\xi \beta_1$, $\nabla_\xi \beta_2$, and $\nabla_\xi \langle \beta_1, \beta_2 \rangle$. It is formulated in terms of an invariant of certain kinds of differential systems on symplectic manifolds.

<u>Proposition</u> 4.18: Let (X, ω) be a symplectic manifold of dimension $2n$ and let $F \in C_X^\infty\, \mathcal{G}_{n+k}(X)$ satisfy $F^\perp \subseteq F$. Suppose that $[C_X^\infty F^\perp, C_X^\infty F] \subseteq C_X^\infty F$. Then there is a unique $\chi_F \in C_X^\infty[(TX)_\mathbb{C}/F]$ such that, if $v_1, \ldots, v_k$, $w_1, \ldots, w_k$ are vector fields in $C_U^\infty F$, U open in X, satisfying $\omega(v_i, v_j) = 0 = \omega(w_i, w_j)$ and $\omega(v_i, w_j) = \delta_{ij}$ for $i,j = 1, \ldots, k$, then $\chi_F = \Sigma_1^k [v_i, w_i]^\wedge$ on U, where $^\wedge$ sends any vector field into its image in $C^\infty[(TX)_\mathbb{C}/F]$. Moreover, if $F = \overline{F}$, χ_F is real.

Proof: There exist such open sets U and vector fields $v_1,\dots,v_k$, $w_1,\dots,w_k$, because our condition is just that the image of $(v_1,\dots,v_k, w_1,\dots,w_k)$ in $C_U^\infty(F/F^\perp)^{2k}$ lies in $C_U^\infty\mathcal{B}(F/F^\perp)$, $F/F^\perp$ being a bundle of symplectic vector spaces. The v's and w's can be chosen to be real if $F = \bar{F}$. It suffices therefore to show the following: if $v_1',\dots,v_k', w_1',\dots,w_k' \in C_U^\infty F$ and satisfy the same conditions, then $\Sigma_i{}_1^k[v_i,w_i]^\wedge = \Sigma_i{}_1^k[v_i',w_i']^\wedge$. We know that there is a matrix valued function on U, $\begin{bmatrix} T_1 & T_2 \\ T_3 & T_4 \end{bmatrix}$, with values in $Sp(n,\mathbb{C})$ such that

$$(4.19)\ (v_1,\dots,v_k, w_1,\dots,w_k) = (v_1',\dots,v_k', w_1',\dots,w_k')\begin{bmatrix} T_1 & T_2 \\ T_3 & T_4 \end{bmatrix} \bmod C_U^\infty F^\perp .$$

Moreover, for $\xi_1,\xi_2 \in C_U^\infty F$ and smooth functions f_1, f_2 on U, we have $[f_1\xi_1, f_2\xi_2]^\wedge = f_1f_2[\xi_1,\xi_2]^\wedge$. Let the entries of T_i be $t_i^{j\ell}$. Then $\Sigma_i{}_1^k[v_i,w_i]^\wedge = \Sigma_{ij\ell}{}_1^k\{t_1^{ji}\,t_2^{\ell i}[v_j',v_\ell']^\wedge + t_3^{ji}\,t_2^{\ell i}[w_j',v_\ell']^\wedge + t_1^{ji}\,t_4^{\ell i}[v_j',w_\ell']^\wedge + t_3^{ji}\,t_4^{\ell i}[w_j',w_\ell']^\wedge\} = \Sigma_{j\ell}{}_1^k\{a_{j\ell}[v_j',v_\ell']^\wedge + b_{j\ell}[v_j',w_\ell']^\wedge + c_{j\ell}[w_j',w_\ell']^\wedge\}$, where $T_1{}^tT_2 = (a_{j\ell})$, $T_3{}^tT_4 = (c_{j\ell})$, and $T_1{}^tT_4 - T_2{}^tT_3 = (b_{j\ell})$. But $T_1{}^tT_2$ and $T_3{}^tT_4$ are symmetric and $T_1{}^tT_4 - T_2{}^tT_3 = I$. Our result follows immediately.

We can now state our main theorem.

<u>Theorem</u> 4.20: With β_1,β_2, F_1,F_2, and ξ as in the paragraph before (4.18), we have

$$\nabla_\xi\langle\beta_1,\beta_2\rangle = \langle\nabla_\xi\beta_1,\beta_2\rangle + \langle\beta_1,\nabla_\xi\beta_2\rangle - \tfrac{1}{2}\langle\omega_*^{F_1\cap\bar{F}_2}\xi,\ X_{F_1+\bar{F}_2}\rangle\langle\beta_1,\beta_2\rangle,$$

where $\omega_*^{F_1\cap\bar{F}_2}$ is as in (2.3) <u>et seq</u>.

Proof: Let U be an open set in X upon which there are defined vector fields $v_1,\dots,v_k$, $w_1,\dots,w_k$, $z_{k+1},\dots,z_n$, $u_{k+1}',\dots,u_n'$, $u_{k+1}'',\dots,u_n''$, such that: (a) $(v_1,\dots,v_k, z_{k+1},\dots,z_n) \in C_U^\infty\mathcal{F}(F_1)$,

(b) $(w_1,\dots,w_k, z_{k+1},\dots,z_n) \in C_U^\infty\mathcal{F}(F_2)$,

(c) $\langle v_i,w_j\rangle = \delta_{ij}$ for $i,j = 1,\dots,k$, where $\langle\cdot,\cdot\rangle = \frac{i}{2\pi\hbar}\,\omega(\cdot,\bar{\cdot})$, as in (2.6),

(d) $\omega(v_i,u_j') = 0$ and $\omega(z_\ell,u_j') = \delta_{\ell j}$ for $i = 1,\ldots,k$ and $\ell,j = k+1,\ldots,n$, and

(e) $\omega(w_i,u_j'') = 0$ and $\omega(z_\ell,u_j'') = \delta_{\ell j}$ for $i = 1,\ldots,k$ and $\ell,j = k+1,\ldots,n$.

Such open sets and vector fields clearly exist.

For any r-tuple of vector fields $(\eta_1,\ldots,\eta_r)$ on U and any subbundle F of $(TX)_C$ we will let $(\eta_1,\ldots,\eta_r)_F$ denote the image of $(\eta_1,\ldots,\eta_r)$ in $C_U^\infty[(TX)_C/F]^r$. Then it is easy to see that

(4.21)
$$\begin{cases} \underline{e}_1 = (\frac{i}{2\pi\hbar}\bar{w}_1,\ldots,\frac{i}{2\pi\hbar}\bar{w}_k, u'_{k+1},\ldots,u'_n)_{F_1} \in C_U^\infty\mathfrak{J}[(TX)_C/F_1], \\ \underline{e}_2 = (\frac{i}{2\pi\hbar}\bar{v}_1,\ldots,\frac{i}{2\pi\hbar}\bar{v}_k, u''_{k+1},\ldots,u''_n)_{F_2} \in C_U^\infty\mathfrak{J}[(TX)_C/F_2], \text{ and} \\ \underline{e} = (v_1,\ldots,v_k, \frac{i}{2\pi\hbar}\bar{w}_1,\ldots,\frac{i}{2\pi\hbar}\bar{w}_k, u'_{k+1},\ldots,u'_n)_{F_1\cap\bar{F}_2} \in \\ \qquad C_U^\infty\mathfrak{J}[(TX)_C/(F_1\cap\bar{F}_2)]. \end{cases}$$

Moreover,

(4.22)
$$\underline{e}_1 = \omega_\#(v_1,\ldots,v_k, z_{k+1},\ldots,z_n) \text{ and}$$
$$\underline{e}_2 = \omega_\#(w_1,\ldots,w_k, z_{k+1},\ldots,z_n).$$

Therefore, applying the canonical isomorphisms of the early part of this section to (3.20) and (3.32), we see that we can, shrinking U if necessary, find $\tilde{\underline{e}}_1, \tilde{\underline{e}}_2 \in C_U^\infty\tilde{\mathfrak{J}}^\flat(X)$ such that $\underline{e}_j = p\circ\tilde{\underline{e}}_j$ and

(4.23)
$$\langle\beta_1,\beta_2\rangle\circ\underline{e} = (\beta_1\circ\tilde{\underline{e}}_1, \beta_2\circ\tilde{\underline{e}}_2).$$

It is easy to see that there exist smooth $k\times k$ matrix valued functions A_1, A_2, and A_3 on U such that

$$(4.24)\quad \begin{cases} (\mathrm{Ad}\,\xi)(v_1,\dots,v_k)_{F_1\cap\overline{F}_2} = (v_1,\dots,v_k)_{F_1\cap\overline{F}_2}\, A_1, \\ (\mathrm{Ad}\,\xi)(w_1,\dots,w_k)_{F_1\cap\overline{F}_2} = (w_1,\dots,w_k)_{F_1\cap\overline{F}_2}\, A_2, \quad \text{and} \\ (\mathrm{Ad}\,\xi)(u'_{k+1},\dots,u'_n)_{F_1+\overline{F}_2} = (u'_{k+1},\dots,u'_n)_{F_1+\overline{F}_2}\, A_3. \end{cases}$$

Here we have used the assumption that F_1 and F_2 are involutive. Note that

$$(4.25)\qquad (u'_{k+1},\dots,u'_n)_{F_1+\overline{F}_2} = (u''_{k+1},\dots,u''_n)_{F_1+\overline{F}_2}.$$

Since ξ is real, (4.24) and (4.25) imply that

$$(4.26)\quad \begin{cases} (\mathrm{Ad}\,\xi)\underline{e}_1 = \underline{e}_1\begin{bmatrix}\overline{A}_2 & * \\ 0 & A_3\end{bmatrix}, \\ (\mathrm{Ad}\,\xi)\underline{e}_2 = \underline{e}_2\begin{bmatrix}\overline{A}_2 & * \\ 0 & A_3\end{bmatrix}, \quad \text{and} \\ (\mathrm{Ad}\,\xi)\underline{e} = \underline{e}\begin{bmatrix}A_1 & 0 & * \\ 0 & \overline{A}_2 & * \\ 0 & 0 & A_3\end{bmatrix}, \end{cases}$$

where $*$ indicates some matrix valued functions.

Applying (4.13), we get

$$(4.27)\quad \begin{aligned} &(\nabla_\xi \langle\beta_1,\beta_2\rangle)\circ\underline{e} = \\ &\xi(\langle\beta_1\circ\beta_2\rangle\circ\underline{e}) - (\mathrm{Re}[\mathrm{Tr}\,A_1 + \mathrm{Tr}\,\overline{A}_2 + \mathrm{Tr}\,A_3])(\langle\beta_1,\beta_2\rangle\circ\underline{e}). \end{aligned}$$

By (4.23) and then (4.7), the first term on the right of (4.27) is

$$\xi(\beta_1\circ\underline{\tilde{e}}_1,\ \beta_2\circ\underline{\tilde{e}}_2) = (\nabla_\xi(\beta_1\circ\underline{\tilde{e}}_1),\ \beta_2\circ\underline{\tilde{e}}_2) + (\beta_1\circ\underline{\tilde{e}}_1,\ \nabla_\xi(\beta_2\circ\underline{\tilde{e}}_2)) =$$

$$((\nabla_\xi\beta_1)\circ\underline{\tilde{e}}_1,\ \beta_2\circ\underline{\tilde{e}}_2) + \tfrac{1}{2}[\mathrm{Tr}\,\overline{A}_2 + \mathrm{Tr}\,A_3](\beta_1\circ\underline{\tilde{e}}_1,\ \beta_2\circ\underline{\tilde{e}}_2) +$$

$$(\beta_1 \circ \underset{\sim}{\tilde{e}}_1, (\nabla_\xi \beta_2) \circ \underset{\sim}{\tilde{e}}_2) + \tfrac{1}{2}[\mathrm{Tr}\, A_1 + \mathrm{Tr}\, \overline{A}_3](\beta_1 \circ \underset{\sim}{\tilde{e}}_1, \beta_2 \circ \underset{\sim}{\tilde{e}}_2),$$

which is just

$$\{\langle \nabla_\xi \beta_1, \beta_2\rangle + \langle \beta_1, \nabla_\xi \beta_2\rangle + [\tfrac{1}{2}\,\mathrm{Tr}\, A_1 + \tfrac{1}{2}\,\mathrm{Tr}\, \overline{A}_2 + \mathrm{Re}\,\mathrm{Tr}\, A_3]\langle \beta_1, \beta_2\rangle\} \circ \underset{\sim}{e}$$

by (4.23) again. Substituting this in (4.27), we get

$$\nabla_\xi \langle \beta_1, \beta_2\rangle = \langle \nabla_\xi \beta_1, \beta_2\rangle + \langle \beta_1, \nabla_\xi \beta_2\rangle - \tfrac{1}{2}[\mathrm{Tr}\, \overline{A}_1 + \mathrm{Tr}\, A_2]\langle \beta_1, \beta_2\rangle.$$

It remains to show that

$$\text{(4.28)} \qquad \mathrm{Tr}\, \overline{A}_1 + \mathrm{Tr}\, A_2 = \langle \omega_*^{F_1 \cap \overline{F}_2}\xi\,,\ \chi_{F_1+\overline{F}_2}\rangle.$$

By (4.24) and (c) above,

$$\text{(4.29)} \qquad \begin{cases} \mathrm{Tr}\, A_1 = \sum_{j=1}^{k} \frac{i}{2\pi\hbar}\, \omega([\xi, v_j], \overline{w}_j) & \text{and} \\[2ex] \mathrm{Tr}\, A_2 = \sum_{j=1}^{k} \frac{i}{2\pi\hbar}\, \omega([\xi, w_j], \overline{v}_j). \end{cases}$$

Therefore

$$\text{(4.30)} \qquad \mathrm{Tr}\, \overline{A}_1 + \mathrm{Tr}\, A_2 = \frac{i}{2\pi\hbar}\sum_{j=1}^{k}\{\omega([\xi, w_j], \overline{v}_j) + \omega([\overline{v}_j, \xi], w_j)\}.$$

Since $d\omega = 0$, we have $\omega([\xi, w_j], \overline{v}_j) + \omega([\overline{v}_j, \xi], w_j) + \omega([w_j, \overline{v}_j], \xi) =$ $\xi\,\omega(w_j, \overline{v}_j) + \overline{v}_j\,\omega(\xi, w_j) + w_j\,\omega(\overline{v}_j, \xi)$, which vanishes because $\omega(\xi, w_j) = 0 =$ $\omega(\overline{v}_j, \xi)$ and $\omega(w_j, \overline{v}_j) = -2\pi i\hbar$. Therefore (4.30) becomes

$$\text{(4.31)} \qquad \mathrm{Tr}\, \overline{A}_1 + \mathrm{Tr}\, A_2 = \frac{i}{2\pi\hbar}\sum_{j=1}^{k}\omega(\xi, [w_j, \overline{v}_j]).$$

Since $\frac{i}{2\pi\hbar}\,\omega(w_j, \overline{v}_\ell) = \delta_{j\ell}$, we have that

$$\text{(4.32)} \qquad \chi_{F_1+\overline{F}_2} = \sum_{j=1}^{k}[w_j, \tfrac{i}{2\pi\hbar}\,\overline{v}_j]^\wedge \quad \text{on } U,$$

and (4.28) follows from the definition of $\omega_*^{F_1+\overline{F}_2}$.

Corollary 4.33: Let β_1, β_2, F_1, and F_2 be as in (4.20). Suppose β_i is covariant constant along F_i and that $\langle \beta_1, \beta_2 \rangle$ is not identically zero. Then $\langle \beta_1, \beta_2 \rangle$ is covariant constant along $F_1 \cap \overline{F}_2$ if and only if $\chi_{F_1+\overline{F}_2} \equiv 0$.

Proof: This is clear because $\omega_*^{F_1+\overline{F}_2}$ maps $F_1 \cap \overline{F}_2$ isomorphically onto $(F_1 + \overline{F}_2)^*$ and because $F_1 \cap \overline{F}_2$ is the complexification of its intersection with TX.

We are therefore led, in the light of the philosophy of [3] and [4], to make the following definition.

Definition 4.34: Let F_1 and F_2 be positive polarizations on X. The pair (F_1, F_2) will be called regular if

(a) $F_1 \cap \overline{F}_2$ has constant fibre dimension,

(b) $X/(F_1 \cap \overline{F}_2)$, the space of leaves of the foliation $(F_1 \cap \overline{F}_2) \cap TX$, is a Hausdorff manifold, and

(c) $\chi_{F_1+\overline{F}_2} \equiv 0$.

A positive polarization F is regular if (F,F) is regular.

Let F be a regular positive polarization and let $\beta \in C_X^\infty \mathcal{E}^{1/2}(X,L)$ satisfy $\rho \circ \beta = F$ and be covariant constant along F. Then $\langle \beta, \beta \rangle \in C_X^\infty \mathcal{D}_{n-k}^{1,b}(X)$, where $\dim X = 2n$ and $\dim(F \cap \overline{F}) = n - k$, satisfies $\rho \circ \langle \beta, \beta \rangle = F \cap \overline{F}$ and is covariant constant along $F \cap \overline{F}$. Now let $\zeta : X \to X/(F \cap \overline{F})$ be the canonical projection sending each point of X into the $F_0 = F \cap TX$ leaf to which it belongs. ζ is a submersion and ζ_* induces an isomorphism $\hat{\zeta}_*$ of $T_x X/F_0$ with $T_{\zeta x}(X/F)$ for every $x \in X$. Let U be a coordinate neighborhood of a point $y_0 \in Y = X/F$, let $\underline{\check{e}} \in C_U^\infty \mathcal{F}(TY)$, and let $\underline{e}$ be the section of TX/F_0 over U given by this isomorphism: $\mathcal{F}(\hat{\zeta}_*)[\underline{e}(x)] = \underline{\check{e}}(\zeta x)$, $x \in \zeta^{-1}U$. Then $(\mathrm{Ad}\, \xi)\underline{e} = 0$ for any $\xi \in C^\infty_{\zeta^{-1}U} F_0$.

But this implies that $\xi(\langle\beta,\beta\rangle \circ \underline{e}) = 0$ for any such ξ. It follows that $\langle\beta,\beta\rangle \circ \underline{e}$ is constant on each F_0 leaf contained in $\zeta^{-1}U$, so that we can define a function $\langle\beta,\beta\rangle^\vee$ on $C_U^\infty\mathfrak{F}(TY)$ by $\langle\beta,\beta\rangle^\vee \circ \check{e} = \langle\beta,\beta\rangle \circ \underline{e}$. By (3.32), $\langle\beta,\beta\rangle^\vee \geq 0$. Clearly $\langle\beta,\beta\rangle^\vee$ has the correct transformation properties for arbitrary choices of U and $\check{\underline{e}}$ to define a positive density $\langle\beta,\beta\rangle^\vee$ on Y. The space of all β as above such that $\langle\beta,\beta\rangle$ is integrable on Y forms a pre-Hilbert space $\mathfrak{H}_0^F$ with norm

$$\|\beta\| = \left(\int_Y \langle\beta,\beta\rangle^\vee\right)^{1/2},$$

whose completion we denote by $\mathfrak{H}^F$.

Now let F_1 and F_2 be regular positive polarizations such that (F_1,F_2) is regular. If $\beta_i \in \mathfrak{H}_0^{F_i}$ then, exactly as above, $\langle\beta_1,\beta_2\rangle$ defines a density $\langle\beta_1,\beta_2\rangle^\vee$ on $X/(F_1 \cap \overline{F}_2)$. If this density is absolutely integrable we can set

$$(\beta_1,\beta_2) = \int_{X/(F_1\cap\overline{F}_2)} \langle\beta_1,\beta_2\rangle^\vee .$$

By (3.32) and the fact that $F_1 \cap \overline{F}_2 = F_2 \cap \overline{F}_1$, we see that $(\beta_2,\beta_1) = \overline{(\beta_1,\beta_2)}$. In favorable cases (β_1,β_2) will be defined for all β_1 in a dense subspace $\mathfrak{K}_1$ of $\mathfrak{H}_0^{F_1}$ and all β_2 in a dense subspace $\mathfrak{K}_2$ of $\mathfrak{H}_0^{F_2}$ in such a way that

$$(\beta_1,\beta_2) = (U\,\beta_1,\beta_2)$$

for all $\beta_j \in \mathfrak{K}_j$ and some unitary $U : \mathfrak{H}^{F_1} \to \mathfrak{H}^{F_2}$. In such cases we will say that $\mathfrak{H}^{F_1}$ and $\mathfrak{H}^{F_2}$ are <u>unitarily related</u>. From this point on, the theory proceeds exactly as in [3] and [4].

<u>Remark</u> 4.35: In the last paragraph of Section 3 of [4], the present author asserted that if F_j is real and if β_j is covariant constant along F_j, then $\langle\beta_1,\beta_2\rangle$ is covariant constant along $F_1 \cap F_2$. This is not true. Corollary 4.33 gives the exact obstruction to this being true. This obstruction

often does not vanish. For example, let X be the complement of the zero section of T^*M, M some manifold. It is easy to find positive polarizations F_1 and F_2 of X such that $F_1 \cap \bar{F}_2$ has as its leaves the rays of X. In that case, $\chi_{F_1+\bar{F}_2} \neq 0$. This sort of example has arisen in a recent attempt by Sternberg and Wolf to construct irreducible representations of $SO(n,2)$ using moving polarizations. In [10], Rothschild and Wolf construct two real polarizations F_1 and F_2 of a nilpotent orbit in the adjoint representation of the split real form of G_2. They show that the representations of that group determined by F_1 and F_2 are disjoint. The present author speculated in [5], p. 12, that the lack of geometrical completeness of the leaves of F_1 and of F_2 lies behind this phenomenon. However, Wolf has calculated (oral communication) $\chi_{F_1+F_2}$ in this case and has shown that it does not vanish. Thus there isn't even a formal intertwining operator given by our method, contrary to the assertion in [5].

REFERENCES

[1] L. Auslander and B. Kostant, Polarization and unitary representations of solvable Lie groups, Invent. Math. 14 (1971), 255-354.

[2] R. Blattner, On induced representations II: infinitesimal induction, Amer. J. Math. 83 (1961), 499-512.

[3] ________, Quantization and representation theory, Proc. Sympos. Pure Math., vol. 26, Amer. Math. Soc., Providence, R.I., 1974, pp. 145-165.

[4] ________, Pairing of half-form spaces, proceedings of the "Colloque Symplectique", Aix-en-Provence 1974, to appear.

[5] ________, Intertwining operators and the half-density pairing, Lecture Notes in Math., vol. 466, Springer-Verlag, Berlin, 1975, pp. 1-12.

[6] J. Dixmier, Représentations induites holomorphes des groupes résolubles algébriques, Bull. Soc. Math. France 94 (1966), 181-206.

[7] I. M. Gel'fand and M. I. Graev, Unitary representations of the real unimodular group (principal non-degenerate series), Izv. Akad. Nauk SSSR. Ser. Mat. 17 (1953), 189-248 (in Russian). Amer. Math. Soc. Translations, Ser. 2, vol. 2, 147-205.

[8] B. Kostant, Quantization and unitary representations, Lecture Notes in Math., vol. 170, Springer-Verlag, Berlin, 1970, pp. 237-253.

[9] _________, Symplectic spinors, Symposia Math., vol. XIV, Academic Press, London, 1974, pp. 139-152.

[10] L. Rothschild and J. Wolf, Representations of semisimple groups associated to nilpotent orbits, Ann. Sci. Ecole Norm. Sup. (4) 1 (1974), 155-174.

[11] I. Satake, Factors of automorphy and Fock representations, Advances in Math. 7 (1971), 83-110.

[12] E. Spanier, Algebraic topology, McGraw-Hill, New York, 1966.

Department of Mathematics
University of California
Los Angeles, CA 90024
U.S.A.

ON COHOMOLOGY GROUPS APPEARING IN GEOMETRIC QUANTIZATION*

Jedrzej Śniatycki
University of Calgary

1. Introduction

Geometric quantization theory provides a framework for a unified treatment of the construction of irreducible representations of Lie groups. On the other hand, it gives a geometric interpretation of the procedure of canonical quantization used in transition from the classical to the quantum description of a physical system. Therefore, it is of interest to study geometric quantization also outside its group theoretical set-up.

Basic objects in geometric quantization are as follows:

(i) a symplectic manifold (X,ω),

(ii) a complex line bundle L over X with a connection ∇, such that ω is the curvature form of ∇, and with an invariant Hermitian metric,

(iii) a complex involutive Lagrangian distribution F on X, called a polarization,

(iv) a bundle $N_F^{\frac{1}{2}}$ of half forms normal to F.

The space of sections of $L \otimes N_F^{\frac{1}{2}}$ covariant constant along F gives the representation space. However, if the integral manifolds of $D = F \cap TX$ are not simply connected, the only smooth section of $L \otimes N_F^{\frac{1}{2}}$ covariant constant along F is the zero section. The Bohr-Sommerfeld quantization conditions define a subset S of X consisting of all integral manifolds Λ of D for which the holonomy group of the canonical flat connection in $(L \otimes N_F^{\frac{1}{2}})|\Lambda$ is trivial. In a sufficiently regular situation connected components of S are submanifolds and there exist non-zero smooth sections of $(L \otimes N_F^{\frac{1}{2}})|S$ covariant constant along F. In quantum mechanical

* Partially supported by the N.R.C. Operating Grant No. A8091.

applications the space $S_F(S)$ of smooth sections of $(L \otimes N_F^{\frac{1}{2}})|S$ covariant constant along F gives rise to the space of wave functions, [6], [7].

It has been suggested by B. Kostant that in the case, when there are no non-zero smooth global sections of $L \otimes N_F^{\frac{1}{2}}$ covariant constant along F, one might be able to use for the representation spaces higher cohomology groups of X with coefficients in the sheaf S_F of germs of smooth sections of $L \otimes N_F^{\frac{1}{2}}$ covariant constant along F, [5]. R.J. Blattner, J. Rawnsley and D.J. Simms showed that one can quantize the one-dimensional harmonic oscillator in the polarization given by the energy levels using the first cohomology group as the space of wave functions. The aim of this paper is to study the cohomology groups $H^m(X,S_F)$ of X with the coefficients in S_F, under some additional conditions imposed on F. The results obtained are summarized in the following theorems.

Theorem 1.1 Let $(X,\omega,F,L \otimes N_F^{\frac{1}{2}})$ satisfy the following conditions: $F = \bar{F}$, for each integral manifold Λ of D, the canonical flat affine connection in Λ is complete, and the space Y of all integral manifolds of D admits a manifold structure such that the canonical projection $\pi: X \longrightarrow Y$ is a fibration admitting local trivializations inducing affine isomorphisms of the fibres. Then $H^m(X,S_F) = 0$ for all $m \neq k$, where k is the rank of the fundamental group of a typical integral manifold of D.

Let $C_F(X)$ denote the ring of complex valued smooth functions on X annihilated by the differentiations in F. The spaces $H^m(X,S_F)$ and $S_F(S)$ are $C_F(X)$ modules.

Theorem 1.2 If in addition to the assumptions of Theorem 1.1 the distribution spanned by the Hamiltonian vector fields in D with periodic orbits is orientable, then $H^k(X,S_F)$ and $S_F(S)$ are isomorphic $C_F(X)$-modules.

The relation of this result to quantization is as follows. The action of $C_F(X)$ on $S_F(S)$ leads to the Bohr-Sommerfeld quantization of functions constant along F. To each $f \in C_F(X)$ there corresponds a linear operator in $S_F(S)$ spectrum of which is determined by the set $f(S)$. Similarly, the action of $C_F(X)$ on $H^k(X,S_F)$ leads to quantization of functions constant along F in terms of linear operators in $H^k(X,S_F)$. The existence of a $C_F(X)$-module isomorphism between $H^k(X,S_F)$ and $S_F(S)$ implies that these two quantizations are equivalent. Thus, one may obtain an equivalent representation of quantum dynamics using elements of $H^k(X,S_F)$ as the wave functions.

In the following section a review of facts in geometric quantization, pertinent to the formulation and the proof of Theorems 1.1 and 1.2 is given. The actual proofs of these theorems are divided into a series of lemmas and propositions contained in section 3.

Throughout this paper all manifolds are assumed to be real of class C^∞ connected and paracompact. All differentiable maps are assumed to be of class C^∞.

2. Elements of geometric quantization

A symplectic manifold is a pair (X,ω), where X is a differentiable manifold and ω is a closed non-degenerate differential 2-form on X. To each complex valued function f on a symplectic manifold (X,ω) there corresponds a unique complex vector field ξ on X such that $\xi \lrcorner \omega = df$, where $\lrcorner$ denotes the left interior product, called the Hamiltonian vector field of f. A polarization of (X,ω) is a complex involutive Lagrangian distribution F on X, that is, a complex involutive distribution F such that ω restricted to F vanishes identically and $\dim_{\underline{C}} F = \frac{1}{2} \dim X$. If the complex conjugate $\bar{F}$ of F coincides with F, the polarization F is a complexification of a real involutive Lagrangian distribution $D = F \cap TX$, and each integral manifold of D has a canonically defined flat affine connection [12].

Throughout this paper it is assumed that the following conditions are satisfied

(i) $F = \bar{F}$,

(ii) for each integral manifold Λ of D the canonical flat connection in Λ is complete,

(iii) the space Y of all integral manifolds of D has a manifold structure such that the canonical projection $\pi : X \to Y$ is a fibration admitting local trivializations which induce affine isomorphisms on the fibres.

Under these conditions each integral manifold Λ of D is isomorphic to $\underline{T}^k \times \underline{R}^{n-k}$, where $\underline{T}^k$ denotes a k-torus and k is the rank of the fundamental group of Λ, and $n = \dim Y$. The Hamiltonian vector fields in D with periodic orbits span a k-dimensional involutive distribution $K \subseteq D$ invariant under the Hamiltonian vector fields in D. There is a unique density κ on K, invariant under the Hamiltonian vector fields in D, assigning to each integral k-torus of K the total volume 1. For any coordinate system $(U, q_1, \ldots, q_n)$ in Y, the affine structures of integral manifolds of D projecting to points in U are defined by n linearly independent Hamiltonian vector fields of the functions $\pi^* q_1, \ldots, \pi^* q_n$ on $\pi^{-1}(U)$. There is an open covering $\{U_i\}$ of Y and a family of diffeomorphisms $\tau_i \colon \pi^{-1}(U_i) \longrightarrow U_i \times \underline{T}^k \times \underline{R}^{n-k}$, where $\underline{T}^k$ denotes the k-torus, such that $\pi \circ \tau_i = pr_1$ and, for each integral manifold Λ of D contained in $\pi^{-1}(U_i)$, $\tau_i | \Lambda$ defines an affine isomorphism of Λ onto $\underline{T}^k \times \underline{R}^{n-k}$.

Let L be a complex line bundle over X with a connection ∇, such that ω is the curvature form of ∇, and with an invariant Hermitian metric. Such a line bundle exists if and only if ω defines an integral de Rham cohomology class, [3]. Since, for each integral manifold Λ of D, $\omega | \Lambda = 0$ the restriction $L | \Lambda$ of L to Λ has a flat connection.

Let BF denote the linear frame bundle for F, that is, elements of BF are ordered bases in F. BF is a principal $GL(n, \underline{C})$ fibre bundle over X. Let $ML(n, \underline{C})$ denote the $n \times n$ complex metalinear group. It is a double covering of $GL(n, \underline{C})$ with the covering map $\rho \colon ML(n, \underline{C}) \longrightarrow GL(n, \underline{C})$. A metalinear frame bundle for F is a principal $ML(n, \underline{C})$ fibre bundle $\widetilde{BF}$ over X together with a map $\tau \colon \widetilde{BF} \longrightarrow BF$ such that the following diagram

$$\begin{array}{ccc} \widetilde{BF} \times ML(n, \underline{C}) & \longrightarrow & \widetilde{BF} \\ \Big\downarrow \tau \times \rho & & \Big\downarrow \tau \\ BF \times GL(n, \underline{C}) & \longrightarrow & BF \end{array}$$

in which the horizontal arrows denote the group actions, commutes.

A metalinear frame bundle $\tilde{B}F$ exists if and only if the corresponding characteristic class of BF in $H^2(X,\underline{Z}_2)$ vanishes, [1], [4]. Let χ: $ML(n,\underline{C}) \longrightarrow C$ denote the unique holomorphic square root of the complex character Det $\circ$ ρ: $ML(n,\underline{C}) \longrightarrow \underline{C}$ such that $\chi(1) = 1$. A bundle of half forms relative to F is a fibre bundle $N_F^{\frac{1}{2}}$ over X associated to $\tilde{B}F$ with a typical fibre $\underline{C}$ on which $ML(n,\underline{C})$ acts by multiplication by $\overline{\chi(a)}^{-1}$. A section ν of $N_F^{\frac{1}{2}}$ can be identified with a function ν: $\tilde{B}F \longrightarrow \underline{C}$ such that $\nu(b) = \overline{\chi(a)}\nu(ba)$ for each $b \in \tilde{B}F$ and each $a \in ML(n,\underline{C})$. There exists a family $\{\nu_j\}$ of local nowhere zero sections of $N_F^{\frac{1}{2}}$ such that their domains cover X and the corresponding transition functions are constant along F. Hence, for each integral manifold Λ of D, the restriction of $N_F^{\frac{1}{2}}$ to Λ has a flat connection, [8].

The Bohr-Sommerfeld set of a system $(X,\omega,F,L \otimes N_F^{\frac{1}{2}})$ is the subset S of X consisting of all integral manifolds Λ of D for which the holonomy group of the flat connection in $(L \otimes N_F^{\frac{1}{2}})|\Lambda$ is trivial. Each connected component of S is a submanifold of X with codimension k and its projection to Y is a submanifold of Y with the same codimension. A section of $L \otimes N_F^{\frac{1}{2}}$ is said to be covariant constant along F if its restriction to each integral manifold of D is covariant constant, and the sheaf of germs of sections of $L \otimes N_F^{\frac{1}{2}}$ covariant constant along F is denoted by S_F. If U is a submanifold of Y such that the space $S_F(\pi^{-1}(U))$ of sections of $(L \otimes N_F^{\frac{1}{2}})|\pi^{-1}(U)$ covariant constant along $F|\pi^{-1}(U)$ does not vanish, then $U \subseteq \pi(S)$. Moreover $S_F(S) = \oplus S_F(S_i)$, where S_i's are connected components of S and each $S_F(S_i) \neq 0$, [9].

3. Proofs

For each positive integer m, let $\wedge^m F^*$ denote the bundle of complex valued m-linear alternating forms on the fibres of F. A section φ of $\wedge^m F^*$ restricted to an integral manifold Λ of D can be identified with a complex valued differential m-form on Λ. Exterior differentiation of differential forms induces exterior differentiation d_F of sections of $\wedge^m F^*$ such that $(d_F\varphi)|\Lambda = d(\varphi|\Lambda)$, for each section φ of $\wedge^m F^*$ and each integral manifold Λ of D. Let C denote the sheaf of complex valued functions on X, C_F the sheaf of complex valued functions on X constant along F and, for each $m = 0,1,\ldots$, F^m the sheaf of sections of $\wedge^m F^*$. The sequence $0 \longrightarrow C_F \longrightarrow C \longrightarrow F^1 \longrightarrow \ldots \longrightarrow F^n \longrightarrow 0$, where $0 \longrightarrow C_F \longrightarrow C$ denotes the inclusion and the remaining arrows denote d_F, is a fine resolution of C_F. The sheaf S_F of sections of $L \otimes N_F^{\frac{1}{2}}$ covariant constant along F is a locally free sheaf of C_F modules, while the sheaf S of sections of $L \otimes N_F^{\frac{1}{2}}$ and the sheaves C, $F^1,\ldots,F^n$, are sheaves of C_F modules. Hence, $S_F \otimes F^m \cong S \otimes F^m$ and the sheaf homomorphisms $1 \otimes d_F$: $S_F \otimes F^m \longrightarrow S_F \otimes F^{m+1}$ induce sheaf homomorphisms ∇_F: $S \otimes F^m \longrightarrow S \otimes F^{m+1}$. Taking the tensor product of the fine resolution of C_F given above with S_F we obtain the following proposition due to B. Kostant.

Proposition 3.1 The sequence

$$0 \longrightarrow S_F \longrightarrow S \longrightarrow S \otimes F^1 \longrightarrow \ldots \longrightarrow S \otimes F^n \longrightarrow 0 ,$$

where $0 \longrightarrow S_F \longrightarrow S$ is the inclusion and the remaining arrows denote ∇_F, is a fine resolution of S_F.

For each open set V in X and each m, $Z^m(V) = \mathrm{Ker}\,(\nabla_F\colon S \otimes F^m(V) \longrightarrow S \otimes F^{m+1}(V))$, $B^m(V) = \nabla_F(S \otimes F^{m-1}(V))$ and $H^m(V) = Z^m(V)/B^m(V)$ are $C_F(V)$ modules. Proposition 3.1 and the abstract de Rham theorem, c.f. [13], imply that the m'th cohomology group $H^m(X,S_F)$ is isomorphic to $H^m(X)$. Since the operator ∇_F involves differentiation in the direction of D only, one may expect that the vanishing of cohomology groups $H^m(X,S_F)$ can be studied locally in Y.

Proposition 3.2 Let $\{U_i\}$ be a locally finite open covering of Y, and $H^m(\pi^{-1}(U_i)) = 0$ for all i. Then $H^m(X,S_F) = 0$.

Proof. Let $\{f_i\}$ be a partition of unity in Y subordinated to the covering $\{U_i\}$, then $\{\pi^* f_i\}$ is a partition of unity in X subordinated to the locally finite covering $\{\pi^{-1}(U_i)\}$ of X. For each $\psi \in Z^m(X)$, $\psi|\pi^{-1}(U_i) \in Z^m(\pi^{-1}(U_i))$ and, since $H^m(\pi^{-1}(U_i)) = 0$, $\psi|\pi^{-1}(U_i) \in B^m(\pi^{-1}(U_i))$. Hence, for each i, there exists a $\psi_i \in S \otimes F^{m-1}(\pi^{-1}(U_i))$ such that $\psi|\pi^{-1}(U_i) = \nabla_F\psi_i$. Then, $\psi = \sum_i \pi^* f_i\psi = \sum_i \pi^* f_i\ \psi|\pi^{-1}(U_i) = \sum_i \pi^* f_i\ \nabla_F\psi_i = \nabla_F \sum \pi^* f_i\psi_i$, and $\psi \in B^m(X)$. Hence, $Z^m(X) = B^m(X)$ which implies that $H^m(X) = 0$. But $H^m(X,S_F)$ is isomorphic to $H^m(X)$, and therefore $H^m(X,S_F) = 0$. Q.E.D.

We know that, for each point $y \in Y$, there exists an open neighbourhood U of y and a trivialization $\tau\colon \pi^{-1}(U) \longrightarrow U \times \underline{T}^k \times \underline{R}^{n-k}$ such that, for each integral manifold Λ of D contained in $\pi^{-1}(U)$, $\tau|\Lambda$ defines an affine isomorphism of Λ onto $\underline{T}^k \times \underline{R}^{n-k}$. In the following we shall use τ to identify $\pi^{-1}(U)$ with $U \times \underline{T}^k \times \underline{R}^{n-k}$.

Lemma 3.3 For each $y_0 \in Y$ we can choose a neighbourhood U of y_0 admitting a trivialization $\tau\colon \pi^{-1}(U) \longrightarrow U \times \underline{T}^k \times \underline{R}^{n-k}$ and a nowhere vanishing section $\lambda \otimes \nu$ of $L \otimes \bar{N}_F^{\frac{1}{2}}\,|\pi^{-1}(U)$ such that the section B of

$F^*|\pi^{-1}(U)$ defined by $\nabla_F(\lambda \otimes \nu) = 2\pi i\ \lambda \otimes \nu \otimes \beta$ can be expressed as $\beta = \sum p_r\ \Theta_r$, where $\Theta_1,\ldots,\Theta_k$ are the harmonic 1-forms on $\underline{T}^k$ such that the i's period of Θ_j is δ_{ij}, and $p_1,\ldots,p_k$ are real valued indpendent functions on U.

<u>Proof</u>. There exists a coordinate neighbourhood U of y_0 in Y with coordinate functions $(q_1,\ldots,q_n)$ admitting a trivialization τ: $\pi^{-1}(U) \longrightarrow U \times \underline{T}^k \times \underline{R}^{n-k}$ such that the induced projection $\pi^{-1}(U) \longrightarrow \underline{T}^k \times \underline{R}^{n-k}$ is a Lagrangian fibration of $(\pi^{-1}(U),\ \omega|\pi^{-1}(U))$ invariant under the actions of the Hamiltonian vector fields $\xi_1,\ldots,\xi_n$ of $\pi^* q_1,\ldots,\pi^* q_n$, respectively. Let $\eta_1,\ldots,\eta_n$ be vector fields on $\pi^{-1}(U)$ tangent to the fibres of $\pi^{-1}(U) \longrightarrow \underline{T}^k \times \underline{R}^{n-k}$ such that, for each $i = 1,\ldots,n$, $\pi_*\eta_i = \dfrac{\partial}{\partial q_i}$. Then $\omega(\eta_i,\eta_j) = \omega(\xi_i,\xi_j) = 0$, $\omega(\xi_i,\eta_j) = \delta_{ij}$, $[\eta_i,\xi_j] = [\xi_i,\xi_j] = 0$, and $\eta_1,\ldots,\eta_n$ preserve ω which implies that $d(\eta_i \lrcorner\, \omega) = 0$, for $i = 1,\ldots,n$. Hence, $\omega|\pi^{-1}(U) = d(\sum_i\ (\pi^* q_i)\eta_i \lrcorner\, \omega)$. Since $\omega|\pi^{-1}(U)$ is the curvature form of $L|\pi^{-1}(U)$, the first Chern class of $L|\pi^{-1}(U)$ vanishes, and so $L|\pi^{-1}(U)$ is trivial.

We use the trivialization τ to identify $\pi^{-1}(U)$ with $U \times \underline{T}^k \times \underline{R}^{n-k}$. Let $\tilde{\lambda}$ be a nowhere vanishing section of $L|\pi^{-1}(U)$. Without loss of generality we may assume that $\tilde{\lambda}$ is covariant constant in the direction of $\underline{R}^{n-k}$ in $U \times \underline{T}^k \times \underline{R}^{n-k}$. The 1-form $\tilde{\alpha}$ on $\pi^{-1}(U)$ defined by $\nabla\tilde{\lambda} = 2\pi i\,\tilde{\lambda} \otimes \tilde{\alpha}$ satisfies $d\tilde{\alpha} = \omega|\pi^{-1}(U)$ and is invariant under vector fields in the direction of $\underline{R}^{n-k}$. Let us write $\underline{T}^k$ as a product of k circles, $\underline{T}^k = T_1\times\ldots\times T_k$, and, for each $i = 1,\ldots,k$, let Θ_i be the harmonic 1-form on T_i with period 1. For each $(y,\underline{t},\underline{z}) \in U \times \underline{T}^k \times \underline{R}^{n-k}$, the integral of $\tilde{\alpha}$ over the circle T_i passing through $(y,\underline{t},\underline{z})$ is independent of

$(\underline{t},\underline{z}) \in \underline{T}^k \times \underline{R}^{n-k}$. Hence, $\tilde{\alpha}$ defines k functions $\tilde{p}_1,\ldots,\tilde{p}_k$ on U such that $\tilde{\alpha} - \sum \tilde{p}_i\, \theta_i$ is invariant under vector fields in the direction $\underline{R}^{n-k}$ and, for each $(y,\underline{z}) \in U \times \underline{R}^{n-k}$, $\tilde{\alpha} - \sum \tilde{p}_i\, \theta_i$ restricted to $\{y\} \times \pi^k \times \{z\}$ is a closed 1-form with zero periods, and the de Rham theorem implies that it is exact. Hence, there exists a function f on $U \times \underline{T}^k \times \underline{R}^{n-k}$, constant in the direction of $\underline{R}^{n-k}$, such that, for each $y \in U$, the restrictions of $(\tilde{\alpha} - \sum \tilde{p}_i\, \theta_i)$ and df to $\{y\} \times \underline{T}^k \times \underline{R}^{n-k}$ coincide. Let λ be the nowhere vansihing section of $L|\pi^{-1}(U)$ given by $\lambda = \exp(-2\pi i f)\tilde{\lambda}$. Then, $\nabla\lambda = 2\pi i\, \lambda \otimes \alpha$, where $\alpha = \tilde{\alpha} - df$ and, for each $y \in U$, $(\alpha - \sum \tilde{p}_i\, \theta_i)|\pi^{-1}(y) = 0$. Moreover, the functions $\tilde{p}_1,\ldots,\tilde{p}_k$ are real since the connection in L preserves a Hermitian form, and they are independent since $d\tilde{\alpha} = d\alpha = \omega|\pi^{-1}(U)$ is a non-singular form.

The Hamiltonian vector fields $\xi_1,\ldots,\xi_n$, trivialize the bundle of linear frames of $F|\pi^{-1}(U)$. Hence, there exists a family $\{\nu_r\}$ of nowhere vanishing local sections of $N_F^{\frac{1}{2}}|\pi^{-1}(U)$ such that each ν_r is covariant constant along F and the corresponding transition functions take on values ± 1. Without loss of generality we may assume that the domains W_r or ν_r are of the form $W_r = U \times V_r \times \underline{R}^{n-k}$. Using the transition functions corresponding to sections $\{\nu_r\}$ we can introduce a flat connection ∇ in $N_F^{\frac{1}{2}}|\pi^{-1}(U)$ such that, for each $y \in U$, it gives rise to the original flat connection in $N_F^{\frac{1}{2}}|\pi^{-1}(y)$. Hence $N_F^{\frac{1}{2}}|\pi^{-1}(U)$ is trivial and we may choose a nowhere zero section $\hat{\nu}$ such that, for each r, $\hat{\nu}|W_r = f_r \nu_r$ where f_r are functions constant along $U \times \underline{R}^{n-k}$ in $W_r = U \times V_r \times \underline{R}^{n-k}$. Let $\hat{\alpha}$ be the 1-form on $\pi^{-1}(U)$ defined by $\nabla\hat{\nu} = 2\pi i\, \nu \otimes \hat{\alpha}$ and, for each $i = 1,\ldots,k$, $\hat{p}_i$ the integral of $\hat{\alpha}$ along the i'th circle T_i in

$U \times \underline{T}_1 \times \ldots \times \underline{T}_k \times \underline{R}^{n-k}$. Since, for each $i = 1,\ldots,k$, $\exp(2\pi i\hat{p}_i) = \pm 1$, it follows that $\hat{p}_i$ is constant on $\pi^{-1}(U)$ and $2\hat{p}_i \in \underline{Z}$. By the same argument as before we can construct a nowhere vanishing section ν of $N_F^{\frac{1}{2}}|\pi^{-1}(U)$ such that $\nabla\nu = 2\pi i\ \nu \otimes \alpha^1$ where, for each $y \in U$, $(\alpha^1 - \sum \hat{p}_i\ \theta_i)|\pi^{-1}(y) = 0$.

Let λ and ν be the sections of $L|\pi^{-1}(U)$ and $N_F^{\frac{1}{2}}|\pi^{-1}(U)$, respectively, constructued above. Then, $\lambda \otimes \nu$ is a nowhere zero section of $L \otimes N_F^{\frac{1}{2}}|\pi^{-1}(U)$, $\nabla_F(\lambda \otimes \nu) = 2\pi i\ \lambda \otimes \nu \otimes (\alpha + \alpha^1)|F = 2\pi i\ \lambda \otimes \nu \otimes (\sum(\tilde{p}_i + \hat{p}_i)\theta_i)$, and $p_1 = \tilde{p}_1 + \hat{p}_1, \ldots, p_k = \tilde{p}_k + \hat{p}_k$ are real valued independent functions on U.

Q.E.D.

For each $\psi \in S \otimes F^m(\pi^{-1}(U))$ there exists a unique $\varphi \in F^m(\pi^{-1}(U))$ such that $\psi = \lambda \otimes \nu \otimes \varphi$. Moreover, $\nabla_F\psi = \lambda \otimes \nu \otimes (d_F\varphi + 2\pi i\ \beta \wedge \varphi)$. Hence, we obtain the following corollary.

Corollary 3.4 $H^m(\pi^{-1}(U)) = 0$ if and only if, for each $\varphi \in F^m(\pi^{-1}(U))$ such that $d_F\varphi + 2\pi i\ \beta \wedge \varphi = 0$, there exists $\tilde{\varphi} \in F^m(\pi^{-1}(U))$ satisfying $d_F\tilde{\varphi} + 2\pi i\ \beta \wedge \tilde{\varphi} = \varphi$.

The identification $\tau: \pi^{-1}(U) \longrightarrow U \times \underline{T}^k \times \underline{R}^{n-k}$ defines a direct sum decomposition for $F^1(\pi^{-1}(U))$ as follows. The elements of $F^1(\pi^{-1}(U))$ can be identified with complex valued 1-forms on $\underline{T}^k \times \underline{R}^{n-k}$ depending on parameters in U. Let $F^{1,0}(\pi^{-1}(U))$ denote the subspace of $F^1(\pi^{-1}(U))$ consisting of forms vanishing on vectors in the direction of $\underline{R}^{n-k}$ and $F^{0,1}(\pi^{-1}(U))$ the subspace of $F^1(\pi^{-1}(U))$ consisting of forms vanishing on vectors in the direction $\underline{T}^k$. Then, $F^1(\pi^{-1}(U)) = F^{1,0}(\pi^{-1}(U)) \oplus F^{0,1}(\pi^{-1}(U))$. Similarly, one defines subspaces $F^{p,q}(\pi^{-1}(U))$ of $F^{p+q}(\pi^{-1}(U))$ consisting of $p + q$ forms expressible as sums of exterior products of p forms in

$F^{1,0}(\pi^{-1}(U))$ and q forms in $F^{0,1}(\pi^{-1}(U))$. Note that $F^{p,q}(\pi^{-1}(U)) = 0$ unless $0 \le p \le k$ and $0 \le q \le n-k$. Hence $F^m(\pi^{-1}(U)) = \oplus_{i=0}^{s} F^{r+i,m-r-i}(\pi^{-1}(U))$, where $r = \max\{0, m+k-n\}$ and $s = \min\{k-r, m-r\}$. Clearly, $\beta \in F^{1,0}(\pi^{-1}(U))$ and $d_F = \partial_1 + \partial_2$ where ∂_1 denotes the exterior differentiation in the direction of $\underline{T}^k$, $\partial_1 : F^{p,q}(\pi^{-1}(U)) \longrightarrow F^{p+1,q}(\pi^{-1}(U))$, ∂_2 denotes the exterior differentiation in direction of $\underline{R}^{n-k}$, $\partial_2 : F^{p,q}(\pi^{-1}(U)) \longrightarrow F^{p,q+1}(\pi^{-1}(U))$, and $\partial_1\partial_1 = \partial_2\partial_2 = \partial_1\partial_2 + \partial_2\partial_1 = 0$.

Lemma 3.5 Let $\varphi \in F^m(\pi^{-1}(U))$ be such that $(d_F + 2\pi i\ \beta\wedge)\varphi = 0$. Then, there exists $\hat{\varphi} \in F^{m,0}(\pi^{-1}(U))$ such that $(d_F + 2\pi i\ \beta\wedge)\hat{\varphi} = 0$ and $\varphi = \hat{\varphi} + (d_F + 2\pi i\ \beta\wedge)\tilde{\varphi}$ for some $\tilde{\varphi} \in F^{m-1}(\pi^{-1}(U))$.

Proof Let $\varphi = \sum \varphi^{p,q}$, $\tilde{\varphi} = \sum \tilde{\varphi}^{p,q}$, where $\varphi^{p,q}, \tilde{\varphi}^{p,q} \in F^{p,q}(\pi^{-1}(U))$. Then, the equation $\varphi = \hat{\varphi} + (d_F + 2\pi i\ \beta\wedge)\tilde{\varphi}$ can be decomposed into a sequence of equations relating forms in the same space $F^{p,q}(\pi^{-1}(U))$, which can be written in the order of increasing p as follows,

$$\partial_2\tilde{\varphi}^{r,m-r-1} = \varphi^{r,m-r}$$

$$\partial_2\tilde{\varphi}^{r+1,m-r-2} = \varphi^{r+1,m-r-1} - (\partial_1 + 2\pi i\ \beta\wedge)\tilde{\varphi}^{r,m-r-1}$$

$$\cdots$$

$$\partial_2\tilde{\varphi}^{r+s,m-r-s-1} = \varphi^{r+s,m-r-s} - (\partial_1 + 2\pi i\ \beta\wedge)\tilde{\varphi}^{r+s-1,m-r-s} - \delta_{r+s,m}\hat{\varphi}\ .$$

where $r = \max\{0, m+k-n\}$ and $r + s = \min\{k, m\}$.

The equations $(d_F + 2\pi i\ \beta\wedge)\varphi = 0$ and $(d_F + 2\pi i\ \beta\wedge)\hat{\varphi} = 0$ imply that ∂_2 of the right hand side of each of the equations above vanishes. Since ∂_2 is exterior differentiation in $\underline{R}^{n-k}$ one can solve successively these equations by means of integration technique used in the proof of Poincaré, Lemma, cf. [11]. If $r + s = m$ then the term on the left hand side of the last equation in the series vanishes identically, and this equation gives rise to an equation defining $\hat{\varphi}$

$$\hat{\varphi} = \varphi^{m,0} - (\partial_1 + 2\pi i\ \beta\wedge)\tilde{\varphi}^{m-1,0}$$

Then $\hat{\varphi}$ satisfies $(d_F + 2\pi i\ \beta\wedge)\hat{\varphi} = 0$ as a consequence of the sequence of the preceeding equations. If $r + s = k < m$ then all the equations in the sequence can be solved for $\tilde{\varphi}^{p,q}$, $\hat{\varphi}$ does not appear at all and it is equal to zero. Q.E.D.

Corollary 3.6 $H^m(X,S_F) = 0$ for all $m > k$.

Proof. If $m > k$ then $F^{m,0}(\pi^{-1}(U)) = 0$ and, by Lemma 3.5, each $\varphi \in F^m(\pi^{-1}(U))$ satisfying $(d_F + 2\pi i\ \beta\wedge)\varphi = 0$ is of the form $\varphi = (d_F + 2\pi i\ \beta\wedge)\tilde{\varphi}$ for some $\tilde{\varphi} \in F^{m-1}(\pi^{-1}(U))$. Hence, by Corollary 3.4 $H^m(\pi^{-1}(U)) = 0$. Since Y can be covered by open sets U such that the condition of Lemma 3.3 are satisfied, it follows, by Proposition 3.2, that $H^m(X,S_F) = 0$. Q.E.D.

Consider the case $m < k$. Lemma 3.5 implies that it suffices to study forms $\varphi \in F^{m,0}(\pi^{-1}(U))$ satisfying the equation $(d_F + 2\pi i\ \beta\wedge)\varphi = 0$, which can be written as a pair of equations $(\partial_1 + 2\pi i\ \beta\wedge)\varphi = 0$ and $\partial_2\varphi = 0$. The condition $\varphi \in F^{m,0}(\pi^{-1}(U))$ implies that φ can be treated as a complex valued differential m-form on the k-torus $\underline{T}^k$ depending on parameters in $U \times \underline{R}^{n-k}$. Since the harmonic forms $\Theta_1,\ldots,\Theta_k$ on $\underline{T}^k$ form a basis of the module of complex valued forms on $\underline{T}^k$, it follows that φ can be expressed as $\varphi = \sum_{i_1\ldots i_m}\varphi_{i_1,\ldots,i_m}\Theta_{i_1}\wedge\ldots\Theta_{i_m}$, where the coefficients $\varphi_{i_1,\ldots,i_m}$ are complex valued functions on $U \times \underline{T}^k \times \underline{R}^{n-k}$. The condition $\partial_2\varphi = 0$ implies that the coefficients $\varphi_{i_1,\ldots,i_m}$ are independent of the variables in $\underline{R}^{n-k}$ and can be treated as complex valued functions on $U \times \underline{T}^k$. Similarly, according to Lemma 3.3, we can treat β as a harmonic form on $\underline{T}^k$ depending on parameters in U and we have $\beta = \sum p_i\,\Theta_i$.

Since the functions $p_1,\ldots,p_k$ are independent in U we can choose such a coordinate system in U that $p_1,\ldots,p_k$ are the first k coordinates. Wihtout loss of generality we may identify U with its image in $\underline{R}^n$ under the coordinate maps and assume that it is of the form of a product $U = U^k \times U^{n-k}$ where U^k is an open set in $\underline{R}^k$ with coordinates $p_1,\ldots,p_k$, and U^{n-k} is an open set in $\underline{R}^{n-k}$ with coordinates $q_1,\ldots,q_{n-k}$. Since $\underline{R}^k$ is the universal covering space of $\underline{T}^k$ we shall use coordinates $t_1,\ldots,t_k$ in $\underline{R}^k$ and Fourier series in the variables $t_1,\ldots,t_k$ to represent functions on $\underline{T}^k$. In order to simplify the notation, we shall use the vector notation for points in the Euclidean space i.e., we shall write $\underline{t} = (t_1,\ldots,t_k) \in \underline{R}^k$, $\underline{q} = (q_1,\ldots,q_{n-k}) \in \underline{R}^{n-k}$, $\underline{p} = (p_1,\ldots,p_k) \in \underline{R}^k$. Moreover, we shall use $\underline{n} = (n_1,\ldots,n_k) \in \underline{Z}^k$ to denote k-tuples of integers appearing in the Fourier series expansion, and put $\underline{n} \cdot \underline{t} = \sum n_i t_i$. With this notation the coefficients $\varphi_{i_1},\ldots,{}_{i_m}$ of φ can be rewritten as $\varphi_{i_1}\ldots,{}_{i_m}(\underline{p},\underline{q},\underline{t}) = \sum \varphi^{\underline{n}}_{i_1,\ldots,i_m}(\underline{p},\underline{q})\exp(-2\pi i\, \underline{n} \cdot \underline{t})$, and so $\varphi(\underline{p},\underline{q},\underline{t}) = \sum \varphi^{\underline{n}}(\underline{p},\underline{q})\exp(-2\pi i\, \underline{n} \cdot \underline{t})$, where $\varphi^{\underline{n}}(\underline{p},\underline{q}) = \sum \varphi^{n}_{i_1,\ldots,i_m}(\underline{p},\underline{q})\theta_{i_1}\wedge\ldots\wedge\theta_{i_m}$. The equation $(\partial_1 + 2\pi i\, \beta\wedge)\varphi = 0$ takes on the form of an infinite sequence of algebraic equations $(\sum(p_i - n_i)\theta_i)\wedge\ \varphi^{\underline{n}}(\underline{p},\underline{q}) = 0$, for each $\underline{n} \in \underline{Z}^k$, since the harmonic forms $\theta_1,\ldots,\theta_k$ on $\underline{T}^k$ correspond to the differentials $dt_1,\ldots,dt_k$ in $\underline{R}^k$.

Lemma 3.7 Let U be a neighbourhood of $\underline{0}$ in $\underline{R}^k \times \underline{R}^{n-k}$ and $f:U \longrightarrow \wedge^m\underline{R}^k$ a smooth map such that, for each $(\underline{p},\underline{q}) \in U$, $\underline{p} \wedge f(\underline{p},\underline{q}) = 0$. If $0 < m < k$, there exists a smooth map $g:U \longrightarrow \wedge^{m-1}\underline{R}^k$ such that $f(\underline{p},\underline{q}) = \underline{p} \wedge g(\underline{p},\underline{q})$ for all $(\underline{p},\underline{q}) \in U$.

Proof. If f is a homogeneous polynomial of degree r in $\underline{p}$, $f(\underline{p},\underline{q}) = f_{i_1\ldots,i_m}{}^{j_1\ldots j_r}(\underline{q})p_{j_1}\ldots p_{j_r}\underline{e}^{i_1}\wedge\ldots\wedge\underline{e}^{i_m}$, where $(\underline{e}^1,\ldots,\underline{e}^k)$ is the canonical basis in $\underline{R}^k$ and summation over repeated indices is understood, the equation $\underline{p}\wedge f(\underline{p},\underline{q}) = 0$ implies that $p_{[i_0}f_{i_1\ldots i_m]}{}^{j_1\ldots j_r}p_{j_1}\ldots p_{j_r} =$ where $[\ldots]$ denotes antisymmetrization with respect to the indices in the bracket. Using permutation symbols $e_{i_1\ldots i_k}$ equal 1 if the permutation $(1,\ldots,k) \longrightarrow (i_1,\ldots,i_k)$ is even, -1 if it is odd and 0 otherwise, (e-systems of Ref. [10]), and their properties, one can verify that in this case $f(\underline{p},\underline{q}) = \underline{p}\wedge g(\underline{p},\underline{q})$ where $g(\underline{p},\underline{q}) = g_{i_1,\ldots,i_{m-1}}{}^{j_1\ldots j_{r-1}}(\underline{q})$ $p_{j_1},\ldots,p_{j_{r-1}}\underline{e}^{i}\wedge\ldots\wedge\underline{e}^{i_{m-1}}$ and $g_{i,\ldots,i_{m-1}}{}^{j_1,\ldots,j_{r-1}}(\underline{q})$ is proportional to $f_{i_1\ldots,i_m}{}^{j_1\ldots j_r}$ contracted in the indices i_m and j_r. Hence, if f is a sum of homogeneous polynomial in $\underline{p}$ or a formal series of homogeneous polynomials in $\underline{p}$, then $f(\underline{p},\underline{q}) = \underline{p}\wedge g(\underline{p},\underline{q})$ where g is the sum of homogeneous polynomials in $\underline{p}$ or a formal series of homogeneous polynomials in $\underline{p}$, respectively. Therefore, for any smooth map $f\colon U \longrightarrow \wedge^m\underline{R}^k$ satisfying $\underline{p}\wedge f(\underline{p},\underline{q}) = 0$ we can apply the reasoning above to the Taylor series of $f(\underline{p},\underline{q})$ around $(0,\underline{q})$. By the Borel extension lemma, c.f. [2], there exists a smooth map $\tilde{g}\colon U \longrightarrow \wedge^{m-1}\underline{R}^k$ such that $f(\underline{p},\underline{q}) - p\wedge\tilde{g}(p,q) = \tilde{f}(\underline{p},\underline{q})$ vanishes to infinite order at $\underline{p} = 0$. Moreover $\underline{p}\wedge\tilde{f}(\underline{p},\underline{q}) = 0$ and we get $0 = \underline{p}\,\lrcorner\,(\underline{p}\wedge\tilde{f}(\underline{p},\underline{q})) = (\underline{p})^2\,\tilde{f}(\underline{p},\underline{q}) - \underline{p}\wedge(\underline{p}\,\lrcorner\,\tilde{f}(\underline{p},\underline{q}))$. Since $\tilde{f}$ vanishes to infinite order at $p = 0$, $g(\underline{p},\underline{q}) = \tilde{g}(\underline{p},\underline{q}) + (\underline{p}\,\lrcorner\,\tilde{f}(\underline{p},\underline{q}))/(\underline{p})^2$ is smooth and $f(\underline{p},\underline{q}) = \underline{p}\wedge g(\underline{p},\underline{q})$ for all $(\underline{p},\underline{q}) \in U$.

Q.E.D.

It follows from Lemma 3.7, that for $s < m < k$, the equations $\sum(p_i - n_i)\theta_i \wedge \varphi^{\underline{n}}(\underline{p},\underline{q}) = 0$ imply $\varphi^{\underline{n}}(\underline{p},\underline{q}) = \sum(p_i - n_i)\theta_i \wedge \tilde{\varphi}^{\underline{n}}(\underline{p},\underline{q})$, for some $(m-1)$-forms $\tilde{\varphi}^{\underline{n}}(\underline{p},\underline{q})$, and therefore the equation $(\partial_1 + 2\pi i\ \beta\wedge)\varphi = 0$ implies $\varphi = (\partial_1 + 2\pi i\ \beta\wedge)\tilde{\varphi}$ where $\tilde{\varphi}(\underline{p},\underline{q},\underline{t}) = \sum \tilde{\varphi}^{\underline{n}}(\underline{p},\underline{q})\exp(-2\pi i\ \underline{n}\cdot\underline{t})$. Hence, we obtain the following corollary.

Corollary 3.8 $H^m(X,S_F) = 0$ for $m < k$.

Proof. From Lemma 3.5, Lemma 3.7 and the discussion above it follows that if $0 < m < k$ and $\varphi \in F^m(\pi^{-1}(U))$ satisfies $(d_F + 2\pi i\ \beta\wedge)\varphi = 0$ there exists $\tilde{\varphi} \in F^{m-1}(\pi^{-1}(U))$ such that $\varphi = (d_F + 2\pi i\ \beta\wedge)\tilde{\varphi}$. Hence, by Corollary 3.4, $H^m(\pi^{-1}(U)) = 0$. Since Y can be covered by open sets U such that the conditions of Lemma 3.3 are satisfied, it follows, by Prop. 3.2 that $H^m(X,S_F) = 0$. Vanishing of $H^0(X,S_F)$ for $k > 0$ is obvious

Q.E.D.

Corollaries 3.6 and 3.8 imply the statement of Theorem 1.1.

Let us consider now the case $m = k$. The equation $(\partial_1 + 2\pi i\ \beta\wedge)\varphi = 0$ is always satisfied for $\varphi \in F^{k,0}(\pi^{-1}(U))$, and the equation $(\partial_1 + 2\pi i\ \beta\wedge)\tilde{\varphi} = \varphi$ for $\tilde{\varphi} \in F^{k-1,0}(\pi^{-1}(U))$ can be rewritten in terms of the Fourier series decompositions of $\tilde{\varphi}$ and φ as $2\pi i(\sum(p_j - n_j)\theta_j) \wedge \tilde{\varphi}^{\underline{n}}(\underline{p},\underline{q}) = \varphi^{\underline{n}}(\underline{p},\underline{q})$, which implies that $\varphi^{\underline{n}}(\underline{n},\underline{q}) = 0$. Conversely, if $\varphi^{\underline{n}}(\underline{n},\underline{q}) = 0$ then $\varphi^{\underline{n}}(\underline{p},\underline{q})$ factors through $(\underline{p} - \underline{n})$, and there exists a smooth form $\tilde{\varphi}^{\underline{n}}(\underline{p},\underline{q})$ such that $\varphi^{\underline{n}}(\underline{p},\underline{q}) = 2\pi i(\sum(p_j - n_j)\theta_j) \wedge \tilde{\varphi}^{\underline{n}}(\underline{p},\underline{q})$. Since the intersection of the Bohr-Sommerfeld set S with $\pi^{-1}(U)$ is characterized by the condition $(\underline{p},\underline{q},\underline{t}) \in S \Leftrightarrow \underline{p} \in \underline{Z}^k$, we obtain the following proposition.

Proposition 3.9 Let $\psi \in S \otimes F^k(X)$ be such that $\nabla_F\psi = 0$ and $\psi|S = 0$. Then there exists $\tilde{\psi} \in S \otimes F^{k-1}(X)$ such that $\psi = \nabla_F\tilde{\psi}$.

Proof. Let $\{U_i\}$ be a locally finite covering of Y such that each U_i satisfies the conditions of Lemma 3.3, and $\{f_i\}$ a partition of unity subordinated to $\{U_i\}$. For each U_j, $\psi|\pi^{-1}(U_j) = \lambda_j \otimes \nu_j \otimes \varphi_j$, and by Lemma 3.5 $\varphi_j = \hat{\varphi}_j + (d_F + 2\pi i\ \beta_j\wedge)\tilde{\varphi}_j$, where $\hat{\varphi}_j \in F^{k,0}(\pi^{-1}(U_j))$ and $\tilde{\varphi}_j \in F^{k-1}(\pi^{-1}(U))$. Since $\varphi_j|S \cap \pi^{-1}(U_j) = 0$ it follows that $\varphi_j(\underline{n},\underline{q}) = 0$, and therefore the Fourier coefficients of φ_j satisfy $\varphi_j^{\underline{n}}(\underline{n},\underline{q}) = 0$. Hence, $\hat{\varphi}_j^{\underline{n}}(\underline{n},\underline{q}) = 0$ and there exists φ_j^* such that $\hat{\varphi}_j = (\partial_1 + 2\pi i\ \beta_j\wedge)\varphi_j^*$, which implies that $\varphi_j = (d_F + 2\pi i\ \beta_j\wedge)(\varphi_j^* + \tilde{\varphi}_j)$ or, equivalently, $\psi|\pi^{-1}(U_j) = \nabla_F\ \tilde{\psi}_j$ where $\tilde{\psi}_j = \lambda_j \otimes \nu_j \otimes (\varphi_j^* + \tilde{\varphi}_j)$. Therefore, $\psi = \sum_j \pi^* f_j \psi = \sum_j \pi^* f_j\ \psi|\pi^{-1}(U_j)$ $= \sum_j \pi^* f_j\ \nabla_F\tilde{\psi}_j = \nabla_F \sum_j \pi^* f_j \tilde{\psi}_j$.

Q.E.D.

For each connected component S_0 of the Bohr-Sommerfeld set S, let $H_{S_0}{}^k(X,S_F)$ denote the subspace of $H^k(X,S_F)$ consisting of all elements which can be represented by $\psi \in S \otimes F^k(X)$ such that $\nabla_F\psi = 0$ and $\psi|(S - S_0) = 0$. By Proposition 3.9, for any two different components S_0 and S_1 of S, $H_{S_0}{}^k(X,S_F) \cap H_{S_1}{}^k(X,S_F) = 0$. Moreover, each $\psi \in S \otimes F^k(X)$ satisfying $\nabla_F\psi = 0$ can be represented as a sum $\psi = \sum_i \psi_i$, indexed over connected components S_i of S, such that $\psi_i|(S - S_i) = 0$ and $\nabla_F\psi_i = 0$. Hence, we have the following corollary.

Corollary 3.10 $H^k(X,S_F) = \oplus H_{S_i}{}^k(X,S_F)$, where the sum is taken over all connected components of the Bohr-Sommerfeld set S.

Since both $H^k(X,S_F)$ and $S_F(S)$ are direct sums of subspaces indexed by connected components of S, it suffices to prove, for each connected component S_0 of S, that $H_{S_0}{}^k(X,S_F)$ is isomorphic to $S_F(S_0)$ in order to conclude that $H^k(X,S_F)$ and $S_F(S)$ are isomorphic. Let S_0 be a connected component of S and $\lambda_0 \otimes \nu_0$ a nowhere vanishing section of $(L \otimes N_F^{\frac{1}{2}})|S_0$. Each class $[\psi] \in H_{S_0}{}^k(X,S_F)$ is uniquely determined by $\psi|S_0$ which can be expressed as $\psi|S_0 = \lambda_0 \otimes \nu_0 \otimes \varphi$, where $\varphi \in F^k(S_0)$ and $d_F\varphi = 0$. Moreover, two d_F-closed forms φ and φ' in $F^k(S_0)$ give rise to the same class $[\psi]$ in $H_{S_0}{}^k(X,S_F)$ if and only if $\varphi - \varphi' = d_F\varphi''$ for some $\varphi'' \in F^{k-1}(S_0)$. Hence, we have a linear isomorphism between $H_{S_0}{}^k(X,S_F)$ and the space of equivalence classes of d_F-closed forms in $F^k(S_0)$ with the equivalence relation given by $\varphi \sim \varphi'$ if and only if $\varphi - \varphi'$ is d_F-exact. Clearly, this isomorphism commutes with multiplication by functions on X constant along F.

The distribution $K \subseteq D$ spanned by the Hamiltonian vector fields in D with periodic orbits has a canonically defined density κ invariant under the action of Hamiltonian vector fields in D and giving total volume 1 to each integral k-torus of K. If K is orientable, the density κ together with an orientation of X give rise to a k-form $\tilde{\kappa}$ on K. Let $\varphi_0 \in F^k(S_0)$ be an extension of $\tilde{\kappa}|S_0$ to a k-form on $F|S_0$ such that $d_F\varphi_0 = 0$; any two such extensions differ by a d_F-exact form.

Lemma 3.11 Each form $\varphi \in F^k(S_0)$ satisfying $d_F\varphi = 0$ can be represented as $\varphi = h\varphi_0 + d_F\varphi'$, where h is a complex valued function on S_0 constant along $F|S_0$ and $\varphi' \in F^{k-1}(S_0)$.

Proof. Let $\{U_i\}$ be a locally finite open covering of $\pi(S_0)$ such that, for each i, there exists a trivialization $\tau_i:\pi^{-1}(U_i) \longrightarrow U_i \times \underline{T}^k \times \underline{R}^{n-k}$. For any form $\varphi \in F^k(S_0)$ with $d_F\varphi = 0$, there exists by Lemma 3.5 a family $\{\varphi_i\}$ of forms in $F^k(S_0)$ satisfying $d_F\varphi_i = 0$, support $\varphi_i \subseteq \pi^{-1}(U_i)$, $\varphi_i|\pi^{-1}(U_i) \in F^{k,0}(\pi^{-1}(U_i))$ and a form $\tilde{\varphi} \in F^{k-1}(S_0)$ such that $\varphi = \sum_i \varphi_i + d_F\tilde{\varphi}$. If φ_i' is a d_F-closed form in $F^k(S_0)$ with support in $\pi^{-1}(U_i)$ such that it agrees with φ_i on $K|\pi^{-1}(U_i)$ then by the same argument as in Lemma 3.5 $\varphi_i' = \varphi_i + d_F\varphi_i''$ for some $\varphi_i'' \in F^{k-1}(S_0)$. Therefore, the class of φ_i is uniquely determined by its restriction to $K|S_0$. Let g_i be the function on S_0 defined by $\varphi_i|(K|S_0) = g_i\tilde{\kappa}$. The support of g_i is contained in $\pi^{-1}(U_i)$ and in the trivialization of $\pi^{-1}(U_i)$ given by τ_i we have $\varphi_i = g_i\theta_1 \wedge\ldots\wedge\theta_k$. Since $\partial_2\varphi_i = 0$, it follows that g_i is constant in the direction of $\underline{R}^{n-k}$ in $\pi^{-1}(U_i) \simeq U_i \times \underline{T}^k \times \underline{R}^{n-k}$. For $(y,\underline{t},\underline{z}) \in U_i \times \underline{T}^k \times \underline{R}^{n-k}$ the integral of φ_i over the torus $\{y\} \times \underline{T}^k \times \{\underline{z}\}$ through $(y,\underline{t},\underline{z})$ depends only on $y \in U_i$ and therefore it defines a function h_i on S_0 with support in $\pi^{-1}(U_i)$ constant along $F|S_0$. The form $h_i\theta_1 \wedge\ldots\wedge\theta_k$ on $\pi^{-1}(U_i)$ extends to a d_F closed form $\hat{\varphi}_i$ on S_0 with support in $\pi^{-1}(U_i)$. Moreover, all periods of $\varphi_i - \hat{\varphi}_i$, treated as k-forms on $\underline{T}^k \times \underline{R}^{n-k}$ depending on parameters in U_i, vanish and by the de Rham theorem there exists a k-1 form $\check{\varphi}_i$ depending on parameters in U_i, such that $\varphi_i - \hat{\varphi}_i = d_F\check{\varphi}_i$. Since supports of φ_i and $\hat{\varphi}_i$ are contained in $\pi^{-1}(U_i)$ we can extend $\check{\varphi}_i$ to a globally defined form on S_0 with support in $\pi^{-1}(U_i)$. Therefore, $\varphi = \sum \hat{\varphi}_i + d_F(\tilde{\varphi} + \sum \check{\varphi}_i)$. Moreover, $(\sum \hat{\varphi}_i)|(K|S_0) = h\tilde{\kappa}$ where $h = \sum h_i$ is constant along $F|S_0$, and so $\sum \hat{\varphi}_i$ and $h\varphi_0$ agree on $K|S_0$ which implies that $\sum \hat{\varphi}_i = h\phi_0 + d_F\varphi''$ for some $\varphi'' \in F^{k-1}(S_0)$. Hence, putting

$\varphi' = \tilde{\varphi} + \sum \check{\varphi}_i + \varphi''$, we have $\varphi = h\varphi_0 + d_F\varphi'$ which prove the Lemma Q.E.D.

Each element of $S_F(S_0)$ is a product of the section $\lambda_0 \otimes \nu_0$ by a function $h \in C_F(X)$. The mapping $S_F(S_0) \longrightarrow H_{S_0}{}^k(X,S_F)$ associating to $h\, \lambda_0 \otimes \nu_0$ the class in $H_{S_0}{}^k(X,S_F)$ defined by $h\, \lambda_0 \otimes \nu_0 \otimes \varphi_0$ is an isomorphism of $C_F(X)$ modules which implies the existence of a $C_F(X)$-module isomorphism between $S_F(S)$ and $H^k(X,S_F)$. This completes the proof of Theorem 1.2.

Acknowledgments

The author would like to thank B. Kostant and V. Guillemin for their interest in this work and helpful discussions on subjects studied in this paper. Stimulating discussions with J. Dodziuk and R. Melrose are gratefully acknowledged.

References

[1] R. J. Blattner, Quantization and representation theory, Proc. Sympos. Pure Math., vol. 26, Amer. Math. Soc., Providence, R.I., 1973, pp. 147-165.

[2] M. Golubitsky and V. Guillemin, Stable mappings and their singularities, Graduate Texts in Math., 14, Springer-Verlag, New York, 1973.

[3] B. Kostant, Quantization and unitary representations, Lecture Notes in Math., vol. 170, Springer-Verlag, Berlin, 1970, pp. 87-208.

[4] B. Kostant, Symplectic spinors, Conv. di Geom. Simp. e Fis. Mat., INDAM, Rome, 1973, to appear in Symp. Math. Series, Academic Press.

[5] B. Kostant, On the definition of quantization, to appear in proceedings of Coll. Int. du C.N.R.S. Géométrie symplectique et physique mathématique, Aix-en-Provence, 1974.

[6] D. J. Simms, Geometric quantisation of the harmonic oscillator with diagonalised Hamiltonian, Proc. 2nd Int. Coll. on Group Theor. Methods in Physics, Nijmegen, 1973.

[7] D. J. Simms, Metalinear structures and a geometric quantisation of the harmonic oscillator, to appear in proceedings of Coll. Int. du C.N.R.S. Géométrie symplectique et physique mathématique, Aix-en-Provence, 1974.

[8] J. Śniatycki, Bohr-Sommerfeld conditions in geometric quantization, Reports in Math. Phys., vol. 7, (1974), pp. 127-135.

[9] J. Śniatycki, Wave functions relative to a real polarization, to appear in Int. J. of Theor. Phys.

[10] I. S. Sokolnikoff, Tensor analysis theory and applications to geometry and mechanics of continua, 2nd ed., John Wiley, New York, 1964.

[11] F. W. Warner, Foundations of differentiable manifolds and Lie groups, Scott, Foresman and Co., Glenview, Illinois, 1971.

[12] A. Weinstein, Symplectic manifolds and their Lagrangian submanifolds, Advances in Math., vol. 6, 1971, pp. 329-346.

[13] R. O. Wells, Jr., Differential analysis on complex manifolds, Prentice Hall, Englewood Cliffs, N.J., 1973.

GEOMETRIC QUANTIZATION AND FEYNMAN PATH INTEGRALS FOR SPIN

K. Gawędzki

Department of Mathematical Methods of Physics, Warsaw, University

This would not be a new observation that the routine quantization procedures extensively used in mechanics (and field theory): the canonical quantization and the pathe-integral prescriptions are interrelated. The Kostant-Souriau geometric generalization of the first procedure provides however an opportunity to understand this interrelation more deeply. The point is that the Kostant-wise quantized observables possess a kernel representation which can be viewed as an infinitesimal version of a path integral formula. We are not going to present a detailed exposition of the idea, this will be given elsewhere. Instead we show how it works in one of the simplest cases: in the case of a pure-spin mechanical system, i.e. a subsystem describing only the spin degrees of freedom of a non-relativistic particle with spin. As a result we obtain a path-integral formulation of a simple dynamics for the spin system which, in our opinion, has never been achieved so far.

We proceed according to the following plan:

i. first we sketch the Kostant-Souriau quantization scheme for the case under consideration (see [2] , [3] , [4]),

ii. then we present the kernel representation of the geometrically quantized observalbes,

iii. basing on the latter we show how to obtain the path-integral expressions.

The phase space of the pure spin system is the unit sphere S^2 in R^3. The symplectic structure of this phase space, the basic structure of the classical mechanical system, is provided by s times the area form of the unit sphere (s > 0 is interpreted as the total spin of the system). This classical mechanical system has rotation group SO(3) as its symmetry group. The Hilbert space of the geometricly quantized system will be a space of sections of a bundle with connection L (see the lecture of D.J. Simms in the present volume - we do not take the half-form bundle factor as it is not necessary in this case). The bundle L exists only for s half-integer or integer. According to the general prescription we must still choose a polarization.

By stereographic projection S^2 can be identified with $P^1(\mathbb{C})$ - the one-dimensional projective plane. We shall keep doing this identification henceforth. $P^1(\mathbb{C})$ carries a natural holomorphic structure and the subbundle of holomorphic vectors provides us with the only rotation invariant positive polarization F of $P^1(\mathbb{C}) \cong S^2$. Now the Hilbert space H of states of the quantized system is the space of sections of L covariantly constant over F.

If we parametrize $P^1(\mathbb{C}) \setminus \{\infty\}$ by the complex coordinate z in the natural way and properly trivialize L over $P^1(\mathbb{C}) \setminus \{\infty\}$ then the sections (elements of H) turn out to be represented by polynomials in $\bar{z}$ of degree at most 2s . Henceforth the space of such polynomials will be identified with H .

We are interested in quantization of functions which are orthogonal projections onto rotation axis of S^2 . These are classical observables whose hamiltonian flows are flows of rotations of S^2 around the axis. If the restriction of such a function to $P^1(\mathbb{C}) \setminus \{\infty\}$ is represented by a function f of the complex variable z and if $\hat{f}$ denotes the geometrically quantized observable then the following kernel expression for $\hat{f}$ can be derived (see [1]):

$$(\hat{f}\psi)(z') = \int K(z';z) \frac{s+1}{s} f(z) \psi(z) d^2z \tag{1}$$

where

$$K(z';z) = \frac{2s+1}{\pi} \frac{(1+\bar{z}'z)^{2s}}{(1+|z|^2)^{2(s+1)}} , \tag{2}$$

$\psi \in H$ is a polynomial of degree at most 2s in $\bar{z}$, and d^2z denotes the Lebesque measure on $\mathbb{C}$.

The kernel K(.;.) has the following reproducing property, crucial in the path integral construction:

$$\int K(z';z) \psi(z) d^2z = \psi(z') \tag{3}$$

for $\psi \in H$.

We shall treat (1) as an infinitesimal version of a path integral formula. In order to build a global version let us state some definitions.

$$X^z := \{\sigma \in C_{pw}([0,t],C) \ : \sigma(t) = z\}$$

where $C_{pw}([0,t],C)$ is the space of piece-wise continuous mappings from $[0, t]$ to $\mathbb{C}$ possessing left and right hand side limits at each point;

$$\pi = (t_o,\ldots,t_k), \qquad 0 = t_o < t_1 < \ldots < t_{k-1} < t_k = t;$$

Π is the set of all πs directed by inclusion; $\mathcal{G}_\pi$ will denote the linear space of all functions G on X^z of the form

$$G(\sigma) = g(\sigma(t_o),\ldots,\sigma(t_{k-1}))$$

for a measurable g on $\mathbb{C}^k$ such that

$$|g(z_o,\ldots,z_{k-1})| \le C_g\ (1+|z_o|^2)^{s+1/2}\ (1+|z_1|^2)^{1/2}\ldots(1+|z_{k-1}|^2)^{1/2}$$

Let us define a family of linear functionals μ_π^z on $\mathcal{G}_\pi$ by

$$\langle G, \mu_\pi^z\rangle := \int g(z_o,\ldots,z_{k-1}) \prod_{j=0}^{k-1} K(z_{j+1};z_j)\, d^2z_j \tag{4}$$

with $z_k := z$. We shall symbolicly write

$$\langle G, \mu_\pi^z\rangle =: \int G(\sigma)\, d\, \mu_\pi^z\, (\sigma) \tag{5}$$

For $\pi' > \pi$ $\mathcal{G}_{\pi'} \supset \mathcal{G}_\pi$ and one can show that for $G \in \mathcal{G}_\pi$

$$\langle G,\mu_\pi^z\rangle = \langle G,\mu_{\pi'}^z\rangle \tag{6}$$

(here the reproducing property (3) enters).

Thus we can define μ^z on $\mathcal{G} = \bigcup_{\pi\in\Pi} \mathcal{G}_\pi$, i.e. on all cylindric functions on X^z. Now for f corresponding to a projection onto a rotation axis of S^2 let

$$G_k(\sigma) := \prod_{j=0}^{k-1} \left(1 + \frac{s+1}{is}\, f\left(\sigma\left(\frac{t}{k}\, j\right)\, \frac{t}{k}\right)\right)$$

Then by (1) and (3)

$$\int G_k(\sigma)\ \psi(\sigma(0))\ d\mu^z(\sigma) = \left[(1+\frac{t}{ik}\ \hat{f})^k\psi\right](z) \tag{7}$$

and consequently

$$\int G_k(\sigma)\ \psi(\sigma(0))\ d\mu^z(\sigma) \xrightarrow[k\to\infty]{} \left[\exp(\frac{t}{i}\ \hat{f})\ \psi\right](z)$$

Pointwise on X^z

$$G_k \xrightarrow[k\to\infty]{} G$$

where

$$G(\sigma) = \exp(\frac{s+1}{is}\int_0^t f(\sigma(u))\ du) \tag{8}$$

What is needed to complete our construction is an extension $\tilde{\mu}^z$ of μ^z to a functional on a class of functions broader then that of cylindric functions, such that the following expression should hold

$$\left[\exp(\frac{t}{i}\ \hat{f})\psi\right](z) = \int \exp(\frac{s+1}{is}\int_0^t f(\sigma(u))\ du)\ \psi(\sigma(0))d\tilde{\mu}^z(\sigma) \tag{9}$$

This can be achieved by defining for example for G on X^z and each

$$G_\pi\ (\sigma) := G(\sigma_\pi)$$

with

$$\sigma_\pi(u) = \begin{cases} \sigma(t_o) & \text{for}\quad u \in [t_p, t_1[\\ \sigma(t_{k-1}) & \text{for}\quad u \in [t_{k-1}, t[\end{cases}$$

and calling G $\tilde{\mu}^z$-integrable if there exists

$$\lim_{\pi\in\Pi} \int G_\pi\ d\mu^z =: \int G\ d\tilde{\mu}^z$$

Now one can easily show that (9) holds.

Remark 1. If one applies (9) to a constant function equal to s^2 the constant operator $s(s+1)$ is obtained giving proper quantization of the observable usually called the square of angular momentum (sum of squares of functions generating rotation flows around perpendicular axes). This cannot be obtained in the original geometric quantization which quantizes function 1 to the identity operator.

Remark 2. Our choice of class of paths belonging to X^z has a technical character rather and is a result of the actual choice of the procedure of extension of μ^z to $\tilde{\mu}^z$.

Remark 3. Our strategy to work in coordinates was not indispensable. We could take X^z as $C_{pw}([0,t], P^1(\mathbb{C}))$, $\mathfrak{G}$ as a space of cylindric mappings from X^z to L such that the diagram

$$X^z \longrightarrow L \xrightarrow{\pi} P^1(\mathbb{C}) \qquad \text{(curved arrow } X^z \to P^1(\mathbb{C}):\ \sigma \longmapsto \sigma(0))$$

is commutative (π being the bundle projection), and we could define μ^z as a linear mapping from $\mathfrak{G}$ to L_z .

Remark 4. All the above considerations can be repeated with only slight modifications for a general kahlerian manifold, or even (in a less complete form however) for the general case of symplectic manifold with a reproducing kernel (see [1]).

References

[1] . K. Gawędzki, Fourier-like kernels in geometric quantization, to appear in Diss. Math.

[2] . B. Kostant, in Lecture Notes in Mathematics, 170 (1970), Springer, Berlin, p. 87.

[3] . P. Rénouard, Varietés symplectiques et quantification, Thesis, 1969, Orsay.

[4] . J.M. Souriau, Structures des systèmes dynamiques, 1970, Dunod, Paris.

V. FOCK, 40 YEARS LATER

Enrico Onofri
Istituto di Fisica, Sezione Teorica,
Università di Parma, 43100-Parma (Italy)
and
Istituto Nazionale di Fisica Nucleare,
Sezione di Milano

Introduction

The title refers to Fock's paper appeared in 1935[1]; it was shown there how Schroedinger equation for the three-dimensional hydrogen atom can be transformed in momentum space into an integral equation over S^3 with a SO(4)-invariant kernel. As an application of modern Quantisation Theory, we are now able to derive this result directly from the classical Hamiltonian system. The result is valid in any number of dimensions. A partial result was announced in Ref.(2); more complete details can be found in Ref.(3).

1 - The regularized Kepler problem as a $SO_\circ(2,n)$-homogeneous space

We start from the Hamiltonian model of the regularized Kepler motion[4] in (n-1) degrees of freedom

$$\Omega_{2(n-1)} = T\, S^{n-1} - S^{n-1} \tag{1}$$

Attempts to quantise such a manifold through non-invariant polarizations are being pursued (Sternberg and Wolf, Rawnsley). A different approach was proposed in Ref.(5) and successfully applied in Ref.(2,3). The first step is to embed the Hamiltonian flow (taking the eccentric anomaly as time parameter) as a one-parameter subgroup in a global symplectic action of $SO_\circ(2,n)$ on $\Omega_{2(n-1)}$. $\Omega_{2(n-1)}$ can be identified, then, with an orbit in the co-adjoint representation of $SO_\circ(2,n)$[6]. The quantisation problem is now identified with the problem of constructing a unitary irreducible representation of the conformal group in n-dimensions.

2 - "Smoothing out the vertex of the cone"

Let S and $M_{\mu\nu} \varepsilon \underline{o(2,n)}$ ($\mu,\nu = 1,2,\ldots,n$) be the generators of compact rotations in the planes (1,2) and ($\mu+2,\nu+2$) respectively, and Z_μ, W_μ the non-compact generators in the planes ($1,\mu+2$) and ($2,\mu+2$). Let $\flat$ denote the duality given by the Killing form; an element in $\underline{o(2,n)^*}$ is given by

(2) $$\omega = s\, S^\flat + \sum_{\mu<\nu} m_{\mu\nu} M^\flat_{\mu\nu} + \sum z_\mu Z^\flat_\mu + \sum_\mu w_\mu W^\flat_\mu$$

The Kepler manifold is identified with the orbit through the point $\omega_0 = S^\flat + M^\flat_{12} + Z^\flat_1 + W^\flat_2$ and it is characterized by the following equations:

(3) $$\begin{cases} s^2 + \sum m^2_\mu - \sum (z^2_\mu + w^2_\mu) = 0 \\ s\, m_{\mu\nu} = z_\mu w_\nu - z_\nu w_\mu \\ \sum z^2_\mu - \sum w^2_\mu = \sum z_\mu w_\mu = 0 \end{cases}$$

The Hamiltonian flow is given by $\exp\{\phi S\}$.

Let us consider a neighborhood of this orbit, namely the family of orbits $\mathcal{O}_\ell$ through the points $\omega_\ell = \ell S^\flat$ ($\ell > 0$) which are characterized by the following equations:

(4) $$\begin{cases} s^2 + \sum_{\mu<\nu} m^2_{\mu\nu} - \sum_\mu (z^2_\mu + w^2_\mu) = \ell^2 \\ s\, m_{\mu\nu} = z_\mu w_\nu - z_\nu w_\mu \end{cases}$$

This corresponds to "smooth out" the vertex of a three-dimensional cone by considering the nearby two-sheeted hyperboloids with the same asymptotics (this is exactly the picture for $SO_0(2,1)$-orbits).

What we gain is that $\mathcal{O}_\ell$ is diffeomorphic to a homogeneous bounded domain $\mathcal{D}$ in C^n. The mapping μ: $\mathcal{O}_\ell \to C^n$ is explicitly given by

(5) $$\zeta = (\zeta_1,\ldots,\zeta_n) = \frac{i}{2s}\left(\sigma + \frac{\sigma\,\sigma'}{2s(s+\ell) - \sigma\bar{\sigma}'}\,\bar{\sigma}\right)$$

where $\sigma_\mu = z_\mu + i\, w_\mu$, $\sigma = (\sigma_1,\ldots,\sigma_n)$ and the prime denotes the transpose. The $SO_0(2,n)$ action is given by holomorphic transformations in terms of ζ; in particular ζ is a vector under SO(n) and $\exp\{\phi S\} = e^{i\phi}\zeta$.

3 - Quantisation of $\mathcal{O}_\ell$ and Fock's equation

The quantisation of $\mathcal{O}_\ell$ is easily obtained in terms of Hilbert spaces of holomorphic functions $\psi(\zeta)$ in $\mathcal{D}$ such that

$$\|\psi\|^2 = N \int_{\mathcal{D}} |\psi(\zeta)|^2 (1 - 2\zeta\bar{\zeta}' + |\zeta\zeta'|^2)^{\ell-\frac{1}{2}n-1} \, \natural \tag{6}$$

is finite; N is a normalization constant and

$$\natural = \prod_\mu \frac{\overline{d\zeta_\mu} \wedge d\zeta_\mu}{2i} \tag{7}$$

$SO_0(2,n)$ transformations are represented by unitary operators

$$(U_g\psi)(\zeta) = (j_g^{-\frac{1}{2}-(\ell-1)/n}\,\psi)(g^{-1}\zeta) \tag{8}$$

j_g being the Jacobian of the transformation $\zeta \to g\zeta$. The representation is irreducible.

The exponent $-\frac{1}{2}-(\ell-1)/n$ is quite peculiar; the term $-\ell/n$ comes from usual prequantization; the term $-\frac{1}{2}$ is characteristic of symplectic spinors, then the theory in its present formulation (7) would suggest an exponent $-(\frac{1}{2}+\ell/n)$. However I was forced to introduce a term $+1/n$ to obtain sensible results, and this term has not yet a simple geometrical interpretation.

Now we should take the limit $\ell\to 0$; a difficulty arises at this point, since the space of finite-norm-holomorphic-functions is trivial if $\ell<\frac{1}{2}n$! A way out is given by the following remark: instead of considering the norm given by Eq.(6), we can equivalently consider the Hilbert space H_ℓ of holomorphic functions in $\mathcal{D}$ axiomatically defined by (see Ref.(8)):

i) H_ℓ is the linear span of $\{U_g\psi_0 \mid g \in SO_0(2,n)\}$ with $\psi_0(\zeta) = 1$;

ii) for a fixed value of ξ the function

$$K_\xi(\zeta) = K(\zeta,\bar{\xi}) = (1 - 2\zeta\bar{\xi}' + \zeta\zeta'\overline{\xi\xi'})^{-\ell-\frac{1}{2}n+1} \tag{9}$$

belongs to H_ℓ;

iii) for every $\psi \in H_\ell$ it holds $\psi(\zeta) = \langle K_\zeta|\psi\rangle$, in particular

$$\langle K_\zeta|K_\xi\rangle = K(\zeta,\bar{\xi}) \qquad \text{(reproducing kernel)} \tag{10}$$

For $\ell>\frac{1}{2}n$ H_ℓ can be realized with a norm given by Eq.(6); if $0<\ell<\frac{1}{2}n$, H_ℓ exists and is non-trivial, but the norm is not given by Eq.(6). By direct computation it is easily shown that for $\ell=0$ a realization of H_0 is given by functions on the sphere S^{n-1} such that

$$\langle\phi_1|\phi_2\rangle = \frac{\Gamma(\frac{1}{2}n-1)}{2\pi^{\frac{1}{2}n}} \int_{S^{n-1}} \overline{\phi_1(x)}\, \sqrt{-\Delta + (\tfrac{1}{2}n-1)^2}\, \phi_2(x)\, \dot{x} \tag{11}$$

Δ being the Laplacean on S^{n-1} (S^{n-1} enters naturally, since $S^1\ S^{n-1}$ is the Bergman-Silov boundary of $\mathcal{D}$; for more details see Ref.(3)). By inserting the reproducing kernel we find:

$$(12)\qquad \phi(x) = \frac{\Gamma(\frac{1}{2}n-1)}{2\ \pi^{\frac{1}{2}n}} \int_{S^{n-1}} \frac{\sqrt{-\Delta_y + (\frac{1}{2}n-1)^2}\ \phi(y)}{||x-y\ ||^{n-2}}\ \dot{y}$$

Finally we obtain for the eigenfunctions of S belonging to the eigenvalue $\nu+\frac{1}{2}n-1$ $(\nu=0,1,2,...)$

$$(13)\qquad \phi_\nu(x) = \frac{\Gamma(\frac{1}{2}n-1)}{2\ \pi^{\frac{1}{2}n}} (\nu+\tfrac{1}{2}n-1) \int_{S^{n-1}} \frac{\phi_\nu(y)}{||x-y||^{n-2}}\ \dot{y}$$

which is Fock's integral equation[9].

Acknowledgments

I thank warmly Prof.Konrad Bleuler for his kind hospitality at the Conference. I thank J.Rawnsley, D.Simms and J.A.Wolf for useful conversations.

References

1) V.Fock, Z.Physik 98, 145 (1935).
2) E.Onofri, "SO(n,2) singular orbits and their quantisation", Proceedings of the Conference on Symplectic Manifolds and Mathematical Physics, Aix-en-Provence 1974, C.N.R.S.(to appear).
3) E.Onofri, "Dynamical Quantisation of the Kepler Manifold", Università di Parma, preprint IFPR-T-047 (1975).
4) J.Moser, Commun.Pure Appl.Math., XXIII, 609 (1970).
5) E.Onofri, "Quantization Theory for homogeneous Kaehler manifolds", Università di Parma, IFPR-T-038 (1974).
6) E.Onofri and M.Pauri, J.Math.Phys., 13(4), 533 (1972).
7) R.Blattner, B.Kostant, D.Simms and J.M.Souriau, Lectures given in this Conference.
8) A.L.Carey, "Induced Representations, reproducing kernels and the conformal group", Mathematical Institute, Oxford preprint (1974).
9) M.Bander and C.Itzykson, Rev.Mod.Phys., 38(2) 330 (1966).

INTERPRETATION GEOMETRIQUE DES ETATS QUANTIQUES

Jean-Marie SOURIAU
Université de Provence
et
Centre de Physique Théorique, CNRS
Marseille
France

§1. ETATS STATISTIQUES CLASSIQUES

L'ensemble des mouvements d'un système dynamique possède une structure de variété symplectique : ce fait est connu depuis la fin du XVIIIème siècle, même s'il a un peu été noyé par la scolastique.

En effet, dès 1788, J.L. Lagrange ("Mécanique Analytique") construisait les "crochets" et "parenthèses" qui portent son nom ; ses calculs -décrits dans le langage actuel- montrent que ce sont les composantes covariantes (resp. contravariantes) d'une 2-forme inversible σ dans des coordonnées locales quelconques de l'espace des mouvements (donc que σ est un invariant intégral absolu des équations du mouvement, au sens d'E. Cartan) ; Lagrange montrait de plus qu'il existe des coordonnées dans lesquelles ces composantes sont constantes (que la forme σ est plate). C'est cette structure (variété munie d'une 2-forme inversible et plate) que l'on appelle aujourd'hui "symplectique".

Un état statistique classique est simplement une loi de probabilité définie sur l'espace des mouvements U : il associe à tout ouvert Ω de U un nombre $p(\Omega)$ que l'on interprète comme la probabilité pour que le mouvement observé du système, dans l'état donné, appartienne à Ω. Cette interprétation exige de l'application p un certain nombre de propriétés classiques : p doit être une mesure positive de masse 1 .

Les états statistiques considérés par Boltzmann correspondent au cas où p est complétement continue ; alors p se caractérise par une fonction ρ définie sur U ("fonction de distribution") ; si on rapporte ρ à l'espace des conditions initiales à l'instant t ("espace de phases"), ρ vérifie nécessairement une équation d'évolution, qui n'est autre que l'équation de Liouville ; dans le cas de la théorie cinétique des gaz, cette équation fournit, par un processus bien connu d'approximations, l'équation de Boltzmann.

Les mathématiciens se partagent sur la manière d'exposer la théorie de la mesure ; au lieu de considérer la fonction p (définie sur certaines parties de U), on peut considérer la fonctionnelle linéaire m (définie sur certaines fonctions de U) telle que

$$m(f) = p(\Omega)$$

si f est la fonction caractéristique de Ω .
m jouit alors des propriétés suivantes :

(1.1)
- m est linéaire sur l'espace vectoriel $\mathcal{B}(U)$ des fonctions continues bornées sur U.
- $m(f) \geqslant 0$ si f est une fonction positive ($f(x) \geqslant 0$, $\forall x \in U$)
- $m(\mathbb{1}) = 1$ si $\mathbb{1}$ est la fonction unité ($\mathbb{1}(x) = 1$, $\forall x \in U$)

que l'on peut prendre comme axiomatique des lois de probabilités de U -donc ici des états statistiques. L'ensemble Prob(U) des lois de probabilité de U ainsi défini est convexe (vérification élémentaire). On peut montrer (si U est séparée) que les points extrêmaux de ce convexe (les éléments de Prob(U) qui n'appartiennent à aucun segment de droit inclus dans Prob(U)) sont les "fonctions" de Dirac $\delta(a)$ associées aux points $a \in U$

(1.2) $$\delta(a)(f) = f(a) \qquad \forall f \in \mathcal{B}(U)$$

Dans un tel état, la probabilité pour que le mouvement soit a vaut 1 : on peut donc identifier les mouvements (ou "états classiques") avec les états statistiques extrêmaux.

- On appelle variable dynamique toute fonction réelle

(1.3) $$g : (x \mapsto u)$$

continue sur U.

A tout état statistique m, et à toute variable dynamique, on peut associer une loi de probabilité μ de R :

(1.4) $$\mu(f) = m(f \circ g) \qquad \forall f \in \mathcal{B}(\mathbb{R})$$

μ s'interprête comme la loi de probabilité de la variable u dans l'état m (ou plus brièvement, comme spectre de u) ; on notera que l'état statistique est caractérisé par l'ensemble des spectres des diverses variables dynamiques (si $f \in \mathcal{B}(U)$, $m(f)$ est la valeur moyenne de la variable dynamique $f(x)$ dans l'état m).

Définitions :

Soit K un ensemble convexe. Nous noterons Conv(K) l'ensemble des bijections f de K dans K qui vérifient

(1.5)
$$f(sx + [1-s]\,y) = s\,f(x) + [1-s]f(y)\ , \qquad \forall s \in [0,1], \forall x,y \in K$$

Conv(K) est un groupe.
Si G est un groupe, nous appellerons action convexe de G sur K tout morphisme de G dans Conv(K).

Théorème (élémentaire) :

Soit Homeo(U) le groupe des homéomorphismes de U (bijections bicontinues). Si $a \in$ Homeo(U), $m \in$ Prob(U), $f \in \mathcal{B}$ (U), nous poserons

(1.6)
$$\underline{a}(m)(f) = m(f \circ a).$$

Alors $a \mapsto \underline{a}$ est une action convexe effective de Homeo(U) sur Prob(U).

Bien entendu, on obtient par ce moyen une action convexe sur Prob(U) de tout sous-groupe de Homeo(U) ; notamment du groupe Diff(U) des difféomorphismes de U, et du groupe Sympl(U) des symplectomorphismes de U (difféomorphismes respectant la forme de Lagrange σ).

La propriété des états statistiques classiques de fournir une action convexe de Sympl(U) se retrouvera au niveau des états quantiques (§ 5 ci-dessous).

§2. FONCTIONS DE TYPE POSITIF

La plupart des résultats figurant dans ce paragraphe sont bien connus, donc énoncés avec peu d'explications.

Définition :

(2.1) Soit G un groupe ; F une application de G dans $\mathbb{C}$.
On dit que F est de type positif (synonyme : $F \gg 0$, $F \in \mathcal{P}(G)$) si:
Pour tout entier n, pour toute application

$$j \mapsto (z_j, a_j)$$

de $\{1,2,\ldots n\}$ dans $\mathbb{C} \times G$, on a

$$\sum_{j,k} \bar{z}_j z_k F(a_j^{-1} \times a_k) \geq 0 \qquad (\times : \text{loi du groupe } G).$$

Il est clair que la matrice dont les éléments sont les nombres $F(a_j^{-1} \times a_k)$ est une matrice positive ; elle possède donc la symétrie hermitienne, et son déterminant est positif. En choisissant $n = 2$, on en tire :

(2.2) Si $F \gg 0$

$$F(e) \geq 0 \qquad (e = \text{élément neutre de } G)$$

$$F(a^{-1}) = \overline{F(a)} \qquad \forall a \in G$$

$$|F(a)| \leq F(e) \qquad \forall a \in G .$$

Les fonctions de type positif sont donc bornées, et atteignent leur norme maximum au point e.

(2.3) Appelons $\mathcal{P}_0(G)$ l'ensemble des fonctions $F \gg 0$ sur G qui vérifient

$$F(e) = 1 .$$

(fonctions normées). On note que :

(2.4) Toute fonction $\gg 0$ est le produit d'une fonction normée par un nombre ≥ 0.

(2.5) $\mathcal{P}(G)$ (resp. $\mathcal{P}_0(G)$) est convexe.

(2.6) Si b est un <u>automorphisme</u> de G, et si on pose

$$b(F)(a) = F(b(a)) \qquad \forall F \in \mathcal{P}(G), \forall a \in G$$

$b \mapsto \underline{b}$ est une <u>action convexe</u> du groupe des automorphismes sur $\mathcal{P}(G)$ (resp. sur $\mathcal{P}_0(G)$).

Soit $F \in \mathcal{P}_0(G)$; en choisissant, dans la définition (2.1) $n = 3$, en prenant les (z_j, a_j) égaux à $(F(b)-F(a),e),(1,a),(-1,b)$, il vient :

$$|F(b)-F(a)|^2 \leqslant 2 - 2\,\mathrm{Re}(F(a^{-1} \times b))$$

d'où résulte

(2.7) $$|F(a) - F(b)| \leqslant \sqrt{2\,|F(e) - F(a^{-1} \times b)|} \quad ;$$

Ceci montre, si G est un <u>groupe topologique</u>, que toute fonction $F \gg 0$, qui est <u>continue au point</u> e est <u>uniformément continue</u> sur G ; ces fonctions forment encore un convexe, qui sera noté $\mathcal{P}'(G)$ ($\mathcal{P}'_0(G)$ pour celles qui sont normées).

En choisissant toujours $n = 3$, et en développant le déterminant de la matrice d'éléments $F(a_j^{-1} \times a_k)$, on constate que :

(2.8) Si $F \in \mathcal{P}_0(G)$, $a,b \in G$

$$|F(axb) - F(a)F(b)| \leqslant \sqrt{1-|F(a)|^2}\ \sqrt{1-|F(b)|^2}\ .$$

(2.9) Il en résulte que l'image réciproque H de $\mathbb{T}$ (groupe multiplicatif des nombres complexes de module 1) par F est un sous-groupe de G, et que

$$F(axb) = F(a) \times F(b) \qquad \forall a \in G\ , \ \forall b \in H$$

en particulier F induit un <u>caractère</u> sur H (un morphisme de H dans $\mathbb{T}$).

<u>Théorème</u> :

(2.10) Si $F \gg 0$, sa conjuguée $\overline{F}$ $(a \mapsto \overline{F(a)})$ est $\gg 0$.

(2.11) Si F, F' sont $\gg 0$, le produit $F\,F'$ $(a \mapsto F(a)\,F'(a))$ est $\gg 0$.

(2.10) est trivial ; (2.11) se montre facilement à l'aide de deux lemmes : si M

une matrice positive, Z quelconque, $Z^*.M.Z$ est positive ; si M et M' sont positives, $\mathrm{Tr}(MM') \geq 0$.

Théorème (Bochner) :

(2.12) Soit E un espace vectoriel de dimension finie ; E^* son dual. Si $m \in \mathrm{Prob}(E)$ (notation (1.1)), si $p \in E^*$, on pose

$$\mathcal{F}(m)(p) = m(q \mapsto e^{ipq}) ;$$

abs $\mathcal{F}$ est une bijection de $\mathrm{Prob}(E)$ sur l'ensemble $\mathcal{P}'_0(E^*)$ [fonctions $\gg 0$ normées continues du groupe additif E^*].

$\mathcal{F}(m)$ est la transformée de Fourier de la loi de probabilité m ; on l'appelle aussi fonction caractéristique.

-Soit m une loi de probabilité sur une variété U ; considérons la fonctionnelle non linéaire F :

$$F(f) = m(x \mapsto e^{if(x)}) \qquad \forall f \in \mathcal{C}(U) \tag{2.13}$$

$\mathcal{C}(U)$ désignant l'ensemble de toutes les fonctions réelles continues sur U .

Il est immédiat que :

(2.14)
a) $F(k) = e^{ik}$ si k est une fonction constante ($k(x) = k \ \forall x \in U$) ;

b) F est une fonction de type positif sur le groupe additif $\mathcal{C}(U)$;

c) F est continue, en ce sens que l'application

$$y \mapsto F(f_y)$$

est continue si $y \mapsto f_y$ est une application d'une variété auxiliaire V dans $\mathcal{C}(U)$, et si $f_y(x)$ est fonction continue du couple (y,x).

En particulier, si $t \in \mathbb{R}$, $f \in \mathcal{C}(U)$, la fonction continue

$$t \mapsto F(tf) \tag{2.15}$$

est la fonction caractéristique de la loi de probabilité μ de la variable dynamique $f(x)$ (Cf. (1.4)), et définit donc complètement celle-ci (th. de Bochner) ; ainsi F définit le spectre de toutes les variables dynamiques, donc la mesure m elle-même (§1). C'est pourquoi nous appellerons F fonctionnelle caractéristique de la loi de probabilité m.

Théorème (Gelfand-Naimark-Segal) :

(2.16) Soit G un groupe ; F une application de G dans $\mathbb{C}$; Alors :

a) [$F \gg 0$]

$\Updownarrow$

b) Il existe un espace préhilbertien (resp. hilbertien) H , une représentation unitaire $a \mapsto \underline{a}$ de G dans H , un vecteur $\psi \in H$, tels que

$$F(a) = \langle \psi, \underline{a}(\psi) \rangle \quad \forall a \in G .$$

L'implication $\Uparrow$ est triviale ; dans le sens $\Downarrow$, on considère l'espace H' des fonctions f de G nulles en dehors d'un fini, muni de la semi-norme $\| f \| = \sqrt{\sum_{a,b} \overline{f(a)}\, f(b)\, F(a^{-1}xb)}$; G agit unitairement sur H' ; H est le quotient de H' par l'ensemble des éléments de semi-norme nulle (resp. son complété) ; ψ la classe de la fonction $a \mapsto \delta(a,e)$.

Le théorème reste vrai si on ajoute à (2.16b) la condition

(2.16') L'espace vectoriel engendré par les $\underline{a}(\psi)$ (lorsque a parcourt G) est dense dans H ;

Cette condition a l'avantage de fixer les éléments H , $a \mapsto \underline{a}$ de la représentation à une équivalence unitaire près.

(2.17) Si U est une variété séparée, on sait construire un espace préhilbertien sur lequel Diff(U) agit unitairement et effectivement (à savoir l'ensemble des semi-densités C^∞ à support compact) ; (2.16 $\Uparrow$) permet d'en déduire une famille séparante de fonctions de type positif sur Diff(U). Pour toute action C^∞ d'un groupe de Lie G sur U , on obtient une famille de fonctions à la fois $\gg 0$ et C^∞ sur G .

Théorème :

(2.18) Soit F une fonction de type positif normée sur un groupe de Lie G ; supposons F de classe C^2 ; soit $\mathcal{G}$ l'algèbre de Lie de G . Alors

a) $F(\exp(Z)) = 1 + i\, M(z) - \frac{1}{2}\left[M(Z)^2 + V(Z,Z) \right] + o(\|Z\|^2)$ $[Z \in \mathcal{G}]$

M étant une forme linéaire réelle sur $\mathfrak{g}$, V une forme bilinéaire symétrique réelle ; on a de plus :

b) $V(Z,Z) \geqslant 0 \qquad \forall\, Z \in \mathfrak{g}$

c) $V(Z,Z)\ V(Z',Z') - V(Z,Z')^2 \geqslant \frac{1}{4}\ M([Z,Z'])^2 \qquad \forall\, Z,Z' \in \mathfrak{g}$,

$[Z,Z']$ étant le crochet de Lie dans $\mathfrak{g}$.

Ce résultat se vérifie à l'aide de (2.2), (2.8) et de la formule de Campbell-Haussdorff.

§3. QUANTOMORPHISMES

Soit Y une variété connexe séparée de dimension impaire 2n+1 (figure 1). On appelle structure de contact la structure définie sur Y par la donnée d'une 1-forme C^∞ ϖ dont le feuilletage caractéristique se réduit à 0 . On en déduit que la dérivée extérieure σ de ϖ a partout le rang 2n ; le feuilletage caractéristique de σ est donc composé de courbes.

Si ces courbes sont compactes (ce sont alors des courbes fermées), le calcul des variations montre que la circulation de ϖ sur chacune d'elles est constante; nous dirons que Y est une variété quantique si cette constante est celle de Planck (notée $2\pi\hbar$).

Il existe alors sur Y un champ de vecteurs $\xi \mapsto I(\xi)$, défini par

(3.2) $$\begin{cases} I(\xi) \in \ker(\sigma) \\ \varpi(I(\xi)) = \hbar \end{cases}$$

Désignons par

$$\exp(tI)(\xi_0)$$

la solution de l'équation différentielle

$$\frac{d\xi}{dt} = I(\xi)$$

qui prend la valeur ξ_0 pour $t = 0$; cette solution est une fonction périodique de période 2π ; par suite l'application

(3.3) $$e^{it} \mapsto \exp(tI)$$

définit une action différentiable libre du tore $\mathbb{T}$ (notation (2.9)) sur Y . Il existe alors une variété symplectique U , et une submersion P de Y sur U , de sorte que :

(3.4) [L'image réciproque de chaque point x de U est une orbite de $\mathbb{T}$;
L'image réciproque de la forme symplectique de U est la dérivée extérieure de ϖ .

Si U se trouve être l'espace des mouvements d'un système dynamique (Cf. §1), nous dirons que Y est une pré-quantification du système.

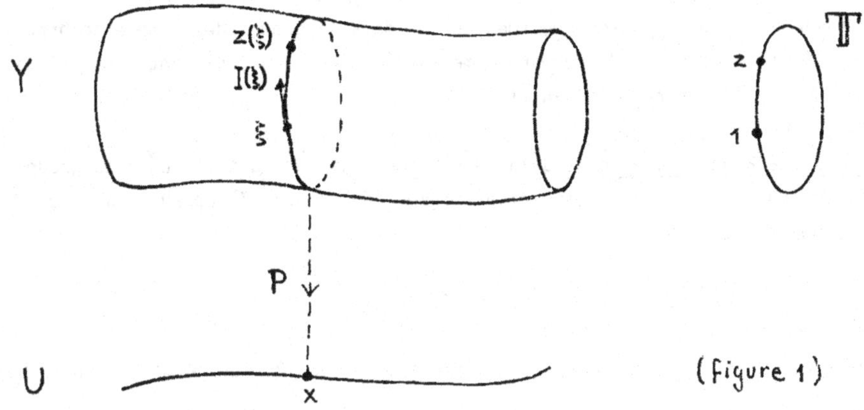

(figure 1)

(3.5) Il semble que la possibilité de préquantifier la variété des mouvements d'un système dynamique doive être considérée comme une loi de la nature ; ainsi l'espace U des mouvements d'une particule à spin (de masse m positive ou nulle) s'identifie à une orbite du groupe de Lorentz-Poincaré dans sa représentation co-adjointe, munie de sa structure symplectique canonique (voir (III)) ; la théorie de l'homologie montre que U n'est préquantifiable que si le spin s de la particule est un multiple entier de $\hbar/2$; c'est bien ce qu'on constate expériemntalement (exemple : le spin d'un neutrino est $\hbar/2$, celui d'un photon $\hbar$).

(3.6) Dans le cas où il existe plusieurs préquantifications essentiellement distinctes, l'une d'entre elles est significative physiquement (dans le cas d'un système de particules indiscernables, il existe deux préquantifications, correspondant respectivement aux cas des bosons et des fermions).

(3.7) Nous appellerons quantomorphismes les difféomorphismes de Y qui respectent la forme ϖ ; tout quantomorphisme respecte aussi le champ de vecteurs I (3.2), et commute donc avec l'action du tore (3.3). Il en résulte qu'il se projette sur U selon un symplectomorphisme ; ce qui définit un morphisme

(3.8) $$\mathrm{Quant}(Y) \longrightarrow \mathrm{Symp}(U)$$

du groupe des quantomorphismes de Y dans le groupe des symplectomorphismes de U . Le noyau de ce morphisme est le tore $\mathbb{T}$ (agissant selon (3.3)) ; $\mathbb{T}$ est le centre de Quant(Y).

Un groupe à un paramètre de quantomorphismes est une action C^∞ du groupe additif $\mathbb{R}$ sur Y , par quantomorphismes ; son générateur Z est un champ de vecteur C^∞ de Y qui vérifie

(3.9) $$\exp(tZ) \in \mathrm{Quant}(Y) \quad \forall t \in \mathbb{R}$$

on dira que Z est un quantomorphisme infinitésimal de Y. On constate que le nombre

(3.10) $$u = \frac{1}{\hbar}\, \varpi(Z(\xi))$$

est fonction différentiable de x seulement, et vérifie

(3.11) $$\sigma(\delta\xi, Z(\xi)) = \hbar\, \delta u$$

pour toute dérivation δ définie sur Y . Dans le cas particulier du vecteur $Z = I$, générateur du tore, on trouve $u = 1$ (3.2).

On notera que les formules (3.10) et (3.11) définissent réciproquement le champ de vecteur Z si l'on connaît la variable dynamique u (définition (1.3)), si bien que nous pourrons noter celui-ci

(3.12) $$Z_u$$

Réciproquement, si on se donne une variable dynamique différentiable u , le vecteur Z_u défini par (3.10) et (3.11) est un quantomorphisme infinitésimal à la seule condition qu'il soit complet.
Le calcul montre que :

(3.13) $$[Z_u, Z_{u'}] = \hbar \; Z_{[u,u']}$$

où $[Z_u, Z_{u'}]$ est le crochet de Lie des champs de vecteurs $Z_u, Z_{u'}$, $[u,u']$ le crochet de Poisson des variables dynamiques u et u'.

§4. ETATS QUANTIQUES

Nous allons chercher une définition convenable des états quantiques en partant de l'idée suivante : ce sont des objets géométriques dont les états statistiques classiques (§1) constituent une "approximation" valable seulement si la constante de Planck $h = 6,6262 \times 10^{-27}$ g cm^2 s^{-1} peut être "négligée".

Si nous négligeons $\hbar$ dans la formule (3.13), on arrive à la conclusion (fausse) que l'application

$$(4.1) \qquad u \mapsto \exp(Z_u)$$

est un isomorphisme du groupe additif des variables dynamiques avec le groupe des quantomorphismes (en fait, cette application n'est pas partout définie, elle n'a aucune raison d'être injective, ni surjective, et elle ne peut pas être un isomorphisme, puisque l'un des groupes est commutatif, et pas l'autre).

Si cependant on le croyait un instant, et si F était la fonctionnelle caractéristique (2.13, 2.14) d'un état statistique classique, F serait la composée avec ce morphisme d'une fonction Φ de type positif sur Quant(Y) :

$$(4.2) \text{ ??? } \qquad F(u) = \Phi\left(\exp(Z_u)\right) \qquad \forall u \in \mathcal{E}(U) \; ;$$

l'application

$$(4.3) \text{ ??? } \qquad t \mapsto \Phi\left(\exp(tZ_u)\right)$$

serait la transformée de Fourier de la loi de probabilité de u dans l'état correspondant (Cf. (2.15)) ;

La formule (2.14a) :

$$F(k) = e^{ik} \text{ si } k \text{ est une constante}$$

montre que l'on aurait :

$$\text{??? } \qquad \Phi\left(\exp(Z_k)\right) = e^{ik} \; ;$$

comme il se trouve que $\exp(Z_k)$ est l'action, sur Y, de l'élément $z = e^{ik}$ du tore (Cf. le §3), cette formule s'écrirait

$$(4.4) \text{ ??? } \qquad \Phi(z) = z \qquad \forall z \in \mathbb{T}$$

Mais rien ne nous empêche d'utiliser ces résultats faux comme définition axiomatique : nous appellerons "états quantiques" les fonctions Φ : Quant(Y) $\to \mathbb{C}$ vérifiant :

(4.5)
a) $\Phi(z) = z \quad \forall z \in \mathbb{T}$

b) Φ est $\gg 0$

c) Φ est continue, en ce sens que $\Phi \circ H$ est continue chaque fois que H est une application différentiable d'une variété dans Quant(Y) ([1]).

Bien entendu, ces objets ne pourront pas être associés à un état statistique par la formule (4.2), et devront donc être étudiés directement.

Par exemple, si u engendre un groupe à un paramètre de quantomorphismes par l'application

$$t \mapsto \exp(tZ_u)$$

on constate -grâce à la définition (4.5c) choisie pour la continuité de Φ et au théorème de Bochner (2.12)- que l'application

$$t \mapsto \Phi\left(\exp(tZ_u)\right)$$

est la transformée de Fourier d'une loi de probabilité μ de la droite réelle :

(4.6) $$\Phi\left(\exp(tZ_u)\right) = \int_{-\infty}^{+\infty} e^{ist} d\mu(s) ;$$

par analogie avec (4.3), nous interpréterons μ comme résultat (aléatoire) de la mesure de u dans l'état quantique Φ (nous dirons simplement spectre de u).

Il ne reste plus qu'à tester les conséquences de l'axiomatique (4.5) et de la règle d'interprétation (4.6).

(4.7) Notons d'abord que l'ensemble des états quantiques n'est pas vide : la méthode des semi-densités permet de vérifier que l'ensemble des fonctions vérifiant (4.5) est séparant sur le groupe Quant(Y) : si a,b $\in$ Quant(Y) et si $a \neq b$, il existe un état quantique Φ tel que $\Phi(a) \neq \Phi(b)$.

([1]) Une application H d'une variété V dans le groupe des difféomorphismes d'une variété V' sera dite différentiable si $(x,x') \mapsto H(x)(x')$ est une application C^∞ de $V \times V'$ dans V'. En raison des axiomes a) et b), Φ vérifie l'inégalité (2.7) ; il suffit donc, pour qu'elle soit continue, qu'elle le soit en l'élément neutre.

(4.8) Si k est une variable dynamique constante, le spectre de k dans n'importe quel état quantique Φ est concentré au point k : il résulte en effet de (4.5a) que

$$\Phi\left(\exp(tZ_k)\right) = e^{itk}$$

et il est clair que $t \mapsto e^{itk}$ est la transformée de Fourier de la mesure de Dirac $\mu = \delta(k)$. Par linéarité, on en déduit que le spectre de u+k est translaté par k du spectre de u (dans n'importe quel état).

(4.9) Par contre, si une variable dynamique u prend ses valeurs dans un ensemble fermé $E \subset \mathbb{R}$ non réduit à un point, il peut arriver que le spectre de u ne soit pas supporté par E (un contre-exemple est facile à construire à l'aide d'une semi-densité) ; cet effet tunnel montre bien la différence entre un état quantique et un état statistique.

Soient p et q deux variables dont le crochet de Poisson vaut 1 ; la formule (3.13) montre que le crochet de Lie $[Z_p, Z_q]$ vaut $\hbar$ I ; si λ, μ, ν sont des nombres réels, la formule (2.18a) donne , s'il existe, le développement limité au second ordre de $F(\exp Z_{\lambda p+\mu q+\nu})$; on constate que $M(Z_p)$ et $M(Z_q)$ sont respectivement les valeurs moyennes de p et q, $V(Z_p,Z_p)$ et $V(Z_q,Z_q)$ les variances de ces mêmes variables (carrés de leurs écarts quadratiques moyens Δp, Δq). La formule (2.18c) donne alors

(4.10)
$$\Delta p \; \Delta q \geqslant \frac{\hbar}{2}$$

c'est la relation d'incertitude de Heisenberg, sous sa forme la plus précise (voir par exemple (II)).

Considérons plus généralement un groupe "quantodynamique", c'est-à-dire un sous-groupe G de Quant(Y) qui possède une structure de groupe de Lie telle que le plongement de G dans Quant(Y) soit différentiable ([1]). Tout état quantique Φ induit sur G une fonction de type positif Φ_G qui est continue. Elle est associée, par le théorème de Gelfand-Naimark-Segal (2.16, 2.16b'), à une représentation unitaire de G sur un certain hilbertien :

(4.11)
$$a \mapsto \underline{a} \;;$$

([1]) Au sens défini en (4.5) . Il revient au même de dire que G est un groupe de Lie qui agit effectivement et différentiablement sur Y par quantomorphismes.

on vérifie facilement que la continuité de Φ_G entraîne la continuité de cette représentation (4.11).

Les résultats connus sur les représentations unitaires continues des groupes localement compacts permettent alors de faire des prédictions sur les spectres des variables dynamiques associées aux sous-groupes à un paramètre de G ("moments"). Dans de très nombreux exemples, elles sont bien vérifiées (action du groupe O(4) dans le cas des mouvements képlériens ; action du groupe de Poincaré dans le cas d'une particule libre sans spin (ou de spin $n\hbar$, $n \in \mathbb{Z}$) ; du revêtement à deux feuillets du groupe de Poincaré dans le cas d'une particule de spin $\hbar/2$; etc). Les relations de commutation associées sont d'ailleurs l'un des outils fondamentaux de la mécanique quantique.

§5. LE PROBLEME DE L'OSCILLATEUR HARMONIQUE

Il se trouve pourtant que ces prédictions tombent en défaut dans l'un des cas les plus simples : celui de l'oscillateur harmonique à une dimension.

Comme pour tout système conservatif, le groupe des translations temporelles est engendré par la variable dynamique $E/\hbar$, E désignant l'énergie ; en utilisant le fait que tous les mouvements sont périodiques de même période T , on établit que $\exp(T\, Z_{E/\hbar})$ est l'identité sur Y ; par conséquent le spectre μ de la variable $E/\hbar$ vérifie (Cf.(4.6)) :

$$\int_{-\infty}^{+\infty} e^{isT}\, d\mu(s) = 1 \tag{5.1}$$

ce qui n'est possible (μ étant une mesure positive de masse 1) que si μ est supporté par l'ensemble des points où $e^{isT} = 1$; ce qui donne pour le spectre de l'énergie les valeurs discrètes

$$\frac{2n\pi\hbar}{T} = n\, h\nu \qquad (n \in \mathbb{Z}\ ;\ \nu = \tfrac{1}{T}) \ ; \tag{5.2}$$

or les valeurs effectivement observées sont données par la formule

$$(n + \tfrac{1}{2})\,\hbar \qquad n \in \mathbb{N} \ . \tag{5.3}$$

On peut lever cette difficulté en admettant qu'une constante additive dans l'énergie est inobservable (les sauts quantiques ne permettent de mesurer que les différences de valeurs de l'énergie) ; mais cette formule empirique (5.3) fournit cependant une indication intéressante : elle conduit en effet à remplacer la formule (5.1) par la formule "opposée" :

$$\int_{-\infty}^{+\infty} e^{isT}\, d\mu(s) = -1 \tag{5.4 ???}$$

ou encore (Cf.(4.6)) :

$$\Phi(\exp(T Z_{E/\hbar})) = -1 \ . \tag{5.5 ???}$$

Ces formules suggèrent de modifier la définition des états quantiques en y introduisant des notions homotopiques : l'application

$$t \mapsto \exp(t\, Z_{E/\hbar}) \ , \tag{5.6}$$

lorsque t parcourt $[0,T]$, définit un lacet différentiable dans Quant(Y) ; la relation (5.5) pourra prendre un sens si on définit Φ sur un revêtement convenable ; elle fournira une condition imposée à tous états.

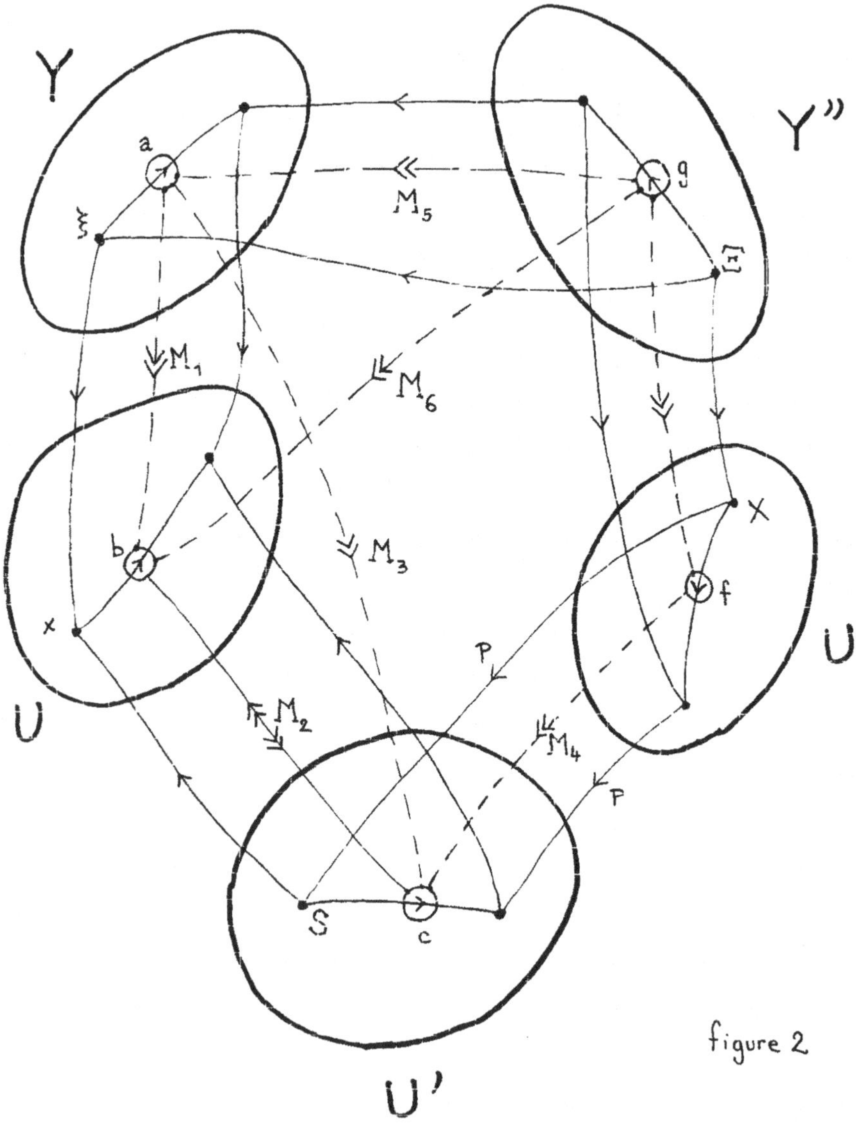

figure 2

Pour réaliser ce programme, effectuons quelques constructions.

1°) Nous avons remarqué (3.8) que la projection sur U (figure 2) d'un quantomorphisme a définit un symplectomorphisme b , et que

(5.7) $$M_1 \ (a \mapsto b)$$

est un morphisme de groupe. Il est de plus différentiable, en ce sens que son composé avec une application différentiable dans Quant(Y) (au sens (4.5)) est une application différentiable dans Sympl(U).

2°) Soit U' l'espace fibré des repères canoniques ([1]) de U (figure 2). U' est un espace fibré principal de base U , dont le groupe structural est le groupe Sp(n) des matrices symplectiques ([1]) d'ordre n . On montre que Sp(n) est connexe ; U' est donc connexe si Y l'est.

Tout symplectomorphisme b de U transforme un repère canonique au point x en un repère canonique au point b(x) ; on peut donc relever b par un difféomorphisme c de U' ;

(5.8) $$M_2 \ (b \mapsto c)$$

est un morphisme injectif différentiable de groupe ; par composition, on obtient donc un morphisme différentiable

(5.9) $$M_3 (a \mapsto c) \quad = \quad M_2 \circ M_1$$

de Quant(Y) dans Diff(U').

3°) Soit U" le revêtement universel de U', P la projection de U" sur U'. Les difféomorphismes f de U" qui vérifient

---

([1]) Un repère S au point x est dit canonique si la matrice des composantes du tenseur symplectique σ dans le repère S est égale à

$$J = \left(\begin{array}{c|c} & \begin{smallmatrix} 1 & & \\ & \ddots & \\ & & 1 \end{smallmatrix} \\ \hline \begin{smallmatrix} -1 & & \\ & \ddots & \\ & & -1 \end{smallmatrix} & \end{array} \right) ;$$

une matrice M est symplectique si elle transforme un repère canonique en repère symplectique, c'est-à-dire si $M^{-1} = J\,M'\,J^{-1}$, M' étant la transposée de M .

(5.10) $$P(X) = P(Y) \Rightarrow P\big(f(X)\big) = P\big(f(Y)\big)$$

forment un groupe ; ils se projettent sur U' suivant un difféomorphisme c :

$$c \circ P = P \circ f ;$$

(5.11) $$M_4(f \mapsto c)$$

est un morphisme <u>différentiable et surjectif</u> de ce groupe sur $\mathrm{Diff}(U')$.

4°) Soit Y'' l'ensemble des couples $\Xi = (\xi, X)$ de $Y \times U''$ qui se projettent en un même point x de U (figure 2). $\xi \mapsto x$ et $X \mapsto x$ étant des submersions, Y'' est une variété.

Considérons l'ensemble G des couples (a,f) :

(5.12) $$a \in \mathrm{Quant}(Y), \quad f \text{ vérifie } (5.10), \quad M_3(a) = M_4(f).$$

A cause de l'injectivité de M_2 , G est un groupe de difféomorphismes de Y'' ;

(5.13) $$M_5(a,f) \mapsto a$$

est un morphisme différentiable surjectif de G sur $\mathrm{Quant}(Y)$, dont le noyau s'identifie à l'ensemble des $f \in \mathrm{diff}(U'')$ vérifiant

(5.14) $$P \circ f = P$$

c'est-à-dire au <u>groupe fondamental d'homotopie</u> de U' ;

(5.15) $$M_6 = M_1 \circ M_5$$

est un morphisme différentiable de G dans $\mathrm{Sympl}(U)$, dont le noyau est l'ensemble des

(5.16) $$\Big\{ (z,k) \;/\; z \in \mathbb{T}, \; k \in \text{groupe d'homotopie de } U' \Big\}.$$

Dans tous ces résultats, nous n'avons pas utilisé le fait qu'il s'agissait d'un oscillateur harmonique, mais seulement que Y était une <u>variété connexe</u> préquantifiant une variété symplectique U .

Dans le cas d'un oscillateur harmonique, quelle que soit d'ailleurs sa dimension, la linéarité des équations du mouvement classique montre que U possède une structure d'espace vectoriel symplectique ; il en résulte que U' est un espace fibré au-dessus de U, et que le groupe d'homotopie de U' est isomorphe à celui des fibres, donc à celui du groupe $Sp(n)$, donc à $\mathbb{Z}$ (voir (IV)). Nous pourrons donc écrire le groupe (5.16) sous la forme

$$\left\{ (z, L^p) \;/\; z \in \mathbb{T} \;,\; p \in \mathbb{Z} \right\} \tag{5.17}$$

L désignant un générateur du groupe d'homotopie de U' ([1]).

Nous pourrons choisir l'axiomatique suivante pour les états quantiques :

(5.18)
a) $\Phi(z, L^p) = z \times [-1]^p$;

b) Φ est une fonction de type positif sur le groupe G ;

c) Φ est continue ([2]).

La compatibilité de ces axiomes peut s'établir en construisant des semi-densités sur Y'', en utilisant le fait que le groupe (5.17) est dans le centre de G, et que la condition (5.18a) fait coïncider, sur ce groupe, Φ avec un caractère.

Soit G_0 l'ensemble des éléments de G dont l'image par M_6 est une transformation affine de U ; soit $\Phi_\circ$ la fonction induite sur G_0 par un état Φ.

On vérifie que G_0 possède une structure de groupe de Lie ; évidemment les axiomes (5.18) induisent les conditions :

(5.19)
a) $\Phi_\circ(z, L^p) = z \times [-1]^p$;

b) $\Phi_\circ$ est une fonction de type positif sur G_0 ;

c) $\Phi_\circ$ est continue.

Or il se trouve (voir (IV)) que l'ensemble des solutions Ψ à carré

([1]) Dans le cas de l'oscillateur à 1 dimension, on peut joindre l'élément $(1, L)$ de G à l'élément neutre par un chemin différentiable tracé dans G, qui se projette sur $\mathrm{Quant}(Y)$ par le lacet (5.6).

([2]) Définition analogue à (4.5) ; nous utilisons le fait que G est un groupe de difféomorphismes de Y''.

sommable de l'équation de Schrödinger de l'oscillateur harmonique est un espace de représentation unitaire de G_0 , et que les fonctions Φ_o associées à chacune d'elles par la construction de Gelfand-Naimark-Segal vérifient ces axiomes (5.19). D'où les problèmes suivants :

- Ces fonctions sont elles prolongeables au groupe G tout entier par des solutions Φ de (5.18) ? Peut-on déterminer un tel prolongement (par exemple en demandant que Φ soit un point extrêmal du convexe (5.19) ?
- Comment étendre cette définition des états à des systèmes dynamiques non linéaires ? Peut-on ainsi établir un lien avec les diverses équations d'onde ?

Principe de correspondance.

Soit $g \in G$; si Φ est un état quantique (5.18), posons

(5.20) $$g(\Phi)(g') = \Phi(g \circ g' \circ f^{-1}) \qquad \forall g' \in G ;$$

on définit ainsi une action convexe (1.5) de G sur les états (voir (2.6)).

Si $g \in \mathrm{Ker}(M_6)$, il résulte de (5.17), (5.18a), (2.9) que $g(\Phi) = \Phi$; par conséquent la formule (5.20) définit une action convexe de $M_6(G) = \mathrm{Sympl}(U)$. Les états quantiques sont donc des objets de géométrie symplectique (au sens de Félix Klein), au même titre que les états statistiques (voir (1.6)) et que les états classiques ; ainsi se manifeste, au niveau de l'axiomatique (5.18), le principe de correspondance entre les mécaniques quantique et classique.

REFERENCES DU TEXTE

(I) J.L. LAGRANGE,
"Mécanique Analytique".
Ré-édition A. Blanchard (1965).

(II) L.I. SCHIFF,
"Quantum Mechanics".
Mac Graw Hill (1955).

(III) J.M. SOURIAU,
"Structures des Systèmes Dynamiques".
Dunod (1969).

(IV) J.M. SOURIAU,
"Construction Explicite de l'Indice de Maslov. Applications".
4th International Colloquium on Group Theoretical Methods in Physics", Univ. of Nijmegen (1975).

GEOMETRIC STRUCTURE OF QUANTIZATION

Jerzy Kijowski

Institute of Mathematical Methods in Physics, University of Warsaw
ul. Hoża 74, 00-682 Warszawa, Poland

1. Introduction

Geometric quantization is an attempt to find the geometrical structure of quantum theories. Till now we understand this structure only in the case of non-relativistic quantum mechanics in flat, euclidean space. In this case the representation of canonical commutation relations: $[p_i, q^j] = i\hbar\delta^j_i$, $[p_i, p_j] = 0 = [q^i, q^j]$, has to be found. The evolution operator is now a sum of kinetic and potential energy:

$$H = \frac{1}{2m} p^2 + V(q)$$

But already quantum mechanics in curved space E (with non-relativistic structure of space-time: $E \times R^1$ where R^1 is a time-axis) can not be formulated this way. Either momenta do not commute or kinetic energy (Laplace-operator) is not equal to $\frac{1}{2m} \sum (p_i)^2$. Quantum dynamics can be however formulated in terms of Schrödinger equation (in position representation !) and we can keep the probabilistic interpretation of wave function.

The situation is much worse in relativistic mechanics where we understand even less.

The popular opinion is that quantization is a representation of Poisson algebra (or its part) in Hilbert space. On the other hand it is not clear to what extent the multiplication of classical observables has to correspond to multiplication of operators. The existence of many such representations (called "procedures of quantization") increases the confusion in this domain.

The present paper proposes another approach to quantization. It is easy to show that on the level of quantum "statics" (description of quantum states) our formulation is equivalent to Souriau-Kostant theory (cf. [7], [8], [10], [11]). The main difference between Kostant's and our point of view consists in a different description of dynamics (see sections 5 and 6). This description leads us to discriminate between "quantizable" and "non-quantizable" theories. The main conclusion is that there is no sense in asking "what the quantum homologue of the classical quantity $x^{10} \cdot p^{8} \cdot e^{x\cos xp}$ is". We use the Occam's razor and quantize only those observables which can be measured. We show (at least on the heuristic level) that our approach can be used for the field theory even in curved space-time whereas the standard formulation of Axiomatic Quantum Field Theory fails because of the lack of Poincaré-group and spectral-condition (all attempts at "quantizing the gravity " show that we do not even imagine clearly what it should be). Some rigorous results in this direction which have been already obtained are not sufficient and need further developement.

2. Complete sets of commuting observables

The canonical structure of both classical mechanics and classical field theory can be summarized as follows. There is a phase-bundle F over the basis M. In (non-relativistic) mechanics M is the 1-dimensional time-axis and F is the bundle of phase-spaces: one for each time. The bundle F (éspace d'evolution in Souriau's language, cf. [11]) can be parametrized by coordinates (t, q^i, p_j). In the field theory M is 4-dimensional space-time and F is the bundle of field-strengths and its derivatives over M. Solutions of the mechanical equations (field-equations) are sections of F. The bundle F has a structure of so called "multi-phase space" which enables to define the symplectic 2-form ω in the space $\mathcal{P}$ of all solutions (cf. [4], [5]).

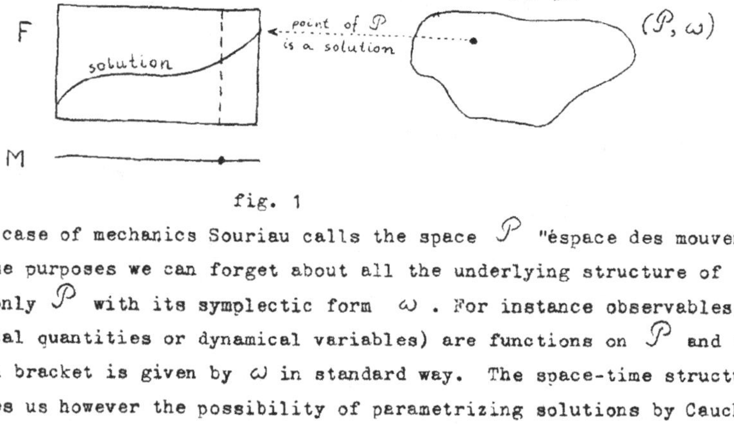

fig. 1

In the case of mechanics Souriau calls the space $\mathcal{P}$ "éspace des mouvements". For some purposes we can forget about all the underlying structure of F, using only $\mathcal{P}$ with its symplectic form ω. For instance observables (physical quantities or dynamical variables) are functions on $\mathcal{P}$ and their Poisson bracket is given by ω in standard way. The space-time structure of F gives us however the possibility of parametrizing solutions by Cauchy-data at given time $t \in M$ (in mechanics) or on given space-like surface $\Sigma \subset M$ (in field theory).

Examples:

1) In 1-particle mechanics coordinates $(x^i(t), p_i(t))$ give for every t the coordinate chart in $\mathcal{P}$ if for each solution of dynamical equations functions $x^i(t)$ and $p_i(t)$ take the value of positions and momenta at time t. Using these coordinates we can represent the symplectic form ω as follows:

$$\omega = dp_i(t) \wedge dx^i(t) \qquad (1)$$

2) Take for simplicity the classical free electrodynamics without currents. For given space-like surface $\Sigma \subset M$ every solution of field equations can be parametrized by Cauchy-data on Σ, i.e. electric and magnetic sourceless fields: (B,E), $\operatorname{div} B = \operatorname{div} E = 0$. Using these coordinates we can represent the symplectic form ω as follows. For two vectors $\underset{1}{X} = (\underset{1}{\delta} B, \underset{1}{\delta} E)$, $\underset{2}{X} = (\underset{2}{\delta} B, \underset{2}{\delta} E)$ we have

$$\omega(\underset{1}{X},\underset{2}{X}) = \int_{\Sigma} \underset{1}{\delta} E^i(x)\, \underset{2}{\delta} A_i(x) - \underset{2}{\delta} E^i(x)\, \underset{1}{\delta} A_i(x) \; d^3x \qquad (2)$$

where $\underset{1}{\delta} A$ and $\underset{2}{\delta} A$ are arbitrary vector fields fulfilling equations:

$$\underset{1}{\delta} B = \operatorname{rot} \underset{1}{\delta} A \quad , \quad \underset{2}{\delta} B = \operatorname{rot} \underset{2}{\delta} A$$

The right-hand side of (2) does not depend on the choice of fields δA provided they vanish sufficiently quickly in infinity. Usually one takes $B,E \in L^2$ and $\delta A \in H^1$.

The crucial role in quantum physics is played by complete sets of commuting observables. The corresponding notion on classical level is "complete set of observables with vanishing Poisson bracket". In both classical and quantum case such a set spans the commuting algebra of observables. The condition of completness means that this algebra is maximal. In classical case it implies (under some regularity conditions) that the symplectic space $\mathcal{P}$ splits into the family Λ of lagrangian submanifolds and our algebra consists of all such functions which are constant on surfaces belonging to Λ. If by $Q_\Lambda = \mathcal{P}/\Lambda$ we denote the quotient space (space of fibres) our algebra is canonically isomorphic to the (commutative) algebra of all functions on Q_Λ. Complete set of commuting observables is thus the set of its generators, i.e. any coordinate chart on Q_Λ.

The lagrangian foliation of symplectic manifold is an example of Kostant's polarization. For our purposes we need even more restrictions concerning global topologies of $\mathcal{P}$, fibres of Λ and of quotient space Q_Λ. We assume namely that all of them admit global coordinate-charts i.e. they are homeomorphic to corresponding vector spaces. We shall use in the sequel Kostant's term "polarization" only in this restricted meaning (the corresponding results for so called complex polarizations are not ready as yet). Polarization plays thus the role of complete set of commuting observables.

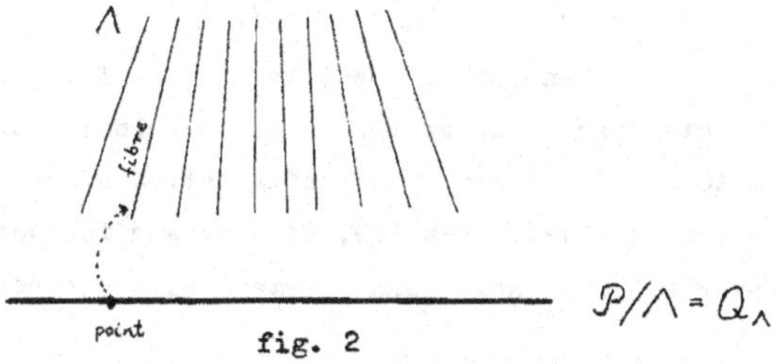

fig. 2

Examples:

1) For givem t in mechanics Λ composed of all fibres
$$\left\{(x^i(t),p_i(t)):\ x^i(t) = \text{const}\right\}$$
is a polarization and $\mathcal{P}/\Lambda = Q_\Lambda$ is a configuration space at the time t with coordinates $(x^i(t))$.

2) For given space-like surface $\Sigma \subset M$ in electrodynamics Λ composed of all fibres $\left\{(B,E):\ B(x) = \text{const}(x)\right\}$ is a polarization. The corresponding configuration space $Q_\Lambda = \mathcal{P}/\Lambda$ is the space of all sourceless vector fields B on Σ.

3. Quantum states and generalized Galilei transformations

One of the most important lessons which quantum physics gives us at its present stage of developement is that all (even most sophisticated) measuring systems permit us to localize our microobjects at most "up to polarization". It means that for every maximal measuring system there exists a polarization Λ such that our system measures the probability of finding our microobjects on fibres of Λ. The result of the series of such experiments is a probability measure ϱ on quotient space Q_Λ. In standard formulation of quantum mechanics ϱ is the square of modulus of the wave function. The density ϱ contains only part of the information about our quantum state. The rest of the information, which prescripts the behaviour of the quantum state with respect to other measuring systems, is contained in the phase of the wave function. The naive way of understanding the geometric meaning of wave function would thus be the following:
$$\Psi_\Lambda = f\cdot\sqrt{\varrho} \tag{3}$$
where $f: Q_\Lambda \to R^1$, $|f| \equiv 1$ is the phase and ϱ is a probability measure. The square root $\sqrt{\varrho}$ is a half-density (and not a half-form) on Q_Λ (in classic Schouten's language half-density is a positive density of weight $\frac{1}{2}$). The set of all pairs $(\varrho, f) =: f\cdot\sqrt{\varrho}$ (complex-valued half-densities), when ϱ and f have Lebesque-measurable coefficients, form in a natural way Hilbert space. But this naive point of view fails. To show this consider non-relativistic one particle mechanics in Newtonian space. There is no ether. No reference frame is distinguished.

Take now two inertial reference frames which coincide at t=0. It means that
$$x^i(0) = \tilde{x}^i(0) \quad , \quad p_i(0) = \tilde{p}_i(0) + m\cdot v_i \tag{4}$$
where coordinates (x,p) are taken with respect to the first and $(\tilde{x},\tilde{p})$ - to the second coordinate frame. The vector $\underline{v}$ is a velocity of second observer with respect to the first one. The configuration space $Q_\Lambda = \mathcal{P}/\Lambda$, where Λ is a foliation $x^i(0)=\text{conts}=\tilde{x}^i(0)$, is the same for both observers. But wave functions describing the same quantum state with respect to both

observers are not the same! For already in kindergarten we learnt that

$$\Psi_{\Lambda}(x) = \widetilde{\Psi}_{\Lambda}(x) \cdot e^{im\underline{v}\cdot\underline{x}} \tag{5}$$

This phenomenon has a beautiful geometric interpretation. In order to present it observe that each fibre of polarization has a natural flat connection.

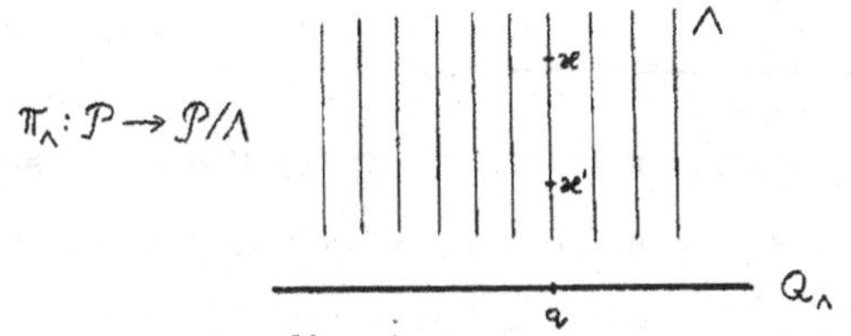

fig. 3

First of all observe that if $\varkappa$ is a point on the fibre q then $T_q(Q_\Lambda)$ (tangent space to the quotient) is itself a quotient:

$$T_q(Q_\Lambda) = T_\varkappa(\mathcal{P}) \,/\, T_\varkappa(q) \tag{6}$$

On the left-hand side q denotes point of Q_Λ and on the right-hand side it denotes a submanifold in $\mathcal{P}$. But the dual space of the quotient is the anihilator. It means that

$$T^*_q(Q_\Lambda) = T_\varkappa(q)^\perp \subset T^*_\varkappa(\mathcal{P}) \tag{7}$$

where $T_\varkappa(q)^\perp$ means the space of co-vectors vanishing on $T_\varkappa(q)$. But $T_\varkappa(q)$ is canonically isomorphic (via symplectic form ω) with its own anihilator since q is lagrangian manifold. It gives us a canonical isomorphism

$$T^*_q(Q_\Lambda) = T_\varkappa(q) \tag{8}$$

The same is true for any other point $\varkappa' \in q$. It means that tangent spaces at different points of q are canonically isomorphic.

$$T_\varkappa(q) = T_{\varkappa'}(q) \tag{9}$$

This isomorphism gives the flat connection on every fibre q. It means that every fibre is a part of an affine space. Now we make the following important assumption: the above connection is complete, i.e. every constant vector field is complete. This assumption is fulfilled if - roughly speaking - different fibres of Λ do not approach each other too quickly in infinity. Under our assumtion every fibre is an affine space. Polarizations satisfying this condition will be called complete.

Now let λ and $\widetilde{\lambda}$ be two global sections of the bundle $\mathcal{P} \to Q_\Lambda$, i.e. two submanifolds transversal with respect to fibres. For each fibre q take the vector from point $q \cap \lambda$ to point $q \cap \widetilde{\lambda}$. By the isomorphism (8) this vector can be treated as a co-vector on Q_Λ at q. It means that λ and $\widetilde{\lambda}$ define a 1-differential form on Q_Λ. This form will be simply denoted $\widetilde{\lambda} - \lambda$.

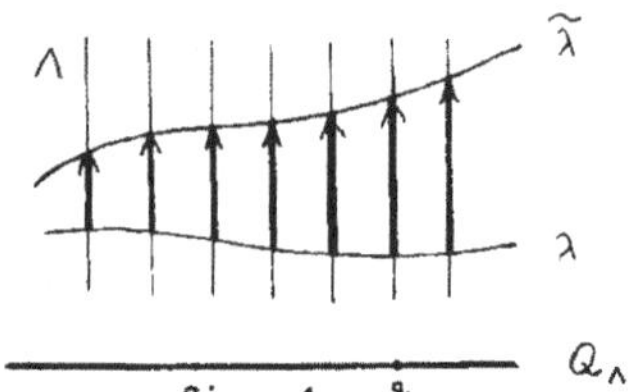

fig. 4

<u>Lemma</u>: If λ and $\tilde{\lambda}$ are lagrangian submanifolds then $d(\tilde{\lambda} - \lambda) = 0$

Using our assumtion about the topology of Q_Λ we see that

$$\tilde{\lambda} - \lambda = d\, S_{\tilde{\lambda}\lambda} \tag{10}$$

where $S_{\tilde{\lambda}\lambda}$ is a function on Q_Λ defined by $(\tilde{\lambda}, \lambda)$ up to an additive constant. Now come back to the formula (5) and observe that

$$m\underline{v}\cdot\underline{x} = S_{\tilde{\lambda}\lambda} \tag{11}$$

where $\lambda = \{p_i = 0\}$, $\tilde{\lambda} = \{\tilde{p}_i = 0\}$.

It suggests that wave function describes the quantum state not only with respect to polarization but also with respect to some transversal lagrangian surface which we shall call in the sequel "the reference frame". The quantum state is thus an equivalence class of triplets and not a pair:

Definition: The quantum state is a class

$$\Omega_\Lambda = [(\varrho, f, \lambda)] \tag{12}$$

where equivalence relation between triplets is following:

$$(\varrho, f, \lambda) \sim (\varrho', f', \lambda') \iff \varrho = \varrho' \ ; \ \varrho\cdot f = \varrho'\cdot f'\cdot e^{iS_{\lambda'\lambda}} \ ; \ S_{\lambda'\lambda} = d(\lambda' - \lambda) \tag{13}$$

For every reference frame λ the function

$$\Psi_{\Lambda,\lambda} = f\cdot\sqrt{\varrho}$$

is called a wave function of the state Ω_Λ with respect to λ.

For different λ' we have

$$\Psi_{\Lambda,\lambda} = \Psi_{\Lambda,\lambda'}\cdot e^{iS_{\lambda'\lambda}} \tag{14}$$

The above formula will be called the generalized Galilei transformation.

Remark:

The space $\mathcal{H}_\Lambda$ composed of all quantum states is a projective Hilbert space and not a Hilbert space. A wave function is always defined up to a constant phase factor e^{ic}, $c \in R^1$, since for $\lambda' = \lambda$ the function $S_{\lambda'\lambda} \equiv c$ satisfies condition (13).

4. Another description of quantum states

There is a beautiful description of those quantum states which correspond to smooth wave functions. It will probably not be very useful for the field theory (ifinite-dimensional case) but I will present it because of its beauty.

For $\Omega_\Lambda = [(\varrho, f, \lambda)]$ take any representant (ϱ, f, λ). The phase $\frac{1}{i}\log f$ is defined up to $n\cdot 2\pi$. It means that the co-vector field $d(\frac{1}{i}\log f)$ is well defined. It can be treated via formula (8) as a field of vertical vectors in the bundle $\mathcal{P} \to Q_\Lambda$. Now we put

$$\mu = \lambda + d(\frac{1}{i}\log f) \tag{15}$$

At each fibre $q \in Q_\Lambda$ the symbol "+" means the parallel translation of the point $\lambda \cap q$ by the vector $d(\frac{1}{i}\log f)$.

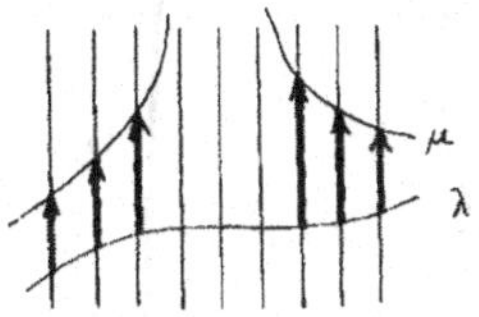

fig. 5

The surface μ is of course lagrangian and transversal to fibres of Λ. It is easy to see that μ does not depend on the representant (ϱ, f, λ) and depends only on the state Ω_Λ. The state can be thus represented by the pair (ϱ, μ). The density ϱ can be treated as a density on μ via canonical projection $\mathcal{P} \to Q_\Lambda$.

Observe however that μ does not need to be a global section of the bundle $\mathcal{P} \to Q_\Lambda$. It is defined only for those points of Q_Λ where the wave function does not vanish (i.e. ϱ does not vanish). The second observation is that for every lagrangian <u>global</u> section λ all the periods of 1-form $\mu - \lambda$ must be equal $n \cdot 2\pi$ because the phase of the wave function must be defined up to $n \cdot 2\pi$. We can thus take the following

Definition: The quantum state (with respect to the polarization Λ) is a pair (ϱ, μ) where 1) μ is a lagrangian section of $\mathcal{P} \to Q_\Lambda$ over some open domain in Q_Λ, 2) periods of μ are integers (times 2π), 3) ϱ is any non-vanishing, normed density on μ.

5. Pairing

In order to have the full quantum theory we must know how to pass from one polarization to another.

Examples

1) In the case of mechanics take the polarization Λ_t defined by $\{x^i(t) = \text{const}\}$ To know the state with respect to Λ_t is, roughly speaking, to know the wave function at the time t. The problem of dynamics: "how to pass from one time to another" is just an example of the general question "how to pass from one polarization to another".

2) In the field theory we can treat Poincaré transformations and even general diffeomorphisms (in General Relativity) as transformations between two polarizations connected with two different space-like surfaces: Σ and its image Σ'.

The transformation from $\mathcal{H}_{\Lambda_1}$ to $\mathcal{H}_{\Lambda_2}$ should be probably some integral operators acting on wave functions. Many people believe that its kernel should depend locally of both polarizations. There is a very natural and simple kernel which can be defined if both polarizations are transversal to each other. In order to define this kernel we must choose two fibres λ_1 and λ_2 belonging respectively to families Λ_2 and Λ_1 (i.e. $\lambda_1 \in \Lambda_2$ $\lambda_2 \in \Lambda_1$ and not vice-versa)

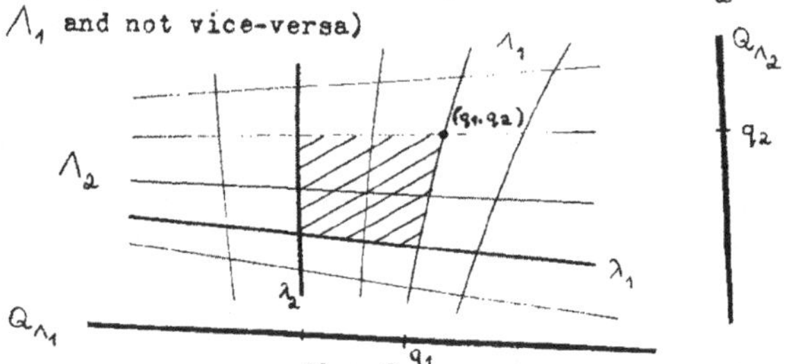

fig. 6

For every $(q_1, q_2) \in Q_{\Lambda_1} \times Q_{\Lambda_2}$ we define a number $k(\lambda_1, q_1, q_2, \lambda_2)$ as a surface of the rectangle spread on those 4 fibres. More precisely: we take 4 points of intersection, we join them with arbitrary lines lying completely in corresponding fibres and take at last any 2-dimensional surface S spread on those lines. We take the orientation compatible with the sequence $\lambda_1, q_1, q_2, \lambda_2$. It can be easily shown that the value

$$k(\lambda_1, q_1, q_2, \lambda_2) = \int_S \omega$$

depend only on 4 fibres. Now we define the kernel

$$K_{\lambda_1 \lambda_2}(q_1, q_2) = \sqrt{|\omega^n|} \cdot e^{ik(\lambda_1, q_1, q_2, \lambda_2)} \tag{16}$$

where $2n = \dim \mathcal{P}$, $\omega^n = \omega \wedge \omega \wedge \cdots \wedge \omega$ is 2n-form on $\mathcal{P}$ and $\sqrt{|\omega^n|}$ is the corresponding half-density. The transformation

$$\mathcal{F}_{\Lambda_2 \Lambda_1} : \mathcal{H}_{\Lambda_1} \longrightarrow \mathcal{H}_{\Lambda_2} \tag{17}$$

is now defined as follows. For any state $\Omega_{\Lambda_1} \in \mathcal{H}_{\Lambda_1}$ take its wave function $\Psi_{\Lambda_1 \lambda_1}$ with respect to λ_1. As a state $\mathcal{F}_{\Lambda_2 \Lambda_1}(\Omega_{\Lambda_1})$ we take such a state whose wave function on Q_{Λ_2} with respect to λ_2 is equal:

$$\Psi_{\Lambda_2 \lambda_2}(q_2) = \int_{Q_{\Lambda_1}} \Psi_{\Lambda_1 \lambda_1}(q_1) \; K_{\lambda_1 \lambda_2}(q_1, q_2) \tag{18}$$

The kernel $K_{\lambda_1 \lambda_2}$ is a half-density on $\mathcal{P}$ i.e. half-density on Q_{Λ_1} and half-density on Q_{Λ_2}. Multiplying it by Ψ which is half-density on Q_{Λ_1} we obtain something which is density on Q_{Λ_1} and half-density on Q_{Λ_2}. After integration over Q_{Λ_1} we thus obtain a half-density on Q_{Λ_2}.

<u>Lemma</u>: The above definition is correct i.e. the state $\mathcal{F}_{\Lambda_2 \Lambda_1}(\Omega_{\Lambda_1})$ does not depend on the choice of λ_1 and λ_2.

The operator $\mathcal{F}_{\Lambda_2 \Lambda_1}$ is called generalized Fourier transformation. There exist its generalizations in the case Λ_1 and Λ_2 are not necessary transversal. (cf. [1], [2], [7], [8]). All of them follow in fact the ideas of

Maslov (cf. [9]). The adequate tool is here the notion of half-form and not half-density (cf. [10]). This line of developement is very important for modern WKB theory and asymptotic solutions of Schrödinger equation (cf. [3]). As a description of fundamental laws of quantum physics these results are however useless because

1) Our Fourier transformations are not (in general) unitary

2) They do not obey group properties of superposition:

$$\mathcal{F}_{\Lambda_2\Lambda_1}\mathcal{F}_{\Lambda_1\Lambda_2} \neq id \quad , \quad \mathcal{F}_{\Lambda_3\Lambda_2}\mathcal{F}_{\Lambda_2\Lambda_1} \neq \mathcal{F}_{\Lambda_3\Lambda_1}$$

3) They have nothing to do (in general) with well established in quantum physics laws of dynamics. In mechanics e.g. operator $\mathcal{F}_{\Lambda_t\Lambda_\tau}$ from position representation at time τ to position representation at time t does not coincide in general with the operator $\exp\{iH(t-\tau)\}$ given by Schrödinger equation.

There are however situations when $\mathcal{F}_{\Lambda_2\Lambda_1}$ is unitary and $\mathcal{F}_{\Lambda_2\Lambda_1}\mathcal{F}_{\Lambda_1\Lambda_2} = id$ namely if Λ_1 and Λ_2 are Heisenberg-transversal.

Definition: **Polarizations Λ_1 and Λ_2 are Heisenberg-transversal** if 1) they are transversal 2) the affine structure given in Q_{Λ_1} by projection of any fibre q_2 along Λ_1 does not depend on q_2.

In this case both Q_{Λ_1} and Q_{Λ_2} are affine spaces and corresponding vector spaces are in canonical duality. Our generalized Fourier transformation (18) coincides with the ordinary Fourier transformation.

Examples

1) Polarizations $\{x^i(t) = \text{const}\}$ and $\{p_i(t) = \text{const}\}$ are Heisenberg-transversal.

2) In linear dynamics (e.g. free particle, harmonic oscilator or constant force field) polarizations $\{x^i(t)=\text{const}\}$ and $\{x^i(\tau)=\text{const}\}$ for $t \neq \tau$ are Heisenberg transversal (with one single exception: when $t - \tau = \frac{n}{2}T$, T is the period of harmonic oscilator)

In all these cases our transformation $\mathcal{F}_{\Lambda_2\Lambda_1}$ agrees perfectly with quantum mechanics. The first idea which comes to mind is to look for such "improved" transformation law $\overline{\mathcal{F}}$ which is transitive and which coincides with $\mathcal{F}$ on Heisenberg-transversal polarizations. But alas such a family of transformations does not exist. The above two requirements about $\overline{\mathcal{F}}$ are incompatible.

6. Quantization as non-flat connection

Let us analyze in details this incompatibility. Take the space $\mathbb{\Lambda}$ of all complete polarizations in $\mathcal{P}$ (the space $\mathbb{\Lambda}$ can be equipped with different topologies and even differentiable structures but the phenomenon which we are going to describe is the same for all reasonable topologies). Take now the bundle $\mathbb{H}$ of all quantum states over $\mathbb{\Lambda}$. The fibre of $\mathbb{H}$ over the point

$\Lambda \in \mathbb{\Lambda}$ is equal $\mathcal{H}_\Lambda$.
A curve $t \longrightarrow \Lambda_t \in \mathbb{\Lambda}$ (1-parameter, smooth family of polarizations) is called a Heisenberg-curve if there is a polarization $\tilde{\Lambda} \in \mathbb{\Lambda}$ such that Λ_t and $\tilde{\Lambda}$ are Heisenberg-transversal for every t. Using operators

$$\mathcal{F}_{\Lambda_\tau \tilde{\Lambda}}\ \mathcal{F}_{\tilde{\Lambda} \Lambda_t} : \mathcal{H}_{\Lambda_t} \longrightarrow \mathcal{H}_{\Lambda_\tau}$$

we can lift every Heisenberg-curve to the bundle $\mathbb{H}$.

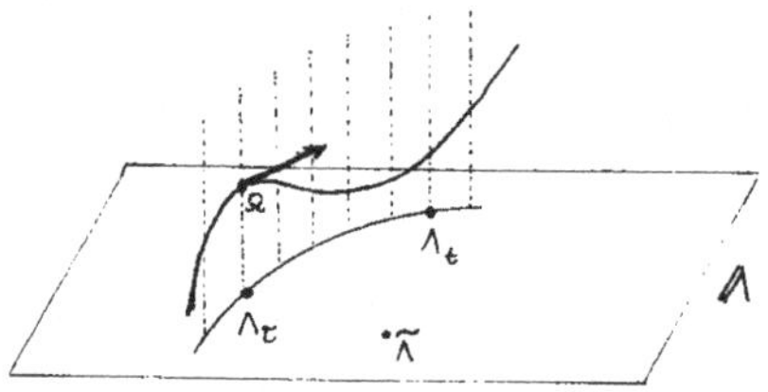

fig. 7

Vectors which are tangent to such lifts will be called Heisenberg-vectors.
Theorem : Heisenberg-vectors span the connection in the bundle $\mathbb{H}$. It means that for every $\Omega \in \mathcal{H}_\Lambda$, $\Lambda \in \mathbb{\Lambda}$ all Heisenberg-vectors attached at Ω span the tangent subspace E_Ω which is transversal to fibres of $\mathbb{H}$ and covers the whole space tangent to $\mathbb{\Lambda}$ at Λ.

The family of all subspaces E_Ω will be called the quantum connection in the bundle $\mathbb{H}$. It is non-flat and that is why we can not find general, transitive law of transformation from $\mathcal{H}_{\Lambda_1}$ to $\mathcal{H}_{\Lambda_2}$. Such a transformation could be obtained by lifting to $\mathbb{H}$ any curve joining Λ_1 and Λ_2, but the result depends not only on Λ_1 and Λ_2 - it depends on the whole curve. The general quantization formula taking into account only the symplectic structure of the space of solutions $\mathcal{P}$ does not exist.

No wonder from the physical point of view. There is no possibility of constructing all measuring systems corresponding to all crazy polarizations which can be thought of. The classical theory which we are going to quantize contains much more structure than $(\mathcal{P}, \omega)$. The space-time structure of F (see fig. 1) enables to distinguish a family $\tilde{\mathbb{\Lambda}} \subset \mathbb{\Lambda}$ of all "position-type" polarizations over all Cauchy-surfaces in M. We must stress that in all classical field theories known to us the notion of "configuration space" is well defined for every space-like surface Σ (see [6])

Examples

1) For mechanics we take as $\tilde{\mathbb{\Lambda}}$ the family of all polarizations $\{x^i(t)=\mathrm{const}\}$ for all times t.

2) For electrodynamics $\tilde{\mathbb{\Lambda}}$ is composed of all polarizations $\{B(x)=\mathrm{const}(x)\}$ for all space-like surfaces Σ.

Definition: The classical theory $(F, M, \mathcal{P}, \omega)$ is called quantizable if the quantum connection in $\mathbb{H}$ is flat when restricted to $\tilde{\mathbb{\Lambda}}$.

Example

Classical, non-relativistic mechanics is always quantizable since the family $\tilde{\Lambda}$ is 1-dimensional. Every connection over 1-dimensional basis is flat. It may be shown that integrating this connection we obtain the result equivalent to Schrödinger theory.

In the field theory in flat space-time we can distinguish smaller family $\tilde{\Lambda}_S$ of polarizations connected only with space-like planes (flat surfaces). The family $\tilde{\Lambda}_S$ is finite-dimensional.

Definition: The field theory is said to be quantizable in the sense of Special Relativity if quantum connection is flat over $\tilde{\Lambda}_S$.

The quantizability in the sense of Special Relativity is connected with the standard Axiomatic Quantum Field Theory approach, where the Poincaré invariance plays the fundamental role. It is very important to investigate the relation between quantizability in general sense and quantizability in the sense of Special Relativity.

Now we see that in general there is no sense of "quantization of classical observables". In order to count the probability of obtaining some experimental result we have the density ς on corresponding polarization and do not need any operators. Such observables which induce groups of canonical transformations <u>preserving</u> $\tilde{\Lambda}$ (like energy, momenta, angular momentum etc.) can however be quantized as generators of 1-parameter groups of transformations. In our approach they are defined always up to an additive constant since we always work in projective Hilbert spaces and there is no canonical way of defining the phase of wave functions. Of course we can also qantize any function φ constant on fibres of some polarization $\Lambda \in \tilde{\Lambda}$ (like position at given time). The corresponding operator in $\mathcal{H}_\Lambda$ consists in multiplying the wave function by φ. We thus see that there are in quantum theories two, essentially different, types of operators: generators of transformations preserving Λ (defined up to additive constant) and "positions" (functions on configuration spaces Q_Λ, $\Lambda \in \tilde{\Lambda}$) which are defined precisely.

Remark 1. Our theory can be used to look for symmetries in quantum mechanics in the following way. Suppose our classical problem has a group of symmetry $\mathcal{G}$ which commutes with time-translations. Take the family $\tilde{\Lambda}_{\mathcal{G}}$ of all polarizations obtained from configuration polarizations $\tilde{\Lambda}$ through the action of elements of $\mathcal{G}$. We say that $\mathcal{G}$ is quantizable if quantum connection is flat over $\tilde{\Lambda}_{\mathcal{G}}$. If it is the case we obtain immediately the (projective) representation of $\mathcal{G}$ integrating quantum connection. This problem is very important for applications and we work on it in Warsaw.

Remark 2. The field X generated by hamiltonian $H = \frac{1}{2m} p^2 + V(x)$ is the sum of two Heisenberg-vectors generated respectively by $\frac{1}{2m} p^2$ and $V(x)$. In order to lift $\tilde{\Lambda}$ to $\mathbb{H}$ we must go step by step, using the Trotter-formula. The convergence of this procedure means on classical level that the field X is complete. On quantum level it means that operator H is self-adjoint. There are serious problems with criterions for self-adjointness of hamiltonians. But all, known to us, examples of hamiltonians which are not self-adjoint correspond to situations when the field X is not complete. There is therefore a conjecture that H is essentially self-adjoint if and only if the corresponding classical field X is complete. This problem is closely related to Feynman integrals.

Remark 3. In the field theory we have as yet very few rigorous statements. It can however be seen that the problem of "good ordering" will be automatically solved in our approach. The mechanism of that is, roughly speaking, the following. The hamiltonian is a sum of two parts which are perfectly self-adjoint (one part depends on "positions" like B in electrodynamics and second on "momenta" like E). But domains of these two parts are completely disjoint and Trotter procedure in Hilbert space diverges. However in projective Hilbert space it converges which enables to integrate quantum connection. We obtain this way the representation (projective) of the group of time-translations. It can be implemented by 1-parameter group of unitary transformations. Any generator of such a group can be called quantum hamiltonian. It is well defined (up to an additive constant) without "substracting infinity".

R E F E R E N C E S

[1] R.J.Blattner, Quantization and Representation Theory, Proceedings of Symposia in Pure Mathematics, vol. XXVI, American Math. Soc. 1973

[2] K.Gawędzki, Fourier like Kernels in Geometric Quantization, to appear in Dissertationes Mathematicae

[3] L.Hörmander, Fourier Integral Operators I, Acta Math. 127 (1971), p. 179

[4] J.Kijowski, A Finite Dimensional Canonical Formalism in the Classical Field Theory, Comm. Math. Phys, 30 (1973) p. 99

[5] J.Kijowski and W.Szczyrba, A canonical Structure of the Classical Field Theory, to appear in Comm. Math. Phys.

[6] J.Kijowski, W.Tulczyjew, Canonical Formalism and Boundary Problems, in preparation

[7] B.Kostant, Quantization and Unitary Representations, Lecture Notes in Math, vol 170, Springer, Berlin 1970

[8] B.Kostant, Symplectic Spinors, Proceedings of Convegno di Geometria Simplettica e Fisica Matematica, INDAM Rome 1973

[9] В.П. Маслов, Теория возмущений и асимптотические методы, Москва 1965

[10] D.J.Simms, Geometric Quantization, Proceedings of Colloque Symplectique, Aix-en-Provence 1974

[11] J.M.Souriau, Structure des Systèmes Dynamiques, Dunod, Paris 1970

THE APPLICATION OF GRADED LIE ALGEBRAS TO INVARIANCE CONSIDERATIONS IN PARTICLE PHYSICS

Y. Ne'eman

Department of Physics and Astronomy
Tel-Aviv University *

Introduction

The application of invariance considerations has recently undergone further conceptual advances with the introduction of Graded Lie Algebras (GLA) as a new symmetry construct, first in Dual Model "supergauges" and then in space-time "supersymmetry". The original algorithm was postulated in Dual models by Neveu and Schwarz (1971) and Ramond (1971) (see also Aharonov, Casher and Susskind, 1971; Gervais and Sakita, 1971). It was designed to improve the physical properties of "spinning strings" as candidate structures for hadrons with their observed spectrum of excitations. Following these developments, J. Wess and B. Zumino (1974a) and simultaneously and independently Volkov and Akulov (1973) introduced transformations in Minkowski space relating states of different Quantum Statistics types. I think the first such idea appeared in a humorous paper by Lipkin on "barbaryons" (1964), which was published upon the initiative of the late L. Rosenfeld. Lipkin showed that you could identify SU(3) octets in which spin replaced I-spin and baryon number took the place of hypercharge. Such an octet would be formed, for example by ($p\uparrow, p\downarrow, K^{*+}\uparrow, K^{*+}\rightarrow, K^{*+}\downarrow, K^{+}, \overline{\Xi^-}\uparrow, \overline{\Xi^-}\downarrow$). In the following year, several such schemes were suggested (Miyazawa et al, 1965; Singh 1965). Exploring the possibility that the neutrino might be a Goldstone particle, I tried to develop this work with some collaborators (Bella, 1973; Joseph, 1972) and describe the composition of the spectrum of hadron states through a symmetry whose supermultiplets would combine bosons and fermions, a symmetry realized either linearly or non-linearly, with a Goldstone-type compensation by a massless fermion. However, the symmetry generator connecting a boson state to a fermion state (an "odd" generator in our subsequent treatment here) is itself a fermion, and its local density can thus only involve an odd number of fermion fields. The fields-cannonical momenta anticommutators therefore do not provide the necessary information required for the evaluation of Lie algebra commutators between two such "odd" generators. We found that there was no way of pursuing this approach while using Lie algebras, except by going over to discrete spaces.

The problem was resolved (Volkov and Akulov, 1973; Wess and Zumino, 1974a) through the introduction of a different algebraic construct, the new supersymmetry, involving Graded Lie Algebras (GLA), in which an anticommutator sometimes appears as the relevant Lie product (between two odd generators).

Graded Lie algebras had previously appeared in the mathematical literature in another context, namely in deformation theory. The first basic mathematical treatment of graded Lie algebras had been provided by Nijenhuis (1955) and then by Frölicher-Nijenhuis (1957). We refer the reader to our recent article (Corwin, Ne'eman, Sternberg, 1975) for a summary of the mathematical work and some new results relating to simplicity criteria and to the enveloping algebra, needed for the development of a representation theory of GLA. That article also contains a review of the main physical applications (supersymmetry), covering the physical results as they appeared in the fall of 1974. In the present paper, we summarize that work and update it. Note that even though we consider the discovery of space-time supersymmetry as a worthwhile result in itself, there are as yet no definite advances in particle physics resulting from supersymmetry. The subject is still in its development phase and has not been applied to the actual phenomenological picture at this stage.

The Mathematical Context

1. Definitions and elementary properties. Let $L = \bigoplus_{c} L_k$ be a graded vector space; in other words L is a vector space, and the most general element of L can be written uniquely as a finite sum of its components, each component lying in one of the vector spaces L_k. The vector spaces are over any field of characteristic different from two and the L_k's will usually be finite dimensional. The index k can range over some given Abelian group as indexing group, however we shall principally be interested in cases where the indexing set is either the group of

integers, $\mathbb{Z}$, or the two element group, $\mathbb{Z}_2$. In what follows we shall adopt the convention that x will denote an element of L_k, that y will denote an element of L_ℓ and that z will denote an element of L_m. We say that L is a graded Lie algebra if we are given a bilinear map, denoted by [,], of $[L \times L] \to L$ such that the following three conditions hold:

$$(1.1) \qquad [L_k, L_\ell] \subset L_{k+\ell}$$

$$(1.2) \qquad [x,y] = -(-1)^k [y,x]$$

and

$$(1.3) \qquad [x,[y,z]] = [[x,y],z] + (-1)^{k\ell}[y,[x,z]] \quad .$$

Here the meaning of the factor $(-1)^n$ is clear for either $\mathbb{Z}$ or $\mathbb{Z}_2$ as indexing group. For more general indexing groups we take it to denote some character of the group with values in the group $\{\pm 1\}$ which must be given as an additional piece of the structure. Condition (1) simply says that the bracket multiplication is consistent with the grading. Condition (2) is the graded version of anticommutativity. Notice that for odd elements it says that "multiplication" is commutative, i.e., the bracket then represents an anticommutator. These are thus our "supergauge" or "supersymmetry" new generators. We see that even elements are Lie algebra generators (i.e., physically, they connect states of similar statistics).

Condition (3) is the graded version of Jacobi's identity. For even x it asserts that left multiplication by x is a derivation of the bracket multiplication, while for odd x it asserts that left multiplication by x is an antiderivation.

We shall call an operator $D: L \to L$ such that $D: L_n \to L_{n+k}$ and

$$(1.4) \qquad D(yz) = (Dy)z + (-1)^{k\ell} yDz$$

a (graded) derivation of degree k. This notion makes sense for any graded algebra A, that is for any graded vector space $A = A_n$ with a bilinear map $A \times A \to A$ such that $A_n \times A_m \to A_{m+n}$. (No associativity or commutativity condition on A is assumed.)

Again notice that L_0 is a Lie algebra in the old fashioned sense and so is the direct sum of all the even L_k's.

We can construct graded Lie algebras as follows: Let $V = \oplus V_n$ be a graded vector space. (For instance, to illustrate a physical application, we might let V be the relevant piece of the Hilbert space of states, graded according to fermion number.) We let $End_k(V)$ consist of those linear maps, x, of V into itself such that $xV_n \subset V_{n+k}$. It is clear that if $x \in End_k(V)$ and $y \in End_\ell(V)$ then the composition xy lies in $End_{k+\ell}(V)$. We define a graded Lie algebra structure on End(V) = $End_k(V)$ by setting

$$(1.5) \qquad [x,y] = x \cdot y - (-1)^{k\ell} y \cdot x \quad .$$

It is obvious that conditions (1) and (2) are satisfied and a straightforward verification shows that condition (3) is satisfied. Similarly, if $A = + A_n$ is any graded associative algebra the above bracket gives a graded Lie algebra called the commutator algebra of A. If A is any graded algebra, associative or not, it is easy to check that the set of graded derivations of A is a Lie subalgebra, Der A, of End A.

It is clear how to define a homomorphism from one graded Lie algebra to another, where they are both indexed by the same group - we require the map to be gradation preserving as well as being a homomorphism of the bracket structure. By a <u>representation</u> of a graded Lie algebra, L, on a graded vector space V we shall mean a homomorphism of L into the graded Lie algebra End(V).

2. <u>An example</u>. Corwin et al. (1975) have presented a large number of mathematical and physical examples. We shall only quote one simple case here, <u>the algebra End(V) for V two dimensional - The Fermi-Dirac anti-commutator</u>. Suppose that $V = V_0 \oplus V_1$ where both V_0 and V_1 are one-dimensional vector spaces over the complex numbers. Then

$$\text{End } V = L_{-1} + L_0 + L_1 \quad .$$

In other words, the algebra is $\mathbb{Z}$ graded with L_i one-dimensional for i = -1,1, two-dimensional for i = 0 and all other L_i trivial. We can define it using real 2 × 2 matrices,

$$e = 1 \qquad x = \tfrac{1}{2}(\sigma_x + i\sigma_y)$$

$$h = \tfrac{1}{2}(\sigma_z + 1) \qquad y = \tfrac{1}{2}(\sigma_x - i\sigma_y)$$

the indexing is given by the eigenvalues of h:

$$h, e \in L_0 \quad , \quad x \in L_1 \quad , \quad y \in L_{-1}$$

with the bracket relations

(2.1) $$[h,x] = x \quad , \quad [h,y] = -y \quad , \quad [x,y] = e$$

all other brackets vanishing. It is easy to check the graded version of the Jacobi identity etc.

One well-known physical realization of this scheme is the Jordan-Wigner quantization (1928; see also Fock, 1932) scheme (which incorporates the Pauli principle) for Fermi-Dirac annihilation and creation operators (we restrict ourselves to one state)

(2.2) $$[b,b^*]_+ = 1 \quad , \quad [b,b]_+ = 0 \quad , \quad [b^*,b^*]_+ = 0$$
$$[N,b^*]_- = b^* \quad , \quad [N,b]_- = -b \quad ,$$

thus

(2.3) $$N \sim h \quad , \quad b^* \sim x \quad , \quad b \sim y \quad , \quad e = 1$$

Note that in this example, the graded vector space $V = V_k$ is the Fermi-Fock space, with V_0 for the vacuum, V_1 for the one fermion state. Indeed, the odd generators b, b^* connect a boson (the vacuum) with a fermion.

Physical Applications

3. Dual Models (strings). We now turn to the "supergauge" transformations of the so-called Dual Models (Neveu and Schwarz, 1971; Gervais and Sakita, 1971). Numerous reviews (Scherk, 1975; Schwarz, 1973; Veneziano, 1974; Rebbi, 1974) will assist readers who happen to be unfamiliar with this relatively recent development of Dispersion theory. It was triggered by Veneziano's (1968) discovery of a crossing-symmetric relativistic strong-interaction amplitude

$$B_4 = \frac{\Gamma(-\alpha(s))\Gamma(-\alpha(t))}{\Gamma(-\alpha(s)-\alpha(t))} + \ldots = \int_0^1 dx\, x^{-\alpha(s)-1}(1-x)^{-\alpha(t)-1} + \ldots$$

satisfying in addition a bootstrap condition in the form of Finite Energy Sum Rules FESR (Dolen, Horn and Schmidt, 1967; Igi and Matsuda, 1967; Logunov, Soloviev and Tavkhelidze, 1967) and possessing appropriate Regge asymptotic behavior. The attempt to unitarize Veneziano's representation has yielded systems, some of which are now developed as non-local, or rather multi-local field theories, instead of the original on-mass shell amplitudes. Nambu, Nielsen, Susskind and others had first replaced the factorized Veneziano model with propagator

$$D(p^2) = -\frac{i\alpha'}{(2\pi)^4}\frac{1}{L_0-1-i\varepsilon} \quad , \quad L_0 - 1 = \alpha'[M^2-p^2] \quad ,$$

$$M^2 = \frac{1}{\alpha'}\sum_{n=1}^{\infty} n\, a_n^{\dagger} a_n - \frac{1}{\alpha'}$$

by a quantized one-space dimensional relativistic string, moving in a larger space-time. The string action is written as

$$S = -\frac{1}{2\pi\alpha'}\int_{\tau_1}^{\tau_2} d\tau \int_0^{\pi} d\sigma\{-\det g\}^{\frac{1}{2}} = -\frac{1}{2\pi\alpha'}\int_{\tau_1}^{\tau_2} d\tau \int_0^{\pi} d\sigma\{(\dot{x}\cdot x')^2 - x'^2\dot{x}^2\}^{\frac{1}{2}}$$

(σ represents a parameter along the string, τ is the proper time). The quantum excitations of the string do reproduce the Veneziano spectrum; the gauge conditions (including the "supergauge") are necessary for the removal of ghost states. However, the system's convergence properties turn out to depend upon the total dimensionality of the space-time, and require a 26 dimensional Minkowski space. The next development adjoined a continuous-spin structure to the linear string, achieving some improved physical features and convergence in 10 dimensions.

We shall only treat here the Neveu-Schwarz (1971) model, which yields a spectrum of physical bosons. It is given by the algebra:

$$(3.1)\qquad [D_{2m}, D_{2n}] = 2(m-n)D_{2m+2n} + \frac{1}{4}(4m^3-m)\delta_{m,-n}I$$

$$(3.2)\qquad [D_{2m}, D_{2n+1}] = (m-2n-1)D_{2m+2n+1}$$

$$(3.3)\qquad [D_{2m+1}, D_{2n+1}] = 2D_{2(m+n+1)} + \frac{1}{2}d(4m^2+4m+\frac{3}{4})\delta_{m,-n}I$$

d is the dimension of the space-time in which we embed the string. Hopefully, d = 4, though this result depends at present upon the introduction of internal degrees of freedom; the actual result is d = 10 for convergence to be ensured.

We see that the grading L_a of the Neveu-Schwarz $L = \oplus L_a$ algebra is a $\mathbb{Z}$ grading, with $D_m \in L_m$ one-dimensional, except for L_0 which contains D_0 and the identity operator I. The index a is given by the commutator

$$(3.4)\qquad [-D_0, L_a] = aL_a$$

Note that (3.1) describes an infinite Lie algebra, connected with the conformal transformations on the two-dimensional (one space, one time) string. (3.3) is an anticommutator by (1.2) and represents the "supergauge" constraints. Eq. (3.2) is a commutator describing the behavior of supergauge generators under the conformal algebra.

The Ramond model has only $\mathbb{Z}_2$ grading.

4. Supersymmetry - the $\mathcal{W}$ and $\mathcal{V}$. The first introduction of a GLA as a supersymmetry of space-time (i.e., as a symmetry containing the Poincaré group) is due to Volkov and Akulov (1973). They adjoined to the Poincaré algebra $\mathcal{P}$ a set of 4 odd generators, behaving like a Majorana spinor. Volkov and Akulov were exploring the

hypothesis that the neutrino's masslessness might indicate that it is a Goldstone particle, necessary to the (non-linear) realization of an exact symmetry of the physical world, a hypothesis suggested earlier by the present author (YN), and which had failed because of the statistics issue (Joseph, 1972; Bella, 1973). They therefore had to introduce conserved generators which do not destroy the vacuum, and which behave like the physical neutrinos under the Poincaré group, i.e., spinors. To preserve Fermi statistics, the new generators were now also required to anticommute. This GLA V thus had 14 generators, was not simple, and had to be realized non-linearly. For ψ the neutrino field,

$$(4.1) \qquad \delta\psi = \zeta + ia(\bar{\zeta}\gamma_\mu\psi)\partial_\mu\psi$$

where a is a constant of dimension (-4), i.e., the fourth power of a length, and ζ is a four-spinor (Majorana) anticommuting parameter. Considering that the neutrino does not seem to fit a Majorana description, the actual physical content would appear rather speculative at this stage, but we shall see that there do exist realizations of this GLA with Weyl neutrinos. Apart from the non-linear realization, the V algebra is indeed the one that has been extensively applied as a supersymmetry, after its reintroduction by Salam and Strathdee (1974a) as the physically interesting subalgebra of the spin-conformal GLA W discovered by Wess and Zumino (1974a). For its explicit construction, see example 2H of Corwin, Ne'eman and Sternberg (1975). Salam and Strathdee even returned to the question of a possible Goldstone role for the neutrino (Salam and Strathdee, 1974c, 1975b).

We now turn to a more detailed study of W.

We preserve the energy-metric of Corwin et al., $g^{00} = 1$, $g^{11} = g^{22} = g^{33} = -1$, and $g^{\mu\nu} = 0$ for $\mu \neq \nu$. The physical algebraic system consists of the conformal algebra SU(2,2) for the even gradation

$$[J^{\mu\nu}, J^{\rho\sigma}] = + i\, g^{\mu\sigma}J^{\nu\rho} - i\, g^{\mu\rho}J^{\nu\sigma} + i\, g^{\nu\rho}J^{\mu\sigma} - i\, g^{\nu\sigma}J^{\mu\rho}$$

$$[J^{\mu\nu}, P^{\lambda}] = - i\, g^{\mu\lambda}P^{\nu} + i\, g^{\nu\lambda}P^{\mu}$$

$$[P^{\mu}, P^{\nu}] = 0$$

$$[J^{\mu\nu},K^{\lambda}] = -i\,g^{\mu\lambda}K^{\nu} + i\,g^{\nu\lambda}K^{\mu}$$

$$(4.2)\qquad [K^{\mu},K^{\nu}] = 0$$

$$[K^{\mu},P^{\nu}] = -2i\,g^{\mu\nu}D - 2i\,J^{\mu\nu}$$

$$[D,P^{\mu}] = i\,P^{\mu}$$

$$[D,K^{\mu}] = -i\,K^{\mu}$$

$$[D,J^{\mu\nu}] = 0$$

where $J^{\mu\nu}$ are the Lorentz group generators, P^{μ} the translations, K^{μ} the pure conformal transformations and D the dilation operator. Note that the indexing for the gradation will be provided by doubling the eigenvalue of D. In the context of adjoining $L_{-1} \oplus L_1$ we shall be led to add a sixteenth (scalar) operator to the even set. This will be E,

$$(4.3)\qquad [E,J_{\mu\nu}] = [E,P^{\mu}] = [E,K^{\mu}] = [E,D] = 0$$

To construct $L_{-1} \oplus L_1$ in terms of physical operators, we use the Dirac matrices with the conventions of Corwin et al. (1975) in a Majorana representation.

The odd generators of $\mathcal{W}$ are Q_α and R_β (α,β = 1..4 in Dirac spinor space) and are Majorana spinors, thus involving two complex or four real functions each. The even-odd Lie brackets are the commutators (Wess and Zumino, 1974a; Corwin, Ne'eman and Sternberg, 1974; Dondi and Sohnius, 1974)

$$[J^{\mu\nu},Q_{\alpha}] = -\frac{1}{2}(\sigma^{\mu\nu})_{\alpha\beta}\,Q_{\beta}$$

$$[J^{\mu\nu},R_{\beta}] = -\frac{1}{2}(\sigma^{\mu\nu})_{\alpha\beta}\,R_{\beta}$$

$$[P^{\mu},Q_{\alpha}] = [K^{\mu},R_{\alpha}] = 0$$

$$[K^{\mu},Q_{\alpha}] = -(\gamma_5\gamma^{\mu})_{\alpha\beta}\,R_{\beta}$$

$$(4.4)\qquad [P^{\mu},R_{\alpha}] = -(\gamma_5\gamma^{\mu})_{\alpha\beta}\,Q_{\beta}$$

$$[D,Q_\alpha] = \frac{i}{2} Q_\alpha$$

$$[D,R_\alpha] = -\frac{i}{2} R_\alpha$$

$$[E,Q_\alpha] = 3i(\gamma^5)_{\alpha\beta} Q_\beta$$

$$[E,R_\alpha] = -3i(\gamma^5)_{\alpha\beta} R_\beta$$

We see that E counts chiral fermions. The odd-odd brackets consist of the anti-commutators,

$$\{Q_\alpha,R_\beta\} = -(\gamma_5\sigma_{\mu\nu}C)_{\alpha\beta} J^{\mu\nu} - i(C)_{\alpha\beta} E + 2i(\gamma_5 C)_{\alpha\beta} D$$

(4.5) $$\{Q_\alpha,Q_\beta\} = -2(\gamma_\mu C)_{\alpha\beta} P^\mu$$

$$\{R_\alpha,R_\beta\} = -2(\gamma_\mu C)_{\alpha\beta} K^\mu$$

The grading will be:

(4.6)

L_{-2}	L_{-1}	L_0	L_1	L_2
K^μ	R_α	$J^{\mu\nu}$	Q_α	P^μ
		D		
		E		

We can also introduce "adjoint" spinors $\bar{Q}_\alpha$ and $\bar{R}_\alpha$. This is especially useful in view of further generalizations in which we shall introduce internal degrees of freedom. For su(n), n $\geq$ 3, the covariant and contravariant representations are not equivalent ($\underline{3}$ and $\underline{3}^*$ in su(3), etc.) and this will require distinguishing between Q_α and $\bar{Q}_\alpha$. The bracket relations are,

$$\{Q_\alpha,\bar{R}_\beta\} = (\gamma_5\sigma_{\mu\nu})_{\alpha\beta} J^{\mu\nu} - 2i(\gamma_5)_{\alpha\beta} D + i\,\delta_{\alpha\beta} E$$

(4.7) $$\{Q_\alpha,\bar{Q}_\beta\} = 2(\gamma_\mu)_{\alpha\beta} P^\mu$$

$$\{R_\alpha,\bar{R}_\beta\} = 2(\gamma_\mu)_{\alpha\beta} K^\mu$$

5. Representations of V. Working with the conformal group as a symmetry implies massless particles (provided the symmetry is not spontaneously broken by a Goldstone boson). We first construct some physically relevant representation of the sub-algebra V corresponding to mass zero. The generators of V are: $J^{\mu\nu} \in L_0$, $Q_\alpha \in L_1$, $P^\mu \in L_2$. For massless particles, the helicity λ (taken here with the same sign as J^{12}) is the only remaining quantum number in the little group of the Poincaré group.

For $p^+ = p^0 + p^3 \neq 0$, $p^- = p^0 - p^3 = 0$, $p^1 = p^2 = 0$ on the states, the little group is generated by $((J^{12}, J^{23} - J^{20}, J^{31} - J^{01}))$.

In the odd set of (4.6), only the Q_α are in the little group. We find that the only two non-vanishing Q_α are Q_1 and Q_4 (in the representation of Corwin et al.). These are not J^{12} (helicity) eigenvectors, and we recombine them into helicity $+\frac{1}{2}$ and $-\frac{1}{2}$ operators.

$$(5.1)\quad \begin{cases} (Q_1 - iQ_4)/\sqrt{2} \equiv x, & (Q_1 + iQ_4)\sqrt{2} \equiv y \\ \{x,y\} = e, & \text{for} \quad 2P^+ \equiv e \\ [h,x] = x, \quad [h,y] = -y & \text{for} \quad 2J^{12} \equiv h \end{cases}$$

This is just the GLA of our example in chapter 2. Its defining 2×2 representation acts on a vector space containing one fermion and one boson state (helicities $\frac{1}{2}$, 0 or any $(\frac{n+1}{2}; n/2)$.

To discuss the representations, we notice that $\alpha_1 = x + y$, $\alpha_2 = -i(x-y)$ define the Clifford algebra C_2,

$$(5.2)\qquad \{\alpha_i, \alpha_j\} = 2\delta_{ij}$$

as can be computed from our defining brackets. It is a $2^2 = 4$ dimensional vector space with basis α_1, α_2, $\alpha_1\alpha_2$, e. Its only irreducible representation is the defining set of 2×2 matrices (see, for example, Boerner, 1963). However, Q_1 and Q_4 are not parity eigenvectors, since under the parity transformation,

(5.3) $Q_\alpha \to \eta_P(\gamma_0)_{\alpha\beta}\, Q_\beta$

and choosing $\eta_P = 1$ here,

$$Q_1 \to -i\, Q_3 \quad , \quad Q_4 \to i\, Q_2$$

To conserve parity, we therefore adjoin a 2-space representing states with $P^- \neq 0$, $P^+ = P^1 = P^2 = 0$. We find,

(5.4) $(Q_3+iQ_2)/\sqrt{2} \equiv x' \quad , \quad (Q_3-iQ_2)\sqrt{2} \equiv y' \quad , \quad 2P^- \equiv e' \quad ,$

$$2J^{12} \equiv h'$$

This time, we pick $n = -1$, getting eigenvalues $(0,-\frac{1}{2})$ or $(-n/2,-(n+1)/2)$ for the helicities λ.

The Fermi states, being helicity eigenstates, have to consist of combinations of the type $(\psi_1 \mp i\psi_4)$ of the real components of a Majorana "neutrino". Parity thus consists of complex-conjugation, leading to the conjugate space. The bosons thus also can be written as $(u \pm iv)$, u a scalar and v a pseudoscalar. We still return to this simplest of all representations when we construct appropriate "superfields", i.e., field representations of supersymmetry (Salam and Strathdee, 1974a). Note however that the findings of Volkov and Soroka (1973) fit within this picture: the massless graviton, with $\lambda = \pm 2$ gets a companion with $\lambda = \pm\frac{3}{2}$.

Actually, there have been to-date very few applications of $\mathcal{W}$. Instead, following Salam and Strathdee (1974a) the non-simple Volkov-Akulov subalgebra $\mathcal{V}$ was used in its linear realizations ("supersymmetry"). However, any Lagrangian which is invariant under that algebra, and which is in addition made invariant under the conformal group (by making all masses and all dimensional couplings vanish) will also be invariant under the $\mathcal{W}$ GLA.

The generators of the subalgebra $\mathcal{V}$ are:

$$J^{\mu\nu} \in L_0$$

$$Q_\alpha \in L_1$$

$$P^{\mu} \quad \varepsilon \quad L_2$$

i.e., the even gradation corresponds to the Poincaré algebra.

Taking here the $M \geq 0$ case, we use the rest-frame to find the "little" GLA.

$$(5.5) \quad \begin{cases} \{Q_\alpha, Q_\beta\} = 2\delta_{\alpha\beta} M \\ \text{or} \\ \{Q_\alpha, \bar{Q}_\beta\} = 2(\gamma_0)_{\alpha\beta} M \end{cases}$$

The first bracket defines C_4, the 4th order Clifford algebra (dimensionality $2^4 = 16$) which is just the algebra of Dirac γ^μ matrices in a Euclidean metric. Its only representation is in 4-dimensional matrices (Boerner, 1963). We thus know that all $M \neq 0$ representations of V will reduce into 4-dimensional subspaces, just as the $M = 0$ ones worked in doubled 2-spaces.

To get the "little" GLA in a more familiar form, we diagonalize γ_0. In our representation $\gamma_0 = -\rho_2$, so that the appropriate unitary operator U will act in ρ space only, with

$$U(-\rho_2)U^{-1} = \rho_3$$

This will rearrange our (Majorana spinor) odd operators

$$Q_\alpha \rightarrow U\, Q_\alpha$$

which has as components

$$(5.6) \quad \begin{aligned} Q_I &= (Q_1 + iQ_3)/\sqrt{2} \\ Q_{II} &= (Q_2 + iQ_4)/\sqrt{2} \end{aligned}$$

and their Hermitian (here just complex) conjugates Q_I^+, Q_{II}^+. We find that J^{23} is now diagonal. The $Q_{A=I,II}$ and Q_A^+ form independent 2-spinors. They fulfil

$$(5.7) \quad \{Q_A, Q_B^+\} = 2\delta_{AB} M \quad , \quad \{Q_A, Q_B\} = 0 \quad , \quad \{Q_A^+, Q_B^+\} = 0$$

These brackets together with the angular momentum commutation relations define the little GLA:

$$J^{ij}, M \quad \varepsilon \quad L_0 \quad (d = 4)$$

$$Q_A \quad \varepsilon \quad L_{-1} \quad (d = 2)$$

$$Q_A^+ \quad \varepsilon \quad L_1 \quad (d = 2)$$

Since γ_0 is diagonal in this representation, the spinors Q_A and Q_A^+ are parity eigenstates with opposite eigenvalues (note that η_P will have to be i or -i here).

An irreducible representation of the little GLA is thus obtained (Salam and Strathdee (1974b)) operating with the Q_A and Q_A^+ on the (2j+1) dimensional carrier space of any representation (j,M) of the (Wigner) little group of the Poincaré group. Because of (5.7), and taking (j_3 corresponds here to the "new" J^{23} direction)

(5.8) $$Q_A|j,j_3,\chi_p,M\rangle = 0$$

in analogy to an annihilation operator, we have 4 possible actions of

Q_A^+: $Q_I^+|j,j_3,\chi_p,M\rangle$, $Q_{II}^+|j,j_3,\chi_p,M\rangle$, $Q_I^+Q_{II}^+|j,j_3,\chi_p,M\rangle$ and $|j,j_3,\chi_p,M\rangle$.

The first two change the spin, j_3, parity, and statistics of the states. The $Q_I^+Q_{II}^+$ action preserves j_3 and j but inverts the parity. We thus have a 4(2j+1) dimensional Fock space, with subspaces $|j,j_3,-\chi_p,M\rangle$, $|j,j_3,\chi_p,M\rangle$, $|j+\frac{1}{2},j_3+\frac{1}{2},\chi_p\eta_p,M\rangle$, $|j-\frac{1}{2},j_3-\frac{1}{2},\chi_p\eta_p,M\rangle$. Notice that fermions and bosons have the same mass.

These rest states are then boosted to any $\vec{p}$ by a Lorentz transformation $U(L_{\vec{p}})$. The action of Q_A and Q_A^+ on the boosted states can be derived from our knowledge of the spinor behavior of the Q_A and Q_A^+ under Lorentz transformations.

It is also sometimes useful to diagonalize $-i\gamma_5$. The chiral projections Q^R and Q^L become each other's hermitian conjugates and fulfil brackets identical with (5.7).

Their graded Lie brackets are in general

$$\{Q^L_a, Q^L_b\} = 0 \quad , \quad \{Q^R_a, Q^R_b\} = 0 \quad ,$$

(5.7)'

$$\{Q^R_a, Q^L_b\} = 2(1 \cdot P^0 + \sigma_3 P^2 - \sigma_1 P^1 - \sigma_2 P^2)_{ab}$$

and for $M \neq 0$ and rest states

$$(5.7) \qquad \{Q^R_a, Q^L_b\} = 2\delta_{ab} M$$

While for the $M = 0$ case, we again get the reduction into 2 subspaces with $P^0 + P^2 \neq 0$, $P^0 - P^2 = P^1 = P^3 = 0$ for the first, and the parity-inverted states for the second subspace. The Q^R_1 and Q^L_1 are in one subspace, and the Q^R_2 and Q^L_2 in the other.

For $M \neq 0$ and rest, Q^R_a and $Q^L_a = Q^{R*}_a$ can thus be treated as annihilation and creation operators in the construction of representations of states or fields. Note that (5.4)-(5.7) are examples of "polarizations" in the context of Kostant's application of Kostant-Souriau methods to GLA as discussed in this conference.

6. <u>Realization on a Grassmann algebra as a generalized (Berezin-Kac) Lie group; Salam-Strathdee superfields.</u> In example 2B of Corwin et al. (1975), a GLA was defined on a graded vector space V which had been generated through exterior multiplication from a vector space W, i.e., a Grassmann algebra. Berezin and Kac (1970) had studied such realizations in the context of a further formal "exponentiation" of a GLA with elements of the Grassmann algebra as parameters (odd and anticommuting elements for the odd generators, even and commuting elements for the Lie algebra). This method had been used in the construction of the supergauges, a fact which had for some time hidden the GLA since one could regard the product of an odd Grassmann element with an odd generator as a Lie generator (the resulting Lie algebra is a solvable one).

We use an N-dimensional vector space (over the complex field) $V(\equiv \Lambda^1 V)$, generating a 2^N dimensional Grassmann algebra $V = \bigoplus_{r=0}^{N} \Lambda^r V$. The basis vectors of V are $v_1, v_2, \ldots v_N$; since the Grassmann algebra is graded-commutative, the elements of V anticommute,

$$v_i \Lambda v_j = (-1) v_j \Lambda v_i$$

We shall write this property as

$$\{v_i, v_j\} = 0 \quad \text{for any } i,j, \; v_{i,j} \in V \tag{6.1}$$

with multiplication thus being defined by the Λ operation.

We shall also use extensively the elements of $\Lambda^2 V \equiv W$, resulting from $v_i v_j$ products. In this case, graded-commutativity ensures that the elements $w_a \in W$ commute. The Minkowski space coordinates are identified with elements of W, $x^\mu \in W = \Lambda^2 V$. If we attach a reflection operation R to the $v_i \in V$, $v_i \to -v_i$, the entire ΛV splits into two parts,

$$\begin{cases} \Lambda V = \Lambda V^{(-)} + \Lambda V^{(+)} \\ \Lambda^r V \subset \Lambda V^{(-)} & \text{if r is odd,} \\ \quad\;\; \subset \Lambda V^{(+)} & \text{if r is even.} \end{cases}$$

"Superfields" as introduced by Salam and Strathdee (1974a) are "local" fields, in the variables $\theta_\alpha \in \Lambda' V$, $x^\mu \in \Lambda^2 V$. θ_α is a Majorana spinor,

$$\theta = \theta^c = C\tilde{\gamma}_0 \theta^* \tag{6.2}$$

which amounts to a true reality condition

$$\theta_\alpha = \theta_\alpha^*$$

in the representation we have used. As to the coordinate, it should be real in any case,

$$x^\mu = (x^\mu)^* \quad .$$

Thus V is at least 4-dimensional. Indeed, a 4-dimensional quasi-Minkowski coordinate (it is not a true Minkowski coordinate since $(x^\mu)^{N+1} = 0$) in

$\Lambda^2 V$ can be constructed from two θ, $\theta' \in V$,

(6.3) $$x^\mu = \bar{\theta}\gamma^\mu\theta'$$

which, by (6.1) is Hermitian and real. Note that

(6.4) $$\bar{\theta}\gamma^\mu\theta' = -\bar{\theta}'\gamma^\mu\theta$$

which can be rewritten as

$$\tilde{\theta_\alpha}(\gamma^0\gamma^\mu)_{\alpha\beta}\,\theta'_\beta = -\tilde{\theta'_\alpha}(\gamma^0\gamma^\mu)_{\alpha\beta}\,\theta_\beta$$

We observe in this expression the (generalized) matrix structure of the Λ operation between two Majorana-like elements of V. It is still antisymmetric, because $\gamma^0\gamma^\mu$ is symmetric; the antisymmetry is thus derived from (6.1)

(6.5) $$\{\theta_\alpha,\theta'_\beta\} = 0$$

and the $\gamma^0\gamma^\mu$ matrices preserve this feature while taking care of the spinor indices.

We now turn to the action of the Q_α on these elements. From

$$\{Q_\alpha,Q_\beta\} = -2(\gamma_\mu C)_{\alpha\beta}\,P^\mu$$

we know that the doubled action of the Q_α represents a translation in W. We can thus guess that Q_α represents such a translation in V, acting in analogy to

$$P_\mu \sim -i\frac{\partial}{\partial x^\mu} \quad , \quad [P_\mu,x^\nu] = -i\,\delta^\nu_\mu$$

As far as its action on V is concerned, $Q_\alpha \sim \Gamma_{\alpha\beta}\frac{\partial}{\partial\theta_\beta}$. Thus Q_α is in V^* or in $\Lambda_1 V$. Note that for $\frac{\partial}{\partial\theta_\alpha}$, an element in V^*, we are in the larger $\oplus\,\Lambda^r_s V$. Thus

(6.6) $$\{\frac{\partial}{\partial v_i},\frac{\partial}{\partial v_j}\} = 0 \quad , \quad \{\frac{\partial}{\partial v_i},v_j\} = \delta_{ij}$$

Q_α will thus bracket with θ_β as $\mathrm{End}_1 V$,

(6.7) $\{Q_\alpha, \theta_\beta\} = i\, C_{\alpha\beta}$

where $C \in V_0$ appears as the appropriate metric for Majorana spinors, so that $\Gamma_{\alpha\beta} = i\, C_{\alpha\beta}$. To obtain an infinitesimal translation by a "constant" parameter $\varepsilon_\alpha \neq \varepsilon_\alpha(\theta)$

(6.8) $\theta_\alpha \rightarrow \theta_\alpha + \varepsilon_\alpha \quad , \quad \{Q_\alpha\, \varepsilon_\beta\} = 0$

with ε_α V, we have to act with $\bar{\varepsilon}_\alpha Q_\alpha$, where we use $\bar{\varepsilon}_\alpha$ rather than ε_α in order to obtain the necessary tensor-construction. Note that exponentiation by $\bar{\varepsilon}_\alpha$ follows the Berezin-Kac (1970) method of generating a generalized Lie group. Integration is defined through

(6.9) $\int dv_i = 0 \; ; \; \int v_i dv_i = 1 \; ; \; \{v_i dv_j\} = \{dv_i, dv_j\} = 0$

Note that $\varepsilon \neq \varepsilon(\theta)$ and $\{Q,\varepsilon\} = 0$, as against (6.6), <u>require additional dimensions in V</u>.

The resulting action is then a commutator bracket, as needed for infinitesimal group action,

(6.10) $i[\bar{\varepsilon}_\alpha Q_\alpha, \theta_\beta] = \varepsilon_\beta$

The action on $x^\mu = \bar{\theta}'\gamma^\mu\theta$ is thus bound to be

(6.11) $i[\bar{\varepsilon}_\alpha Q_\alpha, x^\mu] = \bar{\varepsilon}\gamma^\mu\theta$

Assuming now the existence of a "superfield" $\phi(x_\mu,\theta_\alpha)$, we can use a Taylor series to identify the structure of the infinitesimal operator $\bar{\varepsilon}_\alpha Q_\alpha$,

$$U\phi(x_\mu,\theta_\alpha)U^{-1} = \phi(x_\mu - \bar{\varepsilon}\gamma_\mu\theta, \theta_\alpha - \varepsilon_\alpha) =$$

$$= \phi(x_\mu,\theta_\alpha) - \bar{\varepsilon}\gamma_\mu\theta \frac{\partial}{\partial x^\lambda}\phi(x^\mu,\theta_\alpha) - \varepsilon_\alpha \frac{\partial}{\partial\theta_\alpha}\phi(x_\mu,\theta_\alpha) + O(\varepsilon^2)$$

where the generalized group element is

$$(6.12)\quad \begin{cases} U = 1 - i\,\bar{\varepsilon}_\alpha Q_\alpha \\ (\bar{\varepsilon}_\alpha Q_\alpha)^+ = \bar{\varepsilon}_\alpha Q_\alpha \end{cases}$$

This yields the explicit structure:

$$(6.13)\quad Q_\alpha(\Lambda V) = (iC_{\alpha\beta}\frac{\partial}{\partial\theta_\beta} - i(\gamma_\mu)_{\alpha\beta}\,\theta_\beta\,\frac{\partial}{\partial x_\mu})\Lambda V$$

We now come to one of the difficulties or inconsistencies[+)] of this picture. If we regard $\bar{\varepsilon}Q$ as a Lie group generator, we get

$$(6.14)\quad \bar{\varepsilon}_\alpha Q_\alpha, \bar{\varepsilon}_\beta Q_\beta = \bar{\varepsilon}_\alpha\{Q_\alpha,\bar{Q}_\beta\}\varepsilon_\beta = 2\bar{\varepsilon}_\alpha(\gamma_\mu)_{\alpha\beta}\,\varepsilon_\beta\,P^\mu$$

However, this vanishes, since $\bar{\psi}\gamma_\mu\chi = -\bar{\chi}\gamma_\mu\psi$. Even if we do not sum over the α and β indices, we shall at least have vanishing expressions for $\mu = 0$, since $\gamma_0^2 = 1$. This covers in fact the entire little algebra for $M \neq 0$. We are thus faced with two choices: either the Lie algebra is Abelian, or, as we already noted from (6.8) ε_α and ε'_β have to lie in new subspaces of V, which differ from each other and also do not contain the θ_α. In these new subspaces, we may be able to ensure non-vanishing of the right-hand side. Indeed, the simplest solution is to add 8 dimensions, so as to have different $\bar{\varepsilon}_\alpha$ and ε'_β on the right-hand side. The P^μ are then multiplied by 16 - 4 = 12 new dimensions in $\Lambda^2 V$.

Note that all of this is necessary because the superfield $\phi(x_\mu,\theta_\alpha)$ is acted upon by a Lie group. However, if we allow for finite transformations, θ_α will have "crept" into the new ε_α, ε'_α subspaces, and our efforts will have been to no avail. Moreover, we dare not allow (6.14) to have a vanishing right-hand side since we would then lose the connection with our starting point, in which Q_α acted as the "square-root" of P^μ. We have by all means to recover the GLA bracket's information, even though it will now be supplied by a commutator.

+) We hope to derive a completely consistent approach based upon Kostant's method. The θ_α will be given by functions on the Universal Enveloping Algebra of the GLA; indeed, (6.13) can be regarded as defining a transform, and all nilpotence properties will apply to θ_α without applying to x_μ, the transform of the even P^μ.

Goddard (1974) has shown that it is possible to add only 2 dimensions to V, so that $N \geq 6$, and disconnect the new dimensions from the spinor indices in ε_α. This is the most economical solution. It may have been hinted at by Salam and Strathdee (1974a), but in their solution the number of odd generators would be doubled: $(v_a Q_\alpha)$ with a = 5,6. Rühl and Yunn (1974) have pursued the more direct method and have gotten 26 generators instead of 14 for V. This results from 6 for $J^{\mu\nu}$, 8 for $\bar{\varepsilon}Q$ and $\bar{\varepsilon}'Q$, 12 for $\varepsilon\varepsilon' P_\mu$. Even though a Lie algebra thus replaces the GLA, only infinitesimal transformations of θ_α are allowed. The Lie group is thus physically applied only "very close" to the identity.

We now follow Salam and Strathdee (1974a). Due to the anticommuting properties of θ_α, any function f(θ) must be a polynomial. Since the monomials $\theta_{\alpha_1}\theta_{\alpha_2}\ldots\theta_{\alpha_n}$ have to be completely antisymmetric, expanding $\phi(x^\mu,\theta_\alpha)$ in powers of θ_α is a finite operation terminating at n = 4. The even monomials belong in the $\Lambda V_\theta^{(+)}$, the odd ones in $\Lambda V_\theta^{(-)}$. Altogether, $\phi(x^\mu,\theta)$ is 16-dimensional as long as one does not allow finite transformations in ε_α. Expanding in θ, one gets

$$(6.15)\quad \left\{\begin{aligned} \phi(x,\theta) &= A(x) \\ &+ \bar{\theta}\psi(x) \\ &+ \frac{1}{4}\,\bar{\theta}\theta F(x) + \frac{1}{4}\,\bar{\theta}\gamma\,\theta G(x) + \frac{1}{4}\,(i\bar{\theta}\gamma_5\gamma_\nu\theta)\;A_\nu(x) + \\ &+ \frac{1}{4}\,\bar{\theta}\theta\bar{\theta}\;\chi(x) \\ &+ \frac{1}{32}\,(\bar{\theta}\theta)^2 D(x) \end{aligned}\right.$$

There are altogether (before any subsidiary conditions or equations of motion) 8 spinor and 8 boson components. Foregoing the difficulty about the nilpotence of x^μ, which does not involve (6.15), one finds that A(x), F(x) and D(x) are scalar fields, G(x) is a pseudoscalar and $A_\mu(x)$ an axial vector field. Besides these Bose fields, there are two (Dirac) spinor fields ψ and χ. A "Hermiticity" condition is imposed on the superfield,

$$(6.16)\qquad \phi(x,\theta)^+ = \phi(x,\theta)$$

where + implies besides complex conjugation a reversal of the order of anticommuting factors. The Bose fields then make 8 real components, and the spinors are Majorana spinors. Starting with a pseudoscalar $\phi(x,\theta)$, all parities would be inverted. One can also define $\phi^{\mu}(x,\theta)$, a "vector" superfield, or $\phi_{\alpha}(x,\theta)$, a "spinor" superfield, according to the Poincaré transformation properties

$$(6.17) \qquad \phi'(x',\theta') = \phi(x,\theta) \quad ; \quad \phi'_{\alpha}(x',\theta') = a_{\alpha}^{\beta}(\Lambda)\, \phi_{\beta}(x,\theta) \text{ etc....}$$

The variation of the fields in (6.16) can be found from the equations leading to (6.13). Identifying coefficients in (6.15) one finds,

$$(6.18) \quad \left\{ \begin{aligned} \delta A &= \bar{\varepsilon}\psi \\ \delta\psi &= -\varepsilon\partial\!\!\!/ A + \frac{\varepsilon}{2} F + \frac{\varepsilon}{2}\gamma_5 G + \frac{\varepsilon}{2} i\gamma_5\gamma_{\nu} A^{\nu} \\ \delta F &= \frac{\bar{\varepsilon}}{2}\chi - \bar{\varepsilon}\partial\!\!\!/\psi \\ \delta G &= -\frac{\bar{\varepsilon}}{2}\gamma_5\chi - \bar{\varepsilon}\gamma_5\partial\!\!\!/\psi \\ \delta A_{\nu} &= \frac{\bar{\varepsilon}}{2} i\gamma_5\gamma_{\nu}\chi - \bar{\varepsilon}i\gamma_5\gamma_{\nu}\partial\!\!\!/\psi \\ \delta\chi &= -\varepsilon(\partial\!\!\!/ F+\gamma_5\partial\!\!\!/ G) + \varepsilon i\gamma_5\gamma_{\nu}\partial\!\!\!/ A^{\nu} - \frac{\varepsilon}{2} D \\ \delta D &= -2\bar{\varepsilon}\partial\!\!\!/\chi \end{aligned} \right.$$

Notice that the numbers of fermion and boson components are always equal, as required by our study of the "little" algebra.

In counting components, we did not consider subsidiary conditions.

Indeed, the superfield $\phi(x,\theta)$ is not irreducible. It can be made irreducible by applying a covariant and supersymmetric condition,

$$(6.19) \qquad W_{\mu\nu}\phi = 0$$

where, using the representation $Q(\Lambda V)$

$$(6.20) \qquad W_{\mu\nu} = P_{\mu}W_{\nu} - P_{\nu}W_{\mu} \quad ; \quad W_{\mu} = \frac{1}{2}\varepsilon_{\mu\nu\rho\sigma}P_{\nu}J_{\rho\sigma} + \frac{1}{4}\bar{Q}i\gamma_5\gamma_{\mu}Q$$

This condition cancels 3 fields, which now make 4 fermion and 4 boson components:

$$D = 0 \qquad \chi = 0 \qquad A_{\nu} = 0$$

Covariant and superinvariant conditions can be constructed from powers of $Q(\Lambda V)$, $\bar{Q}(\Lambda V)$ and their chiral projections.

To construct supersymmetric couplings, one utilizes the above method of identifying coefficients of powers of θ. For instance, if

$$\Phi_3(x,\theta) = \Phi_1(x,\theta)\Phi_2(x,\theta)$$

we can identify

$$A_3(x) = A_1(x)A_2(x)$$

$$\psi_3(x) = A_1\bar{\psi}_2 + \bar{\psi}_1 A_2$$

etc.

Note that since the variation of D in (6.18) was only a divergence, the "D_3" component can be used as a Lagrangian density. (Wess and Zumino, 1974a). For the case $\phi_1 = \phi_2$, and $W_{\mu\nu}\phi = 0$ one finds,

$$"D_3" = \varepsilon^{\alpha\beta\gamma\delta}\left(-\frac{1}{4}A\partial^2 A + \frac{i}{2}\bar{\psi}\chi\psi + \frac{1}{4}(\partial_\mu B)^2 + F^2 + G^2\right) =$$

or

$$= \frac{1}{2}(\partial_\mu A)^2 + \frac{1}{2}(\partial_\mu B)^2 + \bar{\psi}i\partial\!\!\!/\psi + 2F^2 + 2G^2 - \frac{1}{2}\partial_\mu(A\partial_\mu A) \tag{6.21}$$

which is indeed an example for a Lagrangian density. The fields F and G have no dynamics, and satisfy equations of motion

$$F = 0 \quad , \quad G = 0$$

Note that the equations of motion for ψ, A, and B reduce the (massless) states to one fermion and one boson.

The $\Phi(x,\theta)$ are reducible. One can also work with chiral projections, by imposing conditions

$$\frac{1}{2}(1\mp i\gamma_5)_{\alpha\beta}\, Q^V_\beta\phi = 0 \tag{6.22}$$

where Q^V stands for the ΛV representation of Q. These superfields are now irreducible. The scalar (i.e., no spinor or vector index on ϕ itself) superfield ϕ_R is then composed of A_-, ψ_R, and F_-. They transform according to

$$\begin{cases} \delta A_\pm = \bar{\varepsilon}\psi_{L,R} \\ \delta\psi_{L,R} = \gamma_{L,R}(F_\pm - i\partial\!\!\!/ A_\pm)\varepsilon \\ \delta F_\pm = -\bar{\varepsilon}i\partial\!\!\!/\psi_{L,R} \end{cases}$$

We identify $\phi_- = \phi_R$, $\phi_+ = \phi_L$, i.e., $\phi_- = (\phi_+)^*$, though one could also have unconnected projections.

We refer the reader to the above-mentioned articles (Ferrara, Wess, Zumino, 1974; Salam and Strathdee, 1974e; see also Nilsson and Tehrakian, 1975; O'Raifeartaigh, 1974) for other examples of superfields, both spinorial ϕ_α, ϕ^μ_α etc. and tensorial ϕ^μ, $\phi^{\mu\nu}$ etc.. Furthermore, Capper (1974) has developed Feynman diagrams reproducing the superfield couplings; these are economical when studying the divergences of multiloop diagrams.

Considering the physical complications involved in the use of the Grassmann algebra substrate, it may be necessary at some stage to possess a formalism producing the field multiplets directly from the GLA. One can use the Q^L, Q^R set. To construct non-unitary irreducible field multiplets (Salam and Strathdee, 1974e) one applies Q^a_R and Q^{a*}_R to a "lowest" representation $\mathcal{D}(j_1,j_2)$ of the proper Lorentz group. Assuming

$$Q^a_R\, \phi(x)_{j_1 j_2} = 0$$

we get 4 submultiplets: 2 from the action of Q^1_L, Q^2_L (in$(\frac{1}{2},0)$) and one from their joint action $\sim(0,0)$, plus the original $\phi(x)_{j_1 j_2}$. The total dimensionality is thus $4(2j_1+1)(2j_2+1)$. One can also have a supermultiplet with inverted parities by starting with

$$Q^a_L\, \phi(x)_{j_1 j_2} = 0$$

These representations are however generally reducible. One can extract pieces by contraction with powers of $\frac{\partial}{\partial x^\mu}$, i.e., graded analogs of subsidiary conditions.

In constructing irreducible representations, it is important to recall that considering as in 2A the boson and fermion states as forming a 2-dimensional graded vector space V, the boson and fermion quantum fields $\phi(x)$ and $\psi(x)$ themselves represent $End_0 V$ and $End_1 V$ operators respectively. Indeed, one may recover the entire (6.18) set, without the ε parameters, by bracketing the Q_α directly with the fields $\psi(x)$, $A(x)$, etc.. Summing up, for G_k a GLA generator,

$$(6.24) \qquad [G_k, \Phi(x)]_{j_1 j_2} = -(-1)^{2(j_1+j_2)k} [\Phi_{j_1 j_2}(x), G_k]$$

7. Inclusion of internal symmetries. Let the indices i, j = 1...n denote an internal symmetry such as the SU(2) of I-spin, or SU(3). We then have, in addition to (4.5) and (4.7) a set (Salam and Strathdee, 1974b)

$$(7.1) \qquad \begin{cases} \{Q_{\alpha i}, Q_{\alpha j}\} = -2\delta_{ij}(\gamma_\mu C)_{\alpha\beta} P^\mu \\ [P_\mu, Q_{\alpha i}] = 0 \end{cases}$$

Restricting the system to rest states, we get a Clifford algebra C_{4n}, whose dimensionality is 2^{4n} and whose matrix representation acts on a 2^{2n} vector space.

$$(7.2) \qquad \{Q_{\alpha i}, Q_{\alpha j}\} = 2\delta_{ij}\,\delta_{\alpha\beta}\, M$$

Thus, for isospin (SU(2)) and assuming that the $Q_{\alpha i}$ transform as an isospinor (n = 2), we find the symmetry realized over a 16-dimensional carrier space (the Clifford algebra will have 256 base elements $Q_{\alpha i}$, $i[Q_{\alpha i}, Q_{\alpha j}]$, etc...). In fact, we can start with any (j,I) multiplet as the lowest state, and construct a representation with 16(2j+1)(2I+1) dimensions. The quantum numbers of the states in the case j = 0, I = 0 are given by the action of the 2n raising operators only; their graded products form a smaller Clifford algebra C_{2n}, whose dimensionality is indeed 2^{2n} (= 16 for I-spin), which will indeed create the 2^{2n} states of the carrier space. This enables us to get their quantum numbers directly: $(\frac{1}{2},\frac{1}{2})$, $\Lambda^2(\frac{1}{2},\frac{1}{2})$, $\Lambda^3(\frac{1}{2},\frac{1}{2})$, $\Lambda^4(\frac{1}{2},\frac{1}{2})$. In this case these are just the 16 matrices of the Dirac-Clifford algebra. They reduce to $(j,I)^P$ multiplets:

$$(0,0)^+ \oplus (\tfrac{1}{2},\tfrac{1}{2})^{\eta} \pm (1,0)^- \pm (0,1)^- + (\tfrac{1}{2},\tfrac{1}{2})^{-\eta} + (0,0)^+$$

Going back to the C_{4n} of (7.2) we note that $\Lambda^2 Q_{\alpha i}$ will form the Lie algebra $SO(8) \supset SO(6) \sim SU(4)$, so that the 16 states can be grouped in SU(4) (Wigner) supermultiplets 1 + 4 + 6 + 4* + 1.

Indeed, we can use a generalization of (4.7)

$$\text{(7.3)} \qquad \begin{aligned} \{Q_{ai},Q^*_{bj}\} &= 2\delta_{ab}\,\delta_{ij}\,M \quad , \quad a,b = 1,2 \\ \{Q_{ai},Q_{bj}\} &= 0 \quad , \quad \{Q^*_{ai},Q^*_{bj}\} = 0 \end{aligned}$$

for rest states. Here we have the same number of odd generators 4n, the results are the same except that $\Lambda^2 Q$ now contains $i[Q_{ai},Q^*_{bj}] = S^{bj}_{ai}$ which is clearly the su(n) algebra, the rest of SO(8) being given by [Q,Q] and $[Q^*,Q^*]$. Note that this "little" GLA now has $Q_{ai} \in L_{-1}$; $Q^*_{ai} \in L_1$; $1, S^{bj}_{ai} \in L_0$.

The (7.3) bracket can be generalized for cases where the representation $\underset{\sim}{n}$ differs from $\underset{\sim}{n}^*$, such as the SU(3) case:

$$\text{(7.4)} \qquad \{Q_{ai},Q^*_{\beta j}\} = 2\delta_{\alpha\beta}\,\delta_{ij}\,M \quad , \quad \alpha = 1..4$$

The Clifford algebra is now C_{8n}, $d = 2^{8n}$, acting on a 2^{4n} dimensional carrier-space. Salam and Strathdee (1974f) have constructed the O(3) case (fitting 6.69) and discussed the totally-antisymmetric features of the multiplets, due to the graded commutativity and filtered structure of the Clifford algebra. It seemed difficult to reconcile with the physical states in the quark model assignments. However, it was soon noted (Wess, 1974) that if one introduces $SU(3)_{color} \otimes SU(3)_{GN}$, the totally antisymmetric representations will indeed contain the observed states whenever the color indices will contract or antisymmetrize to a singlet.

8. Applications of supersymmetry - General symmetry considerations. All supersymmetric models based upon $\mathcal{W}$ or its extension by internal degrees of freedom have in common two simplifying features:

$$(8.1) \qquad [P^{\mu}, Q_{\alpha i}] = 0$$

and

$$(8.2) \qquad H = \sum_{\alpha,i} Q_{\alpha i}^2$$

Conservation is thus guaranteed. In the case of the R_α of (4.5), which do not commute with H, conservation is ensured by

$$(8.3) \qquad \frac{d}{dt}[K^{\mu}, Q_{\alpha}] = (-\gamma_5\gamma^{\mu})_{\alpha\beta}\frac{d}{dt}R_{\beta} = 0$$

These examples can be generalized in the following theorem: "A GLA $\mathcal{G}$ is conserved if its even subalgebra $\mathcal{L}$ (the Lie algebra) is conserved, and if its odd generators $\mathcal{O}$ transform irreducibly under $\mathcal{L}$ and contain at least one non-nilpotent generator $\mathcal{O}_a$."

Clearly, $[\mathcal{O}_a, \mathcal{O}_a] \subset \mathcal{L}$ and doesn't vanish, so that $\frac{d}{dt}\mathcal{O}_a = 0$, leading to $\frac{d}{dt}\mathcal{O}_i = 0$ for all i, through the action of $\mathcal{L}$.

We now discuss the role of the Noether theorem (for recent advances see J. Schwinger, 1951; Orzalesi, 1970; Y. Dothan, 1972; J. Rosen, 1974) in the case of a GLA, and in particular for $\mathcal{W}$. From (6.21) as a Lagrangian we find the conserved (spinor-vector) current,

$$(8.4) \qquad j^{\mu}_{\alpha}(x) = ((\gamma^{\lambda}\partial_{\lambda}(A(x)-B(x)\gamma_5)\gamma^{\mu}\psi(x)))_{\alpha} - 2i(((F(x)+\gamma_5 G(x))\gamma^{\mu}\psi))_{\alpha}$$

and

$$(8.5) \qquad Q_{\alpha} = \int d^3x\, j^0_{\alpha}(x)$$

It has recently been shown by Ferrara and Zumino (1974b) that this current belongs to a GLA ($\mathcal{V}$) supermultiplet which includes the energy momentum tensor and the axial vector current.

The inverse Noether theorem yields either a Lie algebra or a GLA, according to whether the conserved currents (or charges) all have integer spin, or contain a subset with half-integer spin. This results from the same considerations as in the discussions leading to (6.24).

The GLA V and its extensions represent algebras which contain the Poincaré algebra P, or P and F (the $SU(3)_{GN}$ algebra, or even $SU(3)_{GN} \otimes SU(3)_{color}$) as subalgebras. As GLA, they do not come directly under the cases which have been studied and classified by L. O'Raifeartaigh (1965) or under the No-go theorem of S. Coleman and J. Mandula (1967). However, Goddard (1974) has constructed the Lie algebra "equivalent" to V, i.e., having the same vector space as carrier-space for their representations. According to Levi's theorem, any Lie algebra E can be written uniquely as a semi-direct sum

(8.6) $\quad E = \Lambda + \Sigma$

where Λ is semi-simple, and Σ solvable, i.e., for $\Sigma^{(1)} = \Sigma$, $\Sigma^{(n)} = [\Sigma^{(n-1)}, \Sigma^{(n-1)}]$ a commutator bracket, $\Sigma^{(n)} = 0$ for some n. O'Raifeartaigh then proves that there are 4 classes of inclusions of $P \subset E$: $(P = J^{\rho\sigma} \oplus P^{\mu})$

(1) $J^{\rho\sigma} \subset \Lambda$; $P^{\mu} = \Sigma$

(2) $J^{\rho\sigma} \subset \Lambda$; $P^{\mu} \subset \Sigma$, $\vec{\Sigma} - P^{\mu} \neq 0$, $[\Sigma^{\nu}, \Sigma^{\lambda}] = 0$

(example: inhomogeneous isl (6,c), with 72 "translations")

(3) $J^{\rho\sigma} \subset \Lambda$; $P^{\mu} \subset \Sigma$, $\Sigma^{(n)} = 0$

(4) $P \cap \Sigma = 0$

(example: the conformal algebra su(2,2))

Goddard shows that our GLA is equivalent to imposing a solvable class (3) symmetry as a L.A. The O'Raifeartaigh theorem then forbids mass-splitting within a multiplet, if at least one state has a discrete m^2 eigenvalue for $P_{\mu}P^{\mu}|1\rangle$. However, we can deduce the same result directly from (8.1) for V and any extension by F, provided (8.1) holds. Haag and collaborators (1975) have since refined this result and shown that the method we used in chapter 7 is the only allowed one for the inclusion of internal symmetries in a non-trivial way. The supersymmetry cannot be broken linearly.

The Coleman-Mandula (1967) theorem is not applicable. This is because it requires a Hilbert space, and we see in (6.18) that the fields contain Grassman elements $\bar{\varepsilon}$ and ε, even after the extraction of the θ_α monomials. The Hilbert space thus also acquires such a structure, with elements of $V_5 \oplus V_6$ (the additional dimensions in the Grassman space) appearing instead of complex numbers as coefficients of the Fock space states. The physical states are those with complex numbers as coordinates, since only such states can give complex numbers for amplitudes. There is an "inner product" on the entire Fock space, which takes values in $V_5 \oplus V_6$, but reduces to an ordinary inner product on the physical states. The S_α satisfy the hermiticity conditions

$$(8.7) \qquad \langle\Psi_1,\bar{\xi}S\Psi_2\rangle = \langle\bar{\xi}S\Psi_1,\Psi_2\rangle$$

The group thus acts unitarily, preserving this inner product. The Coleman-Mandula (1967) theorem does not apply because (8.7) takes values in $V_5 \oplus V_6$. The conventional Hilbert subspace by itself is not invariant under the group action.

Goddard succeeds in defining a complex-valued inner product in a quadrupled Hilbert space (one each for v_5, v_6, $v_5\Lambda v_6$, 1), but loses positive-definiteness. In either case, the Coleman-Mandula theorem doesn't apply. This is why we have a structure which is not locally isomorphic to just $P + F$.

9. <u>Improved renormalizability in a Yukawa and ϕ^4 interaction.</u> The first example of a supersymmetric interaction was provided by Wess and Zumino (1974b). They added to the free Lagrangian (6.21)

$$(9.1) \qquad L_{free} = \frac{1}{2}(\partial_\mu A)^2 + \frac{1}{2}(\partial_\mu B)^2 + i\bar{\psi}\partial\!\!\!/\psi + 2F^2 + 2G^2$$

a mass term

$$(9.2) \qquad L_m = 2m(FA + GB - \frac{1}{2}\bar{\psi}\psi)$$

and an interaction

$$(9.3) \qquad L_g = g[F(A^2-B^2) + 2GAB - \bar{\psi}(A-\gamma_5 B)\psi]$$

These terms all transform invariantly up to a 4-divergence, under (6.18) as amended through the introduction of the field B. One can also add a term (see δF in (6.23)).

$$(9.4) \qquad L_\lambda = \lambda F$$

A and F are scalar fields, B and G are pseudoscalars, and ψ is a Majorana spinor. F and G are auxilliary and satisfy the equations of motion,

$$- F = \frac{g}{4}(A^2-B^2) + \frac{m}{2} A + \frac{\lambda}{4}$$

$$- F = \frac{g}{2} AB + \frac{m}{2} B$$

Eliminating F and G from the Lagrangian, we find,

$$(9.5) \qquad L = \frac{1}{2}(\partial_\mu A)^2 + \frac{1}{2}(\partial_\mu B)^2 + \bar{\psi}(i\not{\partial}-m)\psi - \frac{1}{2} m^2 (A^2+B^2)$$

$$- \frac{1}{2} gm\, A(A^2+B^2) - \frac{1}{8} g^2 (A^2+B^2)^2 - g\bar{\psi}(A-\gamma_5 B)\psi$$

$$- \frac{\lambda}{2}\left[\frac{\lambda}{4} + mA + \frac{g}{2}(A^2-B^2)\right]$$

which represents a non-linear realization of supersymmetry, corresponding to the elimination of F and G in the linear (6.18). We can regroup the part of the "potential" which involves the A and B fields only,

$$(9.6) \qquad V = - L(A,B) = \frac{m^2}{2} A + \frac{\lambda}{2m}^2 + \frac{m^2}{2} B^2 + g\frac{\lambda}{4}(A^2-B^2) +$$

$$+ \frac{mg}{2} A(A^2+B^2) + \frac{g^2}{8}(A^2+B^2)^2$$

$-L(A,B)$ is the "potential" V whose extrema we shall later study in our search for Goldstone-like solutions. Note that the λF term can be eliminated by a shift in A. Salam and Strathdee (1974e) have shown how to derive (9.5) using the superfield calculus. It results from writing

(9.7) $$L = \frac{1}{8}(\bar{Q}Q)^2(\phi_+\phi_-) - \frac{1}{2}\bar{Q}Q(P(\phi_+) + P(\phi_-))$$

with $\phi_- = \phi_+^*$, $Q = Q(\Lambda V)$, P is a polynomial of order 3. It is apparent that the 4-volume integral vanishes, trivially, so that

$$\delta \int d^4xL = \int d^4x\, \bar{\varepsilon}Q(\Lambda V) = \bar{\varepsilon}\frac{\partial}{\partial\bar{\theta}} d^4xL + \text{surface term} = 0$$

The relevant terms in L are obtained by setting $\theta = 0$, yielding (9.5).

Before we study the effects of renormalization (and disregarding the L_λ term at this stage), we already observe in (9.5) the expected result of a symmetry: A, B, and ψ have related bare masses. The three interactions (ϕ^3, ϕ^4 and the Yukawa term) have related couplings $\frac{1}{2}$ gm, $\frac{1}{8}g^2$, g. Supersymmetry thus does indeed play the role of a symmetry (which we can interpret as a discrete symmetry, using the algebra as a transposition matrix algebra. After elimination of F and G, the conserved current is

(9.8) $$j^\mu_\alpha = (\not\partial(A-\gamma_5B)\gamma^\mu\psi + im(A+\gamma_5B)\gamma^\mu\psi + \frac{i}{2}g(A+\gamma_5B)^2\gamma^\mu\psi)_\alpha \;;\; \partial_\mu j^\mu_\alpha = 0$$

The conservation equation can be checked directly, using the equations of motion and the identity

(9.9) $$\psi(\bar{\psi}\psi) \equiv \gamma_5\psi(\bar{\psi}\gamma)\psi$$

Wess and Zumino (1974b) showed that the theory of (9.5) is less divergent than if the masses and couplings were independent. For instance, in the one-loop approximation, the quadratic divergence of the mass renormalization for A and B cancels out. The logarithmic divergence of the vertex correction to the Yukawa interaction also cancels between the A and B terms, leaving a finite vertex correction.

In its original form, before elimination of F and G, the theory can be regularized (by the method of Pauli and Villars, for instance) without spoiling supersymmetry. Thus, the Ward identities following from (9.8) in perturbation theory are expected to be satisfied. If one uses $L_{\text{free}} + L_m$ as the unperturbed

Lagrangian, one finds as propagators

$$\langle AA\rangle = \langle BB\rangle \sim \Delta_c$$

$$\langle FF\rangle = \langle GG\rangle \sim \Box\Delta_c$$

$$\langle AF\rangle = \langle BG\rangle \sim -m\Delta_c$$

In the one loop approximation, there is only one renormalization needed, a logarithmically divergent wave function renormalization constant Z, common to A, B, ψ, F and G,

$$(9.10)\quad \begin{cases} Z = 1 - 4g^2 I \\ I = -i\int \frac{d^4k}{(2\pi)^4}\frac{1}{(k^2+m^2)^2} = \frac{1}{16\pi^2}\int^{\infty}\frac{d\chi}{\chi} \end{cases}$$

No diagonal mass is generated for either A or B.

The quadratic divergence of the self-energy cancels out and the remaining logarithmically divergent contribution is proportional to $-p^2$. Similarly, the ψ self-energy is proportional to $i\gamma^\mu p_\mu$, and the corrections to the off-diagonal mass terms mFA and mGB cancel. Thus the only mass renormalization is that due to the wave function renormalizations,

$$m_r = mZ$$

Corrections to gFA^2, $-gFB^2$, 2gGAB cancel, and the finite corrections to the Yukawa terms vanish for zero external momenta. One finds

$$g_r = gZ^{3/2}$$

No divergent trilinear or quadrilinear interactions are generated. Iliopoulos and Zumino (1974) and Tsao (1974) have investigated this model in higher orders. For two-loop diagrams they calculated explicitly the various contributions and again found no mass and vertex corrections. They proved to all orders that the theory is renormalized with one single renormalization constant, Z, the wave function renormalization. Note that theories like (9.1) etc. are renormalizable even without supersymmetry (i.e., with arbitrary m_i and g_{ijk}), but supersymmetry has resulted in highly improved renormalizability. There is thus a possibility that some a-priori non-renormalizable model might become renormalizable when supersymmetry is imposed. No such case has been discovered to-date.

The full set of Ward identities corresponding to V supersymmetry has been derived by Iliopoulos and Zumino. They have also adapted a regularization scheme based upon the insertion of higher derivative terms in L, in particular in the kinetic energy term L_{free}. They use the insertion

$$L_\xi = \xi[\frac{1}{2}(\partial_\mu \Box A)^2 + \frac{1}{2}(\partial_\mu \Box B)^2 + i\Box\bar{\psi}\not{\partial}\Box\psi + 2(\Box F)^2 + 2(\Box G)^2$$

L_ξ transforms like L_{free} under the Q_α. It is sufficient to make all diagrams finite, including tadpoles.

Explicit symmetry breaking (in contradistinction to "spontaneous" breaking) is tried by the above authors in the form of a term

$$L_{SB} = cA \tag{9.11}$$

(rather than L_λ, which was invariant under V). L_{SB} is not invariant under V, and breaks current conservation,

$$\partial_\mu j^\mu = c\psi \tag{9.12}$$

However, the entire renormalization program is unaffected, with only finite corrections appearing due to L_{SB}. The masses are now only related by the equation

$$m_A^2 + m_B^2 = 2m_\psi^2 \tag{9.13}$$

derived in the tree approximation. In higher order the equation gets finite corrections.

The L_{SB} term can be eliminated by a simultaneous shift of A and F, $A \to A + a$, $F \to F + f$, with the equations

$$\begin{cases} 4f + 2ma + ga^2 = 0 \\ 2mf + 2gaf + c = 0 \end{cases}$$

which ensure vanishing of linear terms in A or F. Eliminating f we get a cubic equation for a,

$$a(2m+ga)\{(m+ga) + \frac{c}{2}\} = 0$$

Taking the limit $c \to 0$, this has three solutions,

$$a_1 = 0 \quad , \quad a_2 = -\frac{2m}{g} \quad , \quad a_3 = -\frac{m}{g}$$

(a_3 is the "central" value). Taking in $B = \psi = 0$ and $A \to a$ we have a "potential" $-L(a) = V(a)$

$$V(a) = \frac{1}{2}m^2a^2 + \frac{1}{2}gm\,a^3 + \frac{1}{8}g^g a^4 + ca = \frac{1}{2}a^2(m + \frac{1}{2}ga)^2 + ca$$

Our solutions a_i correspond to the stationarity points of V(a). We see that $V(a_1) = 0$, $V(a_2) = -\frac{2mc}{G} \to 0$, $V(a_3) = \frac{1}{8}\frac{m^4}{g^2} - c\frac{m}{g} \to \frac{1}{8}\frac{m^4}{g^2}$ so that a_1 and a_2 produce minima, and a_3 is a maximum. This is unstable, with no possible stabilization through a sign change. From (9.5) we see that (for $c \to 0$, i.e., vanishing of explicit symmetry breaking)

$$m\psi = m + ga \to 0 \quad \text{for } a_3$$

so that this is a "Goldstone spinor" solution, which is however unstable. Notice that one of the two bosons has to be a tachyon, if the other one is massive. Indeed, we have to first order in g

$$m_A^2 = -m^2 - 3gma - \frac{3}{2}g^2a^2 \quad ; \quad \text{for} \quad a_3,\ m_A^2 = \frac{1}{2}m^2$$

$$m_B^2 = -m^2 - 3gma - \frac{1}{2}g^2a^2 \quad ; \quad \text{for} \quad a_3,\ m_B^2 = -\frac{1}{2}m^2$$

Salam and Strathdee (1974c) have investigated directly the idea of a Goldstone spinor in that same Lagrangian, with similar results.

10. Physics results. We shall not review here the large number of articles in which work has continued on the physical applications of supersymmetry. The main results as they stand in mid-1975 are as follows:

1) It has been found that the number of basic couplings (or Lagrangians) is practically limited to two: a coupling involving only "scalar" superfields (i.e., in which the Poincaré spins are 0 and $\frac{1}{2}$) and a coupling in which a "vector" superfield (i.e., involving spins $\frac{1}{2}$ and 1) appears. This is sometimes referred to as a "gauge" coupling, because the J = 1 components can become a Yang-Mills field for Abelian or non-Abelian local gauges.

2) Internal symmetries are included trivially in the gauge case. This implies that the entire ($J = 1 \dotplus J = \frac{1}{2}$) superfield behaves like the regular representation of the internal symmetry, not just the J = 1 field. This is perhaps an important physical clue, since it is the only reason I know which could explain the fact that the lowest baryons are in an SU(3) octet.

3) The study of renormalization in supersymmetric Lagrangians has yielded several examples of improved convergence of theories. Moreover, Zumino (1974b) has shown that in exact supersymmetry, the physical expectation values are given entirely by the tree diagrams. Also, the sum of all vacuum diagrams vanishes identically, which is cosmologically interesting.

4) Spontaneous symmetry breaking works for the internal symmetry, with Higgs-Kibble mechanisms. For supersymmetry itself, it can be broken spontaneously and then requires a massless Goldstone fermion (the neutrino?) as shown by Fayet and Iliopoulos (1974). Rather than a gauge role, as sought by Volkov and collaborators, the neutrino thus seems to fit the Goldstone role we were guessing at originally. P. Fayet (1975) has recently shown that spontaneous supersymmetry breakdown can be achieved whether or not a local gauge is present, and independently of the semi-simplicity of the group. Salam and Strathdee (1975a) have recently classified the various methods of spontaneous breaking of supersymmetry.

5) The inclusion of fermion number N in V or its extensions creates difficulties (Salam and Strathdee, 1975b). It yields values 0, 1, 2 for the right-handed components A_+, ψ_+, F_+ and 2, 1, 0 for the left-handed A_-, ψ_-, F_-. The

exclusion of the F field by a subsidiary condition thus creates a parity-violating set of assignments and an anomaly in N = 2 for the basic spinless left-handed field. The parity difficulty can be turned but the overall situation appears muddled.

6) No actual physical assignments can be tried at this stage. It is as yet not even clear whether the theory should be applied to the fundamental fields (e.g., quarks, and thus fix the composition of the gluons) or the the phenomenological fields.

REFERENCES

Aharonov, Y.,A.Casher and L.Susskind, 1971, Phys. Lett. 35B, 512

Bella,G., 1973, Nuovo Cimento 16A, 143

Berezin,F.A., and G.I.Kac, 197o, Mat.Sbornik (USSR) 82, 124 (English translation 197o, 11, 311)

Boerner,H., 1963, Representations of Groups, North Holland,Pub., Amsterdam, Ch. VIII

Capper, D.M., 1974, Trieste report, IC/74/66

Coleman,S., and J.Mandula, 1967, Phys. Rev. 159, 1251

Corwin,L., Y.Ne'eman and S.Sternberg, 1974, Tel Aviv Univ.report TAUP 448-74

Corwin,L., Y.Ne'eman and S.Sternberg, 1975, Rev.Mod.Phys.

Dolen,R., D.Horn and C.Schmidt, 1967, Phys.Rev.Letters 19, 4o2

Dothan,Y., 1972, Nuovo Cimento A11, 499

Fayet,P. and J.Iliopoulos, 1974, Phys.Lett. 51B, 461

Fayet,P., 1975, PTENS preprint 75/1, Jan.75

Ferrara,S.and E.Remiddi, 1974, Cern report TH 1935; to be pub.

Ferrara,S., J.Wess and B.Zumino, 1974,Phys.Lett. 51B, 239

Ferrara,S.and B.Zumino, 1974, Nucl.Phys. B79, 413

Ferrara,S.and B.Zumino, 1974b, CERN report TH 1947

Fock,V., 1932, Zeit.Physik, 75, 622

Frölicher,A.,and A.Nijenhuis,1957,Proc.Nat.Acad.Sci.USA,43, 239

Gell-Mann,M.,and Y.Ne'eman,1964,The Eightfold Way,W.A.Benjamin pub., N.Y., London.

Gervais,J.L.,and B.Sakita, 1971, Nucl.Phys. B34, 633

Goddard,P.,1974,Berkeley report LBL-3347;to be pub.in Nucl.Phys. B.

Haag,R.,J.T.Lopuszański,and M.Sohnius,1974, Karlsruhe preprint.

Iliopoulos,J.,and B.Zumino, 1974, Nucl.Phys. B76, 31o

Igi,K. and S.Matsuda, 1967, Phys.Rev.Lett. 18, 625

Jordan,P., and E.Wigner, 1928, Zeit.Physik 47, 631

Joseph,A., 1972, Nuovo Cimento 8A, 217

Logunov,A., L.D.Soloviev and A.N.Tavkhelidze,1967,Phys.Lett.24B,181

Miyazawa,H.and H.Sugawara,1965,Prog.Theor.Phys. 34,263

Neveu,A.,and J.H.Schwarz, 1971, Nucl.Phys. B31, 86

Nijenhuis,A.,1955,Proc.Royal Netherlands Acad.of Sci., A58 3.

Nilsson,J.S.,and D.H.Tehrakian, 1975, to be pub.in Nucl.Phys. B.

O'Raiferartaigh,L., 1965, Phys.Rev. 139B, 1o53

O'Raiferartaigh,L., 1974,"Weight-Diagrams for Superfields",DIAS report

Orzalesi,C.A., 197o, Review. of Mod.Phys. 42, 381

Ramond,P., 1971, Phys.Rev. D3, 2415

Rebbi,C., 1974, Phys.Reports 12C, 1

Rosen,J., 1974, Annals of Phys. 82, 54, and 82, 7o

Rühl,W., and B.C.Yunn, 1974,Trier-Kaiserslautern-University report, to be published.

Salam,A., and J.Strathdee, 1974a, Nucl.Phys. B76, 477

Salam,A., and J.Strathdee, 1974b, Nucl.Phys. B8o, 499

Salam,A., and J.Strathdee, 1974c, Phys.Lett., 49B, 465

Salam,A., and J.Strathdee, 1974d, Trieste report IC/74/36, to be pub. in Phys.Lett.

Salam,A., and J.Strathdee, 1974e,Trieste report IC/74/42,to be pub.

Salam,A.,and J.Strathdee,1975a, Trieste report IC/75/38

Salam,A.,and J.Strathdee,1975b, Trieste report IC/75/42

Schwarz,J.H., 1973, Phys.Reports 8C, 269

Schwinger,J., 1951, Phys.Rev. 82, 914

Singh,V., 1965, Phys.Rev.Lett. 15, 271

Tsao,H.S., 1974, Brandeis Univ. report.

Veneziano,G., 1974, Phys.Reports 9C, 199

Volkov,D.V., and V.P.Akulov, 1973,Phys.Lett. B46, 1o9

Volkov,D.V.,and V.A.Soroka,1973,JETP Lett.18,529(English transl.18,312)

Wess,J.and B.Zumino, 1974a,Nucl.Phys. B7o, 39

Wess.J.and B.Zumino, 1974b, Phys.Lett. 49B, 52

Wess,J.and B.Zumino,1974c,CERN report TH 1857;to be pub.in Nucl.Phys.B.

Wess.J., 1974, Lecture at Karlsruhe Summer School.

Woo,G., 1974, Cambridge University report DAMTP 74/12

Zumino,B., 1974, CERN report TH 19o1, Proceedings of the XVII International Conference on High Energy Physics, London.

Zumino,B., 1974b, CERN report TH 1942

---

* This research was supported by a grant from the United States-Israel Binational Science Foundation (BSF), Jerusalem, Israel

Some Recent Results on Supersymmetry

Shlomo Sternberg *)

This paper is divided into two parts. In the first part we give a sketchy descripiton of some results proved within the last year in the field of supersymmetry, several of which are quite striking. In the second part, we give some details showing how the theory of produced representations of Lie algebras, as developed by Blattner, can be modified so as to apply to superalgebras. Two major results are the complete classification of the simple Lie superalgebras, obtained independently by Kaplansky and Kac (with substantial partial results obtained as well as by Nahm, Rittenberg and Sheunert) and the construction of an entire theory of graded differential geometry by Kostant. Since Kostant's paper in this volume describes his results in detail there is no need to present them here. However, the theory of produced representations as given in the second part of this paper can be used by the reader as a sort of introduction to some of Kostant's methods. Indeed, an examination of our methods will show that when we use the universal enveloping algebra of a superalgebra we frequently make use of its structure as a graded Hopf algebra. Kostant replaces this graded Hopf algebra by a larger one which incorporates the group structure associated with the even part of the superalgebra and uses the representation properties of this graded Hopf algebra as one of the key ingredients of his theory.

By a <u>Lie superalgebra</u> we mean a graded algebra $G = \oplus G_i$ (where $i \in \mathbb{Z}$, in which case we talk of a $\mathbb{Z}$ graded superalgebra or $i \in \mathbb{Z}_2$ in which case we talk of a $\mathbb{Z}_2$ graded superalgebra) with an operation [,] which satisfies the axioms

$$[a,b] = -(-1)^{ij}[b,a]; \quad [a,[b,c]] = [[a,b],c] + (-1)^{ij}[b,[a,c]]$$

for $a \in G_i$, $b \in G_j$.

*) Department of Physics and Astronomy, University of Tel Aviv and Department of Mathematics, Harvard University.

Subalgebras, homomorphisms, etc. of Lie superalgebras are assumed to be compatible with the $\mathbb{Z}$ or $\mathbb{Z}_2$ - grading. A Lie superalgebra is called simple if it does not contain nontrivial ideals. In what follows we always exclude the "trivial" case $G = G_0$. In the first part we shall be principally concerned with $\mathbb{Z}_2$ graded superalgebras. For examples see Corwin-Ne'eman-Sternberg [1] where some mathematical and physical applications of these objects are given. Let $V = \oplus V_i$ be a $\mathbb{Z}_2$ - graded space, $\dim V_0 = m$, $\dim V_1 = n$. The algebra End V becomes a $\mathbb{Z}_2$ - graded algebra if we set $(\text{End } V)_i = \{a \in \text{End } V \mid a V_s \subset V_{s+i}\}$. We shall denote the corresponding Lie superalgebra by L(V) or L(m, n), where the bracket is the graded commutator, cf. [1].

We now define the supertrace $T : L(V) \to \mathbb{C}$. We choose a basis in V which is compatible with the $\mathbb{Z}_2$ - grading. Let $\begin{pmatrix} \alpha & \beta \\ \gamma & \delta \end{pmatrix}$ be the matrix of the operator a in this basis. We set $T(a) = \text{Tr}\, \alpha - \text{Tr}\, \delta$.

The bilinear form $(a, b) = T(ab)$ on L(V) has the invariance property: $([a,b], c) = (a, [b, c])$.

Kac [5] calls a Lie superalgebra $G = G_0 \oplus G_1$ over the complex numbers classical if it is simple and the representation of G_0 on G_1 is completely reducible. Kac [5] classifies the finite dimensional simple complex superalgebras. There are two types: the classical superalgebras, and superalgebras which are the graded (and finite dimensional) analogues of the irreducible simple infinite Lie goups of Cartan, see for example Singer-Sternberg [7]. He calls this latter group Cartan Superalgebras. There are six infinite families of classical Lie superalgebras, which Kac denotes by A(n), A(m, n), B(m, n), C(m), D(m, n) and P(m) which we shall describe below. In addition there is a 40 dimensional exceptional superalgebra F(4), a 31 dimensional exceptional superalgebra G(3) and a continuous family of 17 dimensional superalgebras $D(1, 2, \alpha)$ (deformations of D(1, 2)) which are classical. (The 17 dimensional family was missing from Kac's original announcement and was pointed out by Kaplansky.) The following are the infinite families of the classical superalgebras:

In $L(m,n)$ we have the ideal $SL(m,n) = \{a \in L(m,n) \mid T(a) = 0\}$. We set $A(m,n) = SL(m+1, n+1)$ for $m \neq n$, and $A(m,m) = SL(m+1, m+1)/\langle E \rangle$ where E is the unit matrix. The Lie superalgebras $A(m,n)$ are simple for $m, n \geq 0$, $m+n > 0$.

Let f be a non-degenerate bilinear form on the space $V_0 \oplus V_1$ whose restriction to V_0 is skew symmetric, whose restriction to V_1 is symmetric, and such that $f(V_0, V_1) = 0$. In $L(m,n)$ we consider the subalgebra $OSp(m,n) = G_0 \oplus G_1$, where $G_s = \{a \in L_s(m,n) \mid f(a(x), y) + (\sqrt{-1})^s f(x, a(y)) = 0\}$. We set $B(m,n) = OSp(2m, 2n+1)$ $(n \geq 0)$; $C(m) = OSp(2m, 2)$; $D(m,n) = OSp(2m, 2n)$ $(n \geq 2)$. These are simple Lie superalgebras for $m > 0$.

The Lie superalgebra $A(n) = G_0 \oplus G_1$ as a space is the direct sum of two copies of the space of $(n+1)$-order matrices with trace 0; for two elements of G_1 we set: $[a,b] = ab + ba - \frac{2}{n+1} \mathrm{Tr}(ab) E$; in the other cases $[a,b] = ab - ba$. The algebra $A(n)$ is simple for $n \geq 2$.

Let $G_0 = sl_{m+1}$, and let G_{-1} (resp. G_1) be the space of skew symmetric (resp. symmetric) matrices of order $m+1$. We define a simple $\mathbb{Z}$-graded Lie superalgebra structure on the space $P(m) = G_{-1} \oplus G_0 \oplus G_1$ if we set: $[c_1, c_2] = c_1c_2 - c_2c_1$, $c_i \in G_0$, $[a,b] = ab$, $[c,a] = ca + ac^T$ $[c,b] = -c^Tb - bc$ for $a \in G_1$, $b \in G_{-1}$, $c \in G_0$.

As we have seen, the <u>Killing</u> form $(a,b) = T(\mathrm{ad}\, a \cdot \mathrm{ad}\, b)$ on a Lie superalgebra is invariant. Hence, on a simple Lie superalgebra it is either non-degenerate or zero; we consider these cases separately. Kac [5] proves that a Lie superalgebra with non-degenerate Killing form is an orthogonal direct sum of simple algebras, each of which is isomorphic to one of the classical Lie superalgebras $A(m,n)$ for $m \neq n$, $B(m,n)$, $C(m)$, $D(m,n)$ for $n - m \neq 1$, $F(4)$, $G(3)$.

In Corwin-Ne'eman-Sternberg [1] a number of examples were given of complex simple superalgebras which possessed non-trivial reducible representations which were not completely reducible. The problem of complete reducibility has been completely settled by Djokovic-Hochschild [8] who prove that a simple superalgebra has all its finite dimensional representations completely reducible if and only if it is the algebra $OSp(2m, 1)$. A number of mathematicians have proved that $OSp(2m, 1)$ can be characterized as being the only superalgebra with simple even part and non-degenerate Killing form (Kaplansky, Kac, Pais and Rittenberg, Djokovic). Kostant has proved the graded analogue of Ado's theorem. According to a letter from Hoschild to Kaplansky, this result appeared in the unpublished Ph.D. thesis of L. E. Ross, Berkeley 1964. The superalgebra $OSp(2m, 1)$ plays an important role in the study of the metaplectic representation as pointed out by Sternberg-Wolf [12].

We now turn to the Cartan superalgebras. Let $\Lambda = \oplus \Lambda_i$ be an arbitrary $\mathbb{Z}_2$-graded algebra. By the algebra of superdifferentiations of the algebra Λ we mean the subalgebra $\mathrm{Diff}\,\Lambda = \oplus\,\mathrm{Diff}_i\,\Lambda$ in the Lie superalgebra $L(\Lambda)$, where $\mathrm{Diff}_i\,\Lambda = \{a \in L(\Lambda) \mid a(xy) = a(x)y + (-1)^{is} xa(y) \text{ for } x \in \Lambda_s\}$ are the superdifferentiations of degree i.

Now let $\Lambda = \oplus \Lambda_i$ be the Grassmann algebra in n variables $x_1, \cdots, x_n$ with the natural $\mathbb{Z}_2$-grading. We set $W(n) = \mathrm{Diff}\,\Lambda$. The equations $\partial_i(x_j) = \delta_{ij}$ define a superdifferentiation ∂_i; any element in $W(n)$ can be written in the form $\Sigma P_i \partial_i$, $P_i \in \Lambda$.

Two different algebras of differentiable forms can be defined over Λ: Ω and S. The algebra Ω (resp. S) is the algebra over Λ generated by anti-commuting (resp. commuting) differentials $dx_1, \cdots, dx_n$ (resp. $\delta x_1, \cdots, \delta x_n$); we extend the $\mathbb{Z}_2$-grading from Λ if we set $\deg dx_i = 1$ (resp. $\deg \delta x_i = 0$). We define a differential d on Ω (resp. δ on S) to be the superdifferentiation of degree 0 (resp. 1) for which $d(x_i) = dx_i$ and $d^2(x_i) = 0$ (resp. $\delta(x_i) = \delta x_i$ and $\delta^2 = 0$). Any superdifferentiation D of degree s on the algebra Λ extends to a superdifferentiation of the algebra Ω (resp. S) using the condition $D\,dx_i = d\,Dx_i$ (resp. $D\delta = (-1)^s \delta D$).

Let $\omega \in \Omega$ or S; we set $L(\omega) = \{D \in W(n) \mid D\omega = 0\}$. We define the following Lie superalgebras: $S(n) = L(dx_1 \wedge \cdots \wedge dx_n)$; $\widetilde{S}(n) = L((1 + x_1 \cdot \ldots \cdot x_n)\, dx_1 \wedge \cdots \wedge dx_n)$, $n = 2k$; $H(n) = L((\delta x_1)^2 + \cdots + (\delta x_n)^2)'$ (where G' is the commutant of the algebra G). The algebra $S(n)$ is the linear span of the elements $(\partial_i\varphi)\partial_j + (\partial_j\varphi)\partial_i$, $\varphi \in \Lambda$, and $H(n)$ consists of the elements $(\partial_1\varphi)\partial_1 + \cdots + (\partial_n\varphi)\partial_n$, $\varphi \in \Lambda$, where φ does not contain the monomial $x_1 \cdot \ldots \cdot x_n$.

The algebra $W(n)$ is simple for $n > 1$; $S(n)$, $\widetilde{S}(n)$ are simple for $n > 2$: $H(n)$ is simple for $n > 3$. Kac calls these algebras Cartan Lie superalgebras.

(Among the Cartan Lie superalgebras there is no "contact" algebra, which would have to consist of those $D \in W(n)$ which multiply the form $\delta x_n + x_1\,\delta x_1 + \cdots + x_{n-1}\,\delta x_{n-1}$ by an element in Λ. But this algebra does not admit a $\mathbb{Z}_2$ - grading.)

The standard $\mathbb{Z}$ - grading of the algebra Λ induces a $\mathbb{Z}$ - grading of the algebras $W(n)$, $S(n)$, and $H(n)$ of the form $G = \oplus_{i \geq -1} G_i$. The classical Lie superalgebras $A(m,n)$, $C(m)$, and $P(m)$ also admit a unique $\mathbb{Z}$ - grading of this form.

Kac proves that any simple $\mathbb{Z}$ - graded Lie superalgebra $G = \oplus_{i \geq -1} G_i$ is isomorphic to one of the algebras $A(m,n)$, $C(m)$, $P(m)$, $W(n)$, $S(n)$, $H(n)$.

Let $G = G_0 \oplus G_1$ be a simple Lie superalgebra such that the representation of G_0 on G_1 is not irreducible; let L_0 be the maximal subalgebra containing G_0, and let $L_i = \{x \in L_0 \mid [L, x] \subset L_{i-1}\}$ $(i > 0)$. Then the algebra $\operatorname{Gr} L$ for the filtered algebra $L = L_{-1} \supset L_0 \supset L_1 \supset \cdots$ satisfies the above conditions on $\mathbb{Z}$ - gradings, and this implies the classification theorem. Every simple Lie superalgebra is isomorphic either to one of the classical Lie superalgebras or to one of the Cartan Lie superalgebras.

Kac also obtains interesting results on the finite dimensional graded representations of superalgebras, and, in a recent paper [10] obtains the graded analogue of the Weyl character formula and the Kostant formula for the multiplicity of a weight.

In [12] Sternberg and Wolf study examples of what they call "hermitian Lie algebra" structures: $\mathfrak{l}$ is a real Lie algebra, represented by linear transformations of a complex vector space V, and $H: V \times V \to \mathfrak{l}_{\mathbb{C}}$ is an $\mathfrak{l}$ - equivariant hermitian form. Here "hermitian" means that $H(u,v)$ is linear in u and conjugate-linear in v with

$$H(v,u) = \overline{H(u,v)}\ , \quad \bar{\ } = \text{conjugation of } \mathfrak{l}_{\mathbb{C}} \text{ over } \mathfrak{l}, \text{ and}$$

"equivariant" means that

$$[\xi, H(u,v)] = H(\xi u, v) + H(u, \xi v) \quad \text{for } \xi \in \mathfrak{l} \text{ and } u, v \in V \text{ where}$$

$[\ ,\ ]$ is extended as usual from $\mathfrak{l}$ to $\mathfrak{l}_{\mathbb{C}}$.

$2 \operatorname{Im} H: V \times V \to \mathfrak{l}$ is anti-symmetic and $\mathbb{R}$ - bilinear, so one tries to use it to make $\mathfrak{l} + V$ into a Lie algebra by: the usual bracket $\mathfrak{l} \times \mathfrak{l} \to \mathfrak{l}$, the representation $\mathfrak{l} \times V \to V$ (i.e., $[\xi, u] = \xi u = -[u, \xi]$), and $V \times V \to \mathfrak{l}$ given by

$$(*) \qquad [u,v] = 2 \operatorname{Im} H(u,v) = \frac{1}{i}\{H(u,v) - H(v,u)\} \quad \text{for } u, v \in V .$$

This defines a Lie algebra if and only if the Jacobi identity holds, and that is the case just when it holds for any three elements of V:

$$[[u,v],w] + [[u,w],u] + [[w,u],v] = 0 \quad \text{for } u, v, w \in V .$$

In other words, $(*)$ gives a Lie algebra structure on $\mathfrak{l} + V$ just when

$$\{H(u,v)w + H(v,w)u + H(w,u)v\} - \{H(v,u)w + H(w,v)u + H(u,w)v\} = 0 .$$

Similarly $2 \operatorname{Re} H: V \times V \to \mathfrak{l}$ is symmetric and $\mathbb{R}$ - bilinear, so one tries to use it to make $\mathfrak{l} + V$ into a $\mathbb{Z}_2$ - graded superalgebra $\mathfrak{g} = \mathfrak{g}_+ + \mathfrak{g}_-$, $\mathfrak{g}_+ = \mathfrak{l}$ and $\mathfrak{g}_- = V$, by: the usual bracket $\mathfrak{l} \times \mathfrak{l} \to \mathfrak{l}$, the representation $\mathfrak{l} \times V \to V$, and $V \times V \to \mathfrak{l}$ given by

$$(**) \qquad [u,v]_G = 2 \operatorname{Re} H(u,v) = H(u,v) + H(v,u) \quad \text{for } u, v \in V .$$

Again, (**) defines a Lie superalgebra if and only if the graded Jacobi Identity holds, that is the case just when it holds for any three elements of V, and the latter is equivalent to

$$\{H(u,v)w + H(v,w)u + H(w,u)v\} + \{H(v,u)w + H(w,v)u + H(u,w)v\} = 0 \quad .$$

Notice that we obtain both a Lie algebra and a Lie superalgebra, i.e., that both Jacobi Identities are satisfied, just when

$$(***) \qquad H(u,v)w + H(v,w)u + H(w,u)v = 0 \quad \text{for} \quad u,v,w \in V \quad .$$

A basic class of hermitian Lie algebras in which both Jacobi Identities hold are the unitary algebras:

$$\{\underline{u}(k,\ell) \oplus \underline{u}(1)\} \oplus \mathbb{C}^{k,\ell} \quad \text{and} \quad \{\underline{u}(k,\ell)/\underline{u}(1)) \oplus \underline{u}(1)\} \oplus \mathbb{C}^{k,\ell}$$

where $\mathbb{C}^{k,\ell}$ is complex $(k+\ell)$ - space with hermitian scalar product $\langle z,w\rangle = -\sum_1^k z_j \overline{w}_j + \sum_{k+1}^{k+\ell} z_j \overline{w}_j$, and where

$$\underline{u}(k,\ell) = \{\xi : \mathbb{C}^{k,\ell} \to \mathbb{C}^{k,\ell} \text{ linear} : \langle \xi z, w\rangle + \langle z, \xi w\rangle = 0\}$$

is the Lie algebra of its unitary group.

$\underline{u}(k,\ell)$ has complexification $\underline{g\ell}(k+\ell\,;\mathbb{C})$, the Lie algebra of all complex $(k+\ell)\times(k+\ell)$ matrices. Let $*$ denote adjoint relative to $\langle\ ,\ \rangle$, that is $\langle \xi z, w\rangle = \langle z, \xi^* w\rangle$. Then $\underline{u}(k,\ell) = \{\xi \in \underline{g\ell}(k+\ell\,;\mathbb{C}) : \xi^* = -\xi\}$, and $\xi^- = -\xi^*$ is complex conjugation of $\underline{g\ell}(k+\ell\,;\mathbb{C})$ over $\underline{u}(k,\ell)$. Now

$$H_0 : \mathbb{C}^{k,\ell} \times \mathbb{C}^{k,\ell} \to \underline{g\ell}(k+\ell\,;\mathbb{C}) \quad \text{by} \quad H_0(u,v)w = i\langle w,v\rangle u$$

is hermitian and $u(k,\ell)$ - equivariant, for

$$\langle H_0(u,v)w, z\rangle = i\langle w,v\rangle\langle u,z\rangle = i\langle w,v\rangle\overline{\langle z,u\rangle} = -\langle w, H_0(v,u)z\rangle$$

and, for $\xi \in \underline{u}(k,\ell)$,

$$[\xi, H_0(u,v)]w = i\{\langle w,v\rangle\xi u - \langle \xi w,v\rangle u\} = H_0(\xi u,v)w + H_0(u,\xi v)w \quad .$$

As to the Jacobi Identities, note

$$H_0(u,v)w + H_0(v,w)u + H_0(w,u)v = i\{\langle w,v\rangle u + \langle u,w\rangle v + \langle v,u\rangle w\} .$$

The trick here is to give $\underset{\sim}{u}(1)_{\mathbb{C}} = \underset{\sim}{g\ell}(1;\mathbb{C})$ the conjugate of its usual complex structure so that

$$H_1 : \mathbb{C}^{k,\ell} \times \mathbb{C}^{k,\ell} \to \underset{\sim}{g\ell}(1;\mathbb{C}) \quad \text{by} \quad H_1(u,v) = i\langle v,u\rangle$$

is also hermitian. This done,

$$H : \mathbb{C}^{k,\ell} \times \mathbb{C}^{k,\ell} \to \underset{\sim}{u}(k,\ell)_{\mathbb{C}} \oplus \underset{\sim}{u}(1)_{\mathbb{C}} \quad \text{by} \quad H = H_0 \oplus H_1$$

is hermitian and $u(k,\ell)$ - equivariant, and satisfies (***), i.e., satisfies both Jacobi Identities. Thus we have both Lie algebra and Lie superalgebra structures on our unitary algebras $\{\underset{\sim}{u}(k,\ell) \oplus \underset{\sim}{u}(1)\} + \mathbb{C}^{k,\ell}$.

Now let $\underset{\sim}{u}(k,\ell) \oplus \underset{\sim}{u}(1)$ act on $\mathbb{C}^{k,\ell}$ by $(\xi, c)u = \xi(u) + cu$. In effect, this action is the projection

$$\underset{\sim}{u}(k,\ell) \oplus \underset{\sim}{u}(1) \to (\underset{\sim}{u}(k,\ell)/\underset{\sim}{u}(1)) \oplus \underset{\sim}{u}(1) \cong \underset{\sim}{u}(k,\ell) .$$

It gives us ordinary and graded Lie algebra structures on quotient unitary algebras $\underset{\sim}{u}(k,\ell) + \mathbb{C}^{k,\ell}$. In this latter formulation, the hermitian Lie algebra structure is obscured, for H seems not to be hermitian. Here note that $\underset{\sim}{u}(2,2) + \mathbb{C}^{2,2}$ is the spin-conformal algebra of Wess and Zumino which is of some interest in recent physical literature, cf. Corwin-Ne'eman-Sternberg [1] and Sternberg-Wolf [11].

They show how a hermitian Lie algebra is associated to every homogeneous bounded domain. Here the domain $\{m \times m$ complex matrices $Z : I - Z^*Z \gg 0\}$ gives a unitary algebra. They also associate a hermitian Lie algebra to the nilradical in a certain class of parabolic subgroups of classical Lie groups.

They show that the $H_0(u,u)$ for the unitary algebras give the lowest dimensional non-zero co-adjoint orbits for $\underset{\sim}{u}(k,\ell)$, the method generalizing a technique introduced by

Carey and Hannabuss [13] for $\underset{\sim}{u}(2,2)$. For $\underset{\sim}{u}(2,2)$ the maximal parabolic subgroup is the Poincaré group with scale, and these orbits correspond to the zero mass six dimensional orbits in the dual of the Poincaré algebra. They then relate these Hermitian Lie algebras to the orthosymplectic algebra and use them to reduce the metaplectic representation under restriction to $\underset{\sim}{u}(k,\ell)$.

The cohomology and the extension problem for superalgebras has recently been studied by Leites [14] and Tilgner [15].

1. Formal actions of Lie superalgebras. In this section we describe an algebraic construction which provides a model for a Lie superalgebras acting as ("formal power series") transformations. Our method will be to carry over, to the case of Lie superalgebras, the procedures developed for transitive Lie algebras by Guillemin-Sternberg [3] and Blattner [4], the principal algebraic tool being the Lie superalgebra version of the Poincaré-Birkhoff-Witt theorem proved in Corwin-Ne'eman-Sternberg [1]. The idea is roughly the following: Let G be a (classical) Lie group and H a closed subgroup, so that $X = G/H$ is a homogeneous space for G, i.e., a manifold on which G acts transitively. Let R denote the ring of smooth functions on X. The action of G on X induces a representation of G on R. Indeed, in the language of induced representations, it is exactly the representation of G induced from the trivial representation of H. Furthermore, the map $G \to G/H = X$ sending $g \rightsquigarrow gH$ maps R into a subring of the ring of all functions on G, namely onto the subring consisting of those functions which satisfy the identity

$$\varphi(gh) = \varphi(g) \qquad \forall\ h \in H .$$

Now any smooth function, φ, has a power series expansion at the identity element, e, and, if φ is analytic, is determined by this expansion. If D is any (left invariant) differential operator on G, then the power series expansion of φ determines the set of values $(D\varphi)(e)$, and, conversely, these values give the coefficients of the Taylor expansion of φ at e. The set of all left invariant differential operators on G can be identified with the universal enveloping algebra, $U(g)$, where g is the Lie algebra of G. Thus each function, φ, determines a linear functional, $\hat{\varphi}$, on $U(g)$ by the formula $\hat{\varphi}(D) = (D\varphi)(e)$ and, a knowledge of $\hat{\varphi}$ determines the Taylor expansion of φ at e. If we agree to replace, temporarily, the study of functions by the study of their Taylor expansion, we can thus pass to objects which are defined purely in terms of the universal enveloping algebra, which is defined for a Lie superalgebra. In this way we construct an algebraic object associated with a Lie superalgebra L, and a graded subalgebra, K, which plays the role of the "ring of functions" described above. The passage from φ to $\hat{\varphi}$ is a sort of duality, and hence the concept of induced representation must be replaced by its dual notion, that of "produced representation". We refer the reader to Blattner [4] for further motivation of this concept and for the development of the theory for the case of classical Lie algebras. The discussion which follows, will, however be self contained, relying only on the material in [1].

Let A be a graded associative algebra and let B be a subalgebra of A. We assume that A and B are both algebras over some field, which lies as a subalgebra of $B_0 \subset A_0$. In other words we are assuming that $1 \in B_0$. Let F be a (graded) module for B. That is, F is a graded vector space on which B acts as graded endomorphisms. The produced module consists of a pair, (R_F, ρ) where R_F is a graded module for A and $\rho: R_F \to F$ is a homomorphism of B modules, and is characterized by the following universal property: For any graded A module T and any homomorphism $\theta: T \to F$ as B modules, there exists a unique A module homomorphism $\varkappa: T \to R_F$ so that

$$\begin{array}{ccc} R_F & \xrightarrow{\ \rho\ } & F \\ & {}_{\varkappa}\nwarrow \quad \nearrow_{\theta} & \\ & T & \end{array}$$

$\theta = \rho \cdot \varkappa$. It is clear from the standard arguments that (R_F, ρ) is uniquely determined up to isomorphism if it exists. We construct R_F by regarding A as a graded left B module, and setting

$$R_F = \mathrm{Hom}_B(A, F) \quad .$$

(Recall that if V and W are graded vector spaces then $\mathrm{Hom}(V, W)$ becomes a graded vector space where $[\mathrm{Hom}(V, W)]_n$ consists of those maps r such that $r: V_k \to V_{k+n}$ for all k. The space $\mathrm{Hom}_B(A, F)$ consists of those $u: A \to F$ which satisfy $u(ba) = bu(a)$ for all $b \in B$.) We make R_F into an A module by defining au, for

$a \in A$ and $u \in R_F$ by

$$(au)(c) = u(ca) \quad .$$

Then

$$[a(a'u)](c) = (a'u)(ca) = u(caa') = u(c(aa'))$$

so

$$(aa')u = a(a'u)$$

showing that we do get a module structure. We define $\rho: R_F \to F$ by

$$\rho u = u(1) \quad .$$

If T is any A module and $\theta: T \to F$ is a morphism of B modules, we define $\kappa: T \to R_F$ by

$$(\kappa t)(a) = \theta(at) \quad .$$

It is immediate that the required properties hold. We will be interested in the case where $A = U(L)$ is the universal enveloping algebra of some Lie superalgebra, L, and $B = U(K)$ is the universal enveloping algebra of a subalgebra. We refer to [1] for the definition of the universal enveloping algebra and some of its properties. Here we collect some additional properties which will be useful to us:

The algebra $U(L)$ has the structure of a graded associative algebra. The gradation is described as follows: On the tensor algebra $T(L) = k \oplus L \oplus (L \otimes L) + \cdots$ we put the gradation which assigns to the element $X_1 \otimes X_2 \otimes \cdots \otimes X_n$ (where the X_i are all elements of homogenous degrees in L) the degree

$$\deg(X_1 \otimes X_2 \otimes \cdots \otimes X_n) = \deg X_1 + \cdots + \deg X_n \quad .$$

Now

$$U(L) = T(L) / I$$

where I is the ideal generated by all elements of the form

$$X \otimes Y - (-1)^{XY} Y \otimes X - [X, Y]$$

with X and Y homogeneous elements of L (and where here, and in what follows, we use the notation $(-1)^{XY}$ for $(-1)^{\deg X \cdot \deg Y}$ etc.). These elements all are homogeneous (of degree $\deg X + \deg Y$). Since I is generated by homogeneous elements, the gradation on T(L) induces a gradation on U(L) making U(L) into a graded associative algebra.

For any two associative graded algebras, A_1 and A_2, we recall from [1] that their tensor product, $A_1 \otimes A_2$ is again a graded associative algebra with the gradation

$$(A_1 \otimes A_2)_n = \bigoplus_{j+k=n} (A_1)_j \otimes (A_2)_k$$

and multiplication

$$(a_1 \otimes a_2)(c_1 \otimes c_2) = (-1)^{a_2 c_1} a_1 c_1 \otimes a_2 c_2 \quad .$$

In particular, $U(L) \otimes U(L)$ is a graded associative algebra. We claim that there exists a unique "diagonal" homomorphism, $\Delta\colon U(L) \to U(L) \otimes U(L)$ with

$$\Delta(1) = 1 \otimes 1 \quad \text{and} \quad \Delta(X) = X \otimes 1 + 1 \otimes X \quad \text{for} \quad X \in L \quad .$$

Indeed, by the universal property of U(L), it suffices to check that

$$\Delta(X)\,\Delta(Y) - (-1)^{XY} \Delta(Y)\,\Delta(X) = \Delta([X, Y])$$

for X and Y homogeneous elements of L, and this equation is easily checked in view of the definition of the multiplication on $U(L) \otimes U(L)$. The following computation will be useful

<u>Lemma</u> 1.1. <u>Let</u> $m \in U(L)$ <u>be a monomial (i.e., a product of elements of</u> L). <u>Then</u> $\Delta(m)$ <u>is a sum of the form</u>

$$\Delta(m) = \Sigma \pm (a \otimes c + (-1)^{ac} c \otimes a)$$

<u>where</u> a <u>and</u> c <u>are either monomials or</u> 1 .

<u>Proof,</u> by induction on the number of factors in the expression for m . It is true by definition for one factor, since 1 has degree zero. We thus assume it for m and check what happens for Xm , with $X \in L$. Since Δ is a homomorphism, we have merely to expand

$$(X \otimes 1 + 1 \otimes X)(a \otimes c + (-1)^{ac} c \otimes a) = Xa \otimes c + (-1)^{ac+Xc} c \otimes Xa +$$

$$+ (-1)^{ac} (Xc \otimes a + (-1)^{Xa+ac} a \otimes Xc) .$$

Since $(-1)^{ac+Xc} = (-1)^{\deg c \cdot \deg(Xa)}$ and $(-1)^{Xa+ac} = (-1)^{\deg a \cdot \deg(Xc)}$, the lemma is proved.

Let K be a graded subalgebra of L and F_i be graded K (and hence $U(K)$) modules, $i = 1, 2$ or 3 . We can then form the produced modules R_{F_i} . Furthermore, we can form the $U(K) \otimes U(K)$ module $F_1 \otimes F_2$, and hence the produced module

$$\mathrm{Hom}_{U(K)\otimes U(K)} (U(L) \otimes U(L) , F_1 \otimes F_2) ,$$

and we can define the bilinear map

$$R_{F_1} \times R_{F_2} \to \mathrm{Hom}_{U(K)\otimes U(K)} (U(L) \otimes U(L) , F_1 \otimes F_2) , \quad (u , v) \rightsquigarrow u \odot v$$

where

$$u \odot v (a \otimes c) = (-1)^{va} u(a) \otimes v(c) .$$

The diagonal map, $\Delta : U(L) \to U(L) \otimes U(L)$ then induces a map

$$\Delta^* : \mathrm{Hom}_{U(K)\otimes U(K)} (U(L) \otimes U(L) , F_1 \otimes F_2) \to \mathrm{Hom}_{U(K)} (U(L), F_1 \otimes F_2) = R_{F_1 \otimes F_2}$$

where

$$(\Delta^* w)(a) = w(\Delta a) ,$$

and we define

$$u \otimes v = \Delta^*(u \odot v)$$

which defines a bilinear map of $R_{F_1} \times R_{F_2} \to R_{F_1 \otimes F_2}$.

It is immediate that under the natural identification of $(F_1 \otimes F_2) \otimes F_3$ with $F_1 \otimes (F_2 \otimes F_3)$ the product $(u \otimes v) \otimes w$ get identified with $u \otimes (v \otimes w)$ so that the "multiplication" is associative. Furthermore, we claim that for any $X \in L$,

$$X(u \otimes v) = Xu \otimes v + (-1)^{(\deg X)(\deg u)} u \otimes Xv \quad ,$$

so that the elements of L act as graded derivations relative to this multiplication. To check this, we notice that, by definition,

$$[X(u \otimes v)](a) = (u \otimes v)(aX)$$

for any $a \in U(L)$. But

$$\begin{aligned} u \otimes v\,(aX) &= u \odot v((\Delta a)(\Delta X)) \\ &= u \odot v((\Delta a)(X \otimes 1 + 1 \otimes X)) \\ &= [(X \otimes 1)(u \otimes v)](\Delta a) + (-1)^{uX}[(1 \otimes X)(u \otimes v)](\Delta a) \quad , \end{aligned}$$

where the sign in the second term arises from the definition of the action of the tensor product of two algebras on the tensor product of two modules.

Let us now specialize to the situation where F is a graded ring on which K acts as derivations. For example, we could consider the situation where $F = k$ is the ground field, regarded as a graded vector space which is zero in all degrees other than zero, and on which K acts trivially. Multiplication gives a map of $F \otimes F \to F$ which is a morphism of K modules, and hence induces a morphism $R_{F \otimes F} \to R_F$ of L modules. Composing this with the map $R_F \times R_F \to R_{F \otimes F}$ introduced above gives a multiplication map

$$R_F \times R_F \to R_F$$

making R_F into a graded ring on which L acts as derivations.

We claim that if F is graded commutative, then so is R_F. We must show that if u and v are homogeneous elements, then

$$uv = (-1)^{\deg u \cdot \deg v} vu ,$$

i.e., we must show that the above equation holds when evaluated on any element of $U(L)$. It suffices to check what happens when we evaluate on 1 and on monomials. Now

$$uv(1) = u(1)\, v(1) .$$

Since 1 is an element of degreee zero, we have $\deg(u(1)) = \deg u$, $\deg(v(1)) = \deg v$ and hence, from the graded commutativity of F, it follows that

$$\begin{aligned} uv(1) &= u(1)\, v(1) = (-1)^{\deg u(1) \deg v(1)} v(1)\, u(1) \\ &= (-1)^{uv} vu(1) . \end{aligned}$$

Let $\mu : F \otimes F \to F$ denote the multiplication map. Then, for any monomial, m, we have

$$\begin{aligned} uv(m) &= \mu(u \otimes v\,(\Delta m)) \\ &= \Sigma \pm \mu(u \otimes v(a \otimes c + (-1)^{ac}\, c \otimes a)) \\ &= \Sigma \pm ((-1)^{va}\, u(a)\, v(c) + (-1)^{ac+vc}\, u(c)\, v(a)) . \end{aligned}$$

Now $\deg u(a) = \deg u + \deg a$ and $\deg v(c) = \deg v + \deg c$ so

$$(-1)^{uv}((-1)^{ua} v(a)\, u(c) + (-1)^{ac+uc} v(c)\, u(a)) = (-1)^{va} u(a)\, v(c) + (-1)^{ac+vc} u(c)\, v(a)$$

which proves that R_F is graded commutative.

From now on we shall drop the subscript k when discussing the ring R_k, which we shall denote from now on by R. Suppose we choose a basis for the vector space L/K, and representatives in L for each of these basis elements. Then Poincaré-Birkhoff-Witt theorem proved in [1] implies that the monomials which are symmetric in the even representatives and anti-symmetric in the odd representatives form a basis of $U(L)$ as a $U(K)$ module. Hence, for any K module, F, we have the vector space isomorphism

$$R_F = \mathrm{Hom}_{U(K)}(U(L), F) \sim \mathrm{Hom}_k(S(L/K)_{\mathrm{even}} \otimes \Lambda(L/K)_{\mathrm{odd}}, F)$$

giving R_F the structure of a space of "formal power series in commuting and anti-commuting variables". This is true, in particular, for the ring R . To analyze this structure further, we wish to introduce a filtration on the ring R . This filtration is induced from a corresponding filtration on the algebra U(L) , which we shall discuss in some detail in the next section.

2. In this section we discuss filtrations on $U(L)$ and the induced Poisson bracket structure on $\mathrm{gr}\, U(L)$. We will treat the cases of $\mathbb{Z}_2$ graded algebras and $\mathbb{Z}$ graded algebras separately. Suppose first that L is $\mathbb{Z}_2$ graded. The algebra $T(L)$ is graded by tensor degree, and this induces a filtration on $U(L)$. The elements of filtration degree k are those sums of terms in $U(L)$ which can be written as products of elements of L with at most k factors; we denote the space of such elements by $U_k(L)$. We set $\mathrm{gr}_k\, U(L) = U_k(L)/U_{k-1}(L)$ and $\mathrm{gr} U(L) = \bigotimes_0^\infty \mathrm{gr}_k\, U(L)$. The Poincaré-Birkhoff-Witt theorem proved in [1] asserts that

$$\mathrm{gr}\, U(L) \sim S(L_{even}) \otimes \wedge(L_{odd}) \quad .$$

Here the isomorphism is as $\mathbb{Z}_2$ graded algebras, where the $\mathbb{Z}_2$ gradation on $\mathrm{gr}\, U(L)$ is inherited from $U(L)$ (and has nothing to do with the k in the decomposition of $\mathrm{gr}\, U(L) = \oplus\, \mathrm{gr}_k\, U(L)$), while the gradation on $S(L_{even}) \otimes \wedge(L_{odd})$ is inherited from the $\mathbb{Z}_2$ gradation on $T(L)$. We define a Poisson bracket on $\mathrm{gr}\, U(L)$ as follows: Let $x \in \mathrm{gr}_k\, U(L)$ and $y \in \mathrm{gr}_\ell\, U(L)$ be homogeneous relative to the $\mathbb{Z}_2$ gradation with $\deg x = c$ and $\deg y = d$ (so that c and $d \in \mathbb{Z}_2$). Choose $\underline{x} \in U_k(L)$ and $\underline{y} \in U_\ell(L)$ so that $\mathrm{gr}\, \underline{x} = x$ and $\deg \underline{x} = c$, and $\mathrm{gr}\, \underline{y} = y$ and $\deg \underline{y} = d$. Then, in view of the fact that $\mathrm{gr}\, U(L)$ is graded commutative, $\underline{x}\underline{y} - (-1)^{cd}\, \underline{y}\underline{x}) \in U_{k+\ell-1}(L)$. We set

$$\{x, y\} = \mathrm{gr}_{k+\ell-1}\, (\underline{xy} - (-1)^{cd}\, \underline{yx}) \quad .$$

It is easy to check that this definition is independent of the choice of $\underline{x}$ and $\underline{y}$. It is clear from the definition that $\{\ ,\ \}$ is a graded derivation in y for fixed x relative to the graded commutative multiplication on $\mathrm{gr}\, U(L)$ and that $\{\ ,\ \}$ makes $\mathrm{gr}\, U(L)$ into a Lie superalgebra. The subspace $\mathrm{gr}_1\, U(L)$ is a Lie superalgebra isomorphic to L. Since (k and) L generate $\mathrm{gr}\, U(L)$ as an associative algebra, the Poisson bracket is characterized by the property that $\{x, y\}$ is a derivation in y, is graded antisymmetric, and, on $gr_1\ U(L)$ reduces to the Lie bracket. We shall see below that $gr\ U(L)$, with its associative multiplication and Poisson bracket satisfies an important universal property.

In case that L is $\mathbb{Z}$ graded, we can refine the above construction so as to give the algebra $gr\ U(L)$ the structure of a $\mathbb{Z}$ graded algebra. For this purpose we will put a new gradation on the algebra $T(L)$ as follows: On the graded vector space L , assign a new gradation, Deg , by shifting the old gradation by two, i.e., set

$$\text{Deg}\ X = i+2 \quad \text{if} \quad X \in L_i \quad .$$

Then use this gradation on $T(L)$, so that

$$\text{Deg}\ X = \deg X + 2k \quad \text{if} \quad X \in T_k(L)$$

is a tensor of degree K (and, in particular, $\text{Deg}\ 1 = 0$) . This induces a filtration on $U(L)$, and the terms of filtration order i will be denoted by $F_i\ U(L)$. Thus, if $\underline{x} \in U_k(L)$ and $\deg \underline{x} = r$, then $\underline{x} \in F_{r+2k}\ U(L)$. Conversely, if $\underline{x} \in F_{r+2k}\ U(L)$ and $\deg \underline{x} = r$, then $\underline{x} \in U_k(L)$. If $\underline{x}$ and $\underline{y}$ are elements of $F_{r+2k}\ U(L)$ such that $\deg \underline{x} = \deg \underline{y} = r$, and such that $\underline{x} - \underline{y} \in F_{r+2k-1}$, then, since $\deg(\underline{x} - \underline{y}) = r$, we see that $\underline{x} - \underline{y} \in F_{r+2k-2}$ (so that the filtration drops by two on elements of fixed degree) and that $\underline{x} - \underline{y} \in U_{k-1}(L)$. Let F^r_m denote the space of elements of $F_m\ U(L)$ which are homogeneous of degree r . We have shown that

$$\begin{aligned} F^r_{r+2k} / F^r_{r+2k-1} &\sim (gr_k\ U(L))^r \\ &\sim [S(L_{even}) \otimes \wedge(L_{odd})]^r_k \quad , \end{aligned}$$

where the superscript, r , on the right denotes the subspace of degree r relative to the gradation induced on $S(L_{even}) \otimes \wedge(L_{odd})$ from deg . It follows easily that

$$\oplus (F_m / F_{m-1}) \sim S(L_{even}) \otimes \wedge(L_{odd}) \quad ,$$

as a graded associative algebra when the right hand side carries the Deg gradation. We are now precisely in the situation described in [1] section 2J : The algebra $U(L)$ is filtered (by F) and the filtration is consistent with the $\mathbb{Z}_2$ gradation (obtained by reducing deg mod 2) in the sense that even elements are filtered by even degrees and odd elements by odd degrees, with the filtration dropping by two on homogeneous elements, and with the multiplication induced on the graded algebra $\oplus F_m / F_{m-1}$ graded commutative. We know from [1] that this induces a Poisson bracket on $\oplus F_m / F_{m-1}$, where the Poisson bracket $\{x, y\}$ acts as graded derivations in y. This Poisson bracket is defined by

$$\{x, y\} = \underline{x}\underline{y} - (-1)^{\deg x \cdot \deg y} (\underline{y}\underline{x}) / F_{m+n-4}$$

where $x \in F_m / F_{m-1}$ and $y \in F_n / F_{n-1}$ are homogeneous elements with $\underline{x}$ and $\underline{y}$ homogeneous representatives. This Poisson bracket makes $\operatorname{gr} U(L)$ into a $\mathbb{Z}$ graded superalgebra, where F_m/F_{m-1} is given the Lie algebra gradation $m-2$. Let us denote this gradation by $\underline{\deg}$. Thus, if $x \in \operatorname{gr}_k U(L) = U_k(L)/U_{k-1}(L)$ and $\deg x = r$ and $\deg x = r$, then

$$\underline{\deg}\, x = 2k + r - 2 \quad .$$

In particular, L can be characterized as the subspace spanned by those homogeneous elements which satisfy

$$\underline{\deg}\, x = \deg x$$

and forms a subalgebra relative to $\{\ ,\ \}$. The fact that k and L generate $\operatorname{gr} U(L)$ as an associative algebra now implies that the Poisson bracket just defined coincides with the Poisson bracket we introduced above using just the $\mathbb{Z}_2$ gradation.

As an illustration of this construction, consider the following Lie superalgebra

$$L_i = \{0\} \quad i \neq -1, -2 ,$$

$$L_{-1} = V , \text{ a finite dimensional vector space carrying a symmetric bilinear form, } Q ,$$

$$L_{-2} = k \cdot z , \text{ a one dimensional vector space generated by } z$$

with the bracket relations

$$[u, v] = 2Q(u, v) z \quad , \quad u, v \in L_{-1}$$

and

$$[z, u] = 0 \quad .$$

Now $U(L) = T(L)/I$, where I is the ideal generated by $z \otimes u - u \otimes z$ and $u \otimes v + v \otimes u - 2Q(u,v) z$. Let J denote the ideal generated by $z \otimes u - u \otimes z$. Then

$$T(L) / J = k[z] \otimes T(V)$$

where $k[z]$ is the polynomial ring in the generator z . So $T(L)/I = (T(L)/J) / \{u \otimes v + v \otimes u - 2Q(u,v) z\}$ and thus

$$U(L) = C_Q (V[z])$$

is just the Clifford algebra of the module $V[z] = k[z] \otimes V$ over $k[z]$ relative to the quadratic form zQ . The elements z^k all have filtration degree zero, and the filtration, F , on $U(L)$ coincides with the standard filtration on the Clifford algebra as described, for example, in [1]. Thus we see that the Poisson bracket on $\operatorname{gr} U(L)$ generalizes the Clifford Poisson bracket as introduced in [1], and our construction answers the question raised at the end of section 4 of [1]. As pointed out by Kostant [2], the algebra L can be thought of as a graded analogue of the Heisenberg algebra. In this sense, the Clifford algebra (obtained by "specializing" z to 1) is the analogue of the Weyl algebra, i. e., the algebra of the quantum mechanical commutation relations. As Kostant points out, if we take $k = \mathbb{C}$ and dim V to be even, then by a procedure completely analogous to the choice of a polarization in the Heisenberg situation, (i.e., the choice of a maximally isotropic subspace for Q) we obtain the spin representation of the Clifford algebra. We shall discuss this construction later, in terms of a different gradation on L.

We return to general considerations. Suppose that S is a graded commutative algebra, which possesses a Poisson bracket acting as derivations of the multiplicative structure. Let us call such an algebra, together with its Poisson bracket, a Poisson algebra. We claim that the algebra gr $U(L)$ is the "universal" Poisson algebra associated with L in the following sense:

Let R be a graded commutative algebra and let $\xi : L \to \operatorname{Der} R$ be a homomorphism of Lie superalgebras. Suppose that $\varphi : L \to R$ is a graded linear map of even degree such that

$$\xi(X)\varphi(Y) = \varphi([X,Y])$$

for all X and $Y \in L$. Then φ extends to a homomorphism, which we shall also denote by φ of gr $U(L) \to R$. The image, $\varphi(U(L)) \subset R$ has the structure of a Poisson algebra where the Poisson bracket is uniquely determined by

$$\{\varphi(X), \varphi(Y)\} = \varphi([X,Y])$$

for all X and $Y \in L$. Furthermore, $\varphi : \text{gr } U(L) \to \varphi(\text{gr } U(L))$ is a homomorphism of Poisson algebras, i.e.,

$$\varphi(\{a,b\}) = \{\varphi(a), \varphi(b)\}$$

for all a and $b \in$ gr $U(L)$.

Proof. The fact that φ extends to an algebra homomorphism of gr $U(L) \to R$ follows directly from the Poincaré-Birkhoff-Witt theorem. Indeed, $S(L_{\text{even}}) \otimes \wedge(L_{\text{odd}})$ is the universal enveloping algebra of L^{tr}, where L^{tr} denotes the vector space L made into a trivial Lie superalgebras by setting all brackets equal to zero. By the universal property of this enveloping algebra we get the desired algebra homomorphism. We now define the Poisson bracket on $\varphi(\mathrm{gr}\ U(L))$ as follows : for $u = \varphi(X) \in \varphi(L)$ and $v \in \varphi(\mathrm{gr}\ U(L))$ we set

$$\{u,v\} = \xi(X)v \quad ,$$

and this is a derivation in v . This is well defined, since, if $u = \varphi(X) = 0$ then $\xi(X)\varphi(Y) = \varphi([X, Y]) = \pm \varphi([Y, X]) = \pm \xi(Y)\varphi(X) = 0$. The set of $u \in \varphi(L)$ generate $\varphi(\mathrm{gr}\ U(L))$ as a graded commutative algebra. Hence, if we can define a Poisson bracket on $\varphi(\mathrm{gr}\ U(L))$, it will be uniquely determined by the condition that $\{u,v\}$ is a derivation in v for fixed u , and a skew derivation on u for fixed v , i.e., to satisfy

$$\{u_1 u_2, v\} = u_1\{u_2, v\} + (-1)^{u_2 v}\{u_1, v\}u_2 \quad .$$

In proving that the Poisson bracket exists, we might as well assume that $R = \varphi(\mathrm{gr}\ U(L))$. Let $\overline{R} = S(\varphi(L_{even}) \otimes \wedge \varphi(L_{odd})$ so that $\overline{\varphi} : \overline{R} \to R$, is a surjective homomorphism with kernel I, say, where $\overline{\varphi}$ is the obvious homomorphism.

We obtain a well defined map of $\overline{R} \times \overline{R} \to R$ by the requirement that $\{\overline{u}, \overline{v}\}$ be a derivation in $\overline{v}$ and a skew derivation in $\overline{u}$. Thus, for example

$$\{\overline{u}, \overline{v}_1 \overline{v}_2\} = \{\overline{u}, \overline{v}_1\}\overline{\varphi}(\overline{v}_2) + (-1)^{\overline{u}\overline{v}_1}\overline{\varphi}(\overline{v}_1)\{\overline{u}, \overline{v}_2\} \quad ,$$

etc. For any $\overline{u}$ and $\overline{v}$, if we expand $\{\overline{u}, \overline{v}\}$ in $\overline{v}$ according to the above formula using the expression of $\overline{v}$ as a sum of monomials we will obtain a sum of products, each involving a factor of the form $\{\overline{u}, \varphi(Y)\}$. If $\overline{u} \in I$ it follows from the fact that $\{\overline{u}, \varphi(Y)\}$ is a skew derivation in $\overline{u}$, that $\{\overline{u}, \varphi(Y)\} = \pm\{\varphi(Y), \overline{u}\} = 0$ since $\{\varphi(Y), \overline{u}\} = \xi(Y)\overline{\varphi}(u)$ and $\xi(Y)$ is a derivation of R . Similarly, if $\overline{u}$ is arbitrary and $\overline{v} \in I$ we get $\{\overline{u}, \overline{v}\} = 0$. Thus $\{\ ,\ \}$ is defined as a map of $R \times R \to R$, and is a skew derivation in its first variable and a derivation in its second. We must check the remaining properties of the Poisson bracket. For $u = \varphi(X)$ and $v = \varphi(Y)$ we have

$$\{u, v\} = \xi(X)\,\varphi(Y)$$
$$= \varphi([X, Y])$$
$$= (-1)^{XY}\,\varphi([Y, X])$$

so

$$\{u, v\} = (-1)^{\mathrm{degu}\cdot\mathrm{degv}}\,\{v, u\} \quad .$$

Using the fact that $\{u, v\}$ is a derivation in its second variable and a skew derivation in its first, one concludes that the above formula holds for $u \subset \varphi(L)$ and v arbitrary in $\varphi(\mathrm{gr}\ U(L))$ and then for u arbitrary as well. Similarly, one verifies the Jacobi identity

$$\{u, \{v, w\}\} = \{\{u, v\} w\} + (-1)^{\mathrm{degu}\,\mathrm{degv}}\,\{v, \{u, w\}\}$$

first for all three elements in $\varphi(L)$, which follows from the definition, then for w arbitrary, using the derivation properties, then for v arbitrary and then finally for u arbitrary.

To show that the map $\varphi : \mathrm{gr}\ U(L) \to \varphi(U(L))$ is a homomorphism of Poisson algebras, we need only check behaviour of the Poisson bracket, since the multiplicative structure has already been accounted for. By construction, $\varphi(\{X, Y\}) = \{\varphi(X), \varphi(Y)\}$ for X and Y in L. Since (1 and) L generate $\mathrm{gr}\ U(L)$ as a commutative graded algebra, and in view of the derivation properties of $\{\quad\}$, we conclude that φ is indeed a homomorphism of Poisson algebras.

As an example of an application of the above proposition, suppose we pick a $\lambda \in L^*$, and define its isotropy subalgebra, K_λ as

$$K_\lambda = \{Y \in L \mid \langle [Y, X], \lambda \rangle = 0 \quad \text{for all} \quad X \in L\} \quad .$$

It follows immediately from the Jacobi identity that K_λ is a graded Lie subalgebra. Let R^λ be the ring associated to the subalgebra K_λ by the method of the preceeding section, so that

$$R^\lambda = \mathrm{Hom}_{U(K_\lambda)}(U(L), k)$$

is a graded commutative algebra. The algebra L acts as derivations of this ring, and we shall denote this action by ξ so that

$$(\xi(X)u)(a) = u(aX)$$

for any $X \in L$ and any $a \in U(L)$, where $U \in R^\lambda$. Let us define the map $\varphi : L \to R^\lambda$ by

$$\varphi(X)(a) = \langle (\mathrm{ad}\, a) X, \lambda \rangle \quad .$$

Here L acts on itself via the adjoint representation, and the extension of this representation to U(L) is denoted by ad . To check that $\varphi(X)$ actually lies in R^λ, we observe that if $Yb \in KU(K)$, then

$$\varphi(X)(Yba) = \langle [Y, (\mathrm{ad}\, ba) X], \lambda \rangle = 0 \quad .$$

We now verify that for any X and Z in L, the identity

$$\xi(X)\varphi(Z) = \varphi([X, Y])$$

holds, which will prove that $\varphi(\mathrm{gr}\, U(L)) \subset R^\lambda$ has the structure of a Poisson algebra. Indeed, for any $a \in U(L)$.

$$\begin{aligned}
(\xi(X)\varphi(Z)(a) &= \varphi(Z)(aX) \\
&= \langle (\mathrm{ad}\, aX) Z, \lambda \rangle \\
&= \langle (\mathrm{ad}\, a)(\mathrm{ad}\, X) Z, \lambda \rangle \\
&= \langle (\mathrm{ad}\, a)[X, Z], \lambda \rangle \\
&= \varphi([X, Z])(a) \quad .
\end{aligned}$$

As we shall see, the subalgebra $\varphi(\mathrm{gr}\, U(L)) \subset R^\lambda$ is dense, in the formal power series topology in R^λ, so that the Poisson bracket will extend to all of R^λ making R^λ into a Poisson algebra.

Conversely, suppose that $R = \mathrm{Hom}_{U(K)}(U(L), k)$ is such that there exists a map $\varphi : L \to R$ with $\xi(X)\varphi(Z) = \varphi([X, Z])$ where $\xi(X)$ denotes the derivation of R corresponding to X. We may define $\lambda \in L^*$ by

$$\langle X, \lambda \rangle = \varphi(X)(1) \quad .$$

If $Y \in K$, then

$$\begin{aligned} \langle [Y, X], \lambda \rangle &= \varphi([Y, X])(1) \\ &= (\xi(Y)\varphi(X))(1) \\ &= \varphi(X)(Y) \\ &= 0 \end{aligned}$$

since $\varphi(X) \in \mathrm{Hom}_{U(K)}(U(L), k)$, $Y \in K$ and k is a trivial K module.

Thus

$$K \subset K_\lambda \quad .$$

This implies that we have an injection of $\mathrm{Hom}_{U(K_\lambda)}(U(L), k)$ into $\mathrm{Hom}_{U(K)}(U(L), k)$, i.e., of R^λ into R, and it is easy to check that this injection is a homomorphism of Poisson algebras and of L modules. (If L, K, and K_λ were finite dimensional classical Lie algebras corresponding to the Lie group G and closed subgroups H and H_λ, then the "geometric" analogue of the preceeding assertion would be the G/H is fibered over G/H_λ. Here G/H would be an espace d'evolution for G in the sense of Souriau [8] whose corresponding espace des mouvements is G/H_λ. In particular one recovers the result that the most general symplectic homogeneous manifold for G on which G has a Hamiltonian action is a covering of an orbit of G in L^*.)

For classical Lie groups and algebras, one is interested not only in Hamiltonian actions, but also in symplectic actions. That is one wishes to study symplectic manifolds on which the group acts as a group of symplectic automorphisms. In order to construct the analogous notion for graded Lie algebras, we must begin with a graded formulation of the notion differential form, exterior derivative, etc. Let R be a graded commutative ring. Let $\mathcal{V}$ denote the Lie superalgebra of (graded) derivations of R. (The elements of $\mathcal{V}$ play the role of "vector fields".) The space $\mathcal{V}$ is a graded module for R; where $(r\xi)s = r(\xi s)$ for any r and s in R and ξ in $\mathcal{V}$. We define the "exterior algebra" $\wedge(\mathcal{V})$ as

$$\wedge(\mathcal{V}) = R \otimes T(\mathcal{V})/I$$

where $T(\mathcal{V})$ is the tensor algebra of $\mathcal{V}$ (over the ground field) and I is the ideal generated by the elements

$$r\xi_1 \otimes \xi_2 + (-1)^{jk_1} \xi_1 \otimes r\xi_2$$

$$r \otimes \xi - 1 \otimes r\xi \qquad\qquad r \in R,\ \deg r = j,\ \xi_i \in \mathcal{V}\ \deg \xi_i = k_i$$

$$\xi_1 \otimes \xi_2 + (-1)^{k_1 k_2} \xi_2 \otimes \xi_1 \ .$$

It is immediate that $\wedge(\mathcal{V})$ is a graded module over R, and that the ideal J is a homogeneous ideal with respect to the tensor degree, and so that $\wedge(\mathcal{V}) = \oplus \wedge^k(\mathcal{V})$ in the obvious sense. Any $\xi \in \mathcal{V}$ acts as a graded derivation of $R \otimes T(\mathcal{V})$ and, as is easily checked, preserves the ideal J, and so induces a graded derivation of the algebra $\wedge(\mathcal{V})$. A differential form of exterior degree k is then an element of $\mathrm{Hom}_R(\wedge^k(\mathcal{V}), R)$. Of course, such a form will also have a graded degree, as a graded map between two graded vector spaces. We will denote the value of the form ω on the element v of $\wedge^k(\mathcal{V})$ by $\omega(v)$. The algebra $\mathcal{V}$ acts on the forms according to the usual induced action on Hom; we denote this action by D. Thus the "Lie derivative" $D_\xi \omega$ is defined by

$$(D_\xi \omega)(a) = \xi(\omega(a)) - (-1)^{\deg \xi \cdot \deg \omega} \omega(\xi a)$$

for any $a \in \wedge^k(V)$, where ξa denotes the image of the element a under the derivation, ξ. It follows from general principles (or direct verification) that the map $\xi \rightsquigarrow D_\xi$ is a representation. For each $\eta \in V$ we define the map $i(\eta)$ sending forms of exterior degree k into forms of exterior degree $k-1$ by the formula

$$(i(\eta)\omega)(b) = (-1)^{\deg \eta \cdot \deg \omega} \omega(\eta \wedge b) .$$

It is easy to check that

$$D_\xi i(\eta) = i(\xi,\eta) + (-1)^{\deg \xi \cdot \deg \eta} i(\eta) D_\xi .$$

We then can define the differential operator, d, mapping forms of exterior degree k into forms of exterior degree $k+1$ inductively by setting

$$i(\eta)dr = \eta r \qquad \text{for any } r \in R$$

and requiring that

$$i(\eta)d + di(\eta) = D_\eta .$$

In particular, d has degree zero as a map of graded vector spaces. One then checks that the formula

$$dD_\xi = D_\xi d$$

holds (by induction on exterior degree and direct verification for degree zero). One then verifies that

$$d^2 = 0$$

by direct verification when applied to elements of R and then by induction using the identities

$$\begin{aligned} i(\eta)dd &= D_\eta d - di(\eta)d \\ &= D_\eta d - dD_\eta + ddi(\eta) = ddi(\eta) . \end{aligned}$$

We can now define a presymplectic structure to consist of a graded ring R together with a closed two form, ω. A symplectic vector field is then a $\xi \in V$ such that $D_\xi \omega = 0$.

This is the same as the condition $di(\xi)\omega = 0$. If ξ and η are symplectic vector fields, then

$$\begin{aligned} di(\xi)\, i(\eta)\omega &= D_\xi i(\eta)\omega - i(\xi) di(\eta)\omega \\ &= i([\xi,\eta])\omega \end{aligned}$$

since $di(\eta)\omega = 0 = D_\xi \omega$. Thus $\zeta = [\xi,\eta]$ satisfies the stronger condition

$$i(\zeta)\omega = dr \quad .$$

Vector fields satisfying this condition are called Hamiltonian. Let us call a vector field η isotropic if $i(\eta)\omega = 0$. It is clear that every isotropic vector field is Hamiltonian, and that the set of isotropic vector fields and the set of Hamiltonian vector fields form ideals in the graded Lie algebra of symplectic vector fields. Suppose that r_ξ and r_η are elements of R such that $- dr_\xi = i(\xi)\omega$ and $- dr_\eta = i(\eta)\omega$ for suitable vector fields ξ and η (determined up to isotropic vector fields). Then $- d(r_\xi r_\eta) = i(r_\xi \eta \pm r_\eta \xi)\omega$ so that the set of such functions forms a subalgebra, R_P, of R. It is immediate that the Poisson bracket $\{r_\xi, r_\eta\} = \xi r_\eta$ makes R_P into a Poisson algebra.

3. <u>The filtration on the ring</u> R. We return to the study of the graded commutative algebra $R = \mathrm{Hom}_{U(K)}(U(L), k)$ where K is any graded Lie subalgebra of the graded Lie algebra, L. We begin with the case where L is $\mathbb{Z}_2$ graded and define

$$R_i = \{u \in R \mid u(a) = 0 \quad \text{for all} \quad a \in U_{i-1}(L)\} \quad .$$

For $i \leq 0$, we set $R_i = R$. Notice that

$$R_{-1} = R_0 \supset R_1 \supset R_2 \supset \cdots R_i \supset R_{i+1} \cdots \quad .$$

Notice also that for any monomial, $m \in U_r(L)$ and any $u, v \in R$,

$$uv(m) = \Sigma \pm u(a)\, v(b) \quad \text{where} \quad a \in U_k(L) \quad \text{and} \quad b \in U_\ell(L) \quad \text{with} \quad k + \ell = r \quad .$$

If $r \leq i + j - 1$, then in the above sum we cannot have, simultaneously, $r \geq i$ and $\ell \geq j$. From this it follows that

$$R_i \cdot R_j \subset R_{i+j} \quad .$$

This makes R into a filtered algebra (with a decreasing filtration). Let D denote the Lie superalgebra of all derivations of R, so $D = \mathrm{Der}\, R$. We can filter D by setting

$$D_k = \{d \mid dR_m \subset R_{m+k} \quad \text{for all} \quad m\}, \quad -\infty < k < \infty \quad .$$

It follows immediately that this makes D into a filtered Lie superalgebra. It is clear from the definition of ξ, that for any $X \in L, \xi(X) \in D_{-1}$. We claim that

$$\xi(Y) \in D_0 \quad \underline{\text{if and only if}} \quad Y \in K \quad .$$

<u>Proof.</u> a) Suppose $Y \in K$ and $u \in R_m$. Let a be any element of $U_{m-1}(L)$. Then

$$aY = \pm Ya + b \quad \text{with} \quad b \in U_{m-1}(L)$$

so that

$$\begin{aligned}(\xi(Y)u)(a) &= u(aY) \\ &= \pm u(Ya) + u(b) \\ &= \pm Yu(a) + u(b) \\ &= 0 \quad .\end{aligned}$$

b) Suppose $Z \notin K$, and let $\overline{Z}$ be its image in L/K. Choose $\tau \in \mathrm{Hom}(L/K, \mathbb{k})$ such that $\tau(\overline{Z}) \neq 0$. We can regard $\mathrm{Hom}(L/K, \mathbb{k})$ as the subspace of $\mathrm{Hom}(S(L/K)_{\mathrm{even}} \otimes \wedge (L/K)_{\mathrm{odd}}, \mathbb{k})$ which consists of those linear functions which vanish on all components except L/K. Then under the isomorphism of R with $\mathrm{Hom}(S(L/K)_{\mathrm{even}} \otimes \wedge (L/K)_{\mathrm{odd}}, \mathbb{k})$, the element τ corresponds to some $t \in R$ such that $t(1) = 0$ and $t(Z) \neq 0$. Thus $t \in R_0$ but $\xi(Z)t \notin R_0$, proving that $\xi(Z) \notin D_0$.

The filtration on D induces a filtration on L:

$$L_i = \xi^{-1}(D_i) \quad .$$

Thus, $L_{-1} = L$, and $L_0 = K$. We claim that for all $i \geq 0$,

$$L_{i+1} = \{X \in L_i \,|\, [X, L] \subset L_i\} \quad .$$

That the left side is contained in the right follows from the fact that we have a filtration. Let us prove the reverse inclusion, so suppose X belongs to the right hand side. Let $u \in R_m$ and $a \in U_{m+i-1}$. We know that $\xi(X)\,u(a) = 0$ since $X \in L_i$. But, for any $Y \in L$

$$\xi(X)\,u(aY) = u(aYX) = u(a[Y,X]) \pm u(aXY) \quad .$$

The first term on the right vanishes since $[Y,X] \in L_i$. As for the second term, we have $aX = \pm Xa + b$, where b is a sum of terms of the form $c[Z,X]d$ where $Z \in L$ and where c and d are monomials whose degrees add up to at most $m + i - 2$. Now $[Z,X] \in K$ and we can write $c[Z,X] = \pm [Z,X]c + c'$, so that $c[Z,X]d = [Z,X]cd + c'd$ where $c'd \in U_{m+k-2}$. Thus $u(aXY) = \pm u(XaY) + \Sigma\, u([Z,X]cdY) + \Sigma\, u(c'dY) = 0$.

If I is any ideal of L which is contained in K, then clearly $I \subset L_i$ for all i. Since $\cap L_i = \ker \xi$ we see that

$\ker \xi$ <u>is the largest ideal of L contained in</u> K .

The rest of the theory proceeds as in Blattner [4] and we will not proceed with the details, especially since a more general theory is given by Kostant [2].

References

[1] L. Corwin, Y. Ne'eman, and S. Sternberg, "Graded Lie algebras in mathematics and physics", Reviews of Modern Physics, 47 (1975) pp. 573-603.

[2] B. Kostant, article in this volume.

[3] V. W. Guillemin and S. Sternberg, "An algebraic model of transitive differential geometry", Bull. A. M. S., 70 (1964) 16-47.

[4] R. J. Blattner, "Induced and produced representations of Lie algebras", Trans. Amer. Math. Soc. 144 (1969) 457-474.

[5] V. G. Kac, "Classification of Lie superalgebras", Funksional. Anal. 1. Prilozhen. 9 (1975) 75-76.

[6] I. Kaplansky, "Graded Le Algebras", I and II, Univ. of Chicago preprint 1975.

[7] I. M. Singer and S. Sternberg, "The infinite groups of Lie and Cartan", Journal d'Analyse Mathematique 15 (1965) 1-114.

[8] D. Djokovic and G. Hochschild, "Classification of some 2-graded Lie algebras" and "Semi-simplicity of 2-graded Lie algebras", to appear.

[9] J. M. Souriau, Structure des systèmes dynamiques, Dunod, Paris (1970).

[10] V. G. Kac, "Characters of Typical Representations of Classical Lie Superalgebras", to appear.

[11] S. Sternberg and J. A. Wolf, "Charge Conjugation and Segal's Cosmology", Il Nuovo Cimento 28 (1975) 253-271.

[12] S. Sternberg and J. A. Wolf, "Hermitian Lie algebras and metaplectic representations", to appear.

[13] Carey and Hannabuss, to appear.

[14] D. A. Leites, "Cohomology of Lie superalgebras", Funksional. Anal. 1. Prilozhen. 9.

[15] H. Tilgner, "Extensions of Lie graded algebras", to appear.

[16] M. Scheunert, W. Nahm, and V. Rittenberg, "Classification of all simple graded Lie algebras whose Lie algebra is reductive", Journal of Mathematical Physics, 17 (1976), 1626 and 1640

GRADED MANIFOLDS, GRADED LIE THEORY, AND PREQUANTIZATION

by Bertram Kostant

Introduction

.1. Associated to any C^∞ (smooth) manifold X is the commutative algebra, $C^\infty(X)$, of all smooth real valued functions on X. One may reverse the emphasis, however, and, in the spirit of, say, algebraic geometry, or commutative Banach algebras, regard $C^\infty(X)$ as the primary object. All of the fundamental concepts of differential geometry, such as tangent vector, differential operators and forms, vector bundles, distributions (in both senses), de Rham theory, etc. involve constructions which are directly related to $C^\infty(X)$.

In the course of pursuing the main goal of this paper (stated below) one of the things that we shall do is to extend the notion of differentiable manifold, Lie group and, in general, deal with the situation which arises when $C^\infty(X)$ is enriched to include anti-commuting elements. Instead of $C^\infty(X)$ one has a graded commutative algebra $A(X)$. (Graded throughout nearly all of the paper means graded with respect to $\mathbb{Z}_2$). $C^\infty(X)$ is not a subalgebra of $A(X)$ — so that one is not dealing with the sections of a vector bundle — but is in fact a quotient of $A(X)$ and one has an exact sequence

$$0 \longrightarrow A^1(X) \longrightarrow A(X) \longrightarrow C^\infty(X) \longrightarrow 0$$

where $A^1(X)$ is the set of nilpotent elements in $A(X)$.

Remark. There is, of course, a precedent, in algebraic geometry, of including nilpotent elements in the structure sheaf of a variety. However in that case the rings in question are still commutative in the usual sense whereas here $A(X)$ is graded commutative.

Supported in part by NSF grant number MCS76-09177.

Graded manifolds are defined in Section 2 and graded Lie groups in Section 3. Also, in these sections as well as further on, we will track the course of concepts in differential geometry (such as those mentioned in the first paragraph) in the new graded setting. The topics are not chosen at random but are designed so that we have all that is needed to focus in on the graded symplectic theory. In fact starting in the middle of Section 4 we follow a course parallel to that in [8] and end up with prequantization and the "orbit method" for graded Lie groups.

.2. Perhaps the main stimulus for this work comes from our conviction that a graded Lie algebra (GLA) is an important mathematical object and must be taken seriously. There have been a number of developments in the last few years which have contributed to this conviction. One of these developments is the remarkable classification theory due, independently, to Kaplansky [6] and G. Kac [5] of the simple, finite dimensional graded Lie algebras. The classification is not unlike Cartan's classification for the ungraded case. A novelty is a double parameterization. Interestingly, also in the list, as Kac observes, are finite dimensional analogues of certain of Cartan's infinite dimensional simple Lie algebras. Also we have been impressed by the rich supply of examples in the paper [2] of Corwin, Neeman and Sternberg and the application to physical theories (super symmetry). Among other developments (mainly to physical theories) there is also the clearly important usages of GLA's in deformation theory (see e.g. [11]) and in topology (see e.g. [12]).

Our primary objective in this paper is to develop the representation theory of graded Lie algebras or more fully, graded Lie groups (when that has been defined) along the lines of the orbit method for ordinary Lie groups as presented in [8]. This is founded in differential geometry and utilizes symplectic structures, Hamiltonian formalism, integrality conditions, line bundles with connection and prequantization. What we have done here is to develop these concepts for the graded case and to show indeed that these methods work for graded Lie groups. Hopefully at some later point, polarization and other aspects of geometric quantization (see e.g. [9], [13] and [14]) will be done in the graded case.

.3. One of the attractive aspects of graded manifolds is that it puts symmetric and skew-symmetric structures on a more or less equal footing. For example a graded symplectic structure has a "Riemannian" component. More precisely such a structure induces a symmetric inner product on the odd tangent spaces. Without going into details this may be already glimpsed in considering the co-adjoint orbit defined by an even linear functional f on a graded Lie algebra $\underline{g} = \underline{g}_0 + \underline{g}_1$. If $x, y \in \underline{g}_0$ then $[x,y] = -[y,x]$ whereas if $x, y \in \underline{g}_1$ then $[x,y] = [y,x]$. Thus $B_f(x,y) = \langle f,[x,y]\rangle$ defines an alternating bilinear form on $\underline{g}_0$ and a symmetric bilinear form on $\underline{g}_1$. In fact if $\underline{g}_f \subseteq \underline{g}$ is the isotropy subalgebra of $\underline{g}$ at f then the even and odd parts of the tangent space to the orbit through f, at f, are, respectively, $\underline{g}_0/(\underline{g}_f)_0$ and $\underline{g}_1/(\underline{g}_f)_1$ and B_f induces a non-singular alternating bilinear form and a non-singular symmetric bilinear form on these spaces, respectively. Now the orbit method for ordinary Lie algebras appears in its most transparent form when $\underline{g}$ is the Heisenberg Lie algebra. In that case (as first developed by Kirillov, [7]) the orbit method methods yields the Stone-von Neumann theorem. In the graded case one has a generalization of the Heisenberg Lie algebra, a special case of which occurs when $[\underline{g},\underline{g}] = \underline{g}_0$ is 1-dimensional.

Remark. This GLA makes an appearance in any graded symplectic manifold where the underlying manifold X reduces to a single point. It then appears as a Lie subalgebra of $A(X)$ under Poisson bracket.

As is well known from Clifford algebra theory this graded Lie algebra also (up to trivial scalar modifications) has a unique irreducible module -- namely the spin module associated with the Clifford algebra corresponding to the bilinear form B_f on $\underline{g}_1$. An expectation of a general representation theory for graded Lie groups would be that it should yield this Clifford algebra representation for this case as well as reproduce the older theory for ordinary Lie groups. But in fact this is the case for the graded prequantization theory developed here. Indeed the prequantized operator $\nabla_{\xi_\varphi} + 2\pi i\varphi$ is the sum of a differentiation and a multiplication operator which are related by a bilinear form. In case the underlying manifold reduces to point (see Remark above) this formula reduces to a well known formula $(i(x) + \varepsilon(x))$ given for example in Chevalley's book [1], p. 38, for the construction of the spin representation. This particular unifying aspect of the theory provided another motivation for our efforts here.

.4. The proofs of many of the propositions are not given or only indicated. For the most part this is done when it is felt that the proofs are straightforward or require only techniques of the ungraded theory. An exception is the proof of the graded Darboux theorem, which we don't really use and the completion of the proof of (the existence of quotient structures) Theorem 3.9. These will be given elsewhere.

We express our thanks to Professor K. Bleuler for the kindness he has shown us during the conference in Bonn, July 1975 when many of the results stated here were announced.

Contents

1. Algebraic Preliminaries

1.1. All gradings in this paper, unless mentioned otherwise, will be with respect to the ring $\mathbb{Z}_2 = \{0,1\}$. All vector spaces are over a field K. For the most part K will be either $\mathbb{R}$ or $\mathbb{C}$. However initially K is an arbitrary field except that $\text{char } K \neq 2$.

A vector space V over such a field K is a graded vector space if one has fixed subspaces V_0 and V_1, called, respectively, the even and odd parts of V and

$$V = V_0 + V_1$$

is a direct sum. The subspaces V_0 and V_1 are also referred to as the homogeneous components of V. A similar terminology is used for individual elements. An element $v \in V$ is called even if $v \in V_0$ and odd if $v \in V_1$. It is called homogeneous if v is either even or odd. If $v = v_{(0)} + v_{(1)} \in V$ where $v_{(i)} \in V_i$ then $v_{(0)}$ is the even component and $v_{(1)}$ the odd component of v, and the elements $v_{(0)}$, $v_{(1)}$ are called the homogeneous components of v. Also if $v \in V_i$ then i is called the degree (a degree if $v = 0$) of the homogeneous element v and if $v \neq 0$ we put

$$|v| = i. \tag{1.1.1}$$

We adopt the convention that if $v \in V$ and the notation $|v|$ is used we are tacitly assuming (without necessarily mentioning it) that $0 \neq v$ is homogeneous.

A subspace $W \subseteq V$ is called graded if $W = W \cap V_0 + W \cap V_1$. In such a case W is a graded vector space where $W_0 = W \cap V_0$ and $W_1 = W \cap V_1$.

If V and W are graded vector spaces then $V \otimes W = V \otimes_K W$ is graded where if $v \in V$, $w \in W$ are homogeneous then $v \otimes w$ is homogeneous and

$$|v \otimes w| = |v|\,|w|. \tag{1.1.2}$$

Also $\text{Hom}(V,W) = \text{Hom}_K(V,W)$ is graded where $\alpha \in \text{Hom}(V,W)$ is homogeneous of degree $|\alpha|$ if

$$\alpha(V_i) \subseteq W_{i+|\alpha|} \quad \text{for } i = 0,\ 1.$$

In particular $\text{End } V$ is graded.

Remark 1.1. We wish to call attention here to some terminology that will be used throughout the paper. If $\alpha \in \mathrm{Hom}(V,W)$ is described as a homomorphism of graded vector spaces we will understand that $|\alpha| = 0$. This will in particular be the case if V and W have additional structures such as graded algebras. Thus, for example, if α is a homomorphism of graded algebras then $|\alpha| = 0$.

One notes that K itself is graded where $K_0 = K$ and $K_1 = 0$.

All algebras B under consideration here will have a unity $1 \in B$. An algebra B is a graded algebra if B is a graded vector space such that $B_i B_j \subseteq B_{i+j}$ and such that $1 \in B_0$. For example, if V is a graded vector space End V is a graded algebra.

By a left module V for the graded algebra B we mean that V is a left module in the usual sense but that V is also a graded vector space and $B_i \cdot V_j \subseteq V_{i+j}$. A right module is defined similarly.

Two elements x, y in a graded algebra B are graded commutative if any homogeneous component of x is graded commutative with any homogeneous component of y. If x and y are homogeneous then they are graded commutative if either one is zero or otherwise

$$xy = (-1)^{|x|\,|y|} yx . \tag{1.1.3}$$

A graded algebra B is graded commutative (i.e. B is a graded commutative algebra) if one has (1.1.3) for $x, y \in B$.

1.2. Now if V is a left module for a graded commutative algebra B then V inherits a right module structure where we define

$$v \cdot b = (-1)^{|b|\,|v|} b \cdot v. \tag{1.2.1}$$

Similarly a left module structure is defined by a right module structure. Hence if B is a graded commutative algebra we need speak of only modules and not a left or right module for B. We note then that if V and W are B-modules, then so is $V \otimes_B W$ and the grading is given by (1.1.2). We note also by iteration, the tensor algebra $T_B(V)$ over V over B is then well defined as a bigraded $(\mathbb{Z} \oplus \mathbb{Z}_2)$ algebra where $B = T_B^0(V)$. The graded symmetric algebra of V

over B is the quotient $S_B(V)$ of $T_B(V)$ modulo the ideal I in $T_B(V)$ generated by all elements in $T_B(V)$ of the form $x \otimes y - (-1)^{|x||y|} y \otimes x$ for $x, y \in V$. The exterior algebra $\Lambda_B(V)$ of V over B is defined similarly except that $x \otimes y - (-1)^{|x||y|} y \otimes x$ is replaced by $x \otimes y + (-1)^{|x||y|} y \otimes x$. To see that this is well defined one notes first that there is a well defined B linear map

$$T : V \otimes_B V \longrightarrow V \otimes_B V$$

where

$$T(x \otimes y) = (-1)^{|x||y|} y \otimes x. \tag{1.2.1}$$

It follows that $S_B(V)$ and $\Lambda_B(V)$ are bigraded $(\mathbb{Z} \oplus \mathbb{Z}_2)$ algebras. Also $S_B(V)$ is a graded commutative algebra where, on the other hand, if $u_i \in \Lambda_B^{b_i}(V)_{j_i}$, $i = 1, 2$, then

$$u_1 u_2 = (-1)^{b_1 b_2 + j_1 j_2} u_2 u_1. \tag{1.2.2}$$

In case $B = K$ we drop the subscript B. One notes that if $\hat{V}$ denotes V when the grating is ignored then one has linear isomorphisms

$$S(V) \cong S(\hat{V}_0) \otimes \Lambda\hat{V}_1 \tag{1.2.3}$$

and

$$\Lambda V \cong \Lambda\hat{V}_0 \otimes S(\hat{V}_1) \tag{1.2.4}$$

where the symmetric and exterior algebras on the right sides are the usual ones.

If B and C are graded algebras then one induces a graded algebra structure in $B \otimes_K C = B \otimes C$ so that

$$(b_1 \otimes c_1)(b_2 \otimes c_2) = (-1)^{|c_1||b_2|} b_1 b_2 \otimes c_1 c_2. \tag{1.2.5}$$

Thus elements of the form $b \otimes 1$ and $1 \otimes c$ are graded commutative.

Remark 1.2. If one defines graded algebra structures on $\Lambda\hat{V}_i$ and $S(\hat{V}_i)$ so that $\hat{V}_i$ is graded according to i then it is clear that (1.2.3) is an algebra isomorphism. To make (1.2.4) an algebra isomorphism it is only necessary to make the elements of $\hat{V}_0$ and $\hat{V}_1$ anti commute and retain the given algebra structures in $S(\hat{V}_1)$ and $\Lambda\hat{V}_0$. Note that $S(\hat{V}_1)$ is commutative in the usual sense but it is not a graded commutative algebra.

1.3. A graded vector space $\underline{g} = \underline{g}_0 + \underline{g}_1$ together with a bilinear operation $[x,y]$ on $\underline{g}$ such that $[x,y] \in \underline{g}_{|x|+|y|}$ is called a graded Lie algebra (GLA) if

(1.3.1)

$$(1) \quad [x,y] = -(-1)^{|x||y|}[y,x]$$

$$(2) \quad (-1)^{|x||z|}[[x,y],z] + (-1)^{|y||x|}[[y,z],x] + (-1)^{|z||y|}[[z,x],y] = 0$$

for $x, y, z \in \underline{g}$. One defines subalgebra and ideal in the obvious way noting that a graded Lie subalgebra $\underline{h}$ is always assumed to be a graded subspace and hence has the structure of a graded Lie algebra.

Remark 1.3.1. If $\underline{g}$ is a graded Lie algebra then $\underline{g}_0$ is an ordinary Lie algebra and $\underline{g}_1$ is naturally a $\underline{g}_0$ module. The novelty is that $[x,y] = [y,x] \in \underline{g}_0$ for $x, y \in \underline{g}_1$. In particular $[x,x] \in \underline{g}_0$ need not be zero.

Henceforth we reserve $[\ ,\ ]$ for the bracket structure of a GLA. If $\underline{h}$ and $\underline{g}$ are GLA's then a homomorphism $\pi : \underline{h} \longrightarrow \underline{g}$ is a linear map of degree zero such that $\pi[x,y] = [\pi(x),\pi(y)]$ for $x, y \in \underline{h}$. If B is any graded (associative) algebra then B also has the structure of a GLA where $[x,y] = xy - (-1)^{|x||y|}yx$ for $x, y \in B$. In particular if V is a graded vector space then End V has the structure of a GLA. If $\underline{g}$ is a GLA then a representation of $\underline{g}$ on V is a homomorphism

$$\pi : \underline{g} \longrightarrow \text{End } V$$

of GLA's. If π is understood we refer to V as a $\underline{g}$-module.

Remark 1.3.2. If $\underline{g}$ is a GLA one notes that $\text{ad} : \underline{g} \longrightarrow \text{End } \underline{g}$ is a representation where $\text{ad } x(y) = [x,y]$.

If W is a graded vector space with a bilinear operation Q such that $Q(x,y) \in W_{|x|+|y|}$ then an operator $\alpha \in (\text{End } W)_i$ is called a derivation of (W,Q), or a derivation of (W,Q) of degree i, if upon writing $Q(x,y) = x \circ y$ one has

$$\alpha(x\circ y) = \alpha x \circ y + (-1)^{(x)i} x \circ \alpha y$$

for $x, y \in W$. An operator $\alpha \in \operatorname{End} W$ is called a derivation of (W,Q) if its homogeneous components are derivations. The space $\operatorname{Der}(W,Q)$ is easily seen to be a graded Lie subalgebra of $\operatorname{End} W$. If Q is understood we will simply write $\operatorname{Der} W$ for $\operatorname{Der}(W,Q)$ and note then that such a bilinear operation defines a graded Lie subalgebra

$$\operatorname{Der} W \subseteq \operatorname{End} W.$$

<u>Remark 1.3.3.</u> If $\underline{g}$ is a GLA then $\operatorname{ad} \cdot x \in \operatorname{Der} \underline{g}$ for any $x \in \underline{g}$ and hence

$$\operatorname{ad} : \underline{g} \longrightarrow \operatorname{Der} \underline{g}$$

is a homomorphism of GLA's.

The case where B is a graded commutative algebra deserves special consideration. Indeed this will play a major role in this paper. Mimicking the situation of the Lie algebra of all smooth vector fields on a manifold one notes

Remark 1.3.4. If B is a graded commutative algebra then $\operatorname{Der} B$ is a B-module where if $\xi \in \operatorname{Der} B$ and $f, g \in B$ then $f\xi \in \operatorname{Der} B$ where

(1.3.2) $$(f\xi)(g) = f(\xi g).$$

2. <u>Graded Manifolds</u>

2.1. Let X be an m-dimensional C^∞ manifold, not necessarily connected. A presheaf A on X is a correspondence which assigns to each open subset $U \subseteq X$ some abstract set $A(U)$ such that (1) if $V \subseteq U$ is open there is a map $\rho_{U,V} : A(U) \longrightarrow A(V)$, called the restriction map, and (2) the restriction maps satisfy the condition $\rho_{V,W} \circ \rho_{U,V} = \rho_{W,U}$ if $W \subseteq V \subseteq U$. We will use the notation $\rho_{U,V}$ for the restriction map, no matter what the presheaf in question is. A presheaf A is called a sheaf if in addition the following 2 further conditions are satisfied. (3) If $U = \bigcup_{i\in\Lambda} U_i$ is an open covering of an open set and $f, g \in A(U)$

then $\rho_{U,U_i}(f) = \rho_{U,U_i}(g)$ for all $i \in \Lambda$ implies $f = g$ and (4), if $h_i \in A(U_i)$ is given for each $i \in \Lambda$ such that $\rho_{U_i,U_i\cap U_j}(h_i) = \rho_{U_j,U_i\cap U_j}(h_j)$ for all $i,j \in \Lambda$ then there exists (unique by (3)) $h \in A(U)$ such that $\rho_{U,U_i}(h) = h_i$.

Example 2.1. Familiar examples of sheaves on X are the cases (a) where $A(U)$ equals $C^\infty(U)$, the commutative algebra of all real-valued C^∞ functions on U, (b) $A(U)$ equals Der $C^\infty(U)$, the Lie algebra of all smooth vector fields on U, (c) $A(U)$ equals Diff $C^\infty(U)$ the algebra of all smooth differential operators on U and (d) $A(U)$ equal $\Omega(U)$, the algebra of all smooth differential forms on U.

Generally for any sheaf A that will be considered here $A(U)$ will have some ring, Lie algebra or GLA structure. This may be specified by saying that A is a sheaf of rings, Lie algebras or GLA's etc. In any case we expect it will be clear what the algebra is and it will always be assumed that the restriction maps $\rho_{U,V}$ are morphisms of that algebra structure. Furthermore if $A(U)$ has the underlying structure of a graded vector space it will always be assumed that the $\rho_{U,V}$ are maps of graded vector spaces so that they are of degree zero. In any case $A(U)$ will have an underlying additive structure. In particular we can make the following definition. If $f \in A(U)$ and $V \subseteq U$ is open we say that f vanishes on V if $\rho_{U,V}(f) = 0$. One has also the notion of support. If $f \in A(U)$ we define sup f, the support of f to be the complement in U of the set of all $p \in U$ such that f vanishes in some neighborhood of p. It is clear that supp f is a closed subset of U.

2.2. Now assume that A is a sheaf of graded (over $\mathbb{Z}_2$) commutative algebras over $\mathbb{R}$ on X and that for any open set $U \subseteq X$ there is a homomorphism of graded algebras

$$A(U) \longrightarrow C^\infty(U), \qquad f \longmapsto \tilde{f} \tag{2.2.1}$$

which commute with restriction maps.

In saying that the map (2.2.1) is a homomorphism of graded algebras we of course assume that $C^\infty(U)$ is graded so that $C^\infty(U) = (C^\infty(U))_0$. Thus one has

(2.2.2) $$\tilde{f} = 0 \quad \text{if} \quad f \in (A(U))_1 .$$

The identity element in $A(U)$ is denoted by 1_U. By assumption (see §1.1) $1_U \in A(U)_0$. Since by assumption identities map into identities under homomorphisms, $\tilde{1}_U$ is necessarily the identity function on U.

A subalgebra $C(U) \subseteq (A(U))_0$ will be called a function factor (of $A(U)$) if $1_U \in C(U)$ and the map

(2.2.3) $$C(U) \longrightarrow C^{\infty}(U), \qquad f \longmapsto \tilde{f}$$

is an algebra isomorphism.

2.3. If $s, t \in A(U)_1$ then one necessarily has

(2.3.1) $$st = -ts.$$

Let $s_i \in A(U)_1$, $i = 1,\ldots,n$. We will say the s_i are algebraically independent if the product $s_1 \ldots s_n \neq 0$. A subalgebra $D(U) \subseteq A(U)$ will be called an exterior factor of $A(U)$ if $D(U)$, for some n, is generated by 1_U and n algebraically independent odd elements. In such a case it is clear that $D(U)$ is a graded subalgebra of $A(U)$ and the number n is uniquely determined since one clearly has $\dim D(U) = 2^n$.

Now if U is a non-empty open set then $(C(U),D(U))$ will be said to be splitting factors for $A(U)$ if $C(U)$ is a function factor of $A(U)$, $D(U)$ is an exterior factor of $A(U)$, and the map

(2.3.2) $$C(U) \otimes D(U) \longrightarrow A(U) \qquad f \otimes w \longmapsto fw$$

is a linear isomorphism. An open set $U \subseteq X$ will be said to be an A-splitting neighborhood of odd dimension n if there are splitting factors $(C(U),D(U))$ for $A(U)$ such that $\dim D(U) = 2^n$. One immediately has

<u>Lemma 2.3.</u> *If U is an A-splitting neighborhood of odd dimension n then n is unique.*

2.4. An m-dimensional manifold X together with a sheaf A of graded commutative algebras and homomorphisms (2.2.1) is called a graded manifold of dimension (m,n) if any non-empty open set can be covered by A-splitting neighborhoods of odd dimension n. This is denoted by saying that (X,A) is a graded manifold of dimension (m,n).

Now let (X,A) be a graded manifold of dimension (m,n). We will regard (X,A) as fixed for the remainder of §2. Let $U \subseteq X$ be an open set and let $A^1(U)$ be the set of all nilpotent elements in $A(U)$. It follows easily that $A^1(U)$ is a graded ideal since by (2.3.1) one has

$$A(U)_1 \subseteq A^1(U). \tag{2.4.1}$$

Remark 2.4.1. If U is an A-splitting neighborhood then $A^1(U)$ is in fact the ideal generated by $A(U)_1$.

The following proposition implies that $A^1(U)$ is the Jacobson radical of $A(U)$ (recall that $C^\infty(U)$ is a semi-simple ring).

Proposition 2.4.1. *For any open set $U \subseteq X$ the map $f \longmapsto \tilde{f}$ induces an exact sequence*

$$0 \longrightarrow A^1(U) \longrightarrow A(U) \longrightarrow C^\infty(U) \longrightarrow 0. \tag{2.4.2}$$

Remark 2.4.2. Note that if $A(U)$ has a function factor $C(U)$ then $C(U)$ splits the exact sequence (2.4.2). Conversely every splitting of (2.4.2) defines a function factor of $A(U)$. That is by Proposition 2.4 we may characterize a function factor of $A(U)$ as a subalgebra $C(U) \subseteq A(U)_0$ where $1_U \in C(U)$ and such that one has the linear space direct sum

$$A(U) = C(U) \oplus A^1(U). \tag{2.4.3}$$

The proof of Proposition 2.4.1 is obvious locally. The case for an arbitrary open set U follows from the next lemma which asserts that a partition of unity exists for graded manifolds.

Lemma 2.4. *Let $U = \bigcup_{j\in\Gamma} V_j$ be any covering of an open set U. Then there exists a locally finite refinement $U = \bigcup_{i\in\Lambda} U_i$ and elements $f_i \in A(U)_0$ such that* supp $f_i \subseteq U_i$ *and such that* $\sum_{i\in\Lambda} f_i = 1_U$.

To prove Lemma 2.4 one first observes that if W is any open set and $h \in A(W)$ is such that $\tilde{h} = 0$ then h is nilpotent. In fact, from the sheaf properties of A one certainly has that $h^{n+1} = 0$. This, however, implies that if $g \in A(W)$ is such that $\tilde{g}$ is the constant function 1 on W then g is invertible in $A(W)$. (Indeed, writing $g = 1_W + h$ one has $\tilde{h} = 0$ and hence h is nilpotent). As a consequence, one proves that if $A(W)$ has a function factor

$C(W)$, then for any $f \in C(W)$,

$$\text{supp } f = \text{supp } \tilde{f}. \qquad (2.4.4)$$

Now by the local existence of function factors we may find a locally finite refinement $U = \bigcup_{i \in \Lambda} U_i$ of the given covering such that $A(U_i)$ admits a function factor $C(U_i)$. Let $1 = \sum_{i \in \Lambda} \varphi_i$ be a partition of unity for functions in U where $\text{supp } \varphi_i \subseteq U_i$. Now let $g_i \in C(U_i)$ be such that $\tilde{g}_i = \varphi_i$. But then $\text{supp } g_i \subseteq U_i$ by (2.4.4) so that we can regard $g_i \in A(U)_0$ and hence, by the local finiteness, the sum $g = \sum g_i$ is a well defined element in $A(U)_0$ where $\tilde{g} = 1$. But then, from above, g is invertible. Finally one puts $f_i = g_i g^{-1}$ proving the lemma.

Another consequence of (2.4.4) is

Proposition 2.4.2. If $C(U)$ is a function factor in $A(U)$ then for any open set $V \subseteq U$ there exists a unique function factor $C(V)$ in $A(V)$ such that

$$\rho_{U,V}(C(U)) \subseteq C(V). \qquad (2.4.5)$$

Furthermore if V is an A-splitting neighborhood of, say, odd dimension k then V is an A-splitting neighborhood of odd dimension k for any non-empty open set $V \subseteq U$. Moreover, if $(C(U),D(U))$ are splitting factors for $A(U)$ then $(C(V),D(V))$ are splitting factors for $A(V)$ where $C(V)$ is given by (2.4.5) and $D(V) = \rho_{U,V} D(U)$.

It follows from Proposition 2.4.2 that if U is an A-splitting neighborhood of odd dimension k then $k = n$. Henceforth in such a case we will simply refer to U as an A-splitting neighborhood. One notes also that the odd dimension of (X,A) is uniquely defined and we will write $\dim (X,A) = (m,n)$.

2.5. Now if F is a smooth vector bundle over X and $U \subseteq X$ is open we will let $\Gamma(U,V)$ be the space of all smooth sections of F over U. The correspondence $U \longrightarrow \Gamma(U,F)$ is a sheaf. However it is a special sheaf in that

$\Gamma(U,F)$ has the natural structure of a $C^\infty(U)$ module.

Remark 2.5.1. One notes that all the sheaves considered in Example 2.1 are of this form. We wish however to emphasize that A is not of this form. That is, A does not arise from any vector bundle. In particular $A(U)$ is not a $C^\infty(U)$ module and it makes no sense to evaluate $f \in A(U)$ at a point p in U. Of course $A(U)$ admits a $C^\infty(U)$ module structure with every splitting of (2.4.2), but the point is that even if (2.4.2) splits there is no canonical splitting.

Even though A does not arise from a vector bundle one can associate with A another sheaf which does have this property. See Remark 2.6. Let $A^j(U)$ be the jth power of the nilpotent ideal $A^1(U)$. Thus $A^{j+1}(U) \subseteq A^j(U)$. The odd dimension n of (X,A) may be characterized by

Proposition 2.5.1. *For any non-empty open set* $U \subseteq X$ *one has* $A^n(U) \neq 0$ *and* $A^{n+1}(U) = 0$.

One thus has the sequence

$$0 = A^{n+1}(U) \subseteq A^n(U) \subseteq \ldots \subseteq A^1(U) \subseteq A(U) \tag{2.5.1}$$

of ideals in $A(U)$. We will regard $A(U)/A^1(U) = C^\infty(U)$ and hence if

$$\widetilde{A^j}(U) = A^j(U)/A^{j+1}(U) \tag{2.5.2}$$

then $\widetilde{A^j}(U)$ has the structure of a $C^\infty(U)$-module. However $U \longrightarrow \widetilde{A^j}(U)$ has only the structure of a presheaf. For each $p \in X$ let m_p be the maximal ideal in $C^\infty(X)$ given by putting $m_p = \{\varphi \in C^\infty(X) \mid \varphi(p) = 0\}$. One defines a vector space at p by putting

$$F_p^j(A) = \widetilde{A^j}(X)/m_p\widetilde{A^j}(X). \tag{2.5.3}$$

Remark 2.5.1. If $p \in U$ and U is substituted for X in (2.5.3) then the resulting space is canonically isomorphic to $F_p^j(A)$ and may be identified with it.

Now if $F^j(A) = \bigcup_{p \in X} F_p^j(A)$ then $F^j(A)$ has the structure of a smooth vector bundle over X with fiber $F_p^j(A)$ at p having dimension $\binom{n}{j}$.

Projections onto quotients clearly induces a map

(2.5.4) $$\tau_j : A^j(U) \longrightarrow \Gamma(U,(F^j(A))$$

and we recognize that $\Gamma(U,F^0(A)) = C^\infty(U)$ and $\tau_0(f) = \tilde{f}$ for any $f \in A^0(U) = A(U)$.

Now $F^n(A)$ is a real line bundle over X. But if U is open and $s_i \in A(U)_1$, $i = 1,\ldots,n$ then the product $s = s_1 \ldots s_n \in A^n(U)$ since $A(U)_1 \subseteq A^1(U)$. We will say that the s_i are an odd coordinate system in $A(U)$ (or simply in U) if $\tau_n(s) \in \Gamma(U,F^n(A))$ is a nowhere vanishing section of $F^n(A)$ over U. It is clear that if the s_i are an odd coordinate system in U then they are algebraically independent and hence if $D(U)$ is the algebra generated by the s_i and 1_U then $D(U)$ is an exterior factor in $A(U)$.

The following proposition guarantees that an odd coordinate system exists at least locally. It also implies that if $(C(U),D(U))$ are splitting factors then $C(U)$ and $D(U)$ are independent of one another.

Proposition 2.5.2. *Assume U is an A-splitting neighborhood and $(C(U),D(U))$ are splitting factors for $A(U)$. Then any n elements $s_1,\ldots,s_n$ in $D(U)_1$, which, with 1_U, generate $D(U)$, is an odd coordinate system in U. Conversely assume U is any open set such that $A(U)$ has a function factor $C(U)$ and an odd coordinate system $s_1,\ldots,s_n$. Then U is an A-splitting neighborhood. Furthermore $(C(U),D(U))$ are splitting factors for $A(U)$ where $D(U)$ is the subalgebra of $A(U)$ generated by 1_U and s_i.*

Let $U \subseteq X$ be any open set and assume $f_{ij} \in A(U)$, $i, j = 1,\ldots,k$. Regard $\{f_{ij}\}$ as a $k \times k$ matrix with coefficients in $A(U)$. But then $\{\tilde{f}_{ij}\}$ is a $k \times k$ matrix with coefficients in $C^\infty(U)$. The proof of Proposition 2.5.2 makes use of Lemma 2.5 below which in fact is used in a number of places.

Lemma 2.5. *The matrix $\{f_{ij}\}$ is invertible if and only if the matrix $\{\tilde{f}_{ij}\}$ is invertible, i.e. if and only if $\det\{\tilde{f}_{ij}\}$ is an every non-vanishing function in U.*

Using the surjectivity of the map $A(U) \longrightarrow C^\infty(U)$, (see Proposition 2.4.1), the proof of the invertibility of $\{f_{ij}\}$ is just a matrix version of the invertibility of g in the proof of Lemma 2.4. It reduces to the fact that if $h_{ij} \in A(U)$ and $\{\tilde{h}_{ij}\}$ is the zero matrix then $\{h_{ij}\}$ is necessarily nilpotent.

2.6. It will be convenient in this paper to introduce two types of index sets M_d and N_d. First of all $\mathbb{N}$ denotes the set of positive integers and $\mathbb{Z}_+ = \mathbb{N} \cup (0)$. If $d \in \mathbb{N}$ we let M_d denote the set of all sequences $\mu = (\mu_1,\ldots,\mu_k)$ where $\mu_i \in \mathbb{N}$ and $1 \leqslant \mu_1 < \ldots < \mu_k \leqslant d$. The length k of the sequence is denoted by $k(\mu)$ and one of course has

(2.6.1) $$k(\mu) \leqslant d.$$

Next let N_d denote the set of all sequences $\nu = (\nu_1,\ldots,\nu_d)$ where $\nu_i \in \mathbb{Z}_+$. We put

(2.6.2) $$|\nu| = \sum_{i=1}^{d} \nu_i .$$

Now assume that $U \subseteq X$ admits an odd coordinate system $s_1,\ldots,s_n \in A(U)_1$ For any $\mu \in M_n$ let

(2.6.3) $$s_\mu = s_{\mu_1} \ldots s_{\mu_k} .$$

As a consequence of Proposition 2.5.2 one has

Proposition 2.6. Assume $A(U)$ has a function factor $C(U)$ and an odd coordinate system $s_1,\ldots,s_n$. Then any $f \in A(U)$ can be uniquely written

(2.6.4) $$f = \sum_{\mu\in M_n} f_\mu s_\mu$$

where $f_\mu \in C(U)$.

Now put $F(A) = \bigoplus_{j=0}^{n} F^j(A)$. Then the fiber $F_p(A)$ of $F(A)$ at $p \in X$ has the structure of an exterior algebra over $F_p^1(A)$ (regarded as an ungraded vector space). But then with its graded structure $F_p(A)$ is a graded commutative algebra of dimension 2^n where $F_p^j(A) \subseteq (F_p(A))_{j \bmod 2}$. In particular $U \longrightarrow \Gamma(U,F(A))$ is a sheaf $\operatorname{Gr} A$ of graded commutative algebras.

Remark 2.6. It follows easily that $(X,\operatorname{Gr} A)$ is a graded manifold of dimension (m,n). The graded manifold $(X,\operatorname{Gr} A)$ has more structures than (X,A)

in that Gr A arise from a vector bundle on X. One might well ask why not impose the additional structure in the definition of a graded manifold and deal immediately with (X,Gr A). It turns out this additional structure is far too restrictive. It would impose such restrictions on the morphisms of graded manifolds so as to make the development in the paper impossible. This became clear to us when we had to consider the homogeneous spaces of graded Lie groups. An analogy with ordinary differential geometry is the restriction of one's consideration to only linear manifolds. Locally of course every manifold is isomorphic to a linear manifold. Similarly here one notes that if U is an A-split neighborhood then A(U) and Gr A(U) are isomorphic. There is no natural isomorphism but it follows easily from Proposition 2.6 that a choice of a function factor C(U) and an odd coordinate system $s_1 \ldots s_n$ in U sets up an isomorphism

(2.6.5) $$A(U) \longrightarrow \text{Gr } A(U).$$

2.7. Now if U is any open set and $r_1,\ldots,r_m \in A(U)_0$ then we will say that the r_i are an even coordinate system in U if U is a coordinate neighborhood in the usual sense and the functions $\tilde{r}_i \in C^\infty(U)$, $i = 1,\ldots,m$ are a coordinate system in the usual sense. The following result amounts to a sort of implicit function for graded manifolds.

<u>Theorem 2.7.</u> <u>If</u> r_i, $i = 1,\ldots,m$ <u>are an even coordinate system in an open set</u> $U \subseteq X$ <u>then there exists a unique function factor</u> $C(U) \subseteq A(U)$ <u>such that all</u> $r_i \in C(U)$.

We wish to exhibit the function factor C(U) in terms of some fixed function factor which we know exists locally. Assume then that U admits a function factor C(U)'. Thus $A(U) = C(U)' \oplus A^1(U)$ by (2.4.3) and hence we may write

(2.7.1) $$r_i = h_i + z_i$$

$i = 1,\ldots,m$ where the r_i are an even coordinate system in U, $h_i \in C(U)'$ and $z_i \in A^1(U)$. Now let $u_i = \tilde{r}_i \in C^\infty(U)$ so that one also has $\tilde{h}_i = u_i$. Now since C(U)' is isomorphic to $C^\infty(U)$ there exists a unique derivation ∂_i of C(U)' such that ∂_i corresponds to the derivation $\frac{\partial}{\partial u_i}$ of $C^\infty(U)$. For any

$\nu = (\nu_1,\ldots,\nu_m) \in N_m$ let $\partial_\nu = \partial_1^{\nu_1} \ldots \partial_m^{\nu_m} \in \text{End}\, C(U)'$, let $z^\nu = z_1^{\nu_1} \ldots z_m^{\nu_m}$, and let $\nu! = \nu_1! \ldots \nu_m!$. Since $A^{n+1}(U) = 0$ note that $z^\nu = 0$ if $|\nu| \geqslant n+1$. Now for any $g \in C(U)'$ let

(2.7.2)
$$\pi(g) = \sum_{\substack{\nu \in N_m \\ |\nu| \leqslant n}} \frac{1}{\nu!} (\partial_\nu g) z^\nu$$

so that $\pi(g) \in A(U)$. Let $C(U)$ be the image of the map π. Then

(2.7.3)
$$\pi : C(U)' \longrightarrow C(U)$$

is an algebra isomorphism. Moreover $C(U)$ is a function factor and $r_i \in C(U)$. Furthermore for any $g \in C(U)'$ one has

(2.7.4)
$$\tilde{g} = \tilde{\pi}(g).$$

Corollary to Theorem 2.7. *An open set $U \subseteq X$ admits an even coordinate system $r_i \in A(U)_0$ if and only if U is a coordinate neighborhood in the usual sense. Furthermore if $u_i \in C^\infty(U)$, $i = 1,\ldots,m$ are a coordinate system of functions then there exists $r_i \in A(U)_0$, $i = 1,\ldots,m$, (necessarily an even coordinate system) such that*

(2.7.5)
$$\tilde{r}_i = u_i.$$

Furthermore if $C(U) \subseteq A(U)_0$ is the unique function factor which contains the r_i then, the correspondence

$$\{r_1,\ldots,r_n\} \longrightarrow C(U)$$

sets up a bijection between all even coordinate systems in U satisfying (2.7.5) and all function factors in $A(U)$.

2.8. An open set $U \subseteq X$ is called an A-coordinate neighborhood if it is a coordinate neighborhood in the usual sense and $A(U)$ contains an odd coordinate system. If U is an A-coordinate neighborhood the elements $\{r_i,s_j\}$ in $A(U)$, $i = 1,\ldots,m$, $j = 1,\ldots,n$, is called an A-coordinate system if the r_i are an even coordinate system and the s_j are an odd coordinate system.

Remark 2.8.1. If U is an A-coordinate neigborhood note that U is also an A-splitting neighborhood. Furthermore if $\{r_i,s_j\}$ are an A-coordinate system in U and $C(U)$ is the unique function factor which contains the r_i and $D(U)$ is the algebra generated by 1_U and s_j then $(C(U),D(U))$ are splitting factors for $A(U)$.

Now for any $U \subseteq X$ let $\mathrm{Der}\, A(U)$ be the graded Lie algebra of all derivations of $A(U)$. Then as noted in Remark 1.3.4. $\mathrm{Der}\, A(U)$ is an $A(U)$-module. In fact the correspondence $U \longrightarrow \mathrm{Der}\, A(U)$ is a sheaf $\mathrm{Der}\, A$ of A-modules.

We adopt the following convention. A sheaf $U \longrightarrow B(U)$ on X will be called a sheaf of A-modules if $B(U)$ is an $A(U)$ module for any $U \subseteq X$ and the module structures are compatible with the restriction maps.

Now if $\xi \in (\mathrm{Der}\, A(U))_0$ then it follows easily that $A^1(U)$ is stable under ξ and hence ξ induces a derivation $\tilde{\xi}$ (that is, an ordinary vector field on U) of $C^\infty(U) = A(U)/A^1(U)$. The correspondence

$$(\mathrm{Der}\, A(U))_0 \longrightarrow \mathrm{Der}\, C^\infty(U), \qquad \xi \longrightarrow \tilde{\xi} \tag{2.8.1}$$

is in fact a Lie algebra epimorphism and one has (by definition)

$$\tilde{\xi}\tilde{f} = \widetilde{\xi f} \tag{2.8.2}$$

for any $\xi \in (\mathrm{Der}\, A(U))_0$, $f \in A(U)$.

Now assume U is an A-splitting neighborhood and $(C(U),D(U))$ are splitting factors. Let $\mathrm{Der}_{C(U)}\, A(U)$ be the set of all $\xi \in \mathrm{Der}\, A(U)$ which annihilate $C(U)$, i.e. all $\xi \in \mathrm{Der}\, A(U)$ which are $C(U)$ linear. Let $\mathrm{Der}_{D(U)}\, A(U)$ be defined similarly. It is clear that these two subspaces are graded Lie subalgebras. Also they are $A(U)$ submodules. Using (2.6.4) one proves

<u>Proposition 2.8.</u> <u>As linear spaces one has the direct sum</u>

$$\mathrm{Der}\, A(U) = \mathrm{Der}_{D(U)}\, A(U) \oplus \mathrm{Der}_{C(U)}\, A(U). \tag{2.8.3}$$

<u>(Note this is not a Lie algebra direct sum.)</u>

In the case of an A-coordinate neighborhood this becomes much more explicit.

<u>Theorem 2.8.</u> <u>Assume</u> U <u>is an</u> A-<u>coordinate neighborhood with</u> $\{r_i, s_j\}$ <u>as</u> A-<u>coordinates</u>. <u>Thus for any</u> i <u>there exists a unique derivation</u> $\frac{\partial}{\partial r_i} \in \mathrm{Der}\, A(U)$

such that $\frac{\partial r_k}{\partial r_i} = \delta_{ik} 1_U$ and $\frac{\partial s_j}{\partial r_i} = 0$. Also for any j there exists a unique derivation $\frac{\partial}{\partial s_j} \in \text{Der } A(U)$ such that $\frac{\partial s_\ell}{\partial s_i} = \delta_{i\ell} 1_U$ and $\frac{\partial r_i}{\partial s_j} = 0$. Furthermore $\frac{\partial}{\partial r_i} \in (\text{Der } A(U))_0$ and $(\frac{\partial}{\partial s_j}) \in (\text{Der } A(U))_1$ and any derivation $\xi \in \text{Der } A(U)$ can be uniquely written

(2.8.4)
$$\xi = \sum_{i=1}^{m} a_i \frac{\partial}{\partial r_i} + \sum_{j=1}^{n} b_j \frac{\partial}{\partial s_j}$$

where $a_i, b_j \in A(U)$.

Remark 2.8.2. Theorem 2.8 asserts that $\text{Der } A(U)$ is a free $A(U)$-module having $\frac{\partial}{\partial r_i}$ and $\frac{\partial}{\partial s_j}$ as basis. One notes also that the two sums in (2.8.4) are the components of ξ relative to the decomposition (2.8.3). Finally one notes that the $\mathbb{R}$-span $\underline{p}$ of the $\frac{\partial}{\partial r_i}$ and $\frac{\partial}{\partial s_j}$ is a commutative graded Lie algebra of dimension $m + n$. One has $\dim \underline{p}_0 = m$ and $\dim \underline{p}_1 = n$. In particular note that $[\frac{\partial}{\partial s_j}, \frac{\partial}{\partial s_\ell}] = \frac{\partial}{\partial s_j}\frac{\partial}{\partial s_\ell} + \frac{\partial}{\partial s_\ell}\frac{\partial}{\partial s_j} = 0$.

2.9. Now one defines the space of differential operators $\text{Diff}_k A(U) \subseteq \text{End } A(U)$ of degree k of $A(U)$ inductively on k as follows: $\text{Diff}_0 A(U)$ is the space of operators on $A(U)$ obtained by multiplication by elements in $A(U)$. Thus $\text{Diff}_0 A(U) \cong A(U)$ and we may identify $\text{Diff}_0 A(U)$ with $A(U)$. If $\text{Diff}_{k-1} A(U)$ has been defined then

(2.9.1)
$$\text{Diff}_k A(U) = \{\partial \in \text{End } A(U) \mid [\partial, f] \in \text{Diff}_{k-1} A(U) \text{ for all } f \in A(U)\}$$

Here recall that commutation is with respect to the graded structure in $\text{End } A(U)$. Thus if ∂ and f are homogeneous then $[\partial, f] = \partial f - (-1)^{|\partial||f|} f\partial$. One has $\text{Diff}_{k-1} A(U) \subseteq \text{Diff}_k A(U)$ and hence if $\text{Diff } A(U) = \bigcup_{k=0}^{\infty} \text{Diff}_k A(U)$ then $\text{Diff } A(U)$ is a graded subalgebra of $\text{End } A(U)$. Note that

(2.9.2)
$$\text{Diff}_1 A(U) = A(U) \oplus \text{Der } A(U).$$

One also notes that $U \longrightarrow \text{Diff } A(U)$ is a sheaf $\text{Diff } A$ of A-modules.

Now assume that U is an A-coordinate neighborhood and $\{r_i, s_j\}$ an A-coordinate system. For any $\mu = (\mu_1, \ldots, \mu_k) \in M_n$ let $\frac{\partial}{\partial s_\mu} = \frac{\partial}{\partial s_{\mu_1}} \cdots \frac{\partial}{\partial s_{\mu_k}} \in \text{Diff}_{k(\mu)} A(U)$ and for any $\nu = (\nu_1, \ldots, \nu_m) \in N_m$ let

$\frac{\partial^\nu}{\partial r} = (\frac{\partial}{\partial r_1})^{\nu_1} \cdots (\frac{\partial}{\partial r_m})^{\nu_m} \in \mathrm{Diff}_{|\nu|}\ A(U)$. One then has

<u>Proposition 2.9.1.</u> <u>Any element</u> $\partial \in \mathrm{Diff}\ A(U)$ <u>can be uniquely written as a finite sum</u>

(2.9.3) $$\partial = \Sigma\, a_{\mu\nu} \frac{\partial^\nu}{\partial r} \frac{\partial}{\partial s_\mu}$$

<u>where</u> $a_{\mu\nu} \in A(U)$. <u>That is,</u> $\mathrm{Diff}\ A(U)$ <u>is a free</u> $A(U)$-<u>module with basis</u> $\frac{\partial^\nu}{\partial r} \frac{\partial}{\partial s_\mu}$.

Now $A(U)$ has a natural topology for any open set $U \subseteq X$. Indeed for any compact subset $Z \subseteq U$ and any $\partial \in \mathrm{Diff}\ A(U)$ one defines a semi-norm $|\ \ |_{Z,\partial}$ on $\mathrm{Diff}\ A(U)$ by putting, for $f \in A(U)$,

(2.9.4) $$|f|_{Z,\partial} = \sup_{p \in Z} |\widetilde{\partial f}(p)|.$$

<u>Proposition 2.9.2.</u> <u>The vector space</u> $A(U)$ <u>with respect to all the semi norms</u> $|\ \ |_{Z,\partial}$ <u>is a complete locally convex space over</u> $\mathbb{R}$. <u>Furthermore any</u> $\partial \in \mathrm{Diff}\ A(U)$ <u>is a continuous operator on</u> $A(U)$ <u>with respect to this topology</u>.

When we speak of convergence in $A(U)$ it will always be with respect to the topology referred to in Proposition 2.9.2. We will also consider the case for the complexification

(2.9.5) $$A_{\mathbb{C}}(U) = A(U) \otimes_{\mathbb{R}} \mathbb{C}$$

of $A(U)$. One introduces a similar topology and it is again complete with respect to that topology.

<u>Remark 2.9.1.</u> One can be somewhat more explicit with regard to convergences in $A(U)$. Assume U is an A-split neighborhood with $C(U)$ as function factor and $s_1,\ldots,s_n$ an odd coordinate system. Let $f_j \in A(U)$, $j = 1,\ldots,$ be a sequence and let $f \in A(U)$. Now using the notation in (2.6.4) we may write $f_j = \sum_{\mu \in M_n} f_{j\mu} s_\mu$ and $f = \sum_{\mu \in M_n} f_\mu s_\mu$ where $f_{j\mu}, f_\mu \in C(U)$. Now $C(U)$ is isomorphic to $C^\infty(U)$. Furthermore the topology induced on $C(U)$ is equivalent to the topology it inherits from $C^\infty(U)$ using this isomorphism and the usual topology one puts in $C^\infty(U)$ in distribution theory. But now f_j converges to f if and

only if for each $\mu \in M_n$, $f_{j\mu}$ converges to f_μ.

Remark 2.9.2. Another point that should be made is that if $0 \neq f \in A(U)$ one can have $\tilde{f} = 0$ and even $\widetilde{\xi f} = 0$ for all $\xi \in \mathrm{Der}\, A(U)$. However the statement of Proposition 2.9.2 implies that for some $\partial \in \mathrm{Diff}\, A(U)$ one has $\widetilde{\partial f} \neq 0$.

2.10. Now consider the full dual $\mathrm{Hom}(A(X),\mathbb{R}) = A(X)'$ of $A(X)$. Regarding $\mathbb{R} = \mathbb{R}_0$ it follows of course that $A(X)'$ is a graded vector space.

Now if $p \in X$ an element $v \in A(X)'_j$ will be said to be a differentiation of $A(X)$ at p if

(2.10.1) $$v(fg) = (vf)\tilde{g}(p) + (-1)^{|v||f|}\tilde{f}(p)v(g).$$

An element $v \in A(X)'$ will be said to be a differentiation at p if both its homogeneous components are differentiations at p. The tangent space $T_p(X,A)$ of (X,A) at p is defined to be the space of all differentiations of $A(X)$ at p.

If U is any open set let $A|U$ be the restriction of the sheaf A to U. If $p \in U$ one has a natural isomorphism $T_p(U,A|U) \cong T_p(X,A)$ and these spaces will be identified.

Now if $p \in X$ and $v \in T_p(X,A)_0$ then v necessarily vanishes on $A^1(X)$ (see Remark 2.4.1) and hence v induces a differentiation at p of $C^\infty(X) = A(X)/A^1(X)$. That is, it defines an element in $T_p(X)$, the usual tangent space to X at p. Thus we have a map

(2.10.2) $$T_p(X,A)_0 \longrightarrow T_p(X).$$

Proposition 2.10. *One has* $\dim T_p(X,A) = m + n$. *In fact* $\dim T_p(X,A)_1 = n$ *and* $\dim T_p(X,A)_0 = m$. *Moreover* (2.10.2) *is an isomorphism.*

Henceforth we identify $T_p(X,A)_0$ with the usual tangent space $T_p(X)$ by the isomorphism (2.10.2) so one has

(2.10.3) $$T_p(X) = T_p(X,A)_0.$$

2.11. Consider the full dual $(A(X) \otimes A(X))'$ of $A(X) \otimes A(X) = A(X) \otimes_{\mathbb{R}} A(X)$. One has an injection

(2.11.1) $$0 \longrightarrow A(X)' \otimes A(X)' \longrightarrow (A(X) \otimes A(X))'$$

where if $v, w \in A(X)'$ then $v \otimes w$ is regarded as a linear functional on $A(X) \otimes A(X)$ by the formula

$$v \otimes w(f \otimes g) = (-1)^{|w||f|} v(f) w(g) \tag{2.11.2}$$

for $f, g \in A(X)$.

On the other hand one has a map

$$\Delta : A(X)' \longrightarrow (A(X) \otimes A(X))' \tag{2.11.3}$$

defined by the relation

$$\Delta v(f \otimes g) = v(fg) \tag{2.11.4}$$

for $v \in A(X)'$, $f, g \in A(X)$.

Now we will be very much concerned with the subspace $A(X)^*$ of $A(X)'$ defined as the set of all $v \in A(X)'$ such that $v(I) = 0$ for some ideal $I \subseteq A(X)$, where $\dim A(X)/I < \infty$. Now one knows that if $v \in A(X)'$ then $\Delta v \in A(X)' \otimes A(X)'$ if and only if $v \in A(X)^*$. Moreover

<u>Proposition 2.11.1</u>. <u>One has</u>

$$\Delta : A(X)^* \longrightarrow A(X)^* \otimes A(X)^* . \tag{2.11.5}$$

<u>Furthermore</u> (2.11.5) <u>induces on</u> $A(X)^*$ <u>the structure of a graded cocommutative coalgebra</u>.

<u>Remark 2.11.1</u>. See [10] for the theory of graded coalgebras. The statement of Proposition 2.11.1 is equivalent to the statement that if $(A(X)^*)'$ is the dual space to $A(X)^*$ and we express the duality by $\langle v,\lambda\rangle \in \mathbb{R}$ for $v \in A(X)^*$ and $\lambda \in (A(X)^*)'$ then the multiplication is $(A(X)^*)'$ defined by the formula

$$\langle v,\kappa\lambda\rangle = \langle \Delta v,\kappa \otimes \lambda\rangle \tag{2.11.6}$$

for $v \in A(X)^*$, $\kappa, \lambda \in (A(X)^*)'$ induces the structure of a graded commutative algebra on $(A(X)^*)'$. The pairing between tensor products is such that

$$\langle v \otimes w,\kappa \otimes \lambda\rangle = (-1)^{|w||\kappa|} \langle v,\kappa\rangle\langle w,\lambda\rangle . \tag{2.11.7}$$

<u>Remark 2.11.2</u>. This pairing seems to depend on whether we put $A(X)^*$ on the left or the right. However it is independent of this since if v, w, κ, and λ

are homogeneous and $\langle v,\kappa\rangle\langle w,\lambda\rangle \neq 0$ then $(-1)^{|\kappa||w|} = (-1)^{|v||\lambda|}$.

Now for any $p \in X$ let $\delta_p \in A(X)'$ be defined by $\delta_p(f) = \tilde{f}(p)$ for any $f \in A(X)$. Obviously δ_p is an algebra homomorphism and hence if $M_p(A(X))$ is the kernel of δ_p then $M_p(A(X))$ is a maximal ideal in $A(X)$.

Now for $k \in \mathbb{Z}_+$ let

$$A_p^k(X)^* = \{r \in A(X)' \mid v(M_p(A(X)))^{k+1} = 0\}$$

so that $A_p^k(X)^* \subseteq A_p^{k+1}(X)^*$.

<u>Remark 2.11.2</u>. One notes that $\mathbb{R}\delta_p = A_p^0(X)^*$. Also $T_p(X,A) \subseteq A_p^1(X)^*$. In fact one has $A_p^1(X)^* = T_p(X,A) \oplus \mathbb{R}\delta_p$.

Put

(2.11.8) $$A_p(X)^* = \bigcup_k A_p^k(X)^*.$$

Now if $\Delta : C \longrightarrow C \otimes C$ is a graded coalgebra an element $\delta \in C$ is called group-like if $0 \neq \delta \in C_0$ and $\Delta\delta = \delta \otimes \delta$. An element v is called primitive with respect to a group-like element δ if

$$\Delta v = \delta \otimes v + v \otimes \delta.$$

<u>Proposition 2.11.2</u>. <u>One has a direct sum</u>

(2.11.9) $$A(X)^* = \bigoplus_{p\in X} A_p(X)^*.$$

<u>Furthermore</u> $A_p(X)^*$ <u>is a graded subcoalgebra of</u> $A(X)^*$. <u>Also</u> δ_p <u>is a unique group-like element in</u> $A_p(X)^*$ <u>and the tangent space</u> $T_p(X,A)$ <u>is the set of all elements in</u> $A_p(X)^*$ <u>which are primitive with respect to</u> δ_p.

<u>Remark 2.11.3</u>. Note that Proposition 2.11.2 implies that the most general homomorphism $A(X) \longrightarrow \mathbb{R}$ is of the form δ_p for a unique $p \in X$. In fact if we identify p with δ_p so that

(2.11.10) $$X \subseteq A(X)^*$$

then this can be expressed by saying that X is the set of all group-like elements in $A(X)^*$. Also note that (even though we are dealing with algebras over $\mathbb{R}$ and not $\mathbb{C}$) Proposition 2.11.2 implies that the most general maximal ideal of finite codimension in $A(X)$ is uniquely of the form $M_p(A(X))$.

One may also characterize the coalgebra $A(X)^*$ in topological terms. For any $v \in A(X)'$ define the support of v, written $\sup v$, as the complement in X of the set of all $p \in X$ such that there exists a neighborhood U of p such that $v(f) = 0$ for all $f \in A(X)$ satisfying $\sup f \subseteq U$. Now let $\mathrm{Dis}_c\, A(X) \subseteq A(X)'$ be the topological dual of $A(X)$. As in ordinary distribution theory one has

Proposition 2.11.3. *Any* $v \in \mathrm{Dis}_c\, A(X)$ *has compact support.*

Remark 2.11.4. Although we shall not do so one may define $\mathrm{Dis}\, A(X)$, the set of distributions on the graded manifold (X,A). It then follows easily that $\mathrm{Dis}_c\, A(X)$ is exactly the set of the distributions of compact support. The relation of $A(X)^*$ to $\mathrm{Dis}_c\, A(X)$ is given in

Proposition 2.11.4. *One has*

$$A(X)^* \subseteq \mathrm{Dis}_c\, A(X) \subseteq A(X)'.$$

In fact $A(X)^*$ *is the set of all* $v \in \mathrm{Dis}_c\, A(X)$ *such that* $\sup v$ *is a finite set.*

Consistent with Proposition 2.11.4 we will on occasion refer to $A(X)^*$ as the set of distributions of finite support of (X,A).

2.12. Now let $T(X,A) = \bigcup_{p \in X} T_p(X,A)$. Then $T(X,A)$, referred to as the tangent bundle of (X,A), has the structure of a smooth vector bundle over X with fiber $T_p(X,A)$ at p. The space $\Gamma(U,T(X,A))$ of smooth sections of $T(X,A)$ over U will be denoted by $\mathrm{Der}\,(A(U),C^\infty(U))$ since the smooth sections of $T(X,A)$ over U may be identified with the set of all maps $\zeta : A(U) \longrightarrow C^\infty(U)$ such that

$$\zeta(fg) = (\zeta f)\tilde{g} + (-1)^{|\zeta||f|}\tilde{f}\zeta(g) \tag{2.12.1}$$

for $f, g \in A(U)$. If $T(X)$ is the usual tangent bundle then recalling (2.10.3) $T(X)$ is a subbundle of $T(X,A)$ and the space $\mathrm{Der}\, C^\infty(U)$ of all smooth vector fields is contained in $\mathrm{Der}\,(A(U),C^\infty(U))$. In fact as a graded vector space

$$\mathrm{Der}\, C^\infty(U) = (\mathrm{Der}\,(A(U),C^\infty(U)))_0\,. \tag{2.12.2}$$

Now in §2.8 we defined a map $(\mathrm{Der}\,A(U))_0 \longrightarrow \mathrm{Der}\,C^\infty(U)$, $\xi \longrightarrow \tilde{\xi}$. We now observe that this extends to a map

$$\mathrm{Der}\,A(U) \longrightarrow \mathrm{Der}\,(A(U),C^\infty(U)) \tag{2.12.3}$$

again denoted by $\xi \longrightarrow \tilde{\xi}$ where for any $\xi \in \mathrm{Der}\,A(U)$, and $f \in A(U)$ one has

$$\tilde{\xi}f = \widetilde{\xi f}. \tag{2.12.4}$$

Of course $\mathrm{Der}\,(A(U),C^\infty(U))$ is a $C^\infty(U)$ module and with respect to this module structure one has

$$\widetilde{g\xi} = \tilde{g}\tilde{\xi} \tag{2.12.5}$$

for any $g \in A(U)$, $\xi \in \mathrm{Der}\,A(U)$.

Now if $p \in U$ and $\zeta \in \mathrm{Der}\,(A(U),C^\infty(U))$ let $\zeta_p \in T_p(X,A)$ be the value of ζ at p.

<u>Proposition 2.12.1</u>. *Let* $\xi_i \in \mathrm{Der}\,A(U)$, $i = 1,\ldots,k$. *Assume that for each* $p \in U$ *the vectors* $(\tilde{\xi}_i)_p$, $i = 1,\ldots,k$ *span* $T_p(X,A)$ *(so that one must have* $k \geqslant n+m$*). Then for any* $\xi \in \mathrm{Der}\,A(U)$ *there exists* $f_i \in A(U)$ *such that*

$$\xi = \sum_{i=1}^{k} f_i\xi_i.$$

Furthermore if $k = n+m$ *then the* f_i *are unique so that* $\mathrm{Der}\,A(U)$ *is a free* $A(U)$ *module with basis* ξ_i.

The proof of Proposition 2.12.1 follows from Lemma 2.5, Lemma 2.4 (partition of unity) and that if the notation is as in Theorem 2.8 then

$$\left(\widetilde{\frac{\partial}{\partial r_i}}\right)_p, \left(\widetilde{\frac{\partial}{\partial s_j}}\right)_p \quad \text{is a basis of } T_p(X) \text{ and } T_p(X,A)_1 \text{ respectively.} \tag{2.12.6}$$

2.13. Now if U is a non-empty open set then $(U,A|U)$ is a graded manifold of dimension (m,n). But if $p \in U$ then we may clearly identify $((A|U)_p(U))^*$ with $A_p(X)^*$ so that if we write $A(U)^*$ for $(A|U)(U)^*$ then, by Proposition 2.11.2,

$$A(U)^* \subseteq A(X)^*. \tag{2.13.1}$$

<u>Remark 2.13.1</u>. One notes that (2.13.1) is an inclusion of graded cocommutative coalgebras.

On the other hand one of course has $A(U)^* \subseteq A(U)'$, the full dual of $A(U)$, so that the elements of $A(U)^*$ may be applied as linear functionals to elements of either $A(U)$ or $A(X)$. Now $A(U)$ is a left Diff $A(U)$ module and hence $A(U)'$ has the structure of a right Diff $A(U)$ module where if $r \in A(U)'$, $\partial \in \mathrm{Diff}\, A(U)$ then $v \cdot \partial \in A(U)'$ is defined by

$$(v\cdot\partial)(f) = v(\partial f) \tag{2.13.2}$$

for any $f \in A(U)$.

On the other hand if $\partial \in \mathrm{Diff}_k\, A(U)$ and $p \in U$ then for $d \geq k$

$$\partial(M_p(A(U)))^d \subseteq M_p(A(U))^{d-k} \tag{2.13.3}$$

where $M_p(A(U))$ is the kernel of δ_p as applied to $A(U)$. This however implies that if $v \in A_p(X)^*$ then $v \cdot \partial \in A_p(X)^*$. Thus one has

<u>Proposition 2.13.1</u>. $A(U)^*$ <u>is a right</u> Diff $A(U)$ <u>module where if</u> $\partial \in \mathrm{Diff}\, A(U)$, $v \in A(U)^*$ <u>then</u> $v \cdot \partial \in A(U)^*$ <u>is defined by</u> (2.13.2). <u>Furthermore if</u> $p \in X$, $A_p(X)^* \subseteq A(U)^*$ <u>is a submodule</u>.

One notes in particular that if $p \in U$ and $\xi \in \mathrm{Der}\, A(U)$ then

$$\delta_p \cdot \xi = (\tilde{\xi})_p \in T_p(X,A). \tag{2.13.4}$$

Now let $p \in X$ and assume U is an A-coordinate neighborhood with A-coordinates $\{r_i, s_j\}$ where $p \in U$. Let $\frac{\partial^\nu}{\partial r}\frac{\partial}{\partial s_\mu}$ be as in Proposition 2.9.1 so that for $|\nu| + k(\mu) \leq k$ one has $\frac{\partial^\nu}{\partial r}\frac{\partial}{\partial s_\mu} \in \mathrm{Diff}_k\, A(U)$. Then using a partial Taylor expansion around p for the function factor containing the r_i one has

<u>Proposition 2.13.2</u>. <u>The elements</u> $\delta_p \cdot \frac{\partial^\nu}{\partial r}\frac{\partial}{\partial s_\mu}$ <u>for</u> $(\nu,\mu) \in N_m \times M_n$ <u>where</u> $|\nu| + k(\mu) \leq k$ <u>are a basis of</u> $A_p^k(X)^*$.

Now assume $U \subseteq X$ is an arbitrary open set and let E be a graded subalgebra of Diff $A(U)$. For any $k \in \mathbb{Z}_+$ let $E_k = E \cap \mathrm{Diff}_k\, A(U)$. Given $p \in U$ we will say that E is infinitesimally transitive at p if

$$A_p^k(X)^* = \delta_p \cdot E_k \tag{2.13.5}$$

for any $k \geq 0$. In particular if E is infinitesimally transitive at p then

$$A_p(X)^* = \delta_p \cdot E \tag{2.13.6}$$

is a cyclic E-module with δ_p as generator. As a consequence of Proposition 2.12.1 and 2.13.2 one has

Proposition 2.13.3. *Let* E *be a graded subalgebra of* Diff $A(U)$. *Let* $p \in U$. *Assume that for any* $v \in T_p(X,A)$ *there exists* $\xi \in E \cap \text{Der } A(U)$ *such that* $\tilde{\xi}_p = v$. *Then* E *is infinitesimally transitive at* p.

If $Z \subseteq U$ is any set and $E \subseteq \text{Diff } A(U)$ is a graded subalgebra we will say that E is infinitesimally transitive on Z if E is infinitesimally transitive at all points of Z. Obviously Diff $A(U)$ is infinitesimally transitive on U.

2.14. In dealing with a graded manifold (X,A) one finds that the graded coalgebra $A(X)^*$ is easier to deal with than the graded algebra $A(X)$. In effect $A(X)^*$ plays the role of "points". This will be clear when we consider morphisms of graded manifolds in §2.15. Also in the construction of new graded manifolds from old ones (see section 2.18) $A(X)^*$ comes into view much more readily than $A(X)$. This is also true in the case of a graded Lie group. See section 3. In that case $A(X)^*$ is a graded Hopf algebra which one easily constructs from an ordinary Lie group and the universal enveloping algebra of a graded Lie algebra. One may therefore take the point of view that $A(X)^*$ is the primary object. In this section we will deal with the question as to how the sheaf A is determined by the graded coalgebra $A(X)^*$.

Now let (X,A) be a graded manifold and let $p \in X$. Put

(2.14.1) $$C_p^\infty(X)^* = \{v \in A_p(X)^* \mid v(A^1(X)) = 0\}.$$

But then since $A(X)/A^1(X) = C^\infty(X)$ we may regard $C_p^\infty(X)^* \subseteq C^\infty(X)'$ where $C^\infty(X)'$ is the full dual to $C^\infty(X)$. Clearly $\delta_p \in C_p^\infty(X)^*$ and $T_p(X) \subseteq C_p^\infty(X)^*$. In fact

Proposition 2.14.1. *One has* $C_p(X)^* \subseteq (A_p(X)^*)_0$ *is a subcoalgebra of* $A_p(X)^*$. *Furthermore* $C_p(X)^*$ *is the set of all distributions, in the usual sense, whose support is the point* p.

Remark 2.14.1. In the notation of Proposition 2.13.2 note that the elements $\delta_p \cdot \frac{\partial^\nu}{\partial r}$ over all $\nu \in N_m$ are a basis of $C_p(X)^*$.

It follows that if $U \subseteq X$ is open and $C^{\infty}(U)^*$ is the set of all distributions on U, in the usual sense, of finite support them

(2.14.2) $$C^{\infty}(U)^* = \bigoplus_{p \in U} C_p^{\infty}(X)^* \subseteq A(U)^*.$$

Also one has the characterization

(2.14.3) $$C^{\infty}(U)^* = \{v \in A(U)^* \mid v(A^1(U)) = 0\}.$$

Now certainly $C^{\infty}(U)^*$ and $C^{\infty}(U)$ are non-singularly paired. Consequently we may regard $C^{\infty}(U) \subseteq (C^{\infty}(U)^*)'$ where $(C^{\infty}(U)^*)'$ is the full dual of $C^{\infty}(U)^*$. Thus given a linear functional φ on $C^{\infty}(U)^*$ it makes sense to say that φ is a C^{∞} function on U. Now, by Remark 2.9.2, $A(U)$ and $A(U)^*$ are non-singularly paired so that if $(A(U)^*)'$ is the full dual of $A(U)^*$ then

(2.14.4) $$A(U) \subseteq (A(U)^*)'.$$

Furthermore since $A(U)^*$ has the structure of a graded cocommutative coalgebra $(A(U)^*)'$ has the structure of a graded commutative algebra and (2.14.4) is an inclusion of graded commutative algebras. We wish to characterize the subalgebra $A(U)$ of $(A(U)^*)'$. Observe that since $A(U)^*$ has the structure of a right Diff $A(U)$ module then $(A(U)^*)'$ has the structure of a left Diff $A(U)$ module. Of course $A(U)$ is a submodule. The following proposition makes use of Proposition 2.6. In a sense we can regard it as a generalization of the notion of smoothness for ordinary manifolds.

<u>Proposition 2.14.2.</u> <u>Let</u> E <u>be any graded subalgebra of</u> Diff $A(X)$ <u>which is infinitesimally transitive on</u> U (e.g. E = Diff $A(X)$). <u>Then</u> $A(U)$ <u>is the set of all</u> $f \in (A(U)^*)'$ <u>such that the restriction</u> $\partial \cdot f \mid C^{\infty}(U)^*$ <u>is a</u> C^{∞} <u>function on</u> U <u>for all</u> $\partial \in E$.

<u>Remark 2.14.2.</u> One consequence of Proposition 2.14.2 is the observation that the complete structure of a graded manifold (X,A) is determined by the one graded commutative algebra $A(X)$. Indeed given $A(X)$ then first of all, by Remark 2.11.3, X itself is determined by considering the set of all homomorphisms $A(X) \longrightarrow \mathbb{R}$. Furthermore the topology on X is determined from the weak topology on $A(X)'$. Also the C^{∞} structure on X is determined since $A(X)/A^1(X) = C^{\infty}(X)$

where $A^1(X)$ is the set of all nilpotent elements in $A(X)$. Next $A(X)$ determines Diff $A(X)$ and the graded coalgebra $A(X)^*$. But since we know the topology, $A(X)$ also determines the graded subcoalgebras $A(U)^*$. But then the sheaf A is given by Proposition 2.14.2.

2.15. Let (X,A) and (Y,B) be two graded manifolds. As might be expected from Remark 2.14.2 one should be able to describe a morphism of (X,A) to (Y,B) purely in terms of the algebras $A(X)$ and $B(Y)$. Indeed by definition a morphism

(2.15.1) $$\sigma : (X,A) \longrightarrow (Y,B)$$

of graded manifolds is just a homomorphism

(2.15.2) $$\sigma^* : B(Y) \longrightarrow A(X)$$

of graded algebras. If $v \in A(X)^*$ so that v vanishes on an ideal of finite codimension in $A(X)$ it follows that the linear functional $\sigma_* v$ on $B(Y)$ defined by

(2.15.3) $$(\sigma_* v)(g) = v\sigma^*(g)$$

for $g \in B(Y)$ vanishes on an ideal of finite codimension of $B(Y)$ so that $v \longmapsto \sigma_* v$ defines a map

(2.15.4) $$\sigma_* : A(X)^* \longrightarrow B(Y)^*$$

which is a morphism of graded coalgebras. In particular group-like elements map into group-like elements and hence if σ_X is the restriction of σ_* to X then

(2.15.5) $$\sigma_X : X \longrightarrow Y.$$

<u>Remark 2.15.1</u>. Since $A(X)^*$ is non-singularly paired to $A(X)$ one notes that a morphism $\sigma : (X,A) \longrightarrow (Y,B)$ is determined by the map $\sigma_* : A(X)^* \longrightarrow B(Y)^*$. However unlike the case of ordinary manifolds σ is not determined by the point map σ_X. In effect in the case of an ordinary manifold $A(X)$ is a semi-simple ring which is determined by the set of maximal ideals X. This is not the case, in general, for a graded manifold.

<u>Proposition 2.15.1</u>. <u>The map</u> σ_X <u>is a smooth map of</u> C^∞ <u>manifolds</u>. <u>Furthermore for any open set</u> $V \subseteq Y$ <u>there exists a unique homomorphism</u>

(2.15.6) $$\sigma^* : B(V) \longrightarrow A(\sigma_X^{-1}(V))$$

which is compatible with the restriction maps $\rho_{Y,V}$ *and* $\rho_{X,\sigma_X^{-1}(V)}$.

Moreover the map (2.15.2) *is an isomorphism of graded algebras if and only if* (1) σ_X *is a diffeomorphism and* (2) *the map* (2.15.6) *is an isomorphism of graded algebras for all open* $V \subseteq Y$.

As a consequence of Proposition 2.15.1 we are justified in saying that σ is an isomorphism of graded manifolds if σ^* is an isomorphism of graded algebras. Of course given two graded manifolds (X,A) and (Y,B) we say they are isomorphic if there is an isomorphism $\sigma : (X,A) \longrightarrow (Y,B)$.

2.16. Assume $\sigma : (X,A) \longrightarrow (Y,B)$ is a morphism of graded manifolds. Let $p \in X$ and put $q = \sigma_*(p) \in Y$. One easily has

(2.16.1) $$\sigma_* : A_p(X)^* \longrightarrow B_q(Y)^*$$

and in fact for any $k \in \mathbb{Z}_+$

(2.16.2) $$\sigma_* : A_p^k(X)^* \longrightarrow B_q^k(Y)^*$$

Furthermore primitive elements go into primitive elements so that if we let $d\sigma$, referred to as the differential of σ, be the restriction of σ_* to the tangent bundle $T(X,A) \subseteq A(X)^*$ then

(2.16.3) $$d\sigma : T_p(X,A) \longrightarrow T_q(Y,B).$$

As with ordinary manifolds one notes

Remark 2.16. If (2.16.3) has maximal possible rank (i.e. $\min\{\dim T_p(X,A), \dim T_q(Y,B)\}$) then it has this rank for all p' in some neighborhood of p.

The familiar fact that the differential of a map of ordinary manifolds carries a great deal of information about the local behavior of the map is still valid for graded manifolds. Let $q \in V \subseteq Y$ where V is a B-coordinate neighborhood with B-coordinates $\{r_i, s_j\}$. Let $p \subset U \subseteq X$ where U is an A-coordinate neighborhood such that $\sigma_X(U) \subseteq V$.

Proposition 2.16.1. If (2.16.3) *is injective we can choose* U *and* V *such that the restriction of* $\{\sigma^* r_i, \sigma^* s_j\}$ *to* U *contains an* A-*coordinate system for*

U. *If* (2.16.3) *is surjective then we can choose* U *and* V *such that the restriction of* $\{\sigma^* r_i, \sigma^* s_j\}$ *to* U *can be completed to an* A-*coordinate of* U.

The proof of Proposition 2.16.1 relies heavily on Theorem 2.7 and Proposition 2.5.2. Now applying Proposition 2.13.2 one has

Proposition 2.16.2. *If* (2.16.3) *is injective we can choose* U *such that*

(2.16.4) $$\sigma_* : A(U)^* \longrightarrow B(V)^*$$

is injective and such that

(2.16.5) $$\rho_{X,U} A(X) = \rho_{X,U}(\sigma^*(B(Y))).$$

If (2.16.3) *is surjective we can choose* U *and* V *such that* (1) $\sigma_X \mid U$ *is an open map where* $\sigma_X(U) = V$, (2)

(2.16.6) $$\rho_{\sigma_X^{-1} V,U} \circ \sigma^* : B(V) \longrightarrow A(U)$$

is injective and (3) *for all* $p' \in U$ *and* $k \in \mathbb{Z}_+$

(2.16.7) $$\sigma_* : A^k_{p'}(X)^* \longrightarrow B^k_{q'}(X)^*$$

is surjective where $q' = \sigma_* p'$. *In particular* (3) *implies*

(2.16.8) $$\sigma_* : A(U)^* \longrightarrow B(V)^*$$

is surjective.

One can also detect local isomorphisms using (2.16.3).

Theorem 2.16. *Assume that* (2.16.3) *is a linear isomorphism*. *Then we can choose* U *and* V *such that* $\sigma_X : U \longrightarrow V$ *is a diffeomorphism and if* $\tau^* = \rho_{\sigma_X^{-1} V,U} \circ \sigma^*$ *then* $\tau^* : B(V) \longrightarrow A(U)$ *is an isomorphism of graded algebras defining an isomorphism*

(2.16.9) $$\tau : (U, A|U) \longrightarrow (V, B|V)$$

of graded manifolds.

The question of a global isomorphism is then easily dealt with using the differential.

Corollary to Theorem 2.16. *Let* $\sigma : (X,A) \longrightarrow (Y,B)$ *be a morphism of graded manifolds such that* $\sigma_X : X \longrightarrow Y$ *is bijective and*

$$d\sigma : T_p(X,A) \longrightarrow T_{\sigma_*(p)}(Y,B)$$

is a linear isomorphism for all $p \in X$. *Then* σ *is an isomorphism of graded manifolds.*

2.17. If (X,A) and (Y,B) are graded manifolds and

(2.17.1) $$\lambda : A(X)^* \longrightarrow B(Y)^*$$

is a morphism of graded coalgebras then we will say that λ is smooth if $\lambda = \sigma_*$ where $\sigma : (X,A) \longrightarrow (Y,B)$ is a morphism of graded manifolds. Of course if λ is smooth then σ is necessarily unique (see Remark 2.15.1).

By taking the transpose of (2.17.1) one has a map

(2.17.2) $$B(Y) \longrightarrow (A(X)^*)'.$$

The problem of smoothness is only a question of knowing that the image of (2.17.2) lies in $A(X)$.

Remark 2.17.1. If λ is injective and $\lambda(p) = q$ where $q \in Y$ it necessarily follows that p is group-like so that $p \in X$. For a similar reason regarding primitive elements it then follows that if $\lambda(u) = v$ and $v \in T_q(Y,B)$ one must have $u \in T_p(X,A)$. Also if $v \in B_q^k(Y)$ then $u \in A_p^k(X)$.

To detect the injectivity of a smooth map (2.17.1) it is only necessary to restrict one's attention to the tangent bundle.

Proposition 2.17.1. *Let* $\sigma : (X,A) \longrightarrow (Y,B)$ *be a morphism of graded manifolds. Then* $\sigma_* : A(X)^* \longrightarrow B(Y)^*$ *is injective if and only if* (1) $\sigma_X : X \longrightarrow Y$ *is injective and* (2) $\sigma_* : T_p(X,A) \longrightarrow T_{\sigma_*(p)}(Y,B)$ *is injective for all* $p \in X$.

The proof follows from the injectivity of (2.16.4). With regard to the bijectivity of (2.17.1) one has

Proposition 2.17.2. *Let* $\sigma : (X,A) \longrightarrow (Y,B)$ *be a morphism of graded manifolds. Then* $\sigma_* : A(X)^* \longrightarrow B(Y)^*$ *is a bijection if and only if* σ *is an isomorphism of graded manifolds.*

henceforth shall be identified. The following definition is then unambiguous. Let $H \subseteq B(Y)^*$ be any subset such that $X = H \cap Y$ has the structure of a submanifold of Y. We then say that (X,H) or simply H, if the manifold structure on X is understood, defines a graded submanifold of (Y,B) if there exists a graded submanifold (X,A) such that $A(X)^* = H$.

One has a familiar tautology.

Proposition 2.17.4. Let (Y,B) be a graded manifold. Assume $H \subseteq B(Y)^*$ and $X' = H \cap Y$ has the structure of a submanifold of Y. Then H defines a graded submanifold of (Y,B) if and only if there exists a graded manifold (X,A) and a morphism $\sigma : (X,A) \longrightarrow (Y,B)$ such that (1) $\sigma_* : A(X)^* \longrightarrow B(Y)^*$ is injective where $H = \sigma_*(A(X)^*)$ and $\sigma_X : X \longrightarrow X'$ is smooth.

2.18. We wish now to show how new graded manifolds may be constructed from previously given ones.

Assume Y is a C^∞ manifold and $R(Y) \subseteq C^\infty(Y)$ is an algebra of real valued smooth functions on Y. If $R(Y)^*$ denotes the coalgebra of all linear functionals on $R(Y)$ which vanish on an ideal of finite codimension in $R(Y)$ then restriction induces a morphism of coalgebras

$$C^\infty(Y)^* \longrightarrow R(Y)^*. \tag{2.18.1}$$

We will say $R(Y)$ separates $C^\infty(Y)^*$ if the map (2.18.1) is injective.

Remark 2.18.1. It is easy to see that $R(Y)$ separates $C^\infty(Y)^*$ if and only if $R(Y)$ separates the points of Y and for each $q \in Y$ the differentials $(d\varphi)_q$, for $\varphi \in R(Y)$, span the contangent space of Y at q.

We cite three examples of a subalgebra $R(Y)$ which separates $C^\infty(Y)^*$ that we will be concerned with.

Example (1). Assume $Y = V$ is an ungraded real finite dimensional vector space. Put $R(Y) = S(V')$ where $S(V')$ is the symmetric algebra over the dual V' of V so that $R(Y)$ is the algebra of polynomial functions on Y.

Example (2). Assume $Y = X \times Z$ is a product of two manifolds X and Z. Then one has a natural injection $C^\infty(X) \otimes C^\infty(Z) \longrightarrow C^\infty(Y)$. Put $R(Y) = C^\infty(X) \otimes C^\infty(Z)$.

Proof. This follows from the Corollary to Theorem 2.16 and Remark 2.17.1.

Now given graded manifolds (X,A) and (Y,B) and a morphism $\sigma : (X,A) \longrightarrow (Y,B)$ we will say that (X,A,σ) or simply (X,A), if σ is understood, is a graded submanifold of (Y,B) if $A(X)^* \subseteq B(Y)^*$ and $\sigma_* : A(X)^* \longrightarrow B(Y)^*$ is the injection map.

Remark 2.17.2. If (X,A) is a graded submanifold of (Y,B) one notes that X is a submanifold of Y in the usual sense. Also if $p \in X$ then

(2.17.3) $$A_p^k(X)^* \subseteq B_p^k(Y)^*$$

and

(2.17.4) $$T_p(X,A) \subseteq T_p(Y,B).$$

In particular if $\dim(Y,B) = (m,n)$ and $\dim(X,A) = (m_1,n_1)$ then $m \geqslant m_1$ and $n \geqslant n_1$. One also notes that if $U \subseteq Y$ is an open set then $(U,B|U)$ is a graded submanifold of (Y,B).

Proposition 2.17.3. *Assume that (X',A') is a graded submanifold of (Y,B). Assume also that (X,A) is a graded manifold and $\sigma : (X,A) \longrightarrow (Y,B)$ is a morphism of graded manifolds such that $\sigma_*(A(X)^*) \subseteq A'(X')^*$ so that*

(2.17.5) $$\sigma_* : A(X)^* \longrightarrow A'(X')^*$$

is a morphism of graded algebras and one has a map

(2.17.6) $$\sigma_X : X \longrightarrow X'.$$

Then (2.17.5) is smooth if (2.17.6) is a smooth (or just continuous) map of manifolds.

Proof. Taking the transpose of (2.17.5) one has a map $A'(X') \longrightarrow (A(X)^*)'$. But if (2.17.5) is continuous the image of the map is in $A(X)$, since by (2.16.5) every element of $A'(X')$ locally can be obtained by restricting an element of $B(Y)$. Q.E.D.

Now if (X,A) and (X',A') are graded submanifolds of (Y,B) such that $A(X)^* = A'(X')^*$ and the identity map $X \longrightarrow X'$ is smooth then by Proposition (2.17.3) (X,A) and (X',A') are canonically isomorphic graded manifolds and

Example (3). Assume $Y = T^*(Z)$ is the cotangent bundle of a manifold Z. Now the space, $\mathrm{Der}\, C^\infty(Z)$, of all smooth vector fields on Z is a module over $C^\infty(Z)$. Let $S_{C^\infty(Z)}(\mathrm{Der}\, C^\infty(Z))$ be the symmetric algebra of $\mathrm{Der}\, C^\infty(Z)$ over $C^\infty(Z)$. Also let $\pi : Y \longrightarrow Z$ be the bundle projection map. Now one defines an algebra homomorphism

$$\gamma : S_{C^\infty(Z)}(\mathrm{Der}\, C^\infty(Z)) \longrightarrow C^\infty(Y) \tag{2.18.2}$$

where if $b \in Y$ one has $\gamma(\varphi)(b) = \varphi(\pi(b))$ if $\varphi \in C^\infty(Z)$ and $\gamma(\xi)(b) = \langle \xi_{\pi(b)}, b\rangle$ for $\xi \in \mathrm{Der}\, C^\infty(Z)$ where $\xi_p \in T_p(Z)$ is the value of ξ at $p \in Z$. It is not difficult to see that γ is injective. We put $R(Y)$ equal to the image of γ.

Now assume that Y is a manifold, $Q(Y)$ is a graded commutative algebra over $\mathbb{R}$ and one has a homomorphism

$$\psi : Q(Y) \longrightarrow C^\infty(Y) . \tag{2.18.3}$$

Assume also that the image $R(Y)$ separates $C^\infty(Y)^*$ so that one has an injective morphism

$$C^\infty(Y)^* \longrightarrow Q(Y)^*$$

of graded cocommutative coalgebras where $Q(Y)^*$ is the set of all linear functionals on $Q(Y)$ which vanish on an ideal in $Q(Y)$ of finite codimension. For convenience here we identify $C^\infty(Y)^*$ with its image in $(Q(Y)^*)_0$. In particular if $U \subseteq Y$ is any open set then $(C^\infty(U))^* \subseteq Q(Y)^*$.

Now let $\mathrm{Der}\, Q(Y)$ be the graded Lie algebra of all derivations of $Q(Y)$ and let $\mathcal{D} \subseteq \mathrm{End}\, Q(Y)$ be the graded associative algebra of operators on $Q(Y)$ generated by $\mathrm{Der}\, Q(Y)$. We regard $Q(Y)$ as a left $\mathcal{D}$ module. But then it follows that $Q(Y)^*$ is a right $\mathcal{D}$-submodule of the full dual of $Q(Y)$. That is if $v \in Q(Y)^*$, $\partial \in \mathcal{D}$ then $v \cdot \partial \in Q(Y)^*$ where if $h \in Q(Y)$ then $v \cdot \partial(h) = v(\partial h)$. Now for any open set $U \subseteq Y$ let

$$P(U) = C^\infty(U)^* \cdot \mathcal{D} \tag{2.18.4}$$

so that $P(U) \subseteq Q(Y)^*$ is the $\mathcal{D}$-submodule of $Q(Y)^*$ generated by $C^\infty(U)^*$. Furthermore $P(U)$ has also the structure of a graded subcoalgebra of $Q(Y)^*$.

Hence the full dual $P(U)'$ of $P(U)$ has the structure of a graded commutative algebra and also the structure of a left $\mathcal{D}$-module. Now let $B(U)$ be the set of all $f \in P(U)'$ such that $\partial \cdot f \mid C^{\infty}(U)^*$ is a C^{∞} function (see §2.14)) on U for all $\partial \in \mathcal{D}$. It follows easily that $B(U)$ is a graded commutative subalgebra of $P(U)'$ and the correspondence $f \longrightarrow f \mid C^{\infty}(U)^*$ defines a homomorphism

$$B(U) \longrightarrow C^{\infty}(U). \tag{2.18.5}$$

Moreover $U \longrightarrow B(U)$ is a sheaf B of graded commutative algebras compatible with the homomorphisms (2.18.5).

We will now say that the pair $(Q(Y),\psi)$ (see (2.18.3)) generates a graded manifold of dimension (m,n) if (Y,B) is a graded manifold of dimension (m,n). In such a case (Y,B) will be referred to as the graded manifold generated by $(Q(Y),\psi)$.

We will give examples of how some graded manifolds can be constructed. The examples below are "parallel" to Examples (1), (2) and (3) above.

<u>Example I</u>. Assume $V = V_0 + V_1$ is a finite dimensional graded real vector space. Let $V' = V_0' + V_1'$ be the dual space with its obvious gradation. Now if $Y = V_0$ put $Q(Y) = S(V')$ the symmetric algebra over V' on a graded vector space. Thus (see §1.2) one has a graded commutative algebra $S(V') \cong S(\hat{V}_0') \otimes \Lambda\hat{V}_1'$ where $\hat{V}_0'$ and $\hat{V}_1'$ are ungraded vector spaces. Now $S(\hat{V}_0')$ may be regarded as the algebra of polynomial functions on $V_0 = Y$ and as noted in Example (1) this algebra separates $C^{\infty}(Y)^*$. But now one has a natural epimorphism $\psi : S(V') \longrightarrow S(\hat{V}_0') \subseteq C^{\infty}(Y)$. The pair $(S(V'),\psi)$ generates a graded manifold (V_0, A_V) of dimension (m,n) where $m = \dim V_0$, $n = \dim V_1$. One notes that $A_V(V_0) \cong C^{\infty}(V_0) \otimes \Lambda\hat{V}_1'$. We refer to (V_0, A_V) as the graded affine manifold corresponding to V.

<u>Example II</u>. Assume that (X,A) and (Y,B) are graded manifolds of dimensions (m_1,n_1) and (m_2,n_2) respectively. Put $Q(X\times Y) = A(X) \otimes B(Y)$. The epimorphisms $A(X) \longrightarrow C^{\infty}(X)$ and $B(Y) \longrightarrow C^{\infty}(Y)$, $f \longmapsto \tilde{f}$ define an epimorphism

$$\psi : Q(X\times Y) \longrightarrow C^{\infty}(X) \otimes C^{\infty}(Y). \tag{2.18.6}$$

As noted in Example (2) the algebra $C^\infty(X) \otimes C^\infty(Y)$ separates $C^\infty(X\times Y)^*$. Now the pair $(Q(X\times Y,\psi)$ generates a graded manifold $(X\times Y, A\times B)$ of dimension (m_1+m_2, n_1+n_2) which will be referred to as the product graded manifold. We write $(X\times Y, A\times B) = (X,A) \times (Y,B)$.

Remark 2.18.2. One easily shows that

$$A \times B(X\times Y)^* = A(X)^* \otimes B(Y)^* \tag{2.18.7}$$

as graded coalgebras. Furthermore if $h \in A \times B(X\times Y)$ and $w \in B(Y)^*$ then one defines an element $h(\cdot,w) \in A(X)$ by the relation

$$v(h(\cdot,w)) = (v\otimes w)h \tag{2.18.8}$$

for any $v \in A(X)^*$.

Example III. Let (X,A) be a graded manifold of dimension (m,n). Now if $T^*(X)$ is the usual cotangent bundle of X and $\pi : T^*(X) \longrightarrow X$ is the bundle projection then π is an open map and hence we can take the inverse image of A defining in the obvious way a presheaf $\hat{A}$ on $T^*(X)$. For notational convenience we write $\hat{A}(X)$ for $\hat{A}(T^*(X))$ and note that by tautology π induces a graded algebra isomorphism $\hat{\pi} : \hat{A}(X) \longrightarrow A(X)$. This of course extends to an isomorphism of symmetric algebras, in the graded sense,

$$\hat{\pi} : S_{\hat{A}(X)}(\mathrm{Der}\ (\hat{A}(X))) \longrightarrow S_{A(X)}(\mathrm{Der}\ A(X)) \tag{2.18.9}$$

i.e. the symmetric algebra of $\mathrm{Der}\ A(X)$ over the graded commutative algebra $A(X)$. One puts $Q(T^*(X)) = S_{\hat{A}(X)}(\mathrm{Der}\ \hat{A}(X))$. The correspondence $\xi \longmapsto \widetilde{(\hat{\pi}(\xi))}$ for $\xi \in (\mathrm{Der}\ \hat{A}(X))_0$ and $\xi \longmapsto 0$ for $\xi \in (\mathrm{Der}\ \hat{A}(X))_1$ and also $f \longrightarrow \widetilde{\hat{\pi}(f)}$ for $f \in \hat{A}(X)$ defines an epimorphism

$$Q(T^*(X)) \longrightarrow S_{C^\infty(X)}(\mathrm{Der}\ C^\infty(X)). \tag{2.18.10}$$

On the other hand we have defined an epimorphism $\gamma : S_{C^\infty(X)}(\mathrm{Der}\ C^\infty(X)) \longrightarrow R(T^*(X)) \subseteq C^\infty(T^*(X))$ in Example (3). Hence the composite of (2.18.10) and γ defines a homomorphism

$$\psi : Q(T^*(X)) \longrightarrow C^\infty(T^*(X)) \tag{2.18.11}$$

whose image separates $C^{\infty}(T^*(X))^*$. The pair $(Q(T^*(X)),\psi)$ defines a graded manifold $(T^*(X),T^*(A))$ of dimension $(2m,2n)$ which will be referred to as the cotangent bundle of (X,A).

Remark 2.18.3. In general if $(Q(Y),\psi)$ defines a graded manifold (Y,B) one has a homomorphism $Q(Y) \longrightarrow B(Y)$ by definition of $B(Y)$. In Example I this becomes

(2.18.12) $$S(V') \longrightarrow A_V(V_0).$$

In Example II

(2.18.13) $$A(X) \otimes B(Y) \longrightarrow A \times B(X\times Y)$$

and in Example III

(2.18.14) $$S_{\hat{A}(X)}(\text{Der }\hat{A}(X)) \longrightarrow T^*(A)(T^*(X))$$

where $S_{\hat{A}(X)}(\text{Der }\hat{A}(X))$ is isomorphic to the symmetric algebra in the graded sense of $\text{Der }A(X)$ over $A(X)$. One notes also that the maps (2.18.12) and (2.18.13) are certainly injective and by identification with their images we regard (2.18.15

(2.18.15) $$S(V') \subseteq A_V(V_0) \quad \text{and} \quad A(X) \otimes B(Y) \subseteq A \times B(X\times Y).$$

The map (2.18.14) is clearly injective locally since if U is an A-coordinate neighborhood $\text{Der }A(U)$ is free over $A(U)$.

In connection with cotangent bundles note

Remark 2.18.4. If (X,A) is a graded manifold where X reduces to a point then one notes that $A(X) \cong \Lambda V$ where V is some ungraded real vector space of dimension (say) n. Thus $\dim(X,A) = (0,n)$. But $\dim(T^*(X),T^*(A)) = (0,2n)$. In fact one has $T^*(A)(T^*(X)) \cong \Lambda(V \oplus V')$ where V' is the dual space to V. Thus, in effect, the familiar procedure (for example in Clifford algebra theory) of going from ΛV to $\Lambda(V\oplus V')$ from the point of view of graded manifolds is just the operation of taking the cotangent bundle for the case where the base manifold is a point.

Now let $V = V_0 + V_1$ be any finite dimensional real graded vector space and let (V_0, A_V) be the corresponding graded affine manifold. Now note that V_0

is an A_V coordinate neighborhood. In fact $V' \subseteq A_V(V^0)$ (see (2.18.15)) and any homogeneous basis of V' is an A_V-coordinate system. Now let (X,A) be any graded manifold and assume

$$\lambda : V' \longrightarrow A(X) \tag{2.18.16}$$

is a morphism of graded vector spaces.

Proposition 2.18. *There exists a unique morphism* $\sigma : (X,A) \longrightarrow (V_0, A_V)$ *of graded manifolds such that* $\sigma^* \mid V' = \lambda$.

Proof. By Proposition (2.13.2) an element $v \in A_V(V_0)^*$ is uniquely determined by its restriction to $S(V')$. But since σ^* is uniquely determined on $S(V')$ it follows from Remark 2.15.1 that σ itself is unique. To prove the existence of σ it suffices therefore to assume that X is an A-split neighborhood so that one has a function factor $C(X) \subseteq A(X)$ and hence

$$A(X) = C(X) \oplus A^1(X). \tag{2.18.17}$$

Now let

$$\gamma : X \longrightarrow V_0$$

be the smooth map defined so that $\tilde{h}(\gamma(p)) = \widetilde{\lambda(h)}(p)$ for any $p \in X$ and $h \in V'$. Also let

$$\tau : A_V(V_0) \longrightarrow C(X) \tag{2.18.18}$$

be the homomorphism defined so that $\widetilde{\tau(f)} = \tilde{f} \circ \gamma$. Now let r_i, $i = 1,\ldots,d$ be a basis of V_0' and $\{s_j\}$ a basis of V_1' so that $\{r_i, s_j\}$ is an A_V-coordinate system in V_0. Let $C(V_0) \subseteq A_V(V_0)$ be the function factor which contains the r_i so that

$$A_V(V_0) = C(V) \otimes \Lambda V_1'. \tag{2.18.19}$$

To prove the proposition it is clear that we have only to extend $\lambda \mid V_0'$ to a homomorphism $\pi : C(V) \longrightarrow A(X)_0$. Write

$$\lambda(r_i) = g_i + w_i \tag{2.18.20}$$

where $g_i \in C(X)$ and $w_i \in A^1(X)_0$. Clearly $g_i = \tau(r_i)$. Now for any

$\nu = (\nu_1,\ldots,\nu_d) \in N_d$ let $w^\nu = w_1^{\nu_1} \ldots \nu_d^{\nu_d}$. Also let $\nu! = \nu_1! \ldots \nu_d!$ and $\frac{\partial^\nu}{\partial r} = (\frac{\partial}{\partial r_1})^{\nu_1} \ldots (\frac{\partial}{\partial r_d})^{\nu_d} \in \text{Diff}\, A_V(V_0)$. Now for any $f \in C(V)$ define $\pi(f) \in A(X)$ by putting

(2.18.21)
$$\pi(f) = \sum_{\nu \in N_m} \tau\left(\frac{\partial^\nu}{\partial r} f\right) \frac{w^\nu}{\nu!} .$$

The sum is clearly finite since $w^\nu = 0$ for $|\nu|$ greater than the odd dimension of X. One notes that $f \longrightarrow \pi(f)$ defines a homomorphism

(2.18.22)
$$\pi : C(V) \longrightarrow A(X)_0$$

where $\pi(r_i) = \lambda(r_i)$. This proves the proposition. Q.E.D.

3. Graded Lie Theory

3.1. We will now introduce what will be referred to as graded Lie theory. For the most part the graded manifolds with which we will be concerned will arise from graded Lie groups and their homogeneous spaces. The graded submanifolds with which we will be concerned will most often arise as the orbits of graded Lie groups operating on graded manifolds.

Ordinary Lie groups and Lie algebras can be dealt with simultaneously by introducing a suitable cocommutative Hopf algebra. If however we replace this Hopf algebra by a graded cocommutative Hopf algebra then we are in a position to define a graded Lie group. In effect it will turn out that a graded Lie group will be a graded manifold (G,A) where G is an ordinary Lie group and $A(G)^*$ has the structure of a graded Hopf algebra. We recall that for a graded manifold (Y,B) the space $B(Y)^*$ has only the structure of a graded cocommutative coalgebra. For a graded Lie group (G,A) one has in addition an algebra structure on $A(G)^*$ where the two are related so that $A(G)^*$ has the structure of a graded Hopf algebra.

To begin with we first note some properties of graded Lie algebras and graded Hopf algebras.

Let $\underline{g} = \underline{g}_0 + \underline{g}_1$ be a graded Lie algebra over a field K of characteristic zero. One defines the universal enveloping algebra $E(\underline{g})$ of $\underline{g}$ as the tensor algebra $T(\underline{g})$ modulo the ideal in $T(\underline{g})$ generated by all elements in $T(\underline{g})$ of the form $x \otimes y - (-1)^{|x||y|} y \otimes x - [x,y]$ for $x, y \in \underline{g}$. One knows the quotient map

$$T(\underline{g}) \longrightarrow E(\underline{g}) \tag{3.1.1}$$

is injective for $\underline{g}$ and as usual we identify $\underline{g}$ with its image so that $\underline{g} \subseteq E(\underline{g})$. The Poincaré-Birkhoff-Witt theorem is valid for $E(\underline{g})$. (See [2] and [10]). This comes down to the statement that if $x_i \in \underline{g}_0$, $i \in I$, and $y_j \in \underline{g}_1$, $j \in J$, are respectively bases of $\underline{g}_0$ and $\underline{g}_1$ where I and J are well ordered index sets then the set of all elements in $E(\underline{g})$ of the form $x_{i_1}^{d_1} \ldots x_{i_k}^{d_k} y_{j_1} \ldots y_{j_\ell}$ is a basis of $E(\underline{g})$ where $i_1 > \ldots > i_k$ in I, $d_1,\ldots,d_k \in \mathbb{N}$ and $j_1 > \ldots > j_\ell \in J$. In particular if $\underline{g}$ is finite dimensional where $\dim \underline{g}_0 = m$ and $\dim \underline{g}_1 = n$, and $x_1,\ldots,x_m$ is a basis of $\underline{g}_0$, $y_1,\ldots,y_n$ is a basis of $\underline{g}_1$ then

$$x^\nu y_\mu \text{ is a basis of } E(\underline{g}) \text{ for } (\nu,\mu) \in N_m \times M_n \tag{3.1.2}$$

where $\nu = (\nu_1,\ldots,\nu_m) \in N_m$, $\mu = (\mu_1,\ldots,\mu_k) \in M_n$ and $x^\nu = x_1^{\nu_1} \ldots x_m^{\nu_m}$, $y_\mu = y_{\mu_1} \ldots y_{\mu_k}$.

One notes that $E(\underline{g})$ is a graded algebra. In fact using the notation above $x^\nu y_\mu \in (E(\underline{g}))_0$ or $(E(\underline{g}))_1$ according to whether $k(\mu)$ is even or odd.

One also notes that $E(\underline{g})$ is universal with respect to homomorphisms of $\underline{g}$ into graded algebras. That is, if W is a graded (associative) algebra then a linear map $\pi : \underline{g} \longrightarrow W$ of degree zero is a homomorphism if it is a homomorphism of graded Lie algebras where W is given the structure of a graded Lie algebra as in §1.3. If $\pi : \underline{g} \longrightarrow W$ is such a homomorphism then the map extends uniquely to a homomorphism $\pi : E(\underline{g}) \longrightarrow W$ of graded algebras. In particular we note the important case where $W = \operatorname{End} V$ for V a graded vector space.

For purposes which should become clear later we will write the identity element of $E(\underline{g})$ as e rather than 1.

3.2. Now $E(\underline{g} \oplus \underline{g})$ may be naturally identified with $E(\underline{g}) \otimes E(\underline{g})$ (see (1.2.5)) where the element (x,y) in $\underline{g} \oplus \underline{g}$ appears as $x \otimes e + e \otimes y$. The diagonal map $\underline{g} \longrightarrow \underline{g} \oplus \underline{g}$, $x \longrightarrow (x,x)$ is a homomorphism of $\underline{g}$ into

$E(\underline{g}) \otimes E(\underline{g})$ and hence from the universal properties of $E(\underline{g})$ this extends to a homomorphism

(3.2.1) $$\Delta : E(\underline{g}) \longrightarrow E(\underline{g}) \otimes E(\underline{g})$$

giving $E(\underline{g})$ the structure of a graded cocommutative algebra. In fact more than this $E(\underline{g})$ has the structure of a connected graded cocommutative Hopf algebra. We recall the definition of these terms.

Assume E is a graded algebra over K and one has a distinguished algebra homomorphism of graded algebras $1_E : E \longrightarrow K$, referred to as the augmentation map. Assume also that E has the structure of a graded coalgebra with respect to a diagonal map

(3.2.2) $$\Delta : E \longrightarrow E \otimes E$$

and 1_E is the counit. This means that Δ is a coassociative map of graded vector spaces (that is, Δ is a linear map of degree zero such that the two maps $E \longrightarrow E \otimes E \otimes E$ given by $(\Delta \otimes 1) \circ \Delta$ and $(1 \otimes \Delta) \circ \Delta$ are the same) and if $h \in E$ and

(3.2.3) $$\Delta h = \sum_i h_i' \otimes h_i''$$

then

(3.2.4) $$\sum_i 1_E(h_i')h_i'' = \sum h_i' 1_E(h_i'') = h .$$

One then says that $(E, 1_E, \Delta)$ or simply E, if Δ and 1_E are understood, is a Hopf algebra, if Δ is a homomorphism of algebras.

Assume E is a Hopf algebra. One says that E is cocommutative if $T \circ \Delta = \Delta$ where T is given by (1.2.1). See Remark 2.11.1. Also one says that E has an antipode if there is an element $s \in \text{End } E$ called the antipode, such that, using the notation of (3.2.3)

(3.2.5) $$\sum_i s(h_i')h_i'' = \sum_i h_i' s(h_i'') = 1_E(h) .$$

The antipode s, if it exists, is necessarily unique. It then follows easily that $s \in (\text{End } E)_0$. (See [10]).

We recall (see §2.11) an element $0 \neq g \in E$ is called group-like if $g \in E_0$ and $\Delta g = g \otimes g$. One necessarily has $1_E(g) = 1$ for a group-like element.

Furthermore the set G of group-like elements is closed under multiplication and all the elements in G are linearly independent in E. One has $e \in G$ where e is the identity element of E. Moreover if E has antipode s then G is necessarily a group and $s(g) = g^{-1}$ for any $g \in G$.

In §2.11 we defined what we meant by saying that an element $x \in E$ is primitive with respect to a group-like element g. More simply we will say that $x \in E$ is primitive if it is primitive with respect to e. Thus $x \in E$ is called primitive if $\Delta x = x \otimes e + e \otimes x$. The set $\underline{h}$ of all primitive elements in E is a graded subspace, is closed under brackets and hence is a graded Lie algebra. One necessarily has $1_E(x) = 0$ for $x \in \underline{h}$.

Now the dual E' to E is a graded algebra recalling (2.11.6) and (2.11.7). Furthermore the set

(3.2.6) $$M_e(E') = \{f \in E' \mid \langle e,f \rangle = 0\}$$

is a maximal ideal of codimension 1 in E'. Let $E_{(k)}$ be the orthogonal complement in E of the kth power $(M_e(E'))^k$ of $M_e(E')$. One has $E_{(k)} \subseteq E_{(k+1)}$. Put $E_e = \bigcup_{k=1}^{\infty} E_{(k)}$. Then E_e is a Hopf subalgebra of E. Also $\underline{h} \subseteq E_e$. The Hopf algebra E is called connected if $E_e = E$. The definition is less restrictive than the one used in [10] and hence Proposition 3.2 below is more general than similar statements in [10]. For any arbitrary Hopf algebra E we refer to E_e as the connected component of E.

Now as noted above the universal enveloping algebra $E(\underline{g})$ of a graded Lie algebra $\underline{g}$ is a graded connected cocommutative Hopf algebra and $\underline{g}$ is exactly the space of all primitive elements in $E(\underline{g})$. But in fact one has the converse.

Proposition 3.2. *Let E be any graded connected cocommutative Hopf algebra over a field K of characteristic zero. Then E is isomorphic to the universal enveloping algebra $E(\underline{h})$ of its graded Lie algebra $\underline{h} \subseteq E$ of primitive elements.*

The proof of Proposition 3.2 is a graded version of the proof in [15].

3.3. Let $\underline{g}$ be a graded Lie algebra. Then one knows the graded Hopf algebra $E(\underline{g})$ has an antipode. In fact clearly there is a unique element

$s \in \text{End } E(\underline{g})$ such that

(3.3.1)

$$(1) \quad sx = -x \quad \text{for} \quad x \in \underline{g}$$

$$(2) \quad s(uv) = (-1)^{|u||v|} s(v)s(u).$$

One has

Proposition 3.3. *$E(\underline{g})$ has an antipode and the map $s \in \text{End } E(\underline{g})$ satisfying (1) and (2) above is the antipode of $E(\underline{g})$.*

Now if G is any group and $K(G)$ is the group algebra (finite linear combinations) of G over K then $E = K(G)$ is a graded cocommutative Hopf algebra, with antipode, over K, where of course $E_1 = 0$. One defines

$$\Delta : K(G) \longrightarrow K(G) \otimes K(G)$$

so that for $g \in G$, $\Delta(g) = g \otimes g$. Also $s(g) = g^{-1}$ and $1_E(g) = 1$.

Now assume also that $\underline{g}$ is a graded Lie algebra over K and one has a representation $\pi : G \longrightarrow \text{Aut } \underline{g}$ so that $\pi(g)$ is a graded Lie algebra automorphism for any $g \in G$. Then π extends uniquely to a representation

(3.3.2) $$\pi : G \longrightarrow \text{Aut } E(\underline{g})$$

so that G operates as a group of automorphisms of $E(\underline{g})$. Now the smash product

(3.3.3) $$E = K(G) * E(\underline{g})$$

with respect to π, or simply smash product, if π is understood, is a graded cocommutative Hopf algebra with antipode where (1) as a graded vector space $E = K(G) \otimes E(\underline{g})$, (2) as an algebra $K(G)$ and $E(\underline{g})$ are subalgebras but $gug^{-1} = \pi(g)u$ for $g \in G$, $u \in E(\underline{g})$, (3) with respect to the diagonal map Δ, the elements of G are group-like and the elements of $\underline{g}$ are primitive and (4) one has $s(g) = g^{-1}$, $s(x) = -x$ and $1_E(g) = 1$, $1_E(x) = 0$ for $g \in G$, $x \in \underline{g}$.

One sees easily that for the smash product G is exactly the set of all group-like elements in E and $\underline{g}$ is exactly the space of all primitive elements in E. The main difference between the smash product and the Hopf algebra of Proposition 3.2 is that if $G \neq (e)$ the Hopf algebra E is not connected. That is, if we throw away connectedness we are apt to pick up group-like elements. In fact as a generalization (in that grading is permitted here) of our theorem proved in [15], (see also [4]), one has

Theorem 3.3. Let E be any graded commutative Hopf algebra with antipode over an algebraically closed field K of characteristic zero and let G and $\underline{g}$ respectively, be the group of all group-like elements in E and the graded Lie algebra of all primitive elements in E. Then one has a representation $\pi : G \longrightarrow \mathrm{Aut}\, \underline{g}$ of G on $\underline{g}$ so that $gxg^{-1} = \pi(g)x$ for any $g \in G$, $x \in \underline{g}$. Furthermore as a graded Hopf algebra one has an isomorphism

$$E \cong K(G) \ast E(\underline{g})$$

where the smash product is with respect to π.

The proof is just a graded version of the proof of Theorems 13.01 and 8.1.5 in [15].

3.4. Now let $\underline{g} = \underline{g}_0 + \underline{g}_1$ be a finite dimensional graded Lie algebra over the field of real numbers $\mathbb{R}$. Then $\underline{g}_0$ is an ordinary Lie algebra and if we write $(\mathrm{ad}_{\underline{g}}\, x)(y) = [x,y]$ for $x \in \underline{g}_0$, $y \in \underline{g}$ then

(3.4.1) $$\mathrm{ad}_{\underline{g}} : \underline{g}_0 \longrightarrow \mathrm{End}\, \underline{g}$$

is a representation of $\underline{g}_0$ on $\underline{g}$. Now let G be an analytic group (a connected Lie group) whose Lie algebra is $\underline{g}_0$. We will say that $\mathrm{Ad}_{\underline{g}}$ is defined on G if the representation $\mathrm{ad}_{\underline{g}}$ exponentiates to a representation

(3.4.2) $$\mathrm{Ad}_{\underline{g}} : G \longrightarrow \mathrm{Aut}\, \underline{g}\ .$$

For example if G is the simply-connected Lie group whose Lie algebra is $\underline{g}_0$ the $\mathrm{Ad}_{\underline{g}}$ is defined on G.

Now assume that G is a group, $\underline{g}$ is a graded Lie algebra over $\mathbb{R}$ and E is the graded Hopf algebra

$$E = \mathbb{R}(G) \ast E(\underline{g})$$

with respect to some representation $\pi : G \longrightarrow \mathrm{Aut}\, \underline{g}$. We will now say that E has the structure of an L-H algebra (L-H stands for Lie-Hopf) if (1) G has the structure of a (not necessarily connected) Lie group (2) $\underline{g} = \underline{g}_0 + \underline{g}_1$ is a finite dimension graded Lie algebra where $\underline{g}_0$ is the Lie algebra of G. This implies in particular that $\underline{g}_0$ is the tangent space $T_e(G)$ to G at the identity e of G and one has an exponential map

(3.4.3) $$\exp : \underline{g}_0 \longrightarrow G$$

and (3) $Ad_{\underline{g}}$ is defined on the identity component G_e of G and $\pi \mid G_e = Ad_{\underline{g}}$.

Remark 3.4.1. We emphasize that an L-H algebra is more than just an algebra in that the group of group-like elements has the structure of a Lie group G and the even part of the graded Lie algebra of primitive elements is the Lie algebra of G.

If G is a Lie group, $\underline{g}$ is a graded Lie algebra over $\mathbb{R}$ and G operates on $\underline{g}$ as a group of graded Lie algebra automorphisms according to a representation $\pi : G \longrightarrow \mathrm{Aut}\, \underline{g}$ such that conditions (2) and (3) above are satisfied we will let $E(G,\underline{g},\pi)$ be the L-H algebra given by the smash product (3.3.3). If π is understood we will simply write $E(G,\underline{g})$ for this L-H algebra. In particular if G is an analytic group then π is unique and hence is certainly understood.

Now if $E(G,\underline{g})$ is an L-H algebra let $E(G,\underline{g}_0)$ be the L-H algebra obtained by replacing $\underline{g}$ with its even part $\underline{g}_0$. As a Hopf algebra one knows that we can regard

(3.4.4) $$E(G,\underline{g}_0) = C^\infty(G)^*$$

where, we recall, $C^\infty(G)^*$ is the set of distributions on G with finite support. Now one regards the collection of L-H algebras as objects in a category. A morphism

(3.4.5) $$E(G,\underline{g}) \longrightarrow E(H,\underline{h})$$

of L-H algebras is a morphism of Hopf algebras such that the restriction

(3.4.6) $$E(G,\underline{g}_0) \longrightarrow E(H,\underline{h}_0)$$

is induced, in the usual way, by a morphism $G \longrightarrow H$ of Lie groups. The morphism (3.4.5) is an isomorphism if it is setwise bijective. In such a case one knows that the corresponding map of Lie groups $G \longrightarrow H$ is an isomorphism of Lie groups. Of course $E(G,\underline{g})$ and $E(H,\underline{h})$ are isomorphic if there exists an isomorphism (3.4.4).

Remark 3.4.2. If $\underline{g} = \underline{g}_0 + \underline{g}_1$ is an arbitrary finite dimensional real graded Lie algebra note that there always exists a Lie group G such that one may form an L-H algebra $E(G,\underline{g})$. Indeed one can take G equal to the simply connected Lie group whose Lie algebra is $\underline{g}_0$. In this case we refer to $E(G,\underline{g})$ as the simply-connected L-H algebra corresponding to $\underline{g}$.

3.5. We will now define the notion of a graded Lie group.

Let (X,A) and (Y,B) be two graded manifolds. We recall (see §2.17) that a morphism

$$\tau : A(X)^* \longrightarrow B(Y)^*$$

of graded coalgebras is called smooth if $\tau = \sigma_*$ for a morphism

$$\sigma : (X,A) \longrightarrow (Y,B)$$

of graded manifolds.

But now if (X,A) is a graded manifold then $(X\times X, A\times A)$ is a graded manifold (see Example II, §2.18) and $A \times A(X\times X)^* = A(X)^* \otimes A(X)^*$. See (2.18.7). Thus it makes sense to say that a morphism

$$\tau : A(X)^* \otimes A(X)^* \longrightarrow A(X)^*$$

of graded coalgebras is smooth.

Now let (G,A) be a graded manifold of dimension (m,n). Let

(3.5.1) $$\Delta : A(G)^* \longrightarrow A(G)^* \otimes A(G)^*$$

be the diagonal map with respect to which, we recall, $A(G)^*$ is a cocommutative coalgebra. The counit is given by the identity element $1_G \in A(G)$ where $1_G(v) = v(1_G)$ for $v \in A(G)^*$. We will now say that (G,A) has the structure of a graded Lie group if $A(G)^*$ has also the structure of an algebra such that (1), $(A(G)^*, 1_G, \Delta)$ is a Hopf algebra with antipode, s, where (2), the map

(3.5.2) $$A(G)^* \otimes A(G)^* \longrightarrow A(G)^*$$

given by multiplication and the map

(3.5.3) $$s : A(G)^* \longrightarrow A(G)^*$$

given by the antipode are smooth.

Remark 3.5.1. If $A(G)^*$ has the structure of a Hopf algebra with antipode one knows that the maps (3.5.2) and (3.5.3) are morphisms of graded coalgebras. See e.g. [10].

Now assume that (G,A) is a graded Lie group. Then since G is the set of group-like elements in the Hopf-algebra $A(G)^*$ (see Remark 2.11.3) and since $A(G)^*$ has an antipode it follows that G has the structure of a group. Let $e \in G$ be the identity element of $A(G)^*$. Then if

$$\underline{g} = T_e(G,A) \tag{3.5.4}$$

is the tangent space to (G,A) at e then by Proposition 2.11.2 $\underline{g} = \underline{g}_0 + \underline{g}_1$ is the space of primitive elements of $A(G)^*$. (See also §3.2). In particular $\underline{g}$ has the structure of a graded Lie algebra. We refer to $\underline{g}$ as the Lie algebra of the graded Lie group (G,A). Note also that $\underline{g}_0 = T_e(G)$ according to (2.10.3). The following structure result is an easy consequence of Theorem 3.3.

Theorem 3.5. Let (G,A) be graded Lie group. Then G with respect to its manifold and group structure is a Lie group and $\underline{g}_0$, with its Lie algebra structure, as a subalgebra of $\underline{g}$, is the Lie algebra of G. Furthermore $A(G)^*$ with the Lie group structure on G has the structure of an L-H algebra. In fact if $\pi : G \longrightarrow \text{Aut}\, \underline{g}$ is defined by the relation $gxg^{-1} = \pi(g)x$ for $x \in \underline{g}$, $g \in G$, then $\pi \mid G_e = \text{Ad}_{\underline{g}}$ where G_e is the identity component of G and as L-H algebras one has

$$A(G)^* = E(G,\underline{g}). \tag{3.5.5}$$

Remark 3.5.2. As a Hopf algebra the relation (3.5.5) implies that

$$A(G)^* = \mathbb{R}(G) \circledast E(\underline{g}). \tag{3.5.6}$$

Furthermore one easily has

$$A_e(G)^* = E(\underline{g}) \tag{3.5.7}$$

so that, as in an ordinary Lie group, the set of distributions of (G,A) with support at the identity is the enveloping algebra of the Lie algebra $\underline{g}$ of (G,A).

Note also that for any $p \in G$, using the multiplicative structure in $A(G)^*$, we can write

(3.5.8) $$A_p(G)^* = pE(\underline{g}) = E(\underline{g})p$$

and for the tangent space at p one has

(3.5.9) $$T_p(G,A) = p\underline{g} = \underline{g}p.$$

Now if (G,A) and (H,B) are graded Lie groups then a morphism

$$\sigma : (H,B) \longrightarrow (G,A)$$

of graded manifolds is called a morphism of graded Lie groups if

$$\sigma_* : B(H)^* \longrightarrow A(G)^*$$

is a homomorphism of algebras. Also σ is called an isomorphism of graded Lie groups if, in addition, σ_* is bijective.

Remark 3.5.3. Since in any case σ_* is a morphism of coalgebras note that σ is a morphism of graded Lie groups if and only if σ_* is a morphism of Hopf algebras.

It is convenient to reverse the emphasis. Let (G,A) and (H,B) be graded Lie groups. Then a morphism $\tau : B(H)^* \longrightarrow A(G)^*$ of graded Hopf-algebras is called smooth if $\tau = \sigma_*$ where $\sigma : (H,B) \longrightarrow (G,A)$ is a morphism of graded Lie groups. One easily proves

Proposition 3.5.1. *Let (H,B) and (G,A) be graded Lie groups. Then a map $\tau : B(H)^* \longrightarrow A(G)^*$ is a smooth morphism of graded Hopf algebras if and only if τ is a morphism of L-H algebras.*

Remark 3.5.4. One notes also that σ is an isomorphism of graded Lie groups if and only if σ_* is an isomorphism of L-H algebras.

3.6. Now let (G,A) be a graded Lie group. Since (G,A) is also a graded manifold we recall that $A(G)^*$ and the graded commutative algebra $A(G)$ are non-singularly paired. The pairing as in §2.11 will be denoted by $\langle\ ,\ \rangle$. Now for any $w \in A(G)^*$ one can define an operator $R_w \in \mathrm{End}\, A(G)$ by the relation

(3.6.1) $$\langle v, R_w f\rangle = \langle vw, f\rangle$$

for $w, v \in A(G)^*$, and $f \in A(G)$. This uses the smoothness of (3.5.2) and (2.18.8). The map

(3.6.2) $$A(G)^* \longrightarrow \text{End } A(G), \qquad w \longmapsto R_w$$

is a homomorphism of graded algebras and will be referred to as the right regular representation of (G,A) on $A(G)$. The left regular representation of (G,A) on $A(G)$

(3.6.3) $$A(G)^* \longrightarrow \text{End } A(G) \qquad w \longmapsto L_w$$

is the homomorphism of graded algebras defined by

(3.6.4) $$\langle v, L_w f\rangle = (-1)^{|w||v|}\langle s(w)v, f\rangle$$

for $w, v \in A(G)^*$, $f \in A(G)$. One has

(3.6.5) $$[R_u, L_v] = 0$$

for any $u, v \in A(G)^*$ where we recall that commutation is with respect to the graded structure in $\text{End } A(G)$. (See §1.3.)

Again, with this definition of commutation we will say that an operator $\alpha \in \text{End } A(G)$ is left invariant if $[L_u, \alpha] = 0$ for all $u \in A(G)^*$. One defines right invariant similarly.

Now if $u \in E(\underline{g}) \subseteq A(G)^*$ then R_u, $L_u \in \text{Diff } A(G)$ are differential operators. We wish to characterize these differential operators in a fashion similar to the way one knows about in ordinary Lie theory.

Let $\alpha \in \text{End } A(G)$ be any operator and let α^t be its transpose on the full dual $A(G)'$ of $A(G)$. We will say that α admits an $A(G)^*$ transpose if $A(G)^*$ is stable under α^t. It is easy to see that α admits an $A(G)^*$ transpose if and only if for any ideal $I \subseteq A(G)$ of finite codimension there exists another ideal $J \subseteq A(G)$ of finite codimension such that

(3.6.6) $$\alpha(J) \subseteq I.$$

One notes that any operator of the form R_u or L_u, for $u \in A(G)^*$, admits an $A(G)^*$ transpose. More generally any differential operator $\partial \in \text{Diff } A(G)$ or any automorphism of $A(G)$ admits an $A(G)^*$ transpose. However since the set of operators on $A(G)'$ of the form $(L_u)^t$, $u \in A(G)^*$, operates in a simply transitive way on $A(G)^*$ one has

Proposition 3.6.1. *Assume* $\alpha \in \mathrm{End}\, A(G)$ *admits an* $A(G)^*$ *transpose. Then* α *is uniquely of the form* R_u *for* $u \in A(G)^*$ *if and only if it is left invariant. In particular a differential operator* $\partial \in \mathrm{Diff}\, A(G)$ *is of the form* R_u *for* $u \in E(\underline{g})$ *if and only if it is left invariant. The same statement is true if the roles of left and right are reversed.*

Now if $u \in A(G)^*$ one has $R_u \in \mathrm{Der}\, A(G)$ if and only if $u \in \underline{g}$. As a corollary of Proposition 3.6.1 one sees that the Lie algebra $\underline{g}$ of the graded Lie group (G,A) plays the same role as the Lie algebra of an ordinary Lie group.

Proposition 3.6.2. *A derivation* $\xi \in \mathrm{Der}\, A(G)$ *is left invariant if and only if it is of the form* R_x *for* $x \in \underline{g}$. *That is, the map*

(3.6.7) $$\underline{g} \longrightarrow \mathrm{Der}\, A(G), \qquad x \longrightarrow R_x$$

is an isomorphism of the graded Lie algebra $\underline{g}$ *of* (G,A) *onto the graded Lie algebra of all left invariant derivations of* $A(G)$. *Of course left and right can be interchanged.*

Remark 3.6. If (G,A) is a graded Lie group then G itself is an A-splitting neighborhood with splitting factors $(C(G),D(G))$ where

(3.6.8) $$C(G) = \{f \in A(G) \mid L_x f = 0 \text{ for all } x \in \underline{g}_1\}$$

and

(3.6.9) $$D(G) = \{f \in A(G) \mid L_x f = 0 \text{ for all } x \in \underline{g}_0\}.$$

In particular if $C(G)$ is defined by (3.6.8) then the map $f \longrightarrow \tilde{f}$ defines an isomorphism $C(G) \cong C^\infty(G)$, and if $D(G)$ is defined by (3.6.9) then $D(G)$ is an exterior algebra in $\dim \underline{g}_1$ generators and one has

(3.6.10) $$A(G) \cong C(G) \otimes D(G).$$

3.7. Up until now we have not said very much about the existence of graded Lie groups. We will now show that they exist "in abundance".

Let $E(G,\underline{g})$ be an L-H algebra. As a Hopf algebra $E(G,\underline{g}) = \mathbb{R}(G) \ast E(\underline{g})$. As in §3.4 $E(G,\underline{g}_0)$ is the subalgebra of $E(G,\underline{g})$ generated by G and $\underline{g}_0$. As noted in (3.4.4) $E(G,\underline{g}_0)$ can be identified with $C^\infty(G)^*$, the Hopf algebra of all distributions, in the Schwartz sense, of finite support on G. In more explicit terms $C^\infty(G)$ is a left $E(G,\underline{g}_0)$-module where if $a \in G$, $\varphi \in C^\infty(G)$, then $(a\cdot\varphi)(g) = \varphi(ga)$ and if $x \in \underline{g}$ then $(x\cdot\varphi)(g) = \frac{d}{dt}\varphi(g \exp tx)\Big|_{t=0}$. Then if $u \in E(G,\underline{g}_0)$ we identify u with the element in $C^\infty(G)^*$ given by $u(\varphi) = (u\cdot\varphi)(e)$.

Now regard $E(G,\underline{g})$ as a left and right $E(\underline{g})$ module. Let $U \subseteq G$ be any open set and let $E(U,\underline{g})$ be the right $E(\underline{g})$ submodule of $E(G,\underline{g})$ generated by U. One notes that $E(U,\underline{g})$ is also a left $E(\underline{g})$ module. But more than that note that $E(U,\underline{g})$ has the structure of a graded cocommutative coalgebra and hence the dual space $E(U,\underline{g})'$ has the structure of a graded commutative algebra. But also $E(U,\underline{g})'$ has the structure of a left $E(\underline{g})$-module where if $w \in E(\underline{g})$, $v \in E(U,\underline{g})$ and $f \in E(U,\underline{g})'$ one puts $\langle v, w\cdot f\rangle = \langle vw, f\rangle$.

But now $C^\infty(U)^* \subseteq C^\infty(G)^*$ is just the $E(\underline{g}_0)$-submodule of $E(G,\underline{g})$ generated by U. Thus $C^\infty(U)^* \subseteq E(U,\underline{g})$. Now, as in §2.14, $C^\infty(U)$ can be regarded as a subset of the dual of $C^\infty(U)^*$. Let

(3.7.1) $$A(U) = \{f \in E(U,\underline{g})' \mid w\cdot f \mid C^\infty(U)^* \in C^\infty(U) \text{ for all } w \in E(\underline{g})\}.$$

Then $A(U)$ is a graded commutative subalgebra of $E(U,\underline{g})$. The map $f \longmapsto f \mid C^\infty(U)^*$ defines a homomorphism

(3.7.2) $$A(U) \longrightarrow C^\infty(U)$$

and $U \longrightarrow A(U)$ defines a sheaf A of graded commutative algebras on G such that, with respect to the homomorphisms (3.7.2), (G,A) is a graded manifold. Furthermore the pairing of $A(G)$ and the set $E(G,\underline{g})$ induces a bijection

(3.7.3) $$E(G,\underline{g}) \longrightarrow A(G)^*$$

which is a coalgebra isomorphism. Moreover if one carries the Hopf algebra structure on $E(G,\underline{g})$ over to $A(G)^*$ then (G,A) has the structure of a graded Lie group so that $A(G)^*$ has the structure of an L-H algebra. But then (3.7.3) is an isomorphism of L-H algebras. Thus one has

Theorem 3.7. *Let (G,A) be any graded Lie group so that the space $A(G)^*$ of A-distributions of finite support on G has the structure of an L-H algebra (in particular it has the structure of a graded cocommutative Hopf algebra with antipode). Then the correspondences $(G,A) \longrightarrow A(G)^*$ set up a bijection between the set of isomorphism classes of graded Lie groups and the set of isomorphism classes of L-H algebras.*

A graded Lie group (G,A) will be said to be commutative if $A(G)^*$ is a commutative algebra (in the graded sense - see §1.1). A graded Lie group (G,A) will be said to be connected if G is a connected Lie group. We will also refer to a connected graded Lie group as a graded analytic group.

Remark 3.7. Note that if (G,A) is a graded analytic group then it is commutative if and only if its Lie algebra $\mathfrak{g}$ is commutative. Note in particular that this is necessarily the case if $\mathfrak{g}_0 = 0$.

A graded Lie group (G,A) will be said to be simply connected if G is a simply connected Lie group (in our terminology here the also means that G is connected). As a consequence of Theorem 3.7 and Remark 3.4.2 one has the existence theorem.

Corollary to Theorem 3.7. *If $\mathfrak{g} = \mathfrak{g}_0 + \mathfrak{g}_1$ is any finite dimensional real graded Lie algebra there exists a graded Lie group (G,A) such that $\mathfrak{g}$ is the Lie algebra of (G,A). In fact there exists a simply connected graded Lie group (G,A) with this property and such a graded Lie group is unique up to isomorphism.*

Given a finite dimensional real graded Lie algebra $\mathfrak{g}$ we can now speak of the simply connected graded Lie group (G,A) with Lie algebra $\mathfrak{g}$.

3.8. Note that the category of graded Lie groups and the category of L-H algebras are isomorphic. In fact Theorem 3.7 establishes the correspondence between their objects and Proposition 3.5.1 establishes the correspondence between their morphisms. As a consequence one can deal easily with the notion of graded Lie subgroups.

Now if E and F are L-H algebras we will say that F is an L-H subalgebra of E if $F \subseteq E$ and the injection map is a morphism of L-H algebras (see §3.4). One notes that if F_1 and F_2 are L-H subalgebras of E such that $F_1 = F_2$ as sets then, as in ordinary Lie theory, the identity map $F_1 \longrightarrow F_2$ is an isomorphism of L-H algebras. In such a case we can therefore identify F_1 and F_2. Thus given an L-H algebra E and a subset $F \subseteq E$ it makes sense to say that F is an L-H subalgebra of E. More explicitly if $E = E(G,\underline{g}) = E(G,\underline{g},\pi)$ then $F \subseteq E$ is an L-H subalgebra of E if and only if F is of the form $F = E(H,\underline{h}) = E(H,\underline{h},\pi')$ where H is a Lie subgroup of G, $\underline{h}$ is a graded Lie subalgebra of $\underline{g}$ and π' is obtained by restricting π.

Proposition 3.8.2. Given an L-H algebra $E(G,\underline{g})$ the correspondence

$$(H,\underline{h}) \longmapsto E(H,\underline{h}) \tag{3.8.2}$$

defines a bijection of the set of all pairs $(H,\underline{h})$, where H is a Lie subgroup of G and $\underline{h}$ is a graded Lie subalgebra of $\underline{g}$ such that $\underline{h}_0$ is the Lie algebra of H, and the set of all L-H subalgebras of $E(G,\underline{g})$.

Now let (H,B) and (G,A) be graded Lie groups. We will say that (H,B) is a graded Lie subgroup of (G,A) if $B(H)^* \subseteq A(G)^*$ and the injection map $B(H)^* \longrightarrow A(G)^*$ is a smooth morphism of graded Hopf algebras. In such a case one thus has a unique morphism

$$\sigma : (H,B) \longrightarrow (G,A) \tag{3.8.3}$$

of graded Lie groups such that σ_* is the injection map.

Now if (H,B) is a graded Lie subgroup of (G,A) then $B(H)^*$ is an L-H subalgebra of $A(G)^*$ by Proposition 3.8.1. If (H',B') and (H,B) are graded Lie subgroups of (G,A) such that $B(H)^* = B'(H')^*$ then the identity map $B(H)^* \longrightarrow B'(H')^*$ is an isomorphism of L-H algebras and hence it defines, by Proposition 3.8.1, an isomorphism $(H,B) \longrightarrow (H',B')$ of graded Lie groups. We may therefore identify (H',B') and (H,B) if $B(H)^* = B'(H')$. One has

Proposition 3.8.3. Let (G,A) be a graded Lie group so that $A(G)^*$ has the structure of an L-H algebra. Then the correspondence $(H,B) \longmapsto B(H)^*$ sets up a bijection of the set of all graded Lie subgroups of (G,A) and the set of all L-H subalgebras of $A(G)^*$.

For connected graded Lie subgroups all is determined by the graded Lie subalgebras.

Theorem 3.8. *Let (G,A) be a graded Lie group and let g be the Lie algebra of (G,A). Then for any graded Lie subalgebra $h \subseteq g$ there exists a unique connected graded Lie subgroup (H,B) of (G,A) whose Lie algebra is h. Moreover the correspondence $h \longrightarrow (H,B)$ sets up a bijection between the set of all graded Lie subalgebras of g and all connected graded Lie subgroups of (G,A).*

3.9. Let (G,A) be a graded Lie group and let (Y,B) be a graded manifold. Assume that $B(Y)^*$ is an $A(G)^*$ module with respect to a map

(3.9.1) $$A(G)^* \otimes B(Y)^* \longrightarrow B(Y)^*, \qquad u \otimes w \longmapsto u \cdot w$$

such that if $u \in A(G)^*$, $w \in B(Y)^*$,

(3.9.2) $$\Delta(u) = \sum_i u_i' \otimes u_i''$$

and $\Delta(w) = \sum_j w_j' \otimes w_j''$ then

(3.9.3) $$\Delta(u \cdot w) = \sum_{i,j} (-1)^{|u_i''||w_j'|} u_i' \cdot w_j' \otimes u_i'' \cdot w_j'' .$$

We will now say that the map (3.9.1) is an action map and the graded Lie group (G,A) operates or acts on the graded manifold (with respect to (3.9.1)) in case the action map (3.9.1) is smooth. We recall that $A(G)^* \otimes B(Y)^* = A \times B(G\times Y)^*$ with respect to the graded manifold $(G\times Y, A\times B)$. See §2.18.

Now observe that if (G,A) operates on (Y,B) then the graded commutative algebra $B(Y)$ becomes an $A(G)^*$ module with respect to map

(3.9.4) $$\pi : A(G)^* \longrightarrow \text{End } B(Y)$$

where if $u \in A(G)^*$, $f \in B(Y)$, $w \in B(Y)^*$ then

(3.9.5) $$\langle w, u\cdot f\rangle = (-1)^{|u||w|} \langle s(u)\cdot w, f\rangle .$$

We have written $u \cdot f$ for $\pi(u)f$. The map π will be referred to as the coaction representation of (G,A). One observes that if $f, g \in B(Y)$ then

(3.9.6) $$u \cdot fg = \sum (-1)^{|f||u_i''|} (u_i'\cdot f)(u_i''\cdot g).$$

Remark 3.9.1. Note that an action of a graded Lie group (G,A) on a graded manifold (Y,B) is determined by the coaction representation.

Example. Note that the graded Lie group (G,A) operates on the graded manifold (G,A) in two natural ways, (1) where the coaction representation is given by $\pi(u) = R_u$. Here the action map is given by $u \cdot v = (-1)^{|u||v|} vs(u)$. And (2) where $\pi(u) = L_u$. The action map is now given by $u \cdot v = uv$.

Let (G,A) be a graded Lie group. A graded Lie subgroup (H,B) of (G,A) will be called closed if H is a closed subgroup of G. We will now enrich our "supply" of graded manifolds by showing the existence of homogeneous spaces of graded Lie groups as graded manifolds.

Assume that (H,B) is a closed graded Lie subgroup of (G,A). Let

$$\rho : G \longrightarrow G/H, \qquad a \longmapsto aH$$

be the coset projection map. Let $U \subseteq G/H$ be any open set. Put $V = \rho^{-1}(U) \subseteq G$. We now observe that (H,B) operates on the graded manifold $(V,A|V)$. Indeed $A(V)^* \subseteq A(G)^*$. Furthermore if $w \in B(H^*)$, $v \in A(V)^*$, then $vw \in A(V)^*$.

But then for any $w \in B(H)^*$ we define an operator $R_w^V \in \mathrm{End}\, A(V)$ where if $v \in A(V)^*$, $f \in A(V)$ then $\langle vw,f\rangle = \langle v,R_w^V f\rangle$. It is clear that there is an action of (H,B) on $(V,A|V)$ so that $w \longmapsto R_w^V$ is the corresponding coaction representation. Note that the restriction map $\rho_{G,V} : A(G) \longrightarrow A(V)$ is then a (H,B)-module map where A(G) is a $B(H)^*$ module using the right regular representation of (G,A). Now put

$$(3.9.7) \qquad A/B(U) = \{f \in A(V) \mid R_w^V f = \langle w,1_H\rangle f \text{ for all } w \in B(H)^*\} .$$

Using the formula (3.9.6) one sees easily that A/B(U) is a graded commutative subalgebra of A(V). Furthermore the correspondence $U \longrightarrow A/B(U)$ is a sheaf A/B of graded commutative algebras on G/H. Moreover if we regard $C^\infty(U)$ as the algebra of all $f \in C^\infty(V)$ such that $f(gh) = f(g)$ for any $g \in V$, $h \in H$ then $f \longrightarrow \tilde{f}$ defines a homomorphism

$$(3.9.8) \qquad A/B(U) \longrightarrow C^\infty(U)$$

of graded algebras which is compatible with restriction maps on G/H.

The following result provides us with those graded manifolds with which we will be most concerned.

Theorem 3.9. *The sheaf* A/B *on* G/H *together with the homomorphisms* (3.9.7) *define the structure of a graded manifold* $(G/H, A/B)$. *Furthermore if* $\dim (G,A) = (m,n)$ *and* $\dim (H,B) = (m',n')$ *then* $\dim (G/H, A/B) = (m-m', n-n')$.

Since Theorem 3.9 will play such a key role in this paper we wish to give an outline of the proof. The problem of course is to show that if $U \subseteq G/H$ is sufficiently small then $A/B(U)$ has the product structure required for the definition of a graded manifold. If $\bar{e} \in G/H$ is the coset H it suffices to do this where U is a neighborhood of $\bar{e}$. Let $\underline{p}$ be a graded complement of $\underline{h}$ in $\underline{g}$. Choose a homogeneous basis $x_i \in \underline{p}_0$, $y_j \in \underline{p}_1$ of p and let $E(\underline{p}) \subseteq E(\underline{g})$ be the span of all elements of the form $y_\mu x^\nu$ where the notation is as in §3.1. Although $E(\underline{p})$ is of course not, in general, a subalgebra of $E(\underline{g})$ it is however a subcoalgebra of $E(\underline{g})$.

Now if $U \subseteq G/H$ is sufficiently small then one knows that there is a smooth section $\gamma : U \longrightarrow G$ of the fibration $\rho : G \longrightarrow G/H$ such that $\gamma(\bar{e}) = e$. Thus if $V = \rho^{-1}U$ and T is the image of γ then T has the structure of a submanifold of G and the map $T \times H \longrightarrow V$, $(t,h) \longrightarrow th$ is a diffeomorphism. Furthermore U may be chosen sufficiently small so that if $a \cdot \underline{h} = \mathrm{Ad}\, a(\underline{h})$ for $a \in G$ then

(3.9.9) $$\underline{g} = a \cdot \underline{h} + \underline{p}$$

for all $a \in T$ and hence for all $a \in V$. It follows therefore that if $w_k \in \underline{h}_0$ and $z_\ell \in \underline{h}_1$ is a basis of $\underline{h}$ then $w^\mu z_\nu$ is a basis of $E(\underline{h})$ using the notation in §3.1 and the elements

(3.9.10) $$y_\mu x^\nu t h w^{\nu'} z_{\mu'}$$

over all $t \in T$, $h \in H$, $\mu \in M_{\dim \underline{p}_1}$, $\mu' \in M_{\dim \underline{h}_1}$, $\nu \in N_{\dim \underline{p}_0}$ and $\nu' \in N_{\dim \underline{h}_0}$ are a basis of $A(V)^*$. One also notes that the elements $x^\nu t h w^{\nu'}$ are a basis of $C^\infty(V)^*$. Thus if $M(\underline{h}) = M_{\dim \underline{p}_1} \times M_{\dim \underline{h}_1}$ and for each $(\mu,\mu') \in M(\underline{h})$ one puts $A_{\mu,\mu'}(V)^* = y_\mu C^\infty(V)^* Z_{\mu'}$ then

(3.9.11) $$A(V)^* = \sum_{(\mu,\mu')\in M(\underline{h})} A_{\mu,\mu'}(V)^*$$

is a finite direct sum. Now for $(\mu,\mu') \in M(\underline{h})$ let $A_{\mu,\mu'}(V)$ be the set of all $f \in A(V)$ such that f is orthogonal to all the summands in (3.9.11) except the one corresponding to (μ,μ'). Then one proves that

(3.9.12) $$A(V) = \sum_{(\mu,\mu')\in M(\underline{h})} A_{\mu,\mu'}(V)$$

is a direct sum. Indeed $A(V)^*$ is a two sided $E(\underline{g})$ module and one can therefore define $R_u f$, $L_u f \in A(V)$ for any $u \in E(\underline{g})$, $f \in A(V)$ in a manner similar to the case where $f \in A(G)$. Upon considering the matrix for $(R_u)^t$ and $(L_u)^t$ with respect to the basis (3.9.10) one shows that the map

(3.9.13) $$A_{\mu,\mu'}(V) \longrightarrow C^\infty(V)$$

given by $f \longrightarrow \tilde{g}$ where $g = L_{s(y_\mu)} R_{z_{\mu'}} f$ is a linear isomorphism.

Now recall that $C^\infty(U)$ is the set of all $\varphi \in C^\infty(V)$ which are right covariant under the action of H. For any $\mu \in M_{\dim \underline{p}_1}$ let

(3.9.14) $$(A/B)_\mu(U) = \{f \in A_{\mu,0}(V) \mid \widetilde{L_{s(y_\mu)} f} \in C^\infty(U)\}.$$

Then the map $f \longrightarrow \widetilde{L_{s(y_\mu)} f}$ defines a linear isomorphism

(3.9.15) $$(A/B)_\mu(U) \longrightarrow C^\infty(U).$$

Moreover one has $(A/B)_\mu(U) \subseteq A/B(U)$ and in fact

(3.9.16) $$(A/B)(U) = \sum_{\mu\in M_{\dim \underline{p}_1}} (A/B)_\mu(U).$$

Furthermore if $C(U) = (A/B)_0(U)$ then $C(U)$ is a function factor in $A/B(U)$. Finally if $s_\mu \in (A/B)_\mu(U)$ is the element which corresponds to the identity function in $C^\infty(U)$ under the map (3.9.15) and $D(U)$ is the span of s_μ for $\mu \in M_{\dim \underline{p}_1}$ then $D(U)$ is an exterior factor in $A/B(U)$. Finally one notes that U is an A/B splitting neighborhood and $(C(U),D(U))$ are A/B-splitting factors for $A/B(U)$. This establishes that $(G/H,A/B)$ is a graded manifold.

Now since $A/B(G/H) \subseteq A(G)$ the injection map $\tau^* : A/B(G/H) \longrightarrow A(G)$ defines a morphism

(3.9.17) $$\tau : (G,A) \longrightarrow (G/H,A/B)$$

of graded manifolds which will be referred to as the quotient map.

Now consider the left ideal $E(\underline{g})\underline{h}$ in $E(\underline{g})$ generalized by $\underline{h}$. It is clear that the coalgebra structure on $E(\underline{g})$ induces the structure of a graded coalgebra in the quotient space $E(\underline{g})/E(\underline{g})\underline{h}$.

Proposition 3.9.1. *One has* $\tau_G = \rho : G \longrightarrow G/H$ *is the quotient map for the Lie group* G *with respect to the subgroup* H. *On the other hand one has an exact sequence*

(3.9.18) $$0 \longrightarrow E(\underline{g})\underline{h} \longrightarrow E(\underline{g}) \xrightarrow{\tau_*} (A/B)_{\bar{e}}(G/H)^* \longrightarrow 0 .$$

In fact τ_* *induces an isomorphism* $E(\underline{g})/E(\underline{g})\underline{h} \cong (A/B)_{\bar{e}}(G/H)^*$ *of graded coalgebras.*

It is convenient to identify $E(\underline{g})/E(\underline{g})\underline{h}$ with $(A/B)_{\bar{e}}(G/H)^*$. Note then that the graded vector space structure for the tangent space $T_{\bar{e}}(G/H,A/B)$ to $(G/H,A/B)$ at $\bar{e} = \{H\}$ is given by

(3.9.19) $$T_{\bar{e}}(G/H,A/B) = \underline{g}_0/\underline{h}_0 + \underline{g}_1/\underline{h}_1 .$$

Now if the notation is as in (3.9.12) we define a map

(3.9.20) $$\sigma^* : A(V) \longrightarrow A/B(U)$$

as follows: Put $\sigma^*(A_{\mu,\mu'}(V)) = 0$ if $\mu' \neq 0$. On the other hand if $f \in A_{\mu,0}(V)$ let $\sigma^*(f)$ be the unique element in $(A/B)_{\mu}(U)$ such that

(3.9.21) $$\widetilde{L_{s(y_\mu)}} f \mid T = \widetilde{L_{s(y_\mu)}} \sigma^*(f) \mid T .$$

We now observe that σ^* defines a local "section" for the "fibration" $\tau : (G,A) \longrightarrow (G/H,A/B)$.

Proposition 3.9.2. *The map* (3.9.20) *is a morphism of algebras so that it defines a morphism*

$$\sigma : (U,A/B|U) \longrightarrow (V,A|V)$$

of graded manifolds. Moreover σ_* *is injective so that the image is a graded submanifold* (T,A') *of* $(V,A|V)$ *and hence a graded submanifold of* (G,A). *In fact*

$$\tau_* \circ \sigma_* = \text{identity on } A/B(U)^* .$$

Now using Corollary to Theorem 2.16 and the fact that the multiplication map $(G\times G, A\times A) \longrightarrow (G,A)$ is smooth one also has

Proposition 3.9.3. One has a morphism

$$\theta : (U \times H, A/B \times B \mid U \times H) \longrightarrow (V, A|U)$$

of graded manifolds where $\theta_*(u \otimes v) = \sigma_*(u)v$ *for* $u \in A/B(U)^*$, $v \in B(H)^*$. *Furthermore* θ *is an isomorphism of graded manifolds.*

3.10. Now retaining the notation of §3.9 one notes that $A/B(G/H)$ is stable under the operators on $A(G)$ of the form L_u for $u \in A(G)^*$. This defines an action of (G,A) on $(G/H,A/B)$. We may express this as follows:

Proposition 3.10.1. If (G,A) *is a graded Lie group and* (H,B) *is a closed graded Lie subgroup there exists a unique action of* (G,A) *on* $(G/H,A/B)$ *such that*

$$v \cdot \tau_*(w) = \tau_*(vw) \tag{3.10.1}$$

for any $v, w \in A(G)^*$. *Here* $\tau : (G,A) \longrightarrow (G/H,A/B)$ *is the quotient map.*

Assume that (G,A) is a graded Lie group and (G,A) operates on a graded manifold (X',A'). Let $\underline{g}$ be the Lie algebra of (G,A).

Lemma 3.10.1. Let $p \in X'$ *and let* $I_p(G,\underline{g})$ *be the left ideal in* $A(G)^*$ *defined by putting*

$$I_p(G,\underline{g}) = \{u \in A(G)^* \mid u \cdot \delta_p = 0\} \tag{3.10.2}$$

Also let $I_p(\underline{g}) = I_p(G,\underline{g}) \cap E(\underline{g})$, $\underline{h} = I_p(G,\underline{g}) \cap \underline{g}$ *and let* $H = I_p(G,\underline{g}) \cap G$. *Then* $\underline{h}$ *is a graded Lie subalgebra of* $\underline{g}$ *and*

$$I_p(\underline{g}) = E(\underline{g})\underline{h}. \tag{3.10.3}$$

Moreover H *is a closed Lie subgroup of* G *and* $\underline{h}_0$ *is the Lie algebra of* H *so that* $E(H,\underline{h})$ *is an L-H subalgebra of* $E(G,\underline{g})$.

Now let (H,B) *be the graded Lie subgroup of* (G,A) *corresponding to* $E(H,\underline{h})$ *and let* $B_+(H)^* = \{w \in B(H)^* \mid \langle w, 1_H\rangle = 0\}$. *Then*

$$I_p(G,\underline{g}) = A(G)^*B_+(H)^*. \tag{3.10.4}$$

The proof of Lemma 3.10.1 uses the fact that if $I \subseteq E(\underline{g})$ is any graded left ideal such that

$$\Delta I \subseteq E(\underline{g}) \otimes I + I \otimes E(\underline{g}) \tag{3.10.5}$$

then I is primitively generated.

The graded Lie subgroup (H,B) defined in Lemma 3.10.1 will be written as $(G,A)_p$ and will be referred to as the isotropy subgroup of (G,A) at $p \in Z$.

Now assume a graded Lie group (G,A) operates on a graded manifold (X',A'). As an easy application of Lemma 3.10.1 one finds, as in the case of ordinary Lie groups, that the orbits of group-like elements define graded submanifolds.

Proposition 3.10.2. *Let* $p \in X'$ *then the map*

$$\pi : A(G)^* \longrightarrow A'(X')^*, \qquad w \longmapsto w \cdot \delta_p \tag{3.10.6}$$

is a smooth morphism of graded coalgebras. Moreover the image $A(G)^* \cdot \delta_p$ *of* π *defines a graded submanifold* $(G \cdot p, B')$ *of* X'. *In fact if* $(H,B) = (G,A)_p$ *is the isotropy subgroup at* p *then there exists a unique isomorphism*

$$\beta : (G/H, A/B) \longrightarrow (G \cdot p, B') \tag{3.10.7}$$

of graded manifolds such that

$$\pi = \beta_* \circ \tau_* \tag{3.10.8}$$

where $\tau : (G,A) \longrightarrow (G/H,A/B)$ *is the quotient map.*

Assume (G,A) is a graded Lie group and (G,A) operates on a graded manifold (X',A'). We will say that (G,A) operates transitively on (X',A'), or (X',A') with respect to this action, is a homogeneous space for (G,A), if there exists $p \in X'$ such that

$$A'(X')^* = A(G)^* \cdot \delta_p . \tag{3.10.9}$$

Remark 3.10.1. If (G,A) operates transitively on (X',A') note that G operates transitively on X' and

$$A'(X')^* = A(G)^* \cdot \delta_q \tag{3.10.10}$$

for any $q \in X'$.

As a consequence of Proposition 3.10.2 one notes that the quotient spaces of graded Lie groups are its only homogeneous spaces.

Proposition 3.10.3. *Assume* (G,A) *is a graded Lie group and* (H,B) *is a closed graded Lie subgroup. Then with respect to the action of* (G,A) *on* $(G/H,A/B)$ *defined in Proposition* 3.10.1 $(G/H,A/B)$ *is a homogeneous space for* (G,A). *Conversely, assume* (X',A') *is a homogeneous space for* (G,A). *Let* $p \in X'$ *and let* (H,B) *be the isotropy subgroup at* p. *Then the map* (3.10.7) *is an isomorphism*

$$(G/H,A/B) \cong (X',A') \tag{3.10.11}$$

not only of graded manifolds but also of (G,A) *homogeneous spaces.*

3.11. We wish now to consider an example of an action of a graded Lie group, a particular case of which (arising from the coadjoint representation) will concern us later.

Let $V = V_0 + V_1$ be a real finite dimensional graded vector space. As noted in §2.18 one has an associated graded affine manifold (V_0, A_V) of dimension $(\dim V_0, \dim V_1)$. Furthermore (see (2.18.15)) one has

$$S(V') \subseteq A_V(V_0) \tag{3.11.1}$$

where we recall that $V' = V_0' + V_1'$ is the dual space to V and the symmetric algebra $S(V')$ is in the graded sense so that in the notation of §1.2

$$S(V') = \hat{S}(V_0') \otimes \hat{\Lambda}(V_1').$$

Now since $V_0' \subseteq A_V(V_0)$ clearly contains an even coordinate system on V_0 it follows from Theorem 2.7 that there is a unique function factor $C(V_0)$ such that

$$\hat{S}(V_0') \subseteq C(V_0). \tag{3.11.2}$$

Furthermore $\hat{\Lambda V_1'}$ is an exterior splitting factor so that V_0 is an A_V-splitting neighborhood with distinguished splitting factors $(C(V_0), \hat{\Lambda V_1'})$. In particular

$$A_V(V_0) \cong C(V_0) \otimes \hat{\Lambda V_1'}. \tag{3.11.3}$$

One notes in fact that (V_0, A_V) has the structure of a graded commuted Lie group (i.e. $A_V(V)^*$ is a graded commutative Hopf algebra) and the decomposition (3.11.3) is just the decomposition (3.6.10).

Now V_0 is an A_V-coordinate neighborhood. In fact, as noted in §2.18 $V' \subseteq A_V(V_0)$ contains an A_V-coordinate system for V_0. Thus, as one frequently does for ordinary affine manifolds we may identify V with the tangent space $T_0(V_0, A_V)$ to (V_0, A_V) at $0 \in V_0$ so that

$$V = T_0(V_0, A_V) \tag{3.11.3}$$

where for $v \in V$, $f \in V'$ then $\langle v,f \rangle$ is given by the pairing of V and V'. The enveloping algebra $E(V)$ of V is then just the symmetric algebra or, as in the notation of §1.2,

$$E(V) = S(\hat{V}_0) \otimes \hat{\Lambda} V_1 . \tag{3.11.4}$$

Now to avoid a possible ambiguity (a situation which also arises in ordinary affine manifolds) we will occasionally write $\delta(V_0)$ for V_0 where it is to be understood that V_0 is to be regarded as the set of group-like elements in $A_V(V_0)^*$. In particular if $p \in V_0$ then $\delta_p \in \delta(V_0)$ is the corresponding group-like element. Thus if $p \in V_0$ then (see (3.5.9)) the tangent space at p as a graded vector space is given by

$$T_p(V_0, A_V) = \delta_p V_0 + \delta_p V_1 . \tag{3.11.5}$$

One also notes that since (V_0, A_V) is commutative the smash product becomes the tensor product and hence

$$A_V(V_0)^* = \mathbb{R}(\delta V_0) \otimes E(V) \tag{3.11.6}$$

as L-H algebras.

Now let (G,A) be any graded Lie group. By a representation of (G,A) we mean a homomorphism

$$\gamma : A(G)^* \longrightarrow \operatorname{End} V \tag{3.11.7}$$

of graded algebras such that the restriction of γ to $C^\infty(G)^*$ $(= E(G,\underline{g}_0)$ as L-H algebras) is given in the usual smooth way by a Lie group representation of G on V. In particular V is an $A(G)^*$-module. But $V = T_0(V_0, A_V) \subseteq A_V(V_0)^*$. One easily proves

<u>Lemma 3.11.1.</u> <u>There exists a unique</u> $A(G)^*$ <u>module structure on</u> $A_V(V_0)^*$ <u>given by a map</u>

(3.11.8) $$A(G)^* \otimes A_V(V_0)^* \longrightarrow A_V(V_0)^*, \qquad u \otimes v \longrightarrow u \cdot v$$

such that $u \cdot v = \gamma(u)v$ for $v \in V$ and such that

(1) $$u \cdot zw = \sum (-1)^{|z||u_i''|} (u_i' \cdot z)(u_i'' \cdot w)$$

where $u \in A(G)^*$, $z, w \in A_V(V_0)^*$, using the notation (3.9.2), and

(2) $$x \cdot \delta_p = \delta_p\, x \cdot p \in T_p(V_0, A_V)$$

for $x \in \underline{g}$, $p \in V_0$, using the notation of (3.11.5), and

(3) $$a \cdot \delta_p = \delta_{a \cdot p}$$

for $a \in G$.

The point is that (G,A) operates on the graded affine manifold (V_0, A_V).

Proposition 3.11.1. If γ is a representation of a graded Lie group (G,A) on a real finite dimensional graded vector space V then there is a unique action of (G,A) on the graded affine manifold (V_0, A_V) such that (3.11.8) is the action map.

Now let

(3.11.9) $$\pi_\gamma : A(G)^* \longrightarrow \text{End } A_V(V_0)$$

be the coaction representation with respect to the action of (G,A) on (V_0, A_V) given by Proposition 3.11.1. One notes that $S(V')$ (see (3.11.1)) is stable under $\pi_\gamma(A(G)^*)$. In fact $V' \subseteq S(V')$ is $A(G)^*$ stable and one has

(3.11.10) $$\langle v, u \cdot v' \rangle = (-1)^{|u||v|} \langle s(u) \cdot v, v' \rangle$$

for $v \in V$, $v' \in V'$, $u \in A(G)^*$ where we have written $u \cdot v'$ for $\pi_\gamma(u)v'$.

Now let $p \in V_0$. Then if $O = G \cdot p \subseteq V_0$ is the G orbit of p then by Proposition 3.10.2 the orbit $A(G)^* \cdot \delta_p$ defines a graded submanifold (O,B) of (V_0, A_V). Furthermore if we let $(G_p, A_p) = (G,A)_p$ be the isotropy subgroup at p then by Proposition 3.10.2 one has an isomorphism

(3.11.11) $$(O,B) \cong (G/G_p, A/A_p)$$

of graded manifolds. But now since (O,B) is a graded submanifold of (V_0, A_V) one has a natural homomorphism of $A_V(V_0) \longrightarrow B(O)$ of graded commutative algebras. Let $I_O(V_0)$ be the kernel of this homomorphism so that one has an exact sequence

(3.11.12) $$0 \longrightarrow I_O(V_0) \longrightarrow A_V(V_0) \longrightarrow B(O).$$

One characterizes the ideal $I_O(V_0)$ in using the coaction representation π_Y as follows.

Proposition 3.11.2. One has

$$I_O(V_0) = \{\psi \subset A_V(V_0) \mid u \cdot \tilde{\psi}(p) = 0 \quad \text{for all} \quad u \in A(G)^*\}.$$

4. Graded Differential forms, Cohomology and Line bundles

4.1. Now let (X,A) be a graded manifold and let $U \subseteq X$ be an open set. Now $\mathrm{Der}\, A(U)$ is not only a graded Lie algebra but also a module over the graded commutative algebra $A(U)$. With respect to the latter structure we can consider the tensor algebra $T(U)$ of $\mathrm{Der}\, A(U)$ over $A(U)$. See §1.2. One notes of course that $T(U)$ is bigraded (with respect to $\mathbb{Z} \oplus \mathbb{Z}_2$). Let $T^b(U)$ be the homogeneous subspace corresponding to $b \in \mathbb{Z}_+$.

Now let $J(U)$ be the two-sided bigraded ideal in $T(U)$ generated by all elements in $T^2(U)$ of the form $\xi \otimes \eta + (-1)^{|\xi||\eta|}\eta \otimes \xi$ where $\xi, \eta \in \mathrm{Der}\, A(U)$ are homogeneous. Also let $J^b(U) = J(U) \cap T^b(U)$.

Now, regarding $T^b(U)$ as a left $A(U)$ module, an element $\beta \in \mathrm{Hom}_{A(U)}(J^b(U), A(U))$ can be regarded as a b-linear map on $\mathrm{Der}\, A(U)$ with values in $A(U)$. Its value in $\xi_1, \ldots, \xi_b \in \mathrm{Der}\, A(U)$ will be denoted by $\langle \xi_1, \ldots, \xi_b \mid \beta \rangle \in A(U)$. Furthermore $\mathrm{Hom}_{A(U)}(T^b(U), A(U))$ is the set of all b-linear maps on $\mathrm{Der}\, A(U)$ which satisfy the condition

(4.1.1) $$\langle \xi_1, \ldots, f\xi_\ell, \ldots, \xi_b \mid \beta \rangle = (-1)^{|f| \sum_{i=1}^{\ell-1} |\xi_i|} f\langle \xi_1, \ldots, \xi_\ell, \ldots, \xi_b \mid \beta \rangle .$$

Now let $\Omega^b(U,A)$ be the set of all $\beta \in \mathrm{Hom}_{A(U)}(T^b(U), A(U))$ which vanish on $J^b(U)$. The elements β in $\Omega^b(U,A)$ are characterized by the additional condition that

(4.1.2) $$\langle \xi_1, \ldots, \xi_j, \xi_{j+1}, \ldots, \xi_b \mid \beta \rangle = (-1)^{1+|\xi_j||\xi_{j+1}|} \langle \xi_1, \ldots, \xi_{j+1}, \xi_j, \ldots, \xi_b \mid \beta \rangle .$$

One also notes that $\Omega^b(U,A)$ (and also $\mathrm{Hom}_{A(U)}(\mathcal{T}^b(U),A(U))$) are $A(U)$ modules where

(4.1.3) $$\langle \xi_1,\ldots,\xi_b \mid \beta f\rangle = \langle \xi_1,\ldots,\xi_b \mid \beta\rangle f$$

and

(4.1.4) $$\langle \xi_1,\ldots,f\xi_\ell,\ldots,\xi_b \mid \beta\rangle = (-1)^{|f| \sum_{i=\ell}^{b} |\xi_i|} \langle \xi_1,\ldots,\xi_b \mid f\beta\rangle .$$

Also $\Omega^b(U,A)$ is graded $(\mathbb{Z}_2)$ where

$$\langle \xi_1,\ldots,\xi_b \mid \beta\rangle \in A(U)_k$$

for $k = |\beta| + \sum_{i=1}^{b} |\xi_i|$.

One puts $\Omega^0(U,A) = A(U)$ and $\Omega(U,A) = \bigoplus_{b=0}^{\infty} \Omega^b(U,A)$. We will define an algebra structure on $\Omega(U,A)$. In fact $\Omega(U,A)$ will have the structure of a bigraded commutative algebra over $A(U)$. That is, if $\beta_i \in \Omega^{b_i}(U,A)_{j_i}$, $i = 1,2,$ then we will have

(4.1.5) $$\beta_1\beta_2 \in \Omega^{b_1+b_2}(U,A)_{j_1+j_2}$$

and also

(4.1.6) $$\beta_1\beta_2 = (-1)^{b_1b_2+j_1j_2}\beta_2\beta_1 .$$

To define $\beta_1\beta_2$ let $b = b_1 + b_2$ and let $\xi_1,\ldots,\xi_b \in \mathrm{Der}\,A(U)$ be any b homogeneous elements. Also let B be the set $(1,\ldots,b)$ and for any subset $C = (c_1,\ldots,c_d)$, written in increasing order, let ξ_C be the d-tuple $(\xi_{c_1},\ldots,\xi_{c_d})$. Let $\overline{C}$ be the complement of C in B. One defines

(4.1.7) $$|\xi_C| = \sum_{c\in C} |\xi_c|$$

and

(4.1.8) $$|\xi_{C,\overline{C}}| = \sum (1 + |\xi_c||\xi_{\overline{c}}|)$$

where the sum in (4.1.8) is over all pairs $(c,\overline{c}) \in C \times \overline{C}$ such that $c > \overline{c}$. Now let Γ be the set of all subsets $C \subseteq B$ whose cardinality is b_1. Thus $\langle \xi_C \mid \beta_1\rangle$ and $\langle \xi_{\overline{C}} \mid \beta_2\rangle$ are defined for any $C \in \Gamma$. One defines $\beta_1\beta_2$ by the relation

(4.1.9) $$\langle \xi_1,\ldots,\xi_b \mid \beta_1\beta_2\rangle = \sum_{C\in\Gamma} (-1)^{|\xi_{C,\overline{C}}|+|\xi_{\overline{C}}||\beta_1|} \langle \xi_C|\beta_1\rangle\langle \xi_{\overline{C}}|\beta_2\rangle$$

where we recall $|\beta_1| = j_1$ is the $\mathbb{Z}_2$-grading of β_1.

Proposition 4.1. The relations (4.1.5) and (4.1.6) are satisfied.

One establishes the associative law by observing that if $\beta_i \in \Omega^{b_i}(U,A)$, $i = 1,2,3$, are $\mathbb{Z}_2$-homogeneous and $b = b_1 + b_2 + b_3$ then for both $(\beta_1\beta_2)\beta_3$ and $\beta_1(\beta_2\beta_3)$ one has

$$(4.1.10) \qquad <\xi_1,\ldots,\xi_b|\beta_1\beta_2\beta_3> = \sum (-1)^{\sigma(\xi_{C_1,C_2,C_3})} <\xi_{C_1}|\beta_1><\xi_{C_2}|\beta_2><\xi_{C_3}|\beta_3>$$

where the sum is over all triples (C_1,C_2,C_3) of subsets of B such that $B = C_1 \cup C_2 \cup C_3$ is a disjoint union where the cardinalities of C_1, C_2 and C_3 are respectively b_1, b_2 and b_3. Moreover if

$$(4.1.11) \qquad |\xi_{C_1,C_2,C_3}| = \sum (1 + |\xi_c||\xi_{c'}|)$$

where the sum is over all pairs $(c,c') \in B \times B$ such that $(c,c') \in C_1 \times C_2 \cup C_1 \times C_3 \cup C_2 \times C_3$ and $c > c'$ then

$$(4.1.12) \qquad \sigma(\xi_{C_1,C_2,C_3}) = |\xi_{C_1,C_2,C_3}| + |\xi_{C_2}||\beta_1| + |\xi_{C_3}||\beta_2| + |\xi_{C_3}||\beta_1|$$

Remark 4.1. If $\beta = \beta_1 \ldots \beta_b$ where $\beta_i \in \Omega^1(U,A)$ are $\mathbb{Z}_2$-homogeneous then one may give a formula for $<\xi_1,\ldots,\xi_b \mid \beta>$ in terms of the elements $<\xi_i \mid \beta_k> \in A(U)$. This formula appears to establish the existence of some sort of graded determinant. Indeed let S_b denote the permutation group of B and for each $\pi \in S_b$ let

$$(4.1.13) \qquad \Delta_\pi = \{(i,j) \in B \times B \mid i < j,\ \pi(i) > \pi(j)\}\,.$$

Also if $\xi_i \in \operatorname{Der} A(U)$, $i = 1,\ldots,b$ let

$$(4.1.14) \qquad |\xi_\pi| = \sum_{(i,j)\in\Delta_\pi} 1 + |\xi_i||\xi_j|$$

and let

$$(4.1.15) \qquad \sigma(\xi_\pi) = |\xi_\pi| + \sum_{\ell>k} |\xi_{\pi(\ell)}||\beta_k|\,.$$

Then by induction on the number of factors into which β is decomposed one establishes

$$(4.1.16) \qquad <\xi_1,\ldots,\xi_b|\beta_1,\ldots,\beta_b> = \sum_{\pi\in S_b} (-1)^{\sigma(\xi_\pi)} <\xi_{\pi(1)}|\beta_1> \ldots <\xi_{\pi(b)}|\beta_b>$$

4.2. Now if $V \subseteq U$ is an open subset one has a restriction homomorphism $\rho_{U,V} : \Omega(U,A) \longrightarrow \Omega(V,A)$. This is defined so that if $\xi_i \in \operatorname{Der} A(U)$, $i = 1,\ldots,b$, and $\beta \in \Omega^b(U,A)$ then $\rho_{U,V}(\beta) \in \Omega^b(V,A)$ is characterized by

(4.2.1) $$\langle \rho_{U,V}\xi_1,\ldots,\rho_{U,V}\xi_b \mid \rho_{U,V}\beta\rangle = \rho_{U,V}\langle \xi_1,\ldots,\xi_b \mid \beta\rangle$$

It is clear that $U \longrightarrow \Omega(U,A)$ defines a sheaf of bigraded commutative algebras on X. This will be referred to as the sheaf of graded differential forms on (X,A).

Assume $\dim (X,A) = (m,n)$. An open set $U \subseteq X$ is called A-parallelizable if there exists $\eta_\ell \in \operatorname{Der} A(U)$, $\ell = 1,\ldots,m+n$, such that $\eta_\ell \in (\operatorname{Der} A(U))_0$ if $\ell \leq m$ and $\eta_\ell \in (\operatorname{Der} A(U))_1$ if $\ell > m$ and such that every $\xi \in A(U)$ can be uniquely written

(4.2.2) $$\xi = \sum_{\ell=1}^{n+m} f_\ell \eta_\ell$$

where $f_\ell \in A(U)$.

Remark 4.2. Note that by Theorem 2.8 every A-coordinate neighborhood is A-parallelizable.

Now if U is A-parallelizable and the notation is as in (4.2.2) one defines $\alpha_\ell \in \Omega^1(U,A)$ by the condition that

(4.2.3) $$\langle \xi \mid \alpha_\ell\rangle = f_\ell$$

for any $\xi \in \operatorname{Der} A(U)$. Thus

(4.2.4) $$\langle \eta_k \mid \alpha_\ell\rangle = \delta_{k\ell} 1_U .$$

Put $\beta_\ell = \alpha_\ell$ for $1 \leq \ell \leq m$ and $\gamma_\ell = \alpha_{\ell+m}$ for $1 \leq \ell \leq n$. One notes that $\beta_\ell \in \Omega^1(U)$ and $\gamma_\ell \in \Omega^1(U)_1$ and hence by (4.16) one has the anticommutativity

(4.2.5) $$\beta_\ell \alpha_i = - \alpha_i \beta_\ell$$

for $i = 1,\ldots,m+n$ but the commutativity

(4.2.6) $$\gamma_i \gamma_j = \gamma_j \gamma_i .$$

Now for any $\mu = (\mu_1,\ldots,\mu_k) \in M_m$, $\nu = (\nu_1,\ldots,\nu_n) \in N_n$ let

(4.2.7) $$\beta_\mu\gamma^\nu = \beta_{\mu_1} \cdots \beta_{\mu_k}\gamma_1^{\nu_1} \cdots \gamma_n^{\nu_n}.$$

Also let $(M_m \times N_n)_b = \{(\mu,\nu) \in M_m \times N_n \mid k(\mu) + |\nu| = b\}$. One notes then that

(4.2.8) $$\beta_\mu\gamma^\nu \in \Omega^b(U,A)_{|\nu|}$$

for any $(\mu,\nu) \in (M_m N_n)_b$.

Proposition 4.2.1. *Assume U is A-parallelizable and the elements $\beta_\mu\gamma^\nu \in \Omega(U,A)$ are as above. Then $\Omega(U,A)$ is a free A(U)-module with basis $\{\beta_\mu\gamma^\nu\}$, $(\mu,\nu) \in M_m \times N_n$.*

Now if $U \subseteq X$ is any open set one has a map

(4.2.9) $$d : \Omega^0(U,A) \longrightarrow \Omega^1(U,A)$$

where for any $g \in A(U) = \Omega^0(U,A)$, $dg \in \Omega^1(U,A)$ is defined by the relation

(4.2.10) $$\langle\xi \mid dg\rangle = \xi g$$

for any $\xi \in \operatorname{Der} A(U)$.

Now assume that U is an A-coordinate neighborhood. Let $\{r_i, s_j\}$, $i = 1,\ldots,m$, $s = 1,\ldots,n$, be an A-coordinate system in U. Then, by Theorem 2.8, Der A(U) is a free A(U)-module with $\{\frac{\partial}{\partial r_i}, \frac{\partial}{\partial s_j}\}$ as basis. We may then apply Proposition 4.2.1 noting the 1-forms $\{\beta_i, \gamma_j\}$ are just $\{dr_i, ds_j\}$. Thus if $dr_\mu ds^\nu$ is defined similarly for $(\mu,\nu) \in M_m \times N_n$ one has

(4.2.11) $$dr_\mu ds^\nu \in \Omega^{k(\mu)+|\nu|}(U,A)_{|\nu|}$$

and, as an immediate corollary of Proposition 4.2.1 one has

Proposition 4.2.2. *Every $\beta \in \Omega(U,A)$ can be uniquely written*

(4.2.12) $$\beta = \sum_{(\mu,\nu)\in M_m \times N_n} dr_\mu ds^\nu f_{\mu,\nu}$$

where $f_{\mu,\nu} \in A(U)$. In particular $\Omega^b(U,A)$ for any $b \in \mathbb{Z}_+$ is a free A(U)-module and the elements $\{dr_\mu ds^\nu\}$, $(\mu,\nu) \in (M_m \times N_n)_b$, are a basis.

The classic formula for the differential of a functions holds in graded manifolds. However one must be careful to use the right (not the left) A(U)-module structure in $\Omega^1(U,A)$.

Proposition 4.2.3. Let the notation be as above. If $f \in A(U)$ is arbitrary one has

(4.2.13)
$$df = \sum_{i=1}^{m} dr_i \frac{\partial f}{\partial r_i} + \sum_{j=1}^{n} ds_j \frac{\partial f}{\partial s_j} .$$

4.3. Now let $U \subseteq X$ be an arbitrary open set. The map $d : \Omega^0(U,A) \longrightarrow \Omega^1(U,A)$ has $\mathbb{Z}_2$-degree equal to zero. Thus one should not be surprised to see no sign term in

Proposition 4.3.1. Let $f, g \in A(U)$ then

(4.3.1)
$$dfg = (df)g + fdg .$$

Now since $\Omega(U,A)$ is bigraded with respect to $\mathbb{Z} \oplus \mathbb{Z}_2$ it follows that End $\Omega(U,A)$ is also bigraded with respect to $\mathbb{Z} \oplus \mathbb{Z}_2$. Thus $u \in \text{End}\, \Omega(U,A)$ is of bidegree (c,j) if

(4.3.2)
$$u(\Omega^b(U,A)_i) \subseteq \Omega^{b+c}(U,A)_{i+j}$$

for any $(b,i) \in \mathbb{Z} \oplus \mathbb{Z}_2$. If $u \in \text{End}\,(\Omega(U,A))$ is of bidegree (c,j) we will say that u is a derivation of bidegree (c,j) if for any $\alpha \in \Omega^b(U,A)_i$ and $\beta \in \Omega(U,A)$ one has

(4.3.3)
$$u(\alpha\beta) = u(\alpha)\beta + (-1)^{bc+ij}\alpha u(\beta) .$$

Our initial but incorrect view of $\Omega(U,A)$ was that an element in $\Omega(U,A)$ is some sort of symmetric algebra valued differential form on U. One's experience with vector valued differential forms was that in order to define exterior differentiation d one needs an affine connection and then one only has $d^2 = 0$ if the curvature is zero. Therefore the following theorem came as rather a pleasant surprise.

Theorem 4.3. Let $U \subseteq X$ be an arbitrary open set. Then there exists a unique derivation d

$$d : \Omega(U,A) \longrightarrow \Omega(U,A)$$

of bidegree $(1,0)$ such that (a) d on $\Omega^0(U,A)$ is given by (4.2.10) and (b) $d^2 = 0$.

We wish to give an outline of the proof. First of all one observes that it is enough to assume that U is an A-coordinate neighborhood, say, with A-coordinates $\{r_i, s_j\}$. Let $dr_\mu ds^\nu$ be as in (4.2.12). One necessarily has $d(dr_\mu ds^\nu f) = (-1)^{k(\mu)+|\nu|} dr_\mu ds^\nu df$ for $f \in A(U)$. But df is given by (4.2.13). Hence it follows that any d satisfying (a) and (b) in Theorem 4.3 is necessarily unique. Now if $\beta \in \Omega(U)$ then one defines $d\beta$ using (4.2.12) by putting

(4.3.4) $$d\beta = \sum_{(\mu,\nu)\in M_m \times N_n} (-1)^{k(\mu)+|\nu|} dr_\mu ds^\nu df_{\mu,\nu} .$$

Some straightforward calculations then establish that d is indeed a derivation of bidegree (1,0). It remains only to show that $d^2 = 0$. But this readily reduces to prove that $d^2 f = 0$ for $f \in A(U)$. One has

(4.3.5) $$-d^2 f = \sum_{i,k} dr_i dr_k \frac{\partial^2 f}{\partial r_k \partial r_i} + \sum dr_i ds_\ell \frac{\partial^2 f}{\partial s_\ell \partial r_i} + \sum ds_\ell dr_i \frac{\partial^2 f}{\partial r_i \partial s_\ell} + \sum ds_j ds_\ell \frac{\partial^2 f}{\partial s_\ell \partial s_j} .$$

One has $ds_\ell dr_i = -dr_i ds_\ell$. But $\frac{\partial}{\partial s_\ell}\frac{\partial}{\partial r_i} = \frac{\partial}{\partial r_i}\frac{\partial}{\partial s_\ell}$. Thus the second and third sums cancel out. The first sum vanishes for a similar reason. The novelty is that the fourth sum also vanishes but for the opposite reasons. That is $ds_j ds_\ell = ds_\ell ds_j$ but $\frac{\partial}{\partial s_\ell}\frac{\partial}{\partial s_j} = -\frac{\partial}{\partial s_j}\frac{\partial}{\partial s_\ell}$. This establishes Theorem 4.3.

We refer to the operator d as exterior differentiation of graded differential forms.

Other familiar operators on ordinary manifolds have counterparts on graded manifolds. If $U \subseteq X$ is open, $\xi \in \text{Der}\, A(U)$, we will now define the operation $i(\xi) \in \text{End}\, \Omega(U,A)$ of interior differentiation by ξ.

Assume ξ is homogeneous. Let $\beta \in \Omega^{b+1}(U,A)$. Then a b-linear form $i(\xi)\beta$ on Der A(U) may be given by

(4.3.6) $$\langle \xi_1, \dots, \xi_b | i(\xi)\beta \rangle = (-1)^{|\xi| \sum_{i=1}^{b} |\xi_i|} \langle \xi, \xi_1, \dots, \xi_b | \beta \rangle .$$

<u>Proposition 4.3.2. One has</u> $i(\xi)\beta \in \Omega^b(U,A)$ <u>so that</u> $\beta \longrightarrow i(\xi)\beta$ <u>induces a linear map</u> $i(\xi) : \Omega(U,A) \longrightarrow \Omega(U,A)$. <u>Furthermore</u> $i(\xi)$ <u>is a derivation of bidegree</u> $(-1, |\xi|)$.

The proof, although somewhat long, is straightforward from definitions.

For arbitrary $\xi \in \mathrm{Der}\, A(U)$ one puts $i(\xi) = i(\xi_{(0)}) + i(\xi_{(1)})$ where $\xi_{(0)}$ and $\xi_{(1)}$ are the homogeneous components of ξ.

One easily has

Lemma 4.3.1. *For* $f \in A(U)$, $\xi \in \mathrm{Der}\, A(U)$ *and* $\beta \in \Omega(U,A)$, *one has*

(4.3.7) $$i(f\xi)\beta = fi(\xi)\beta .$$

The following is also immediate.

Lemma 4.3.2. *If* $u_i \in \mathrm{End}\, \Omega(U,A)$ *are, respectively, derivations of bidegree* (b_i,j_i), $i = 1,2$ *then*

$$u = u_1 u_2 - (-1)^{b_1 b_2 + j_1 j_2} u_2 u_1$$

is a derivation of bidegree $(b_1 + b_2, j_1 + j_2)$ *of* $\Omega(U,A)$.

Let $\mathrm{Der}\, \Omega(U,A)$ be the set of all $u \in \mathrm{End}\, \Omega(U,A)$ whose bigraded homogeneous components are derivations of that bidegree. It follows from Lemma 4.3.2 that $\mathrm{Der}\, \Omega(U,A)$ is a bigraded Lie algebra.

Now for any $\xi \in \mathrm{Der}\, A(U)$ let

(4.3.8) $$\theta(\xi) = di(\xi) + i(\xi)d.$$

As a consequence of Lemma 4.3.2 one notes

Proposition 4.3.3. *If* $\xi \in \mathrm{Der}\, A(U)$ *is homogeneous then* $\theta(\xi)$ *is a derivation of bidegree* $(0,|\xi|)$ *of* $\Omega(U,A)$.

Consistent with terminology in the ungraded case we refer to $\theta(\xi)$ as Lie differentiation of $\Omega(U,A)$ by $\xi \in \mathrm{Der}\, A(U)$.

Also since $d^2 = 0$ it follows from (4.3.8) that for any $\xi \in \mathrm{Der}\, A(U)$ one has the commutativity

(4.3.9) $$d\theta(\xi) = \theta(\xi)d.$$

The familiar relations between Lie differentiation and interior differentiation now take the following form.

Proposition 4.3.4. *Assume that* $\xi, \eta \in \mathrm{Der}\, A(U)$ *are homogeneous. Then one has the following bracket relations in* $\mathrm{Der}\, \Omega(U,A)$

$$(1)\quad i(\xi)i(\eta) + (-1)^{|\xi||\eta|} i(\eta)i(\xi) = [i(\xi),i(\eta)] = 0$$

$$(2)\quad \theta(\xi)i(\eta) - (-1)^{|\xi||\eta|} i(\eta)\theta(\xi) = [\theta(\xi),i(\eta)] = i([\xi,\eta])$$

$$(3)\quad \theta(\xi)\theta(\eta) - (-1)^{|\xi||\eta|} \theta(\eta)\theta(\xi) = [\theta(\xi),\theta(\eta)] = \theta([\xi,\eta])$$

The relationship between contraction and Lie differentiation becomes

Proposition 4.3.5. *Let* $\beta \in \Omega^b(U,A)$ *and assume that* ξ *and* $\xi_i \in \mathrm{Der}\,A(U)$ *are homogeneous,* $i = 1,\ldots,b$. *Then*

$$\xi\langle\xi_1,\ldots,\xi_b|\beta\rangle = \sum_{i=1}^{b} (-1)^{|\xi|j_{i-1}} \langle\xi_1,\ldots,[\xi,\xi_i],\ldots,\xi_b|\beta\rangle$$

$$+ (-1)^{|\xi|j_b} \langle\xi_1,\ldots,\xi_b|\theta(\xi)\beta\rangle$$

where $j_i = \sum_{k=1}^{i} |\xi_k|$.

The classic formula for the coboundary in de Rham theory becomes

Proposition 4.3.6. *Let the notation be as in Proposition* 4.3.5 *with* $\xi = \xi_{b+1}$. *Then*

$$\langle\xi_1,\ldots,\xi_{b+1}|d\beta\rangle = \sum_{i=1}^{b+1} (-1)^{i-1+j_{i-1}|\xi_i|} \xi_i\langle\xi_1,\ldots,\hat{\xi}_i,\ldots,\xi_{b+1}|\beta\rangle$$

$$+ \sum_{k<\ell} (-1)^{d_{k,\ell}} \langle[\xi_k,\xi_\ell],\xi_1,\ldots,\hat{\xi}_k,\ldots,\hat{\xi}_\ell,\ldots,\xi_{b+1}|\beta\rangle$$

where $d_{k,\ell} = |\xi_k|j_{k-1} + |\xi_\ell|j_{\ell-1} + |\xi_k||\xi_\ell| + k + \ell$.

It is convenient to write out the sum in Proposition 4.3.6 for the case where $b = 1$. Thus if $\beta \in \Omega^1(U,A)$ and ξ_1, $\xi_2 \in \mathrm{Der}\,A(U)$ are homogeneous then

$$(4.3.10)\qquad \langle\xi_1,\xi_2|d\beta\rangle = \xi_1\langle\xi_2|\beta\rangle - (-1)^{|\xi_1||\xi_2|} \xi_2\langle\xi_1|\beta\rangle - \langle[\xi_1,\xi_2]\ \beta\rangle$$

For $b = 2$ one has

Proposition 4.3.7. *Let* $\beta \in \Omega^2(U,A)$. *Then* $d\beta = 0$ *if and only if for every homogeneous* $\xi_i \in \mathrm{Der}\,A(U)$, $i = 1,2,3$ *one has*

$$(-1)^{|\xi_1||\xi_3|}\xi_1\langle\xi_2,\xi_3|\beta\rangle + (-1)^{|\xi_1||\xi_2|}\xi_2\langle\xi_3,\xi_1|\beta\rangle + (-1)^{|\xi_2||\xi_3|}\xi_3\langle\xi_1,\xi_2|\beta\rangle$$

$$= (-1)^{|\xi_1||\xi_3|}\langle[\xi_1,\xi_2],\xi_3|\beta\rangle + (-1)^{|\xi_1|.|\xi_2|}\langle[\xi_2,\xi_3],\xi_1|\beta\rangle$$

$$+ (-1)^{|\xi_2||\xi_3|}\langle[\xi_3,\xi_1]\xi_2|\beta\rangle .$$

4.4. Unlike the algebra of all ordinary differential forms on U the algebra $\Omega(U,A)$ is not the space of all smooth sections of a vector bundle on U. Hence it is not immediately obvious that the correspondence $X \longrightarrow \Omega(X,A)$ is a functor in the category of graded manifolds. Nevertheless it is.

Theorem 4.4. *Assume that*

$$\sigma : (X,A) \longrightarrow (Y,B)$$

is a morphism of graded manifolds. Let $W \subseteq Y$ *be an open set so that if* $U = \sigma_X^{-1} W \subseteq X$ *then*

$$\sigma^* : B(W) \longrightarrow A(U)$$

is a homomorphism of graded commutative algebras. Then σ^* *extends to a unique homomorphism of bigraded commutative algebras*

$$\sigma^* : \Omega(W,B) \longrightarrow \Omega(U,A) \tag{4.4.1}$$

such that σ^* *commutes with exterior differentiation. Moreover the map* (4.4.1) *is compatible with restriction maps to smaller open sets.*

We give some indication of the proof of Theorem 4.4. First it easily suffices to consider the case where W is a B-coordinate neighborhood in (Y,B), say with B-coordinates $\{r_i, s_j\}$. It follows from Proposition 4.2.2 that $\Omega(W,B)$ is just the exterior algebra in the sense of §1.2 of $\Omega^1(W,B)$ over $B(W) = \Omega^0(W,B)$. Thus it suffices only to show that there is a unique map

$$\sigma^* : \Omega\ (W,B) \longrightarrow \Omega\ (U,A)$$

which satisfies the conditions of Theorem 4.4. Using (4.2.13) this however reduces to showing that if $f \in B(W)$ then

$$(4.4.2) \qquad d(\sigma^*(f)) = \sum_i d(\sigma^*(r_i))\sigma^*(\frac{\partial f}{\partial r_i}) + \sum_j d(\sigma^*(s_j))\sigma^*(\frac{\partial f}{\partial s_j})$$

or if $\xi \in \operatorname{Der} A(U)$ then

$$(4.4.3) \qquad \xi\sigma^*(f) = \sum \xi(\sigma^*(r_i))\sigma^*(\frac{\partial f}{\partial r_i}) + \sum \xi(\sigma^*(s_j))\sigma^*(\frac{\partial f}{\partial s_j}).$$

Now if f is a polynomial in the r_i and s_j the equation is obvious. Now assume that (4.4.3) does not hold for some $f \in B(W)$ and $\xi \in \operatorname{Der} A(U)$. Then there exists $p \in W$ and $u \in A_p^k(X)^*$ for some k such that u distinguishes between the left and right sides of (4.4.3). Now if necessary we may replace r_i by $r_i - \tilde{r}_i(\sigma_*(p))1_W$ so that one has $\tilde{\sigma}^*(r_i)(p) = \tilde{\sigma}^*(s_j)(p) = 0$. This does not change either side of (4.4.3). But now, recalling the decomposition of f given in (2.6.4), then, by taking the partial Taylor expansion, we may write $f_\mu = g_\mu + h_\mu$ where g_μ is a polynomial in the r_i and h_μ is in the ideal generated by homogeneous polynomials in the r_i of degree $k+2 - k(\mu)$. This decomposes f into the sum $g + h$. But now one has equality if we substitute g for f in (4.4.3). On the other hand if we substitute h for f in (4.4.3) and apply u then both sides vanish since $u \in A_p^k(X)^*$ and both sides of (4.4.3) are in the ideal of $A(U)$ generated by homogeneous polynomials in the $\sigma(r_i)$, $\sigma(s_j)$ of degree $k+1$. This is a contradiction proving Theorem 4.4.

4.5. Now let $p \in X$ and assume that $p \in U \subseteq X$. The elements of $\Omega(U,A)$ define $A(U)$-valued multilinear forms on $\operatorname{Der} A(U)$ but unlike the case with ordinary differential forms they do not induce multilinear forms on the tangent space $T_p(X,A)$ at p.

Let $\Omega_A^b(p)$ be the linear space of all real valued b-linear forms z on

$$T_p(X,A) = T_p(X) \oplus T_p(X,A)_1$$

such that

$$(4.5.1) \qquad \langle v_1,\ldots,v_j,v_{j+1},\ldots,v_b|z\rangle = (-1)^{1+|v_j||v_{j+1}|}\langle v_1,\ldots,v_{j+1},v_j,\ldots,v_b|z\rangle$$

for all homogeneous $v_i \in T_p(X,A)$.

Note that $\Omega_A^b(p)$ is $\mathbb{Z}_2$-graded so that if z is homogeneous then (4.5.1) vanishes unless

$$(4.5.2) \qquad \sum_{i=1}^{b} |v_i| = |z|.$$

Also given $z \in \Omega_A^b(p)$ let $z \mid T_p(X,A)_j$, $j = 1,2$, be, respectively, the b-linear forms on $T_p(X,A)_0 = T_p(X)$ and $T_p(X,A)_1$ obtained by restricting the v_i to be in $T_p(X,A)_0$ or $T_p(X,A)_1$.

Remark 4.5.1. By (4.5.1) note that $z \mid T_p(X)$ is an alternating b-linear form on $T_p(X)$ and $z \mid T_p(X,A)_1$ is a symmetric b-linear form on $T_p(X,A)_1$.

Let $\Omega_A^0(p) = \mathbb{R}$ and put $\Omega_A(p) = \bigoplus_{b=0}^{\infty} \Omega_A^b(p)$. One defines the structure of a bigraded commutative algebra over $\mathbb{R}$ (i.e. relations (4.1.5) and (4.1.6) are satisfied) on $\Omega_A(p)$ by the same formula as (4.1.9) where the ξ_i are replaced by v_i and β_1, β_2 are replaced by $z_i \in \Omega_A^{b_i}(p)_{j_i}$, $i = 1,2$.

Now let $T_p^*(X,A)$ be the dual space to the tangent space $T_p(X,A)$. One has a canonical isomorphism $T_p^*(X,A) \longrightarrow \Omega_A^1(p)$. Furthermore this clearly extends to a homomorphism of the exterior algebra

$$\Lambda T_p^*(X,A) \longrightarrow \Omega_A(p)$$

which, one readily establishes, is an isomorphism. Henceforth we identify $\Lambda T_p^*(X,A)$ with $\Omega_A(p)$. But for any b one has a smooth vector bundle $\Lambda^b T^*(X,A)$ on X whose fiber at p is $\Lambda^b T_p^*(X,A)$. If $U \subseteq X$ is open let $\Omega_A^b(U)$ be the space of all smooth sections of $\Lambda^b T_p^*(X,A)$ over U. One has $\Omega_A^0(U) = C^\infty(U)$ and $\Omega_A(U) = \bigoplus_{b=0}^{\infty} \Omega_A^b(U)$ has the structure of a bigraded commutative algebra. We recall (see §2.12) that $\mathrm{Der}\,(A(U), C^\infty(U))$ is the space of all smooth sections of $T(X,A)$ over U. Thus if $\gamma \in \Omega_A^b(U)$ and $\zeta_i \in \mathrm{Der}\,(A(U), C^\infty(U))$, $i = 1,\ldots,b$, one has the b-linear form

$$(4.5.3) \qquad \langle \zeta_1, \ldots, \zeta_b \mid \gamma \rangle \in C^\infty(U)$$

and the identity (4.5.1) is satisfied where the ζ_i replaces the v_i and γ replaces z.

We now observe that the map $A(U) \longrightarrow C^\infty(U)$, $f \longmapsto \tilde{f}$ extends to a homomorphism $\Omega(U,A) \longrightarrow \Omega_A(U)$, $\beta \longmapsto \tilde{\beta}$.

Proposition 4.5.1. For any $\beta \in \Omega^b(U,A)$ *an element* $\tilde{\beta} \in \Omega_A^b(U)$ *is uniquely defined by the relation*

$$(4.5.4) \qquad (\langle \xi_1,\ldots,\tilde{\xi}_b|\beta\rangle) = \langle \tilde{\xi}_1,\ldots,\tilde{\xi}_b|\tilde{\beta}\rangle$$

for any $\xi_i \in \mathrm{Der}\, A(U)$, $i = 1,\ldots,b$. *Furthermore the linear extension*

$$(4.5.5) \qquad \Omega(U,A) \longrightarrow \Omega_A(U), \qquad \beta \longmapsto \tilde{\beta}$$

is an algebra homomorphism.

Now for any $p \in X$ and $\gamma \in \Omega_A(X)$ let $\gamma_p \in \Omega_A(p)$ be the value of γ at p. We will also write $\gamma \mid T_p(X,A)$ for γ_p. In fact if $\gamma \in \Omega_A^b(X)$ we will write $\gamma \mid T_p(X,A)_j$, $j = 1,2$, for $\gamma_p \mid T_p(X,A)_j$. Thus (see Remark 4.5.1) one has $\gamma \mid T_p(X)$ is an alternating b-linear form on $T_p(X)$ and $\gamma \mid T_p(X,A)_1$ is a symmetric b-linear form on $T_p(X,A)_1$.

Assume $\omega \in \Omega^2(X,A)$. Then ω defines (see Lemma 4.3.1) an $A(X)$-linear map

$$(4.5.6) \qquad \mathrm{Der}\, A(X) \longrightarrow \Omega^1(X,A), \qquad \xi \longmapsto i(\xi)\omega$$

where we recall

$$(4.5.7) \qquad \langle \eta | i(\xi)\omega\rangle = (-1)^{|\eta||\xi|}\langle \xi,\eta|\omega\rangle$$

for $\xi, \eta \in \mathrm{Der}\, A(X)$. We will say that ω is non-singular if the map (4.5.6) is an isomorphism.

But now if $p \in X$ then $\tilde{\omega} \mid T_p(X)$ is an alternating bilinear form on $T_p(X)$ and $\tilde{\omega} \mid T_p(X,A)_1$ is a symmetric bilinear form on $T_p(X,A)_1$. Furthermore if ω has $\mathbb{Z}_2$-grading zero then also $\tilde{\omega} \in \Omega_A^2(X)_0$ and hence $(\langle v_1,v_2|\tilde{\omega}_p\rangle) = 0$ for $v_j \in T_p(X,A)_j$, by (4.5.2). One has

Proposition 4.5.2. If $\omega \in \Omega^2(X,A)_0$ *then* ω *is non-singular if and only if both the alternating bilinear form* $\tilde{\omega} \mid T_p(X)$ *and the symmetric bilinear form* $\tilde{\omega} \mid T_p(X,A)_1$ *are non-singular at every point* $p \in X$.

4.6. Now for any open set $U \subseteq X$ let $\Omega(U)$ denote the algebra of ordinary differential forms on X. As usual, d will also denote exterior differentiation on $\Omega(U)$.

Now let

(4.6.1) $$\Omega_A(X) \longrightarrow \Omega(X),$$

$\gamma \longrightarrow \gamma \mid T(X)$, be the $C^\infty(X)$-linear map defined so that if γ is $\mathbb{Z}$-homogeneous then the value of $\gamma \mid T(X)$ at $p \in X$ is just $\gamma \mid T_p(X)$. See §4.5. Also let

(4.6.2) $$\kappa : \Omega(X,A) \longrightarrow \Omega(X)$$

be defined by putting $\kappa(\beta) = \tilde{\beta} \mid T(X)$. Hence one has a commutative diagram

(4.6.3) $$\begin{array}{ccc} \Omega(X,A) & \xrightarrow{\ \kappa\ } & \Omega(X) \\ & \searrow & \uparrow \\ & & \Omega_A(X) \end{array} \quad .$$

Proposition 4.6.1. All three maps in (4.6.3) are algebra homomorphisms. Furthermore κ *commutes with exterior differentiation and hence is a mapping co-chain complexes.*

Remark 4.6. In the notation we are using note that the manifold X itself is just (X,C^∞) where C^∞ is the sheaf of C^∞ functions on X. The map $f \longmapsto \tilde{f}$, for $f \in A(X)$, defines a morphism

(4.6.4) $$\sigma : (X,C^\infty) \longrightarrow (X,A)$$

of graded manifolds. One now observes that κ is just σ^*.

We now observe that the Poincaré lemma holds for graded manifolds.

Theorem 4.6. Let (X,A) *be a graded manifold. Then if* $U \subseteq X$ *is a connected open set and* $f \in A(U)$ *then* $df = 0$ *if and only if* $f = \lambda \cdot 1_U$ *for some constant* $\lambda \in \mathbb{R}$.

Furthermore if U, *in addition, is a contractible* A-*coordinate neighborhood and* $\beta \in \Omega^b(U,A)$, *when* $b \geq 1$, *is such that* $d\beta = 0$, *then there exists* $\omega \in \Omega^{b-1}(U,A)$ *such that* $\beta = d\omega$.

We will sketch the proof. The first statement is just an easy consequence of (4.2.13) and (2.6.4). Assume that U is an A-coordinate neighborhood with A-coordinates $\{r_i,s_j\}$. Let $C(U)$ be the function factor which contains the r_i

and $D(U)$ the exterior factor generated by the s_j so that $A(U) = C(U) \otimes D(U)$. See Remark 2.8.1. Now let $\Omega(U,C)$ be the subalgebra of $\Omega(U,A)$ generated by $C(U)$ and dr_i. Let $\Omega(U,D)$ be the subalgebra of $\Omega(U,A)$ generated by $D(U)$ and ds_j. Then both $\Omega(U,C)$ and $\Omega(U,D)$ are stable under exterior differentiation by (4.2.13). Furthermore by (2.6.4) and (4.2.12) one has the tensor product decomposition of cochain complexes

$$\Omega(U,A) = \Omega(U,C) \otimes \Omega(U,D). \tag{4.6.5}$$

Now the map $f \longrightarrow \tilde{f}$ induces an isomorphism $C(U) \longrightarrow C^\infty(U)$. The extension $\kappa : \Omega(U,C) \longrightarrow \Omega(U)$ is then clearly also an isomorphism. Furthermore this is an isomorphism of cochain complexes by (4.2.12). But $\Omega(U)$ is acyclic by the usual Poincaré lemma. Thus $\Omega(U,C)$ is acyclic. On the other hand $\Omega(U,D)$ is easily seen to be the usual Koszul complex with the roles of symmetric and skew-symmetric interchanged. The proof of the acyclicity of the Koszul complex may then be trivially modified to show that $\Omega(U,D)$ is also acyclic. The Kunneth formula applied to (4.6.5) proves that $\Omega(U,A)$ is also acyclic proving Theorem 4.6.

4.7. We will use the following notation with regard to cohomology. If S is a sheaf on X then $H(X,S)$ will denote the usual Cech cohomology with values in S. On the other hand if, say, B is a cochain complex whose coboundary boundary is understood we will write $\mathrm{Coh}(B)$ for the cohomology of B. Thus $\mathrm{Coh}\,(\Omega(X))$ is the usual de Rham cohomology of X. But now we can also consider $\mathrm{Coh}\,(\Omega(X,A))$. We will refer to this as the de Rham cohomology of (X,A).

Now recall (see Lemma 2.4) that one has a partition of unity for the graded manifold (X,A). It follows therefore that $\Omega^b(A)$ is flasque if $\Omega^b(A)$ is the sheaf $U \longrightarrow \Omega^b(U,A)$. But then by Theorem 4.6 exterior differentiation defines

$$\longrightarrow \Omega^b(A) \longrightarrow \Omega^{b+1}(A) \longrightarrow \quad , \tag{4.7.1}$$

a flasque resolution of the constant sheaf. But then the de Rham theorem is still valid and hence one has a natural isomorphism

$$\mathrm{Coh}\,(\Omega(X,A)) \longrightarrow H(X,\mathbb{R}). \tag{4.7.2}$$

On the other hand the usual de Rham theorem yields a natural isomorphism $\mathrm{Coh}\,(\Omega(X)) \longrightarrow H(X,\mathbb{R})$.

The maps are easily related by

<u>Theorem 4.7.</u> <u>Let</u> (X,A) <u>be any graded manifold and let</u> $\bar{\kappa} : \mathrm{Coh}\,(\Omega(X,A)) \longrightarrow \mathrm{Coh}\,(\Omega(X))$ <u>be the map on the de Rham cohomology induced by the cochain map</u> $\kappa : \Omega(X,A) \longrightarrow \Omega(X)$ <u>given by</u> (4.6.2). <u>Then</u> $\bar{\kappa}$ <u>is an isomorphism and one has a commutative diagram of algebra isomorphisms</u>

(4.7.3)
$$\begin{array}{ccc} \mathrm{Coh}\,(\Omega(X,A)) & \xrightarrow{\bar{\kappa}} & \mathrm{Coh}\,(\Omega(X)) \\ & \searrow & \downarrow \\ & & H(X,\mathbb{R}) \end{array} .$$

It will be useful for us to be more explicit about the map (4.7.2) particularly for the case of the second cohomology group $\mathrm{Coh}^2\,(\Omega(X,A))$. Let $\omega \in \Omega^2(X,A)$ be a closed 2-form and let $\{U_i\}$, $i \in \Lambda$, be a contractible covering of X. Since $H^2(U_i,\mathbb{R}) = 0$ it follows that there exists $\alpha_i \in \Omega^1(U_i,A)$ so that $d\alpha_i = \omega$ in U_i. But then $d(\alpha_j - \alpha_i) = 0$ in $U_i \cap U_j$ and hence for similar reasons there exists $f_{ij} \in A(U_i \cap U_j)$ so that $\alpha_j - \alpha_i = df_{ij}$ in $U_i \cap U_j$. But then $d(f_{ij}+f_{jk}-f_{ik}) = 0$ and hence by the same argument (or Theorem 4.6) there exists $r_{ijk} \in \mathbb{R}$ so that

$$f_{ij} + f_{jk} - f_{ik} = r_{ijk} 1_{U_i \cap U_j \cap U_k}$$

in $U_i \cap U_j \cap U_k$. But $\{r_{ijk}\}$ is a Cech cocycle with values in $\mathbb{R}$. Let $[r_{ijk}] \in H^2(X,\mathbb{R})$ be the corresponding class. Also let $[\omega] \in \mathrm{Coh}^2\,(\Omega(X,A))$ be the class of ω. Then

(4.7.4)
$$[\omega] \longmapsto [r_{ijk}]$$

with respect to the map (4.7.2).

4.8. The sheaf A is a sheaf of algebras over $\mathbb{R}$. Let $A_{\mathbb{C}}$ be the complexification of A. Thus $A_{\mathbb{C}}$ is the sheaf $U \longrightarrow A(U) \otimes_{\mathbb{R}} \mathbb{C} = A_{\mathbb{C}}(U)$. If $C^\infty_{\mathbb{C}}(U) = C^\infty(U) \otimes_{\mathbb{R}} \mathbb{C}$ is the algebra of all complex-valued smooth functions on U obviously the map $f \longmapsto \tilde{f}$ extends linearly to an algebra homomorphism

$A_{\mathbb{C}}(U) \longrightarrow C^{\infty}_{\mathbb{C}}(U)$, $f \longrightarrow \tilde{f}$.

Now let L be a sheaf on X of graded $A_{\mathbb{C}}$-modules. Thus for any open set $U \subseteq X$, $L(U)$ is a graded $A_{\mathbb{C}}(U)$-module.

We will say that a non-empty open $U \subseteq X$ is principal for L if $L(U)$ is a free $A_{\mathbb{C}}(U)$-module with only one generator in $L(U)_0$. Such a generator t will be referred to as a basal element for L in U.

Remark 4.8.1. If U is principal for L using the sheaf properties it is not hard to prove that V is principal for L for any non-empty open $V \subseteq U$. Furthermore if $t \in L(U)$ is basal for L in U its restriction $\rho_{U,V}(t) \in L(V)$ is basal for L in V.

Now the sheaf L will be said to be a line bundle sheaf over (X,A) if X can be covered by open sets which are principal for L.

Remark 4.8.2. Note that $A_{\mathbb{C}}$ itself is a line bundle sheaf over (X,A).

Now assume that L is a line bundle sheaf over (X,A). A set $\{(U_i,t_i)\}$, $i \in \Lambda$, will be said to be a local system for L if $\{U_i\}$, $i \in \Lambda$, is an open covering of X by sets which are principal for L and $t_i \in L(U_i)_0$ is a basal element for L in U_i. By assumption a local system for L exists.

Now for any open set $U \subseteq X$ let $A^*_{\mathbb{C}}(U)$ be the set of all $f \in A_{\mathbb{C}}(U)_0$ which are invertible in the algebra $A_{\mathbb{C}}(U)$. The argument in the proof of Lemma 2.4 yields

Lemma 4.8.1. *If* $f \in A_{\mathbb{C}}(U)_0$ *then* $f \in A^*_{\mathbb{C}}(U)$ *if and only if the function* $\tilde{f} \in C^{\infty}_{\mathbb{C}}(U)$ *is nowhere vanishing on* U.

Now let $\{(U_i,t_i)\}$, $i \in \Lambda$, be a local system for L. But then by Remark 4.8.1 there exists uniquely $c_{ij} \in A^*_{\mathbb{C}}(U_i \cap U_j)$ such that

(4.8.1) $$t_i c_{ij} = t_j \qquad \text{in} \quad U_i \cap U_j.$$

Furthermore one has

(4.8.2) $$c_{ii} = 1_{U_i}, \quad c_{ij}c_{ji} = 1_{U_i \cap U_j}$$

and

(4.8.3) $$c_{ij}c_{jk} = c_{ik} \qquad \text{in} \quad A^*_{\mathbb{C}}(U_i \cap U_j \cap U_t).$$

We refer to the c_{ij} as transition elements for L.

But now mimicking the situation with ordinary line bundles one may define an equivalence relation among the line bundle sheaves over (X,A) and introduce $L(X,A)$ the set of equivalence classes of such line bundle sheafs.

If L and L' are line bundle sheaves over (X,A) then we will say that they are equivalent if there is an $A_{\mathbb{C}}(X)$-linear isomorphism

(4.8.4) $$L(X) \longrightarrow L'(X).$$

It is clear then that (4.8.4) induces an $A_{\mathbb{C}}(U)$-linear isomorphism

(4.8.5) $$L(U) \longrightarrow L'(U)$$

for any open set $U \subseteq X$. Given a line bundle sheaf L the corresponding class in $L(X,A)$ will be denoted by [L].

Remark 4.8.3. In the notation above the class [L] as with ordinary line bundles is determined by the set of transition elements $\{c_{ij}\}$. Furthermore given elements $c_{ij} \in A^*_{\mathbb{C}}(U_i \cap U_j)$ satisfying (4.8.2) and (4.8.3) there exists a line bundle sheaf having these c_{ij} as transition elements.

Let $L(X)$, as in [8], be the set of equivalence classes of ordinary complex line bundles on X. Then as with $L(X)$ the set $L(X,A)$ has a natural structure of a group where the identity element is the class $[A_{\mathbb{C}}]$ of A. In fact the correspondence $U \longrightarrow A^*_{\mathbb{C}}(U)$ defines a sheaf $A^*_{\mathbb{C}}$ on X. Furthermore since $A^*_{\mathbb{C}}$ is clearly a sheaf of abelian groups (under multiplication) we may form the cohomology group $H(X,A^*_{\mathbb{C}})$. Now the relations (4.8.2) and (4.8.3) imply that the c_{ij} define a class $[c_{ij}] \in H^1(X,A^*_{\mathbb{C}})$. But now the definition of equivalence of line bundle sheaves on (X,A) clearly implies

Proposition 4.8.1. *One has a natural isomorphism of abelian groups*

(4.8.6) $$L(X,A) \longrightarrow H^1(X,A^*_{\mathbb{C}})$$

where if L *and* c_{ij} *are as above and* $[L] \in L(X,A)$ *is the class corresponding to* L *then* $[L] \longrightarrow [c_{ij}]$ *under the map* (4.8.6).

The isomorphism (4.8.6) enables one to determine $L(X,A)$. Now recall (see Proposition 2.9.2) that if $U \subseteq X$ is open the space $A(U)$ has the structure of a complete locally convex space over $\mathbb{R}$. Clearly, then, also $A_{\mathbb{C}}(U)$ has the structure of a complete locally convex space over $\mathbb{C}$. One notes in particular that if $f \in A_{\mathbb{C}}(U)_0$ then $\varepsilon(f) = \sum_{n=0}^{\infty} \frac{(2\pi i f)^n}{n!} \in A^*_{\mathbb{C}}(U)$ converges absolutely and $f \longmapsto \varepsilon(f)$ defines a homomorphism

$$(4.8.7) \qquad \varepsilon : A_{\mathbb{C}}(U)_0 \longrightarrow A^*_{\mathbb{C}}(U)$$

of abelian groups. The nature of the map (4.8.7) is more transparent in case U admits a function factor $C(U)$. Indeed in such a case we can write $f = g+h$ where $g \in C_{\mathbb{C}}(U) = C(U) \otimes_{\mathbb{R}} \mathbb{C}$ and $h \in A^1_{\mathbb{C}}(U)_0 = A^1(U)_0 \otimes_{\mathbb{R}} \mathbb{C}$. Since $C_{\mathbb{C}}(U)$ is isomorphic to $C^{\infty}_{\mathbb{C}}(U)$ one has $\varepsilon(g) \in C_{\mathbb{C}}(U)$. On the other hand h is nilpotent and hence $\varepsilon(h)$ is given by a finite sum and is of the form $1_U + h'$ where $h' \in A^1_{\mathbb{C}}(U)_0$. On the other hand one has

$$(4.8.8) \qquad \varepsilon(f) = \varepsilon(g)\varepsilon(h).$$

In this connection since elements of the form $1_U + A^1_{\mathbb{C}}(U)_0$ have unique logarithms in $A^1_{\mathbb{C}}(U)_0$ note

Lemma 4.8.2. *One has a bijection*

$$(4.8.9) \qquad \varepsilon : A^1_{\mathbb{C}}(U)_0 \longrightarrow 1_U + A^1_{\mathbb{C}}(U)_0 .$$

But then if U is connected it follows from (4.8.8) that the kernel of ε is just $\mathbb{Z}\cdot 1_U$. Furthermore it also follows that if U is simply-connected then the map is surjective. Thus if $(A_{\mathbb{C}})_0$ is the sheaf $U \longrightarrow A_{\mathbb{C}}(U)_0$ one has

Lemma 4.8.3. *The map ε induces an exact sequence*

$$(4.8.10) \qquad 0 \longrightarrow \mathbb{Z} \longrightarrow (A_{\mathbb{C}})_0 \longrightarrow A^*_{\mathbb{C}} \longrightarrow 0$$

of sheaves.

Now let L be a line bundle sheaf over (X,A) and let c_{ij} be as in (4.8.1) with respect to a contractible covering $\{U_i\}$, $i \in \Lambda$. Since $c_{ij} \in A^*_{\mathbb{C}}(U_i \cap U_j)$ we may find $g_{ij} \in A_{\mathbb{C}}(U_i \cap U_j)_0$ so that $\varepsilon(g_{ij}) = c_{ij}$ and hence if $h_{ijk} = g_{ij} + g_{jk} - g_{ik}$ one has $\varepsilon(h_{ijk}) = 1_{U_i \cap U_j \cap U_k}$ so that

$h_{ijk} = n_{ijk} 1_{U_i \cap U_j}$ where $n_{ijk} \in \mathbb{Z}$ and defines a class $[n_{ijk}] \in H^2(X,\mathbb{Z})$. But since $(A_{\mathbb{C}})_0$ is flasque, taking cohomology, it follows from Proposition 4.8.1 and Lemma 4.8.3 that we have established, as is the case with ordinary line bundles,

Theorem 4.8. *Let (X,A) be a graded manifold and let $L(X,A)$ be the group of equivalence classes of line bundle sheaves over (X,A). Then one has an isomorphism of groups*

(4.8.11) $$L(X,A) \cong H^2(X,\mathbb{Z})$$

where in the notation above $[L] \longrightarrow [n_{ijk}]$.

4.9. Now let L be a line bundle sheaf over (X,A). We will say that (L,∇) is a line bundle sheaf with connection ∇ if for any open set $U \subseteq X$ and $\xi \in \operatorname{Der} A(U)$ one has a linear map, referred to as covariant differentiation by ξ,

(4.9.1) $$\nabla_\xi : L(U) \longrightarrow L(U)$$

where $|\nabla_\xi| = |\xi|$ with respect to the $\mathbb{Z}_2$ grading, and which is compatible with restriction maps to smaller open sets, and also, is such that

(4.9.2) (1) $$\nabla_\xi ft = (\xi f)t + (-1)^{|f||\xi|} f\nabla_\xi t$$

for $f \in A(U)$, $t \in L(U)$ and

(2) the map

$$\operatorname{Der} A(U) \longrightarrow \operatorname{End} L(U)$$

given by $\xi \longrightarrow \nabla_\xi$ is $A(U)$-linear.

Now the complexification of $\operatorname{Der} A(U)$ may be taken to be the complex Lie algebra of derivations, $\operatorname{Der} A_{\mathbb{C}}(U)$. It is clear that by linearity we may take ξ and f in (4.9.1) and (4.9.2) to be in $\operatorname{Der} A_{\mathbb{C}}(U)$ and $A_{\mathbb{C}}(U)$ respectively. Also $\Omega_{\mathbb{C}}(U,A)$ will denote the complexification of $\Omega(U,A)$ and notation of §4.1-4.8 will extend to complexifications by linearity.

Now assume that (L,∇) is a line bundle sheaf with connection over (X,A). Let $U \subseteq X$ be principal for L and let $t \in L(U)_0$ be a basal element. Then for any $\xi \in \operatorname{Der} A_{\mathbb{C}}(U)$ there exists $g \in A_{\mathbb{C}}(U)$ such that $\nabla_\xi t = gt$. But the correspondence $\xi \longmapsto g$ defines an $A_{\mathbb{C}}(U)$-linear map $\operatorname{Der} A_{\mathbb{C}}(U) \longrightarrow A_{\mathbb{C}}(U)$. By definition of $\Omega^1_{\mathbb{C}}(U,A)$ therefore, there exists a unique element $\alpha(t) \in \Omega^1_{\mathbb{C}}(U,A)$ such that

(4.9.3) $$\nabla_\xi t = 2\pi i<\xi|\alpha(t)>t$$

for all $\xi \in \text{Der}\, A_{\mathbb{C}}(U)$. One notes for the $\mathbb{Z}_2$-grading that, since $|\nabla_\xi t| = |\xi|$ it follows that

(4.9.4) $$|\alpha(t)| = |\alpha| = 0$$

and

$$|<\xi|\alpha(t)>| = |\xi| \ .$$

Now if $s \in L(U)$ then clearly s is a basal element if and only if s is of the form ft where $f \in A^*_{\mathbb{C}}(U)$. The analogue of Proposition 1.4.1 in [8] follows in the same way.

Proposition 4.9.1. *If U is principal for L and $t \in L(U)$ is a basal element. Then for any other basal element $ft \in L(U)$, $f \in A^*_{\mathbb{C}}(U)$ one has*

(4.9.5) $$\alpha(ft) - \alpha(t) = \frac{1}{2\pi i}\frac{df}{f} \ .$$

Remark 4.9. Note that there is no ambiguity with regard to which side $\frac{1}{f}$ multiplies df since $\frac{1}{f} \in A_{\mathbb{C}}(U)_0$.

It follows immediately from (4.9.5) that $d(\alpha(ft)) = d\alpha(t)$ since the logarithmic differential $\frac{df}{f}$ is clearly closed (again recall that $|f| = 0$). Since X can be covered by open sets which are principal for L one has

Proposition 4.9.2. *Let (L,∇) be a line bundle sheaf with connection on a graded manifold (X,A). Then there exists a unique graded 2-form $\omega \in \Omega^2_{\mathbb{C}}(X,A)_0$ such that for any open set $U \subseteq X$ which is principal for L and any basal element $t \in L(U)$ one has*

(4.9.6) $$\omega = d\alpha(t) \qquad \text{in } U.$$

Furthermore ω is closed (i.e. $d\omega = 0$).

The 2-form ω is called the curvature of (L,∇) and is written

(4.9.7) $$\omega = \text{curv}\ (L,\nabla).$$

The analogue of Proposition 3.4.4 in [8] is also valid. One uses the formula (4.3.10).

Proposition 4.9.3. *Let* $\omega = \text{curv}\,(L,\nabla)$. *Then for any element* $\xi, \eta \in \text{Der}\, A_{\mathbb{C}}(X)$ *and* $t \in L(X)$ *one has*

$$(4.9.8) \qquad ([\nabla_\xi, \nabla_\eta] - \nabla_{[\xi,\eta]})t = 2\pi i \langle \xi, \eta | \omega \rangle t$$

where we recall commutation is in the graded sense.

4.10. Now one says that two line bundle sheaves with connection, (L,∇) and (L',∇') over (X,A) are equivalent if there exists an equivalence

$$(4.10.1) \qquad L(X) \longrightarrow L'(X)$$

of line bundle sheaves which commutes with covariant differentiation by ξ for any $\xi \in \text{Der}\, A_{\mathbb{C}}(X)$. We will let $L_c(X,A)$ denote the set of equivalence classes and write $[(L,\nabla)]$ for the equivalence class of (L,∇).

Now let (L,∇) be a line bundle sheaf with connection over (X,A) and let $\{(U_i,t_i)\}$, $i \in \Lambda$, be a local system for L. We will then refer to the set $\{(c_{ij},\alpha_i)\}$, $i, j \in \Lambda$, where $c_{ij} \in A^*_{\mathbb{C}}(U_i \cap U_j)$ is defined by $t_i c_{ij} = t_j$ in $U_i \cap U_j$ and $\alpha_i = \alpha(t_i) \in \Omega^1_{\mathbb{C}}(U_i,A)_0$ as the corresponding local data for (L,∇). The c_{ij} satisfy the relation (4.8.2) and (4.8.3) and by (4.9.5) they are related to the α_i by the equation

$$(4.10.2) \qquad \alpha_j - \alpha_i = \frac{1}{2\pi i}\frac{dc_{ij}}{c_{ij}} \quad \text{in } U_i \cap U_j .$$

Conversely if $\{U_i\}$, $i \in \Lambda$, is any open covering of X and a set $\{(c_{ij},\alpha_i)\}$, $i, j \in \Lambda$ is given where $c_{ij} \in A^*_{\mathbb{C}}(U_i \cap U_j)$, $\alpha_i \in \Omega^1_{\mathbb{C}}(U_i,A)_0$ and conditions (4.8.2), (4.8.3) and (4.10.2) are satisfied then one sees easily that there exists a line bundle sheaf with connection (L,∇) with a local system $\{(U_i,t_i)\}$, $i \in \Lambda$, for L such that $\{(c_{ij},\alpha_i)\}$ is the corresponding local data. Moreover (L,∇) is unique up to equivalence. In fact if $\{c'_{ij},\alpha'_i\}$, $i, j \in \Lambda$, is another such set and is the local data for the line bundle sheaf with connection (L',∇') then (L,∇) is equivalent to (L',∇') if and only if there exists $\lambda_i \in A^*_{\mathbb{C}}(U_i)$, $i \in \Lambda$, such that

$$(4.10.3) \qquad \lambda_i c_{ij} \lambda_j^{-1} = c_{ij} \quad \text{in } U_i \cap U_j$$

and

$$(4.10.4) \qquad \alpha_i - \alpha'_i = \frac{1}{2\pi i}\frac{d\lambda_i}{\lambda_i} .$$

Since every line bundle sheaf with connection admits local data with respect to a contractible covering it follows from (4.10.4) that the notion of curvature is an equivalence invariant and hence curv $[(L,\nabla)]$ = curv (L,∇) is well defined. One also notes that $L_c(X,A)$ has the structure of an abelian group where if, with respect to a contractible covering, $\{(c'_{ij},\alpha'_i)\}$ and $\{c''_{ij},\alpha''_i\}$ are the local data for (L',∇') and (L'',∇'') then

(4.10.5) $$[(L,\nabla)] = [(L',\nabla')] + [L'',\nabla'')]$$

where the local data (c_{ij},α_i) for (L,∇) is given by

(4.10.6) $$c_{ij} = c'_{ij}c''_{ij}$$
$$\alpha_i = \alpha'_i + \alpha''_i .$$

The identity element of $L_c(X,A)$ is $[(A_{\mathbb{C}},\nabla^e)]$ where $\nabla^e_\xi f = \xi f$ for $\xi \in \text{Der } A_{\mathbb{C}}(X)$ and $f \in A_{\mathbb{C}}(X)$.

Remark 4.10. Note that in the notation of (4.10.5)

(4.10.7) $$\text{curv } [(L,\nabla)] = \text{curv } [(L',\nabla')] + \text{curv } [(L'',\nabla'')].$$

Now for any closed 2-form $\omega \in \Omega^2_{\mathbb{C}}(X,A)_0$ let $L_\omega(X,A)$ be the set of all $[(L,\nabla)] \in L_c(X,A)$ such that $\omega = \text{curv }((L,\nabla))$. Obviously

(4.10.8) $$L_c(X,A) = \cup\, L_\omega(X,A)$$

is a disjoint union over the set of all closed 2-forms $\omega \in \Omega^2_{\mathbb{C}}(X,A)_0$.

Now given a closed 2-form $\omega \in \Omega^2_{\mathbb{C}}(X,A)_0$, the question as to whether $L_\omega(X,A)$ is empty or not, we now observe has the same answer as in the ungraded case. See Proposition 2.1.1 in [8]. We first observe that the cohomology group $H^1(X,\mathbb{C}^*)$, where $\mathbb{C}^*$ is the multiplicative group of non-zero complex numbers, operates on $L_c(X,\omega)$. Let $\{U_i\}$, $i \in \Lambda$, be a contractible covering of X and assume (L,∇) is a line bundle sheaf with connection over (X,A). Let $\{(c_{ij},\alpha_i)\}$ be the corresponding local data for (L,∇). Let $\{z_{ij}\}$ be a Cech cocycle for the constant sheaf $\mathbb{C}^*$ and let $[z_{ij}] \in H^1(X,\mathbb{C}^*)$ be the corresponding class. One has

Proposition 4.10.1. *The group $H^1(X,\mathbb{C}^*)$ operates as a group of endomorphisms of $L_c(X,A)$ in such a fashion that, in the notation above, one has*

(4.10.9) $$[z_{ij}] \cdot [(L,\nabla)] = [(L',\nabla')]$$

where (L',∇') has local data $\{(c_{ij}z_{ij},\alpha_i)\}$ with respect to the covering $\{U_i\}$, $i \in \Lambda$.

Now consider the cohomology $\mathrm{Coh}\,(\Omega_{\mathbb{C}}(X,A))$ of the complexified de Rham complex $\Omega_{\mathbb{C}}(X,A)$. The isomorphism (4.7.2) clearly extends to an isomorphism

(4.10.10) $$j : \mathrm{Coh}\,(\Omega_{\mathbb{C}}(X,A)) \longrightarrow H(X,\mathbb{C}).$$

Now the injection $\mathbb{Z} \longrightarrow \mathbb{C}$ induces a homomorphism $H(X,\mathbb{Z}) \longrightarrow H(X,\mathbb{C})$. A class $\nu \in \mathrm{Coh}\,(\Omega_{\mathbb{C}}(X,A))$ will be called integral if $j\nu$ lies in the image of $H(X,\mathbb{Z})$.

Proposition 4.10.2. Let (X,A) be a graded manifold and let $\omega \in \Omega^2_{\mathbb{C}}(X,A)_0$ be a closed 2-form. Then $L_\omega(X,A)$ is non-empty if and only if the class $[\omega] \in \mathrm{Coh}^2\,(\Omega_{\mathbb{C}}(X,A))$ is integral. Furthermore in such a case $L_\omega(X,A)$ is stable under the action $H^1(X,\mathbb{C}^*)$. In fact $L_\omega(X,A)$ is a principal homogeneous space for $H^1(X,\mathbb{C}^*)$ so that if a base point in $L_\omega(X,A)$ is fixed the action induces a bijection

(4.10.11) $$H^1(X,\mathbb{C}^*) \longrightarrow L_\omega(X,A).$$

The proof of Proposition (4.10.2) proceeds in the same way as the proof of Proposition 2.1.1 in [8]. We will repeat here that part of the proof showing that if $[\omega]$ is integral then $L_\omega(X,A)$ is not empty. Indeed assume $[\omega]$ is integral. Let $\{U_i\}$, $i \in \Lambda$, be a contractible covering of X. Then there exists $\alpha_i \in \Omega^1_{\mathbb{C}}(U_i,A)_0$ such that $d\alpha_i = \omega$ in U_i and hence $d(\alpha_j-\alpha_i) = 0$ in $U_i \cap U_j$. Thus there exists $f_{ij} \in A_{\mathbb{C}}(U_i \cap U_j)_0$ such that $\alpha_j - \alpha_i = df_{ij}$. But $d(f_{ij}+f_{jk}-f_{ik}) = 0$ in $U_i \cap U_j \cap U_k$ so that one has $z_{ijk} \in \mathbb{C}$ such that $f_{ij} + f_{jk} - f_{ik} = z_{ijk} 1_{U_i \cap U_j \cap U_k}$. But now recalling (4.7.4), the statement that $[\omega]$ is integral implies that the z_{ijk} can be chosen so that $z_{ijk} \in \mathbb{Z}$. But then if $c_{ij} = \varepsilon(f_{ij})$ one has $c_{ij} \in A^*_{\mathbb{C}}(U_i\, U_j)$. But then $\{(c_{ij},\alpha_i)\}$ satisfies the conditions (4.8.2), (4.8.3) and (4.10.2) and hence is the local data for some (L,∇). Clearly $\omega = \mathrm{curv}\,(L,\nabla)$ proving that $L_\omega(X,A)$ is not empty.

4.11. The main definitions we have made in the last few sections are functorial in the category of graded manifolds. Assume

$$\sigma : (Y,B) \longrightarrow (X,A)$$

is a morphism of graded manifolds (Y,B) and (X,A). Let L be a line bundle sheaf over (X,A) and let $\{U_i,t_i\}$, $i \in \Lambda$, be a local system for L. Also let $c_{ij} \in A^*_{\mathbb{C}}(U_i \cap U_j)$ be the transition elements for L. But if $V_i = \sigma_Y^{-1}(U_i)$ then $\{V_i\}$, $i \in \Lambda$, is a covering of Y and $\sigma^* c_{ij} \in B^*_{\mathbb{C}}(V_i \cap V_j)$. Furthermore the set $\{\sigma^* c_{ij}\}$ satisfies the conditions (4.8.2) and (4.8.3) and hence there exists a line bundle sheaf L' on (Y,B) having the $\sigma^* c_{ij}$ as transition elements. Moreover if L has a connection ∇ and $\{(c_{ij},\alpha_i)\}$ is the corresponding local data for (L,∇) then the $\{(\sigma^* c_{ij},\sigma^*\alpha_i)\}$ also satisfies condition (4.10.2). It follows that there is a connection ∇' on L' so that the $\{(\sigma^* c_{ij},\sigma^*\alpha_i)\}$ is local data for (L',∇'). Furthermore σ^* on $\Omega_{\mathbb{C}}(X,A)$ commutes with exterior differentiation proving the last statement in

Proposition 4.11. *Assume $\sigma : (Y,B) \longrightarrow (X,A)$ is a morphism of graded manifolds. Let the notation be as above. Then $[L']$ (resp. $[(L',\nabla')]$) is independent of the choice of transition elements (resp. local data) and depends only on $[L]$ (resp. $[(L,\nabla)]$). Furthermore if we put $[L'] = \sigma^*[L]$ (resp. $[(L',\nabla')] = \sigma^*[(L,\nabla)]$) then*

(4.11.1) $$\sigma^* : \mathcal{L}(X,A) \longrightarrow \mathcal{L}(Y,B)$$

(4.11.2) (resp. $$\sigma^* : \mathcal{L}_c(X,A) \longrightarrow \mathcal{L}_c(Y,B))$$

is a homomorphism of abelian groups. Finally for any closed 2-form $\omega \in \Omega^2_{\mathbb{C}}(X,A)_0$ one has

(4.11.3) $$\sigma^* : \mathcal{L}_\omega(X,A) \longrightarrow \mathcal{L}_{\sigma^*\omega}(Y,B).$$

5. Graded Symplectic Manifolds, Hamiltonian formulas and Coadjoint orbits

5.1. Let (X,A) be a graded manifold and let $\omega \in \Omega^2(X,A)_0$ be a 2-form. We recall from §4.5 that ω is called non-singular if the $A(X)$-linear map $\operatorname{Der} A(X) \longrightarrow \Omega^1(X,A)$, $\xi \longrightarrow i(\xi)\omega$, is an isomorphism. A graded symplectic manifold, denoted by (X,A,ω), is a graded manifold (X,A) together with a closed (i.e. $d\omega = 0$) non-singular 2-form $\omega \in \Omega^2(X,A)_0$.

One notes that the underlying manifold X for a graded symplectic manifold (X,A,ω) is necessarily even dimensional. Recalling the map $\kappa : \Omega(X,A) \longrightarrow \Omega(X)$, (see (4.6.2)), one has

Proposition 5.1. *If (X,A,ω) is a graded symplectic manifold then $(X,\kappa\omega)$ is a symplectic manifold in the usual sense.*

This follows from Propositions 4.5.2 and 4.6.1.

Remark 5.1.1. Note that a graded symplectic structure ω on (X,A) defines more than just a symplectic structure on X. Indeed by Proposition 4.5.2 it induces also a "Riemannian structure" on the odd tangent spaces. That is, for any $p \in X$, by Proposition 4.5.2, $\tilde{\omega} \mid T_p(X,A)_1$ is a non-singular symmetric bilinear form. The signature of the quadratic form is a feature of a graded symplectic manifold that has no parallel in the ungraded case.

Example. Before continuing we wish to observe that graded symplectic manifolds abound. Indeed if X is any manifold one knows that the cotangent bundle $T^*(X)$ has the natural structure of a symplectic manifold. If (X,A) is a graded manifold then (see Example III in §2.18) we defined the cotangent bundle $(T^*(X),T^*(A))$ of (X,A). We assert that $(T^*(X),T^*(A))$ has a natural structure of graded symplectic manifold. As in the ungraded case the 2-form where $\omega = d\alpha$ is exact where $\alpha \in \Omega^1(T^*(X),T^*(A))_0$ is a natural 1-form. To describe α it is enough to consider the case when X is an A-coordinate neighborhood say with A-coordinates $\{r_i,s_j\}$ and observe that the definition is independent of the coordinates. Let $\xi \in \operatorname{Der} T^*(A)(T^*(X))$. Now since (2.18.14) is injective locally we can regard $\hat{r}_i, \hat{s}_j \in \hat{A}(X) \subseteq T^*(A)(T^*(X))$. But also $(\hat{\frac{\partial}{\partial r_i}}), (\hat{\frac{\partial}{\partial s_j}}) \in T^*(A)(T^*(X))$ and one has

$$(5.1.1) \qquad \langle\xi|\alpha\rangle = \sum_i (\xi\hat{r}_i)\,(\hat{\frac{\partial}{\partial r_i}}) + \sum_j (\xi\hat{s}_j)\,(\hat{\frac{\partial}{\partial s_j}})$$

This is of course the analogue of the classical $\sum p_k dq_k$. The p_k are $(\hat{\frac{\partial}{\partial r_i}})$, $(\hat{\frac{\partial}{\partial s_j}})$ and the dq_k are $d\hat{r}_i$, $d\hat{s}_j$.

One notes that the signature of the quadratic form $\tilde{\omega} \mid T_b(T^*(X),T^*(A))_1$, for any $b \in T^*(X)$, is (n,n) where $\dim (X,A) = (m,n)$.

Remark 5.1.2. In Remark 2.18.4 we noted that if, for a graded manifold (X,A), the space X reduces to a point p then $A(X) = \Lambda V$ for a finite dimensional real vector space. Furthermore we have noted that for the cotangent bundle $(T^*(X),T^*(A))$ one has $T^*(A)(T^*(X)) = \Lambda(V \oplus V')$ where V' is the dual space to V. Now one has a tangent space identification

$$T_p(T^*(X),T^*(A))_1 = V \oplus V' \tag{5.1.2}$$

and we note that the quadratic form induced by the natural symplectic structure is the one obtained by the pairing of V and V'.

5.2. Now let (X,A,ω) be a fixed graded symplectic manifold. We will carry over much of the terminology used in connection with ordinary symplectic manifolds to the graded case. A graded vector field $\xi \in \operatorname{Der} A(X)$ will be called locally Hamiltonian if $\theta(\xi)\omega = 0$. By Proposition 4.3.4, (3), the space, $\underline{a}'(X)$, of all locally Hamiltonian graded vector fields is clearly a graded Lie subalgebra of $\operatorname{Der} A(X)$.

Now for any $\xi \in \operatorname{Der} A(X)$ put $\beta_\xi = i(\xi)\omega \in \Omega^1(X,A)$. Thus $\beta_{f\xi} = f\beta_\xi$ by Lemma 4.3.1 and with respect to the $\mathbb{Z}_2$-grading $|\beta_\xi| = |\xi|$. Now the relation (4.3.8) implies

$$\theta(\xi)\omega = d\beta_\xi \tag{5.2.1}$$

and hence, as in the ungraded case, one has

Proposition 5.2.1. *If $\xi \in \operatorname{Der} A(X)$ then ξ is locally Hamiltonian if and only if β_ξ is closed. In particular the correspondence $\xi \longrightarrow \beta_\xi$ sets up a bijection from the graded Lie algebra $\underline{a}'(X)$ of all locally Hamiltonian graded vector fields to the space of all closed 1-forms.*

An element $\xi \in \operatorname{Der} A(X)$ is called Hamiltonian if β_ξ is exact. That is if $\beta_\xi = df$ for some $f \in A(X)$. Let $\underline{a}(X)$ denote the space of all Hamiltonian graded vector fields on (X,A). Obviously $\underline{a}(X) \subseteq \underline{a}'(X)$. But more than that, as in the ungraded case, not only is $\underline{a}(X)$ an ideal in $\underline{a}'(X)$ but

Proposition 5.2.2. One has

$$[\underline{a}'(X),\underline{a}'(X)] \subseteq \underline{a}(X) \subseteq \underline{a}'(X) .$$

The argument goes exactly as in the proof of Proposition 3.2.1 in [8]. We recall two identities used in the proof. One has, as a consequence of Proposition 4.3.4, (2),

(5.2.2) $$\theta(\xi)\beta_\eta = \beta_{[\xi,\eta]}$$

for $\xi \in \underline{a}'(X)$, $\eta \in \mathrm{Der}\, A(X)$ and hence

(5.2.3) $$\beta_{[\xi,\eta]} = di(\xi)\beta_\eta$$

if ξ, $\eta \in \underline{a}'(X)$ by (4.3.8).

Now if ξ, $\eta \in \mathrm{Der}\, A(X)$ are homogeneous then $<\xi|i(\eta)\omega> = (-1)^{|\xi||\eta|}<\eta,\xi|\omega>$ by (4.3.6). But $<\eta,\xi|\omega> = (-1)^{1+|\eta||\xi|}<\xi,\eta|\omega>$. Thus homogeneous or not one has in any case

(5.2.4) $$<\xi|\beta_\eta> = -<\xi,\eta|\omega>.$$

But $<\xi|\beta_\eta> = i(\xi)\beta_\eta$. Thus if ξ, $\eta \in \underline{a}'(X)$ then by (5.2.3) and (5.2.4) one has

(5.2.5) $$\beta_{[\xi,\eta]} = -d<\xi,\eta|\omega> .$$

This together with (5.2.4) implies that for any $\rho \in \mathrm{Der}\, A(X)$, ξ, $\eta \in \underline{a}'(X)$

(5.2.6) $$<\rho,[\xi,\eta]|\omega> = \rho<\xi,\eta|\omega> .$$

Now for any $f \in A(X)$ let $\xi_f \in \underline{a}(X)$ be the Hamiltonian graded vector field defined by the relation $\beta_{\xi_f} = df$. That is

(5.2.7) $$i(\xi_f)\omega = df .$$

Since $|\omega| = 0$ with respect to the $\mathbb{Z}_2$-grading one notes that

(5.2.8) $$|f| = |\xi_f|$$

with respect to the $\mathbb{Z}_2$-grading.

Note that if X is connected then $df = 0$ for $f \in A(X)$, by Theorem 4.6, implies $f \in \mathbb{R}\cdot 1_X$. But then by (5.2.8) and the definition of $\underline{a}(X)$ it follows that the map $f \longmapsto \xi_f$ induces an exact sequence

(5.2.9) $$0 \longrightarrow \mathbb{R}\cdot 1_X \longrightarrow A(X) \longrightarrow \underline{a}(X) \longrightarrow 0$$

of graded linear spaces.

Now let $\xi \in \text{Der } A(X)$ and $g \in A(X)$. Note that

(5.2.10) $$\xi g = - \langle \xi, \xi_g | \omega \rangle$$

by (5.2.4) where $\eta = \xi g$. One now defines a bracket operation in $A(X)$, referred to as Poisson brackets, by putting, for any $f, g \in A(X)$

(5.2.11) $$[f,g] = \xi_f g .$$

Note that if $f, g \in A(X)$ are homogeneous but $h \in A(X)$ is arbitrary

(5.2.12) $$[f,gh] = [f,g]h + (-1)^{|f||g|} g[f,h].$$

On the other hand substituting ξ_f for ξ in (5.2.10) one has

(5.2.13) $$[f,g] = - \langle \xi_f, \xi_g | \omega \rangle .$$

It then follows from (4.1.2) that if f, g are homogeneous

(5.2.14) $$[f,g] = - (-1)^{|f||g|} [g,f] .$$

The following statement contains the basic facts about the existence and properties of Poisson bracket for graded symplectic manifolds.

Theorem 5.2. *$A(X)$ is a graded Lie algebra with respect to the bracket operation $[f,g]$ defined by (5.2.11). Furthermore the map $A(X) \longrightarrow \underline{a}(X)$, $f \longrightarrow \xi_f$ is a homomorphism of graded Lie algebras. That is*

(5.2.15) $$\xi_{[f,g]} = [\xi_f, \xi_g]$$

for any $f, g \in A(X)$ and hence, if X is connected, (5.2.9) is an exact sequence of graded Lie algebras.

Proof. We cannot just quote say Proposition 4.1.1 in [8] for the proof since we are in the graded case and we have to make sure that the signs work out correctly. But they do. Indeed let $f, g \in A(X)$. Then since $\xi_f, \xi_g \in \underline{a}'(X)$ one has $\beta_{[\xi_f, \xi_g]} = - d\langle \xi_f, \xi_g | \omega \rangle$ by (5.2.5). On the other hand $d[f,g] = - d\langle \xi_f, \xi_g | \omega \rangle$ by (5.2.13). Thus $\beta_{[\xi_f, \xi_g]} = d[f,g]$. But then $[\xi_f, \xi_g] = \xi_{[f,g]}$ by definition of $\xi_{[f,g]}$. This establishes the equation (5.2.15).

Now if f, g and $h \in A(X)$ are homogeneous then since $d\omega = 0$ we have by Proposition 4.3.7

$$(5.2.16) \qquad 0 = \sum (-1)^{|f||h|}(\xi_f\langle\xi_g,\xi_h|\omega\rangle - \langle[\xi_f,\xi_g],\xi_h|\omega\rangle)$$

where the sum is over the three cyclic permutations of f, g and h. But now

$$(5.2.17) \qquad -(-1)^{|g||f|}\langle[\xi_g,\xi_h],\xi_f|\omega\rangle = (-1)^{|f||h|}\langle\xi_f,[\xi_g,\xi_h]|\omega\rangle$$

by (4.1.2) since $|g||f| + (|g| + |h|)|f| = |f||h|$. On the other hand

$$(5.2.18) \qquad \langle\xi_f,[\xi_g,\xi_h]|\omega\rangle = \xi_f\langle[\xi_g,\xi_h]|\omega\rangle$$

by (5.2.6). Thus the cyclic sum (5.2.16) implies the cyclic sum

$$(5.2.19) \qquad 2\sum (-1)^{|f||h|}\xi_f\langle\xi_g,\xi_h|\omega\rangle = 0 .$$

But then by (5.2.13) this implies the cyclic sum

$$\sum (-1)^{|f||h|}[f,[g,h]] = 0 .$$

This however together with (5.2.15) implies that $A(X)$ is a graded Lie algebra with respect to Poisson bracket. Since we have already established (5.2.15) this proves the theorem. Q.E.D.

Remark 5.2. Of course all the relations above proved for X are valid for any open subset $U \subseteq X$. In particular we will use the above notation for the case where U is substituted for X.

5.3. Darboux's theorem for ordinary symplectic manifolds is very important since it produces natural coordinates to work with, the p's and q's. Fortunately Darboux's theorem is also valid in the graded case. But in fact it is more interesting in the graded case since as we shall see not only does the alternating part normalize producing p's and q's but the quadratic part also remarkably normalizes (unlike the case of a Riemannian manifold) producing a very special odd coordinate system s_j. The new invariant, the signature of $\tilde{\omega} \mid T_p(X,A)_1$ becomes explicit in this basis.

Now let $\dim (X,A,\omega) = (m,n)$. Since $(X,\kappa\omega)$ is a symplectic manifold $m = 2m_0$ is even. An open set $U \subseteq X$ will be called an A-Darboux coordinate neighborhood for (X,A,ω) if U is an A-coordinate neighborhood which admits

A-coordinates $\{\{p_k,q_\ell\},s_j\}$ where $\{p_k,q_\ell\} \in A(U)_0$, $k, \ell = 1,\ldots,m_0$ is an even coordinate system, $s_j \in A(U)_1$, $j = 1,\ldots,n$ is an odd coordinate system and

(5.3.1) $$\omega = \sum_{k=1}^{m_0} dp_k dq_k + \sum_{j=1}^{n} \frac{\varepsilon_j}{2} (ds_j)^2$$

where ε_j is either $+1$ or -1. Such a coordinate system will be referred to as an A-Darboux coordinate system.

If $\{\{p_k,q_\ell\},s_j\}$ is an A-Darboux coordinate system then it follows immediately from Proposition 4.3.2 that

$$\xi_{p_k} = -\frac{\partial}{\partial q_k}, \quad \xi_{q_k} = \frac{\partial}{\partial p_k}$$

(5.3.2) and

$$\xi_{s_j} = \varepsilon_j \frac{\partial}{\partial s_j} .$$

As a consequence one has the bracket relations

$$[p_i,p_k] = [q_i,q_\ell] = [s_j,p_k] = [s_j,q_\ell] = 0$$

and

(5.3.3) $$[q_k,p_\ell] = \delta_{k\ell} 1_U,$$

and

$$[s_i,s_j] = \varepsilon_j \delta_{ij} 1_U .$$

Also the familiar formula for the Hamiltonian vector field corresponding to a function now becomes

(5.3.4) $$\xi_f = \sum_{k=1}^{m_0} \left(\frac{\partial f}{\partial q_k}\frac{\partial}{\partial p_k} - \frac{\partial f}{\partial p_k}\frac{\partial}{\partial q_k}\right) + \sum_{j=1}^{n} (-1)^{|f|} \varepsilon_j \frac{\partial f}{\partial s_j}\frac{\partial}{\partial s_j}$$

for $f \in A(U)$. Thus if $f, g \in A(U)$

(5.3.5) $$[f,g] = \sum_{k=1}^{m_0} \left(\frac{\partial f}{\partial q_k}\frac{\partial g}{\partial p_k} - \frac{\partial f}{\partial p_k}\frac{\partial g}{\partial q_k}\right) + \sum_{j=1}^{n} (-1)^{|f|} \varepsilon_j \frac{\partial f}{\partial s_j}\frac{\partial g}{\partial s_j} .$$

Remark 5.3.1. Note that the second sum in (5.3.5) is more reminiscent of Riemannian structures than symplectic ones.

The analog of Darboux's theorem for graded manifolds is

Theorem 5.3. *Any graded symplectic manifold* (X,A,ω) *can be covered by A-Darboux coordinate neighborhoods.*

Our proof of Theorem 5.3 is quite long with a large number of steps and will not be given here. We will say in very rough terms how one proceeds. Of course one uses the usual Darboux theorem. This however, considering only an even coordinate system, produces p_k and q_ℓ such that $\widetilde{[p_i,p_k]} = \widetilde{[q_i,q_\ell]} = 0$ and $\widetilde{[q_k,p_\ell]} = \delta_{k\ell}$. The problem here is to get rid of the higher nilpotent terms in $[p_i,p_k]$, $[q_i,q_\ell]$ and $[q_k,p_\ell]$. This is done inductively, recalling the filtration (2.5.1), using a type of Poincaré lemma. After this has been accomplished the exterior factor D generated by the desired odd coordinates s_j is actually uniquely determined. It is given by $D = \{f \in A(U) \mid [p_k,f] = [q_\ell,f] = 0\}$. The problem of extracting the s_j from this algebra is again an inductive process using an alternating version of the Poincaré lemma to get rid of higher nilpotent terms.

There is, of course, a relationship between the graded Poisson bracket structure for (X,A,ω) and the usual Poisson bracket structure of functions for $(X,\kappa\omega)$. If $f, g \in A(X)_0$ then

(5.3.6) $$\widetilde{[f,g]} = [\tilde{f},\tilde{g}].$$

In fact one has

(5.3.7) $$\tilde{\xi}_f = \xi_{\tilde{f}}$$

for $f \in A(X)_0$ where $\xi_\varphi \in \operatorname{Der} C^\infty(X)$ for $\varphi \in C^\infty(X)$ is defined so that $i(\xi_\varphi)\kappa\omega = d\varphi$.

<u>Remark 5.3.2</u>. It follows from (5.3.6) that if $\{\{p_k,q_\ell\},s_j\}$ is an A-Darboux coordinate system in U then $\tilde{p}_k$, $\tilde{q}_\ell \in C^\infty(U)$ is a Darboux coordinate system of functions in the usual sense. In fact Theorem 5.3 can be strengthened in that if $b \in X$ and φ_k, $\psi_\ell \in C^\infty(U)$ is a Darboux coordinate system of functions in some neighborhood U of b then there exists an A-Darboux coordinate neighborhood W with A-Darboux coordinates $\{\{p_k,q_\ell\},s_j\}$ such that $b \in W \subseteq U$ and $\tilde{p}_k = \varphi_k$, $\tilde{q}_\ell = \psi_\ell$ in W.

<u>Remark 5.3.3</u>. It is interesting to consider the Poisson algebra structures on $A(X)$ in case X reduces to a single point. As noted in Remark 2.18.4 the given (associative, commutative) algebra structure on $A(X)$ is of the form ΛV

where V is a real finite dimensional vector space. However V in general is not a distinguished subspace of ΛV. However in the case of the graded symplectic manifold (X,A,ω) we can choose V to be the subspace spanned by A-Darboux coordinates s_i. If $\dim (X,A) = (0,n)$ and the notation is as in (2.6.3) let $[\mu] \subseteq (1,\dots,n)$ be the underlying set for the sequence $\mu \in M_n$. Then for the basis s_μ of ΛV one has

$$(5.3.8) \qquad [s_\mu, s_{\mu'}] = \begin{cases} 0 & \text{if } \operatorname{card}([\mu] \cap [\mu']) \neq 1 \\ \pm s_{\mu''} & \text{otherwise where } [\mu''] \text{ is the symmetric difference of } [\mu] \text{ and } [\mu']. \end{cases}$$

This graded Lie algebra was discovered by Sternberg in connection with filtering Clifford algebras. Also modulo the constants it is by (5.2.9), isomorphic to the $2^n - 1$ dimensional graded Lie algebra $\underline{a}(X)$ which is simple. This quotient occurs in the classification lists of Kac and Kaplansky. In the Kac theory this is one of the finite dimensional simple graded Lie algebras which is an analogue of one of Cartan's infinite dimensional simple Lie algebras.

Remark 5.3.4. Let $\underline{m} = \underline{m}_0 + \underline{m}_1$ be a finite dimensional graded real vector space and assume that B is an alternating, in the graded sense, non-singular bilinear form of $\mathbb{Z}_2$-grading zero on $\underline{m}$. Thus

$$(5.3.9) \qquad \begin{aligned} B(x,y) &= -B(y,x) && \text{for } x, y \in \underline{m}_0 \\ B(u,v) &= B(v,u) && \text{for } u, v \in \underline{m}_1 \\ &\text{and} \\ B(x,u) &= B(u,x) = 0 && \text{for } x \in \underline{m}_0, \ u \in \underline{m}_1 . \end{aligned}$$

Now let $\mathbb{R}z$ be a one dimensional real vector space and put $\underline{n} = \mathbb{R}z \oplus \underline{m}$. Then $\underline{n}$ has the structure of a graded vector space where $\underline{n}_0 = \mathbb{R}z \oplus \underline{m}_0$ and $\underline{n}_1 = \underline{m}_1$. We now define the structure of a graded Lie algebra on $\underline{n}$ by defining $[\mathbb{R}z, \underline{n}] = 0$ and for $x, y \in \underline{n}$ one puts

$$(5.3.10) \qquad [x,y] = B(y,x)z .$$

Any graded Lie algebra isomorphic to a graded Lie algebra of the form $\underline{n}$ will be referred to as a graded Heisenberg Lie algebra. Of course when $\underline{m}_1 = 0$

this reduces to the usual Heisenberg Lie algebra. It is clear from the bracket relation (5.3.3) that if $\{\{p_k,q_\ell\},s_j\}$ are A-Darboux coordinates in some connected open set $U \subseteq X$ and $P \subseteq A(U)$ is the $2m_0 + n + 1$ dimensional subspace spanned by $\mathbb{R}\cdot 1_U$ and p_k, q_ℓ and s_j then P is a graded Lie subalgebra of $A(U)$ and in fact is a graded Heisenberg Lie algebra. Let $\underline{p} \subseteq \underline{a}(U)$ be the graded commutative $2m_0 + n$ dimensional Lie algebra spanned by $\frac{\partial}{\partial p_k}$, $\frac{\partial}{\partial q_\ell}$, and $\frac{\partial}{\partial s_j}$. Observe that the homomorphism $f \longrightarrow \xi_f$ induces the following exact subsequence of (5.2.9) where U is substituted for X

$$0 \longrightarrow \mathbb{R}\,1_U \longrightarrow P \longrightarrow \underline{p} \longrightarrow 0\,. \tag{5.3.11}$$

Thus we find the graded Heisenberg Lie algebra playing a familiar role (modulo the word graded), namely as a central extension of a graded commutative Lie algebra. We will come back to this sequence when we consider prequantization.

5.4. Now let (Y,B,ω) be a graded symplectic manifold so that if Y is connected the exact sequence (5.2.9) becomes $0 \longrightarrow \mathbb{R}\cdot 1_Y \longrightarrow B(Y) \longrightarrow \underline{a}(Y) \longrightarrow 0$. Now let $\underline{g}$ be a finite dimensional real graded Lie algebra. By a Poisson representation of $\underline{g}$ on (Y,B,ω) we mean a homomorphism

$$\lambda : \underline{g} \longrightarrow B(Y) \tag{5.4.1}$$

of graded Lie algebras where the graded Lie algebra structure in $B(Y)$ is given by Poisson bracket.

Now let (G,A) be the simply connected graded Lie group with Lie algebra $\underline{g}$. See Corollary to Theorem 3.7. We will say that λ is integrable if there is an action of (G,A) on (Y,B) with action map (see §3.9)

$$A(G)^* \otimes B(Y)^* \longrightarrow B(Y)^*$$

such that if

$$\pi : A(G)^* \longrightarrow \text{End}\, B(Y)$$

is the corresponding coaction representation of (G,A) then

$$\pi(x) = \xi_{\lambda(x)} \in \underline{a}(Y) \tag{5.4.2}$$

for any $x \in \underline{g}$.

Remark 5.4.1. Note that if λ is integrable then the action of (G,A) on (Y,B) is necessarily unique. That is, as an L-H algebra $A(G)^* = \mathbb{R}(G) * E(\underline{g})$. Since $\pi(x) = \xi_{\lambda(x)}$ for $x \in \underline{g}$ it is clear that π is uniquely determined on $E(\underline{g})$. However, by smoothness and the fact that $\exp x$ commutes with x in $A(G)^*$ it follows that π is uniquely determined on $\mathbb{R}(G)$. That is, if π_1 and π_2 were two possible coaction maps, $f \in B(Y)$ and $v \in B(Y)^*$, then $\varphi(t) = \langle v, \pi_1(\exp tx)\pi_2(\exp(-tx)f\rangle$ has zero derivative for all $t \in \mathbb{R}$ which implies $\pi_1(\exp x)f = \pi_2(\exp x)f$.

Now assume (G,A) is a simply connected graded Lie group. Let $\underline{g}$ be the Lie algebra of (G,A). Assume also that (Y,B,ω) is a graded simplectic manifold and one has a Poisson representation $\lambda : \underline{g} \longrightarrow B(Y)$ of $\underline{g}$ on (Y,B,ω). We will then say that (Y,B,ω) is a Hamiltonian (G,A) space with respect to λ if λ is integrable and (Y,B) is a homogeneous space with respect to the corresponding action of (G,A) on (Y,B).

The main point of this and the following section is to show that if (G,A) is any simply connected (this condition is retained only for simplicity) graded Lie group then, as in the ungraded case, all the orbits of the coadjoint representations are Hamiltonian (G,A) spaces.

Now let (G,A) be a simply connected graded Lie group and let $\underline{g}$ be the Lie algebra of (G,A). Let $\mathrm{ad} : \underline{g} \longrightarrow \mathrm{End}\, \underline{g}$ be the adjoint representation of $\underline{g}$ on itself so that $(\mathrm{ad}\, x)(y) = [x,y]$. Now let

$$\mathrm{ad}' : \underline{g} \longrightarrow \mathrm{End}\, \underline{g}'$$

be the coadjoint representation. Thus $\underline{g}' = \underline{g}_0' + \underline{g}_1'$ is the dual space to $\underline{g}$ and if $f \in \underline{g}'$ and we write $x \cdot f = \mathrm{ad}'\, x(f)$ for any $x \in \underline{g}$ then

$$\langle x \cdot f, y\rangle = (-1)^{|x||f|+1} \langle f, [x,y]\rangle . \tag{5.4.3}$$

Now we adopt the notation of §3.11 where $\underline{g}' = V$ and $\mathrm{ad}' = \gamma$. By Proposition 3.11.1 one thus has an action of the graded Lie group (G,A) on the graded affine manifold $(\underline{g}_0', A_{\underline{g}'})$. We will refer to this as the coadjoint action. Let

$$A(G)^* \otimes A_{\underline{g}'}(\underline{g}_0')^* \longrightarrow A_{\underline{g}'}(\underline{g}_0')^*, \qquad u \otimes v \longrightarrow u \cdot v \tag{5.4.4}$$

be the corresponding action map and

(5.4.5) $$\pi : A(G)^* \longrightarrow \text{End}\, A_{\underline{g}'}(\underline{g}_0')$$

the corresponding coaction representation of (G,A).

Now let $f \in \underline{g}_0'$. (The choice of letter f is so as to be consistent with the notation of [8]). Then as in §3.11 (see also Proposition 3.10.2) if $\delta_f \in A_{\underline{g}'}(\underline{g}\ )^*$ is the group-like element corresponding to f then the orbit $A(G)^*\cdot\delta_f \subseteq A_{\underline{g}'}(\underline{g}_0')^*$ defines a graded submanifold (O,B) of $(\underline{g}_0', A_{\underline{g}'})$ where the underlying manifold $O \subseteq \underline{g}_0'$ is given by

(5.4.6) $$O = G \cdot f$$

with respect to the usual coadjoint action of G on $\underline{g}_0'$. One has

(5.4.7) $$B(O)^* = A(G)^*\cdot\delta_f$$

and (O,B) is a graded homogeneous space for the graded Lie group (G,A). Let

(5.4.8) $$A(G)^* \otimes B(O)^* \longrightarrow B(O)^*, \qquad u \otimes w \longrightarrow u\cdot w$$

be the corresponding action map. Recalling the action map (5.4.4) one clearly has a commutative diagram

(5.4.9) $$\begin{array}{ccc} A(G)^* \otimes A_{\underline{g}'}(\underline{g}_0')^* & \longrightarrow & A_{\underline{g}'}(\underline{g}_0')^* \\ \uparrow & & \uparrow \\ A(G)^* \otimes B(O)^* & \longrightarrow & B(O)^* \end{array}$$

where the vertical maps are injective.

Now let (G_f, A_f) be the isotropy graded Lie subgroup of (G,A) defined by f. By Proposition 3.10.2 one has an isomorphism

(5.4.10) $$(O,B) \cong (G/G_f, A/A_f)$$

of graded homogeneous spaces.

Let $\underline{g}_f = (\underline{g}_f)_0 + (\underline{g}_f)_1$ be the Lie algebra of (G_f, A_f). Then as noted in (3.9.19) the graded structure of the tangent space $T_{\overline{e}}(G/G_f, A/A_f) = \underline{g}/\underline{g}_f$ of $(G/G_f, A/A_f)$ at the coset $\overline{e} = (G_f)$ is given by

(5.4.11) $$T_{\overline{e}}(G/G_f, A/A_f) = \underline{g}_0/(\underline{g}_f)_0 + \underline{g}_1/(\underline{g}_f)_1 \ .$$

Now (5.4.10) induces an isomorphism

(5.4.12) $$\underline{g}/\underline{g}_f \longrightarrow T_f(O,B)$$

where if $x \in \underline{g}$ and the image in $T_f(O,B)$ of the coset $x + \underline{g}_f$ is denoted by $\bar{x} \in T_f(O,B)$ then

(5.4.13) $$\begin{aligned}\bar{x} &= x \cdot \delta_f \\ &= \delta_f\, x \cdot f\end{aligned}$$

where $x \cdot f \in \underline{g}'$. See Lemma 3.11.1, (2). Note that $\underline{g}' \subseteq A_{\underline{g}'}(\underline{g}_0')^*$ is regarded as the tangent space to $(\underline{g}_0', A_{\underline{g}'})$ at the origin. Thus by Lemma 3.10.1 one has $\underline{g}_f = \{x \in \underline{g} \mid x \cdot f = 0\}$. That is

(5.4.14) $$\underline{g}_f = \{x \in \underline{g} \mid \langle f, [x,y]\rangle = 0 \text{ for all } y \in \underline{g}\}.$$

The following is then immediate

Proposition 5.4. Let $\underline{g}$ be a finite dimensional graded Lie algebra and let $f \in \underline{g}_0'$. Let (G,A) be the simply connected graded Lie group corresponding to $\underline{g}$ and let (O,B) be the orbit defined by f with respect to the coadjoint action. Then there exists a unique element $z_f \in \Lambda^2 T_f^*(O,B)$ such that

(5.4.15) $$\langle \bar{x}, \bar{y} \mid z_f\rangle = -\langle f, [x,y]\rangle$$

for any $x, y \in \underline{g}$. Furthermore z_f defines a non-singular alternating bilinear form on the usual tangent space $T_f(O)$ and a non-singular symmetric bilinear form on the odd tangent space $T_f(O,B)_1$.

Now in the notation of §3.11 as applied here $V' = \underline{g}$ so that (see (3.11.1)) if $S(\underline{g})$ is the symmetric algebra over $\underline{g}$ in the graded sense (see §1.2) one has $S(\underline{g}) \subseteq A_{\underline{g}'}(\underline{g}_0')$. To avoid ambiguity rather than regarding $S(\underline{g})$ as being contained in $A_{\underline{g}'}(\underline{g}_0')$ it is more convenient here to regard it as being mapped into $A_{\underline{g}'}(\underline{g}_0')$. For each $z \in S(\underline{g})$ we will denote its image by $\psi^z \in A_{\underline{g}'}(\underline{g}_0')$. Now for any $x \in \underline{g}$ write η^x for $\pi(x)$ (see (5.4.5)) so that $\eta^x \in \operatorname{Der} A_{\underline{g}'}(\underline{g}_0')$. The relation (3.11.10) then implies

(5.4.16) $$\eta^x \psi^y = \psi^{[x,y]}$$

for any $x, y \in \underline{g}$.

Now let

(5.4.17) $$\pi_O : A(G)^* \longrightarrow \operatorname{End} B(O)$$

be the coaction representation of (G,A) corresponding to the action map (5.4.8) Also let $\sigma : (O,B) \longrightarrow (\underline{g}_0', A_{\underline{g}'})$ be the morphism of graded manifolds corresponding to injection so that

(5.4.18) $$\sigma^* : A_{\underline{g}'}(\underline{g}_0') \longrightarrow B(O)$$

is a homomorphism of graded commutative algebras. The commutative diagram (5.4.9) clearly implies

(5.4.19) $$\sigma^*(\pi(u)h) = \pi_O(u)\sigma^* h$$

for any $u \in A(G)^*$, $h \in A_{\underline{g}'}(\underline{g}_0')$. In particular, if for any $z \in S(\underline{g})$ one puts $\varphi^z = \sigma^*\psi^z$ then

(5.4.20) $$\xi^x \varphi^y = \varphi^{[x,y]}$$

by (5.4.16) where $x, y \in \underline{g}$ and we put $\xi^x = \pi_O(x) \in \mathrm{Der}\, B(O)$.

We now have

Theorem 5.4. Let $\underline{g}$ be any finite dimensional real graded Lie algebra and let (G,A) be the corresponding simply connected Lie group. (See §3.7). Let (O,B) be any orbit of (G,A) for the coadjoint action of (G,A). Then there exists a unique graded 2-form $\omega \in \Omega^2(O,B)_0$ such that in the notation above

(5.4.21) $$\varphi^{[x,y]} = -\langle \xi^x, \xi^y \mid \omega\rangle$$

for all $x, y \in \underline{g}$. Moreover ω is non-singular and $d\omega = 0$ so that (O,B,ω) is a graded symplectic manifold.

To prove Theorem 5.4 one first establishes

Lemma 5.4. Let $b \in O$ and let $x_i \in \underline{g}$, $i = 1,\ldots,n+m$ be a basis of $\underline{g}$. Then there exists a neighborhood $b \in U \subseteq O$ of b such that for any $\xi \in \mathrm{Der}\, B(U)$ there exists $\psi_i \in A_{\underline{g}'}(\underline{g}_0')$ such that if $\varphi_i = \sigma^*(\psi_i) \in B(O)$ then

(5.4.22) $$\xi = \sum \varphi_i \xi^{x_i} \quad \text{in } U.$$

Indeed by Proposition 2.12.1 there exists $\rho_i \in B(O)$ such that $\xi = \sum \rho_i \xi^{x_i}$. But now if U is given as in (2.16.5) there exists by (2.16.5), $\psi_i \in A_{\underline{g}'}(\underline{g}_0')$ such that $\rho_i = \sigma^*(\psi_i)$ in U. This proves Lemma 5.4.

Now let the notation be as in Lemma 5.4. Also let $\xi' \in \mathrm{Der}\, B(U)$ and let ψ_i', φ_i' be defined as in Lemma 5.4 with ξ' substituted for ξ. We propose to define ω in U so that if ξ' and the basis x_i is homogeneous one has

$$\langle \xi,\xi' | \omega\rangle = - \sum_{i,j} (-1)^{|\varphi_j'||x_i|} \varphi_i \varphi_j' \varphi^{[x_i,x_j]} \tag{5.4.23}$$

in U. To show this is well defined we have to prove that if $\xi = 0$ then

$$\sum_i \varphi_i \varphi^{[x_i,x_j]} = 0 \tag{5.4.24}$$

for all j. However by (5.4.20) the left side of (5.4.24) is just $\xi\varphi^{x_j}$ in U. But this vanishes since $\xi = 0$. This proves ω is well defined. It is clearly independent of all choices and hence one obtains a global 2-form $\omega \in \Omega^2(O,B)_0$ such that (5.4.23) is satisfied. In particular one has

$$\varphi^{[x,y]} = - \langle \xi^x, \xi^y | \omega\rangle$$

for any $x, y \in \underline{g}$. Now let $f \in O$. Then $\widetilde{\varphi^{[x,y]}}(f) = \langle f,[x,y]\rangle$ and hence, recalling (4.5.4),

$$\langle (\tilde{\xi}^x)_f, (\tilde{\xi}^y)_f | \tilde{\omega}_f\rangle = - \langle f,[x,y]\rangle . \tag{5.4.25}$$

But now by (3.9.5) one has

$$\begin{aligned}(\tilde{\xi}^x)_f &= - x \cdot \delta_f \\ &= - \overline{x} .\end{aligned} \tag{5.4.26}$$

(See (5.4.13)). It follows therefore that

$$\langle \overline{x}, \overline{y} | \tilde{\omega}_f\rangle = - \langle f,[x,y]\rangle \tag{5.4.27}$$

and hence

$$\tilde{\omega}_f = z_f \tag{5.4.28}$$

by Proposition 5.4. It follows then that ω is non-singular by Propositions 4.5.2 and 5.4.

Now by (4.1.2) and (4.3.6) one has

$$\langle \xi^x | i(\xi^y)\omega\rangle = - \langle \xi^x,\xi^y | \omega\rangle \tag{5.4.29}$$

for any $x, y \in \underline{g}$. But since we have established (5.4.21) it follows from (5.4.20) that

(5.4.30) $$i(\xi^y)\omega = d\varphi^y$$

for all $y \in \underline{g}$. In particular $i(\xi^{[x,y]})\omega = d\varphi^{[x,y]}$. However $\theta(\xi^x)$ commute with exterior differentiation in $\Omega(O,B)$ so that $\theta(\xi^x)d\varphi^y = d\varphi^{[x,y]}$ and hence $i(\xi^{[x,y]})\omega = \theta(\xi^x)d\varphi^y$. But then by (5.4.30) one has

(5.4.31) $$i(\xi^{[x,y]})\omega = \theta(\xi^x)i(\xi^y)\omega$$

for all $x, y \in \underline{g}$. However $\xi^{[x,y]} = [\xi^x,\xi^y]$. But then $i(\xi^{[x,y]}) = \theta(\xi^x)i(\xi^y) - (-1)^{|x||y|}i(\xi^y)\theta(\xi^x)$ by Proposition 4.3.4, (2). It follows therefore from (5.4.31) that

(5.4.32) $$i(\xi^y)\theta(\xi^x)\omega = 0$$

for all $x, y \in \underline{g}$. But then by Lemmas 4.3.1 and 5.4 one has

(5.4.33) $$\theta(\xi^x)\omega = 0$$

for all $x \in \underline{g}$. But $\theta(\xi^x) = di(\xi^x) + i(\xi^x)d$. However $di(\xi^x)\omega = d^2\varphi^x = 0$ by (5.4.30). Thus (5.4.33) implies $i(\xi^x)d\omega = 0$ for all $x \in \underline{g}$. Again by Lemmas 4.3.1 and 5.4 this implies $d\omega = 0$ so that (O,B,ω) is a graded symplectic manifold proving the theorem. Q.E.D.

Remark 5.4.2. In the course of the proof of Theorem 5.4 note that we have established the relation

$$<(\tilde{\xi^x})_f, (\tilde{\xi^y})_f|\tilde{\omega}_f> = - <f,[x,y]>$$

for any $f \in O$, $x, y \in \underline{g}$. See (5.4.25). Also we have proved $\tilde{\omega}_f = z_f$, (see (5.4.28)), using the notation of Proposition 5.4, so that $\tilde{\omega}_f$ defines a non-singular alternating bilinear form on the ordinary tangent space $T_f(O)$ of O at f and a non-singular symmetric bilinear form on the odd tangent space $T_f(O,B)_1$ of (O,B) at f.

To avoid confusion we will occasionally write (O,B_O) for (O,B) and ω_O for ω. Also we will refer to (O,B) or (O,B,ω) as a (G,A) orbit of the coadjoint action. If $f \in O$ then (O,B) or (O,B,ω) may be referred to as the (G,A) orbit defined by $f \in \underline{g}_0'$.

5.5. Let the notation be as in Theorem 5.4 and let

$$\lambda : \underset{\sim}{g} \longrightarrow B(O) \tag{5.5.1}$$

be the map given by $\lambda(x) = \varphi^x$.

The coadjoint orbits (O,B) of Theorem 5.4 are Hamiltonian (G,A) spaces.

Theorem 5.5.1. The map λ is a Poisson representation of $\underset{\sim}{g}$ (i.e. $\varphi^{[x,y]} = [\varphi^x,\varphi^y]$). Furthermore λ is integrable and the corresponding action of (G,A) on (O,B) is the same as the given one ((5.4.8)) so that (O,B) is a Hamiltonian (G,A) space with respect to λ.

Proof. By (5.4.30) one has $\xi_{\varphi^x} = \xi^x$. That is

$$\xi_{\lambda(x)} = \xi^x . \tag{5.5.2}$$

But then the relation (5.4.20) implies that $[\varphi^x,\varphi^y] = \varphi^{[x,y]}$ by definition of Poisson bracket. Thus λ is a Poisson representation. The relation (5.5.2) together with the definition of ξ^x, establishes the remainder of the theorem.

Q.E.D.

To avoid ambiguity we will write λ_O for the Poisson representation (5.5.1) corresponding to the orbit (O,B_O,ω_O).

Remark 5.5.1. Let $S^k(\underset{\sim}{g}) \subseteq S(\underset{\sim}{g})$ be the subspace corresponding to $k \in \mathbb{Z}$ for the natural $\mathbb{Z}$-grading of $S(\underset{\sim}{g})$. Also let $E_k(\underset{\sim}{g}) \subseteq E(\underset{\sim}{g})$ denote the filtration of the enveloping algebra $E(\underset{\sim}{g})$ defined in the same manner as in the ungraded case. One has a Birkhoff-Witt theorem (see [2]) which establishes an isomorphism

$$S^k(\underset{\sim}{g}) \simeq E_k(\underset{\sim}{g})/E_{k-1}(\underset{\sim}{g}) \tag{5.5.3}$$

But (5.5.3) also induces a graded Lie algebra structure on $S(\underset{\sim}{g})$. Indeed the bracket in $E(\underset{\sim}{g})$, $[u,v] = uv - (-1)^{|u||v|}vu$, $u, v \in E(\underset{\sim}{g})$, yields the relation

$$[E_k(\underset{\sim}{g}),E_\ell(\underset{\sim}{g})] \subseteq E_{k+\ell-1}(\underset{\sim}{g}) \tag{5.5.4}$$

and this induces the bracket relation

$$[E_k(\underset{\sim}{g})/E_{k-1}(\underset{\sim}{g}),E_\ell(\underset{\sim}{g})/E_{\ell-1}(\underset{\sim}{g})] \subseteq E_{k+\ell-1}(\underset{\sim}{g})/E_{k+\ell-2}(\underset{\sim}{g}) \tag{5.5.5}$$

defining, by (5.5.3), a graded Lie algebra structure on $S(\underset{\sim}{g})$ where

$$[S^k(\underset{\sim}{g}),S^\ell(\underset{\sim}{g})] \subseteq S^{k+\ell-1}(\underset{\sim}{g}) . \tag{5.5.6}$$

But then with regard to orbit (O,B) one has a generalization of the fact that λ_O is a Poisson representation of $\underline{g}$ in $B(O)$. Indeed one has

$$[\varphi^z,\varphi^w] = \varphi^{(z,w]} \tag{5.5.7}$$

for any $z, w \in S(\underline{g})$. This follows easily from (5.2.12) and the fact that λ_O is a Poisson representation.

Now let $\underline{g}$ be a finite dimensional graded Lie algebra and let (G,A) be the corresponding simply connected graded Lie group. Let (X,B_X,ω_X) be a Hamiltonian (G,A) space with respect to a Poisson representation $\lambda_X : \underline{g} \longrightarrow B_X(X)$. Now assume that Y is a covering space for X with respect to a covering map

$$Y \longrightarrow X . \tag{5.5.8}$$

Now X has the structure of an ordinary homogeneous space for G and, since G is simply-connected, the same is true for Y in such a manner that the action of G commutes with the covering map. Now the preimage of the sheaf B_X on X by the map (5.5.8) defines a presheaf on Y. Every presheaf defines a sheaf (see Remark 1.2.3 in [3]). If B_Y is the corresponding sheaf then it is locally isomorphic to B_X so that (Y,B_Y) is a graded manifold and one has a morphism of

$$\sigma : (Y,B_Y) \longrightarrow (X,B_X)$$

of graded manifolds such that the covering map (5.5.8) is just σ_Y. (See (2.15.5)). Furthermore if one defines $\omega_Y = \sigma^*\omega_X$ then (Y,B_Y,ω_Y) is a graded symplectic manifold and $\lambda_Y = \sigma^*\circ\lambda_X$ defines a Poisson representation $\underline{g} \longrightarrow B_Y(Y)$. Since G is simply connected one proves

<u>Proposition 5.5.1.</u> (Y,B_Y,ω_Y) <u>is a Hamiltonian</u> (G,A) <u>space with respect to the Poisson representation</u> λ_Y.

Now if (X_i,B_i,ω_i) are Hamiltonian (G,A) spaces with respect to λ_i, $i = 1, 2$, and $\sigma : (X_1,B_1) \longrightarrow (X_2,B_2)$ is a morphism of graded manifolds we will say $(X_1,B_1,\omega_1,\lambda_1)$ is a covering of $(X_2,B_2,\omega_2,\lambda_2)$ with respect to σ if (1) σ_{X_1} is a covering map, (2) σ locally is an isomorphism of graded symplectic manifolds and, (3) $\sigma^*\circ\lambda_2 = \lambda_1$. One notes of course that, in the notation of Proposition

5.5.1, $(Y,B_Y,\omega_Y,\lambda_Y)$ is a covering of $(X,B_X,\omega_X,\lambda_X)$ with respect to σ.

Remark 5.5.2. If $(X_1,B_1,\omega_1,\lambda_1)$ is a covering of $(X_2,B_2,\omega_2,\lambda_2)$ with respect to σ one notes that $\sigma^* : B_2(Y_2) \longrightarrow B_1(Y_1)$ is a map of $A(G)^*$ modules.

Now let $\underline{g}$ be a finite dimensional real graded Lie algebra and let (G,A) be the corresponding simply connected graded Lie group. Now if the notation is as in Theorem 5.4, then Theorem 5.5 asserts that the orbits (O,B) of the coadjoint action are Hamiltonian (G,A) spaces, $(O,B_O,\omega_O,\lambda_O)$, for (G,A). By Proposition 5.5.1, if Y is a covering space of O then one has a Hamiltonian (G,A) space, $(Y,B_Y,\omega_Y,\lambda_Y)$, which is a covering of the orbit. The following theorem asserts that the coverings of the orbits of the coadjoint action are the only Hamiltonian (G,A) spaces that (G,A) has.

Theorem 5.5.2. *Let $\underline{g}$ be any real finite dimensional graded Lie algebra and let (G,A) be the corresponding graded Lie group. Let $(Y,B_Y,\omega_Y,\lambda_Y)$ be a Hamiltonian (G,A) space. Then there exists a unique orbit (O,B_O) of the coadjoint action of (G,A) and a unique morphism $\sigma : (Y,B_Y) \longrightarrow (O,B_O)$ of graded manifolds such that $(Y,B_Y,\omega_Y,\lambda_Y)$ is a covering of $(O,B_O,\omega_O,\lambda_O)$ with respect to σ.*

Proof. Now

(5.5.9) $$\lambda_Y : \underline{g} \longrightarrow B_Y(Y)$$

is a homomorphism of graded Lie algebras. Let $\rho^x = \lambda_Y(x) \in B_Y(Y)$ for any $x \in \underline{g}$ so that $[\rho^x,\rho^y] = \rho^{[x,y]}$. On the other hand if $\zeta^x = \xi_{\rho^x} \in \underline{a}(Y) \subseteq \mathrm{Der}\, B_Y(Y)$ one has

(5.5.10) $$\zeta^x \rho^y = \rho^{[x,y]} .$$

One notes that $x \longmapsto \zeta^x$ with respect to the coaction representation of (G,A) on $B_Y(Y)$.

Now we can interpret (5.5.9) in another way. Consider the graded affine manifold $(\underline{g}_0',A_{\underline{g}'})$ and recall that $\psi^y \in A_{\underline{g}'}(\underline{g}_0')$ for any $y \in \underline{g}$. (See §5.4.). By Proposition 2.18 one has a unique morphism

(5.5.11) $$\sigma : (Y, B_Y) \longrightarrow (\underline{g}_0', A_{\underline{g}'})$$

of graded manifolds such that

(5.5.12) $$\sigma^*(\psi^y) = \rho^y$$

for any $y \in \underline{g}$. Substituting $[x,y]$ for y in (5.5.12) one has $\sigma^*(\eta^x \psi^y) = \zeta^x \rho^y$. In fact by iteration for any $x_1, \ldots, x_k, y_1, \ldots, y_\ell \in \underline{g}$ one has

(5.5.13) $$\sigma^*(\eta^{x_1} \eta^{x_2} \cdots \eta^{x_k} \psi^{y_1} \cdots \psi^{y_\ell}) = \zeta^{x_1} \cdots \zeta^{x_k} \rho^{y_1} \cdots \rho^{y_\ell}$$

But now if $\tilde{\lambda}_Y : \underline{g}_0 \longrightarrow C^\infty(Y)$ is defined by putting $\tilde{\lambda}_Y(x) = \widetilde{\lambda_Y(x)}$ it follows easily from (5.3.6) that $(Y, \kappa\omega_Y, \tilde{\lambda}_Y)$ is a Hamiltonian G-space in the sense of §5 in [8]. It then follows, by applying Theorem 5.4.1 in [8], that the image of $\sigma_Y : Y \longrightarrow \underline{g}_0'$ is an orbit $G \cdot f = O \subseteq \underline{g}_0'$.

Moreover the map

(5.5.14) $$\sigma_Y : Y \longrightarrow O$$

is smooth.

Now let $b \in Y$. Recalling (see (2.13.2)) that $(B_Y(Y))^*$ is a right $\text{Diff}\, B_Y(Y)$ module and $A_{\underline{g}'}(\underline{g}_0')^*$ is a right $\text{Diff}\, A_{\underline{g}'}(\underline{g}_0')$ module the equation (5.5.13) implies

(5.5.15) $$\sigma_*(\delta_b \cdot \zeta^{x_1} \cdots \zeta^{x_k}) = \delta_{\sigma_*(b)} \cdot \eta^{x_1} \cdots \eta^{x_k}$$

since the image of $S(\underline{g})$ in $A_{\underline{g}'}(\underline{g}_0')$ clearly separates $A_{\underline{g}'}(\underline{g}_0')^*$.

But by definition of (O,B) the right side of (5.5.15) over all k and all choices of the $x_i \in \underline{g}$, spans $B_b(O)^*$. Similarly the left side of (5.5.15) over all such choices, by the homogeneity of (Y,B_Y) spans $(B_Y)_{\sigma_*(b)}(Y)^*$. Thus one has

(5.5.16) $$\sigma_*(B_Y(Y))^* = B(O)^* .$$

But then by Proposition 2.17.3 one has

(5.5.17) $$\sigma : (Y,B_Y) \longrightarrow (O,B_O)$$

is a morphism of graded manifolds. The relation (5.5.12) then becomes

(5.5.18) $$\sigma^*(\varphi^y) = \rho^y$$

and one has

(5.5.19) $$\sigma^*(\xi^x\varphi^y) = \zeta^x\rho^y \ .$$

But (5.5.18) implies $\sigma^* d\varphi^y = d\rho^y$ and (5.5.19) implies

(5.5.20) $$\sigma^* \langle \xi^x | d\varphi^y \rangle = \langle \zeta^x | d\rho^y \rangle .$$

But then by (4.1.16) if $x_i, y_j \in \underline{g}$, $i, j = 1,\ldots,k$, then

(5.5.21) $$\sigma^* \langle \xi^{x_1} \ldots \xi^{x_k} | d\varphi^{y_1} \ldots d\varphi^{y_k} \rangle = \langle \zeta^{x_1},\ldots,\zeta^{x_k} | d\rho^{y_1} \ldots d\rho^{y_k} \rangle .$$

However the $d\varphi^y$ for $y \in \underline{g}$ together with $B(O)$ at least locally generate $\Omega(O,B)$ by Propositions 2.16.1 and 4.2.2 and hence one has

(5.5.22) $$\sigma^* \langle \xi^x, \xi^y | \omega_O \rangle = \langle \zeta^x, \zeta^y | \sigma^* \omega_O \rangle$$

for all $x, y \in \underline{g}$. On the other hand one has

(5.5.23) $$\sigma^* \langle \xi^x, \xi^y | \omega_O \rangle = \langle \zeta^x, \zeta^y | \omega_Y \rangle$$

by (5.5.18) where $[x,y]$ is substituted for y since the right side of (5.5.23) equals $-\rho^{[x,y]}$ and $\langle \xi^x, \xi^y | \omega_O \rangle = -\varphi^{[x,y]}$ by (5.2.13). Thus

(5.5.24) $$\sigma^* \omega_O = \omega_Y$$

by Proposition 2.12.1. Finally if $p \in Y$ and $q = \sigma_* p$ then applying tilda to (5.5.19) one has

(5.5.25) $$\sigma_* \tilde{\zeta}^x_p = \tilde{\xi}^x_q \ .$$

But applying tilda to (5.5.23) yields

(5.5.26) $$\langle \tilde{\xi}^x_q, \tilde{\xi}^y_q | (\tilde{\omega}_O)_q \rangle = \langle \tilde{\zeta}^x_p, \tilde{\zeta}^y_p | (\tilde{\omega}_Y)_p \rangle .$$

But then by Proposition 4.5.2 σ_* induces an isomorphism on tangent spaces. Hence, by Corollary to Theorem 2.16 and (5.5.24) σ is a local isomorphism of graded symplectic manifolds. Thus $(Y,B_Y,\omega_Y,\lambda_Y)$ is a covering of $(O,B_O,\omega_O,\lambda_O)$ with respect to σ since (5.5.18) implies $\sigma^* \circ \lambda_O = \lambda_Y$. The uniqueness of O is guaranteed by Theorem 5.4.1 in [8] and the relation $\sigma^* \circ \lambda_O = \lambda_Y$ guarantees that σ is unique. This proves Theorem 5.5.2. Q.E.D.

6. Prequantization and Representations of Graded Lie Groups

6.1. Let (G,A) be a graded Lie group and let $\underline{g}$ be the Lie algebra of (G,A). Recall that as an L-H algebra $A(G)^* = E(G,\underline{g}) = \mathbb{R}(G) \circledast E(\underline{g})$ where $E(\underline{g})$ is the enveloping algebra of the graded Lie algebra $\underline{g}$.

Now let V be a complete locally convex complex vector space and let $\text{End}_c V$ be the algebra of all continuous operators on V. Assume also that V is graded and that both homogeneous components of V are closed subspaces of V. It is clear then that $\text{End}_c V$ has the structure of a graded algebra. We will now say that a homomorphism of graded algebras

(6.1.1) $$\gamma : A(G)^* \longrightarrow \text{End}_c V$$

is a smooth representation of (G,A) on V if (1) the restriction

$$\gamma : G \longrightarrow \text{End}_c V$$

is a continuous representation of the locally compact group G in the sense of say, Chapter 4 in [16] (the map $G \times V \longrightarrow V$, $(g,r) \longrightarrow \gamma(g)v$ is continuous). (2) Each $v \in V$ is a C^∞-vector in the sense of §4.4 in [16] (the map $G \longrightarrow V$, $g \longrightarrow \gamma(g)v$ is C^∞) and (3) $\gamma(x)v = \frac{d}{dt}\pi(\exp tx)v\Big|_{t=0}$ for each $v \in V$, $x \in \underline{g}_0$.

Example. If (G,A) operates on the graded manifold (X,B) then the corresponding coaction representation $\pi : A(G)^* \longrightarrow \text{End}\, B(X)$ is a smooth representation of (G,A).

A second example will arise from induced representations. This will be defined below.

Let (H,B) be a graded Lie group. Now $\mathbb{C}^*$ can be regarded as a graded Lie group $(\mathbb{C}^*,C^\infty)$ where of course $\dim(\mathbb{C}^*,C^\infty) = (2,0)$. A homomorphism

(6.1.2) $$\chi : (H,B) \longrightarrow (\mathbb{C}^*,C^\infty)$$

of graded Lie groups will be referred to as a character on (H,B).

Given a character χ of (H,B) then the restriction

(6.1.3) $$\chi_* : H \longrightarrow \mathbb{C}^*$$

of $\chi_* : B(H)^* \longrightarrow C^\infty(\mathbb{C}^*)^*$ to H is just a character on H in the usual sense. Conversely one easily has

Lemma 6.1. *Let* $\chi_H : H \longrightarrow \mathbb{C}^*$ *be a character on* H. *Then there exists a character* χ *of* (H,B) *such that* $\chi_H = \chi_* \mid H$ *if and only if the differential of* χ_H *vanishes on* $[\underline{h}_1, \underline{h}_1] \subseteq \underline{h}_0$. *Moreover* χ *is necessarily unique.*

Remark 6.1. Note that χ_* necessarily vanishes on $\underline{h}_1 + [\underline{h},\underline{h}]$.

Now assume that (H,B) is a closed graded Lie subgroup of (G,A) and χ is a character on (H,B).

Now recall the quotient map $\tau : (G,A) \longrightarrow (G/H, A/B)$. See (3.9.17). Let $U \subseteq G/H$ be an open set and let $V = \tau_G^{-1}(U) \subseteq G$. One thus has $A(V)^* B(H)^* \subseteq A(V)^*$. Now let

(6.1.4) $$A(V,\chi) = \{f \in A_{\mathbb{C}}(V) \mid \langle vw,f\rangle = \chi_*(sw)\langle v,f\rangle \quad \text{for all } v \in A(V)^*,\ w \in B(H)^*\}.$$

Now recall (see (3.9.7)) $A/B(U)$ is identified with the set of all $g \in A(V)$ such that $\langle vw,g\rangle = \langle w,1_H\rangle\langle v,g\rangle$ for all $v \in A(V)^*$, $w \in B(H)^*$. It is clear then that if $f \in A(V,\chi)$, $g \in A/B(U)$ then $gf \in A(V,\chi)$. But if we put

(6.1.5) $$L^\chi(U) = A(\tau_G^{-1}(U),\chi)$$

then clearly $U \longrightarrow L^\chi(U)$ is a sheaf L^χ on G/H. Hence L^χ is a sheaf of A/B modules. The following theorem guarantees that the sheaf L^χ is non-trivial. Its proof uses the graded cross-section result, Proposition 3.9.2.

Theorem 6.1. *Let* (G,A) *be a graded Lie group and* (H,B) *a closed graded Lie subgroup. Then* L^χ *defined as above is a line bundle sheaf on* $(G/H,A/B)$ *for any character* $\chi : (H,B) \longrightarrow (\mathbb{C}^*, C^\infty)$.

To prove the theorem it suffices by using translations to show that if $U \subseteq G/H$ is as in the outline of the proof of Theorem 3.8 then U is principal for L^χ. Now recall that, for this open set U, we have defined in Proposition 3.9.3, an isomorphism

(6.1.6) $$\theta : (U\times H, (A/B)\times B \mid U\times H) \longrightarrow (V, A\mid V)$$

of graded manifolds where $V = \tau_G^{-1}(U)$.

Now since $\mathbb{C}^*$ is a Lie group $C^\infty(\mathbb{C}^*)^*$ operates on $C^\infty(\mathbb{C}^*)$ in a natural way (left regular representation). Let $\hat{\chi} \in B_{\mathbb{C}}(H)$ be defined so that for any $\omega \in B(H)^*$,

$$(6.1.7) \qquad \langle w, \hat{\chi} \rangle = \chi_*(sw) \cdot 1 .$$

Now for any $h \in (A/B)_{\mathbb{C}}(U)$ let $F(h) \in A_{\mathbb{C}}(V)$ be the image of $h \otimes \hat{\chi}$ in $A_{\mathbb{C}}(V)$ with respect to the isomorphism (6.1.6). It follows easily from the definition of θ that $F(h) \in A(V,\chi)$ and that every element in $A(V,\chi)$ is uniquely of this form. It follows then that U is principal for L^χ and $F(1_U)$ is a basal element. This proves Theorem 6.1.

Now it is clear that $A(G,\chi)$ is a closed, graded, subspace of $A_{\mathbb{C}}(G)$. Moreover, by (3.6.5), $A(G,\chi)$ is stable under L_u for any $u \in A(G)^*$. Let $(\text{Ind } \chi)(u) = L_u \mid A(G,\chi)$ so that

$$(6.1.8) \qquad \text{Ind } \chi : A(G)^* \longrightarrow \text{End}_c\, A(G,\chi)$$

is a smooth representation of (G,A) on $A(G,\chi)$. We refer to $\text{Ind } \chi$ as the representation of the graded Lie group (G,A) induced by χ.

Remark 6.1. It should be noted that if $H = G$ then $\text{Ind } \chi$ is a finite dimensional representation which will be non-trivial if $\underline{h} \neq \underline{g}$.

6.2. Let (X,B) be a graded manifold and let L be a line bundle sheaf over (X,B). If $U \subseteq X$ is any open subset we now observe that $L(U)$ has the structure of a locally convex complex vector space. By Theorem 4.8 the group $\mathcal{L}(X,B)$ of equivalence classes of all such L is naturally isomorphic to the cohomology group $H^2(X,\mathbb{Z})$. But $\mathcal{L}(X)$, the set of equivalence classes of ordinary complex line bundles, is also naturally isomorphic to $H^2(X,\mathbb{Z})$ so one has a canonical isomorphism

$$(6.2.1) \qquad \mathcal{L}(X,B) \cong \mathcal{L}(X) .$$

One can be very explicit about this isomorphism. The correspondence $U \longrightarrow L(U)/B^1(U)L(U)$ defines a presheaf on (X,B). If $\tilde{L}$ is the corresponding sheaf (see [3], p. 112) then $\tilde{L}(U)$ is a $C^\infty(U)$ module. In fact $\tilde{L}$ defines a complex line bundle $\hat{L}$ so that $\tilde{L}(U) = \Gamma(U,\hat{L})$. Now for any open set $U \subseteq X$ one has a map

(6.2.2) $$L(U) \longrightarrow \tilde{L}(U), \qquad t \longmapsto \tilde{t}$$

such that for any $f \in A(U)$

(6.2.3) $$\widetilde{ft} = \tilde{f}\tilde{t} \ .$$

In particular $B^1(U)L(U)$ is in the kernel of (6.2.2). In fact if U is principal for L one has an exact sequence

(6.2.4) $$0 \longrightarrow B^1(U)L(U) \longrightarrow L(U) \longrightarrow \tilde{L}(U) \longrightarrow 0 \ .$$

Remark 6.2.1. The isomorphism (6.2.1) may be given by $[L] \longrightarrow [\tilde{L}]$ where $[\hat{L}] \in L(X)$ is the class of $\hat{L}$. Note that if $\tilde{c}_{ij} \in B(U_i \cap U_j)$ are transition elements for L with respect to a local system $\{U_i,t_i\}$ then $\tilde{c}_{ij} \in C^\infty(U_i \cap U_j)$ are transition functions for $\hat{L}$ relative to the local system $\{U_i,\tilde{t}_i\}$.

Now if ∇ is a connection in L observe that ∇ induces a connection ∇ in the line bundle $\hat{L}$. Indeed it follows easily from (2.12.5) and (6.2.3) that if $\xi \in (\mathrm{Der}\ B(X))_0$ one may uniquely define $\tilde{\nabla}_{\tilde{\xi}} \in \mathrm{End}\ \tilde{L}(B)$ so that

(6.2.5) $$\tilde{\nabla}_{\tilde{\xi}}\tilde{t} = \widetilde{(\nabla_\xi t)}$$

for any $t \in L(X)$. We will let $(\tilde{L},\tilde{\nabla})$ denote the line bundle $\hat{L}$ with connection $\tilde{\nabla}$ rather than $(\hat{L},\tilde{\nabla})$.

Lemma 6.2. *If $\{U_i,t_i\}$ is a local system for L and $\{c_{ij},\alpha_i\}$ is the corresponding local data for (L,∇) then $\{\tilde{c}_{ij},\tilde{\alpha}_i\}$ is the data for local $(\tilde{L},\tilde{\nabla})$ relative to the local system $\{U_i,\tilde{t}_i\}$.*

This is immediate from (6.2.5). As a consequence one has

(6.2.6) $$\widetilde{\mathrm{curv}\ (L,\nabla)} = \mathrm{curv}\ (\tilde{L},\tilde{\nabla}) \ .$$

Now assume $\omega = \mathrm{curv}\ (L,\nabla)$. Let $L_{\kappa\omega}(X)$ be the set of all equivalence classes of line bundles with connection over X whose curvature is $\kappa\omega$. One knows (see §2.5 in [8]) that $L_{\kappa\omega}(X)$ is a principal $H^1(X,\mathbb{C}^*)$ homogeneous space where the action is defined in a way similar to the action of $H^1(X,\mathbb{C}^*)$ on $L_\omega(X,A)$ given in §4.10. In fact one has

Proposition 6.2.1. *Let* (X,B) *be a graded manifold and let* $\omega \in \Omega^2_{\mathbb{C}}(X,B)_0$. *Assume* $[\omega] \in \mathrm{Coh}^2(\Omega_{\mathbb{C}}(X,B))$ *is integral so that by Theorem* 4.7 *the de Rham class* $[\kappa\omega] \in \mathrm{Coh}^2(\Omega_{\mathbb{C}}(X))$ *is also integral. Then the map*

$$L_{\omega}(X,B) \longrightarrow L_{\kappa\omega}(X) \tag{6.2.7}$$

given by $[(L,\nabla)] \longrightarrow [(\tilde{L},\tilde{\nabla})]$ *is an isomorphism of principal* $H^1(X,\mathbb{C}^*)$ *homogeneous space.*

Now if $k \in \mathbb{Z}_+$ one defines a space of operators $\mathrm{Diff}_k\, L(U)$ on $L(U)$ in a manner similar to the definition of $\mathrm{Diff}_k\, A(U)$ (see §2.13). That is, $\mathrm{Diff}_0\, L(U)$ is the set of operators $\partial \in \mathrm{End}\, L(U)$ which commute, in the graded sense, with any multiplication operator M_f, $f \in A(U)$, on $L(U)$. Inductively, $\mathrm{Diff}_k\, L(U)$ is the set of all $\partial \in \mathrm{End}\,(LU)$ such that $[\partial, M_f] \in \mathrm{Diff}_{k-1}\, L(U)$ for any $f \in A(U)$ where the bracket is defined in the graded sense. One puts $\mathrm{Diff}\, L(U) = \bigcup_k \mathrm{Diff}_k\, L(U)$. Now for any $\partial \in \mathrm{Diff}\, L(U)$ and any $\gamma \in \Gamma(U,\hat{L}')$ of compact support where $\hat{L}'$ is the dual line bundle to $\hat{L}$ one defines a semi-norm $|\ |_{\partial,\gamma}$ where

$$|t|_{\partial,\gamma} = \max_{p\in U} |\langle \tilde{\partial} t, \gamma\rangle (p)| \tag{6.2.8}$$

for any $t \in L(U)$.

Proposition 6.2.2. $L(U)$ *is a complete locally convex space with respect to the semi-norms* (6.2.5).

Remark 6.2.2. One notes that any $\partial \in \mathrm{Diff}\, L(U)$ is a continuous operator on $L(U)$. In particular if (L,∇) is a line bundle sheaf over (X,B) with connection then convariant differentiation ∇_{ξ}, $\xi \in \mathrm{Der}\, A(U)$, is a continuous operator on $L(U)$. Indeed $\nabla_{\xi} \in \mathrm{Diff}_1\, L(U)$ since for any $f \in A(U)$

$$[\nabla_{\xi}, M_f] = M_{\xi f}\,. \tag{6.2.9}$$

6.3. Now assume that (X,B,ω) is a graded symplectic manifold. Thus if X is connected one has an exact sequence of graded Lie algebras

$$0 \longrightarrow \mathbb{R}1_X \longrightarrow B(X) \longrightarrow \underline{a}(X) \longrightarrow 0 \tag{6.3.1}$$

where $B(X)$ is a graded Lie algebra with respect to Poisson bracket.

Now assume that the class $[\omega] \in \mathrm{Coh}^2\ (\Omega(X,B)) \cong H^2(X,\mathbb{R})$ is integral. See §4.10.

Remark 6.3.1. This condition is of course always satisfied if $H^2(X,\mathbb{R}) = 0$. In particular it can always be satisfied locally for any graded symplectic manifold.

Then by Proposition 4.10.2 there exists a line bundle sheaf with connection (L,∇) over (X,B) such that

(6.3.2) $$\omega = \mathrm{curv}\ (L,\nabla).$$

We now observe that $L(X)$ has a natural structure of a $B(X)$ (as a graded Lie algebra under Poisson bracket) module. For any $g \in B(X)$ let $\nu_L(g) \in \mathrm{Diff}_1\ L(X)$ be the differential operator on $L(X)$ defined so that for any $t \in L(X)$

(6.3.3) $$\nu_L(g)t = (\nabla_{\xi_g} + 2\pi i g)t\ .$$

We refer to the map $\nu_L : B(X) \longrightarrow \mathrm{Diff}\ L(X)$ as prequantization (for the graded symplectic manifold (X,B,ω)). A simple but crucial point is that ν_L is a homomorphism of graded Lie algebras.

Proposition 6.3.1. Let (X,B,ω) be a graded symplectic manifold such that $[\omega] \in \mathrm{Coh}^2\ (\Omega(X,A))$ is integral and let (L,∇) be a line bundle sheaf with connection over (X,B) such that $\omega = \mathrm{curv}\ (L,\nabla)$. Then where $B(X)$ is regarded as a graded Lie algebra under Poisson bracket the prequantization map

(6.3.4) $$\nu_L : B(X) \longrightarrow \mathrm{Diff}\ L(X)$$

is a homomorphism of graded Lie algebras.

Proof. If $g, h \in B(X)$ one has $[\nabla_{\xi_g} + 2\pi i M_g, \nabla_{\xi_h} + 2\pi i M_h] = [\nabla_{\xi_g}, \nabla_{\xi_h}] + 2\pi i[\nabla_{\xi_g}, M_h] + 2\pi i[M_g, \nabla_{\xi_h}]$. Now $[g,h] = \xi_g h$. Thus $[\nabla_{\xi_g}, M_h] = M_{[g,h]}$ by (6.2.9). But also

(6.3.5) $$[M_g, \nabla_{\xi_h}] = M_{[g,h]}\ .$$

Indeed $[M_g, \nabla_{\xi_h}] = (-1)^{1+|h||g|}[\nabla_{\xi_h}, M_g] = (-1)^{1+|h||g|} M_{[h,g]}$. But

$[h,g] = (-1)^{1+|h||g|}[g,h]$. This proves

(6.3.6) $$[\nu_L(g),\nu_L(h)] = [\nabla_{\xi_g},\nabla_{\xi_h}] + 4\pi i M_{[g,h]} \cdot$$

But by (4.9.8), $[\nabla_{\xi_g},\nabla_{\xi_h}] = \nabla_{[\xi_g,\xi_h]} + 2\pi i M_{\langle\xi_g,\xi_h|\omega\rangle}$. However $[\xi_g,\xi_h] = \xi_{[g,h]}$ and $\langle\xi_g,\xi_h|\omega\rangle = -[g,h]$ by (5.2.13). This proves

(6.3.7) $$[\nu_L(g),\nu_L(h)] = \nu_L[g,h] .$$

Q.E.D.

The notation above will be retained when a open subset $U \subseteq X$ is substituted for X. Thus if $g \in B(U)$, prequantization is denoted by $\nu_L(g) = \nabla_{\xi_g} + 2\pi i M_g$ and of course Proposition 6.3 is valid when U replaces X. If U is principal for L then the representation $\nu_L : B(U) \longrightarrow \text{Diff } L(U)$ may be replaced by a representation of $B(U)$ on itself. Indeed assume $t \in L(U)$ is a basal element. Put $\alpha(t) = \alpha \in \Omega^1_{\mathbb{C}}(U,B)$.

Remark 6.3.2. One recalls of course that $d\alpha = \omega$ in U. In case $H^1(U,\mathbb{C}) = 0$ which, of course, can be the case locally one notes that by choosing t appropriately one can obtain in this way any one-form $\beta \in \Omega^1_{\mathbb{C}}(U,B)_0$ such that $d\beta = \omega$. Indeed by the isomorphism (4.10.10) we can write $\beta - \alpha = df$ where $f \in B_{\mathbb{C}}(U)_0$. But then $t' = ht$ is a basal element where $h = \varepsilon(2\pi i f)$. See Lemma 4.8.1. Furthermore $\alpha(t') = \beta$ by Proposition 4.9.1.

Now one defines a representation

$$\nu_\alpha : B(U) \longrightarrow \text{Diff } B(U),$$

where $B(U)$ is a graded Lie algebra under Poisson bracket, by the formula

(6.3.8) $$\nu_L(g)ft = (\nu_\alpha(g)f)t$$

for any $f, g \in B(U)$. The notation ν_α is justified since explicitly one has

(6.3.9) $$\nu_\alpha(g)f = (\xi_g + 2\pi i(\langle\xi_g|\alpha\rangle + g))f .$$

We will refer to the correspondence $g \longrightarrow \nu_\alpha(g)$ as coordinate prequantization.

Remark 6.3.3. For the case of ordinary symplectic manifolds we have been concerned with prequantization in connection with representation theory and quantization. An explanation of the form of the operator $\nu_L(g)$ for ordinary symplectic manifolds is given in Theorem 3.3.1 in [8]. Now notice, here, for any $g \in B(U)$ the operator $\nu_\alpha(g)$ on $B(U)$ is the sum of a derivation of $B(U)$ and a multiplication operator on $B(U)$ where the two are related by means of the graded 2-form ω. Now let W be any complex vector space where $\dim W = n < \infty$ and assume (u,v) is a non-singular symmetric bilinear form on W. One has two familiar 2^n-dimensional algebras over W, the Clifford algebra $C(W)$ and the exterior algebra ΛW. In preparation for the construction of the spin representation Chevalley in [1], Chapter 11, p. 38 introduces a $C(W)$-module structure on ΛW as follows: He defines a representation

(6.3.10) $$\nu : C(W) \longrightarrow \mathrm{End}\ \Lambda W$$

where

(6.3.11) $$\nu(w) = i(w) + M(w)$$

for $w \in W$ where $i(w)$ is the derivation of degree -1 on ΛW such that $i(w)u = (w,u)$ for $u \in W$ and $M(w)$ is multiplication on ΛW by w. That is, $\nu(w)$ is a sum of a derivation of ΛW and a multiplication operator and the two are related by (this time) a symmetric bilinear form. We are thus calling attention to the formal similarity of $\nu_\alpha(g)$ and $\nu(w)$.

We will show indeed in the example below that if the manifold X reduces to a point then Chevalley's operator $\nu(w)$ does indeed arise from coordinate prequantization. Thus conceptually for us prequantization for graded symplectic manifolds is a unification of the usual prequantization of functions and Clifford algebra theory. This was one of the features of the graded theory which motivated our work here.

Example. Let (X,B,ω) be a graded symplectic manifold where X reduces to a single point. Thus $\dim (X,B) = (0,n)$. By the graded Darboux theorem, Theorem 5.3, we may find an odd coordinate system $s_i \in B(X)$, $i = 1,\ldots,n$ such that $\omega = \frac{1}{2}\sum_{j=1}^{n} \varepsilon_j (ds_j)^2$ where $\varepsilon_j = \pm 1$. Thus with respect to Poisson bracket one has $[s_i,s_j] = \varepsilon_i\delta_{ij}$ and $\xi_{s_g} = \varepsilon_j \frac{\partial}{\partial s_j}$. See (5.3.2) and (5.3.3). Now let W be the complex span of the s_i. Then regarding W as an ungraded vector space one has

$$B_{\mathbb{C}}(X) = \Lambda W \tag{6.3.12}$$

as graded commutative algebras.

Now since $H^2(X,\mathbb{C}) = 0$ the class $[\omega] = 0$ by the isomorphism (4.10.10) and a line bundle sheaf with connection (L,∇) exists on (X,B) such that $\omega = \text{curv}\,(L,\nabla)$. Furthermore the class $[(L,\nabla)] \in L_\omega(X,B)$ is unique by Proposition 4.10.1. Now if $\alpha \in \Omega^1(X,B)_0$ is given by

$$\alpha = \frac{1}{2}\sum_{j=1}^{n} \varepsilon_j s_j ds_j = -\frac{1}{2}\sum_{j=1}^{n} \varepsilon_j (ds_j) s_j \tag{6.3.13}$$

then $d\alpha = \omega$. But by Remark 6.3.2 there exists a basal element $t \in L(X)$ such that $\alpha(t) = \alpha$. But clearly $\langle \xi_{s_j} | \alpha \rangle = -\frac{s_j}{2}$. Thus one has

$$\nu_\alpha(s_j) = \varepsilon_j \frac{\partial}{\partial s_j} + \pi i M_{s_j} \tag{6.3.14}$$

where M_{s_j} is multiplication in $B_{\mathbb{C}}(X)$ by s_j. But now we may define a symmetric bilinear form (f,g) in W by putting

$$(f,g) = \frac{1}{\pi i}[f,g] \tag{6.3.15}$$

where Poisson bracket is extended linearly from $B(X)$ to $B_{\mathbb{C}}(X)$. One thus has $(s_i,s_j) = \frac{\varepsilon_j}{\pi i}$. Let $C(W)$ be the corresponding Clifford algebra. Comparing (6.3.14) with (6.3.11) one has

<u>Proposition 6.3.2</u>. *Let (X,B,ω) be a graded symplectic manifold where X reduces to a point. Let $W_{\mathbb{R}} \subseteq B(X)$ be the span of a B-Darboux coordinate system and let $W \subseteq B_{\mathbb{C}}(X)$ be its complexification. Let $C(W)$ be the Clifford algebra over W with respect to the bilinear form (6.3.15). On the other hand let $\alpha \in \Omega^1(X,B)_0$ be the 1-form given by (6.3.13) so that $d\alpha = \omega$, and let $g \longrightarrow \nu_\alpha(g)$, $g \in B(X)$, by the corresponding coordinate prequantization. See (6.3.9). Then*

$$\nu_\alpha(g) = \nu(\pi i g) \tag{6.3.16}$$

for any $g \in W_{\mathbb{R}}$ where $w \longmapsto \nu(w)$ is given by (6.3.11) so that ν is the representation of the Clifford algebra $C(W)$ on ΛW given in Chevalley [1], p. 38 (In the notation of [1], $\nu(w) = L_w + \delta_w$).

6.4. Now let $\underline{g}$ be any finite dimensional real graded Lie algebra and let (G,A) be the corresponding simply connected graded Lie group.

Let $f \in \underline{g}_0'$ and let (G_f,A_f) be the isotropy subgroup of (G,A) at f with respect to the coadjoint action of (G,A). Let $\underline{g}_f \subseteq \underline{g}$ be the Lie algebra of (G_f,A_f).

Now let G_f' be the set of all characters

$$\chi : (G_f,A_f) \longrightarrow (\mathbb{C}^*,C^\infty) \tag{6.4.1}$$

of (G_f,A_f) such that

$$\chi_*(\exp x) = e^{2\pi i\langle f,x\rangle} \tag{6.4.2}$$

for any $x \in (\underline{g}_f)_0$. The notation is justified by

Lemma 6.4. *The correspondence* $\chi \longrightarrow \chi_* \mid G_f$ *set up a bijection of* G_f' *and the set of all group characters* δ *on* G_f *such that* $\delta(\exp x) = e^{2\pi i\langle f,x\rangle}$ *for any* $x \in (\underline{g}_f)_0$.

Note that f vanishes on $[\underline{g}_f,\underline{g}]$. In particular then f vanishes on $[(\underline{g}_f)_1,(\underline{g}_f)_1] \subseteq (\underline{g}_f)_0$. The proof of Lemma 6.4 then follows from Lemma 6.1.

The element $f \in \underline{g}_0'$ is called integral if G_f' is not empty.

Remark 6.4. Note that by Lemma 6.4 the integrability of f depends only on the even part $\underline{g}_0$ of $\underline{g}$ and hence is the same as the notion of integrability of f in [8].

Now let (O,B_O) be the (G,A) orbit of f so that, in the notation of §5.4, (O,B_O,ω_O) is a graded symplectic manifold which, in the notation of §5.5, is a Hamiltonian (G,A) space with respect to the Poisson representation $\lambda_O : \underline{g} \longrightarrow B_O(O)$. As a (G,A) homogeneous space one has the isomorphism

$$(O,B_O) \cong (G/G_f,A/A_f).$$

For notational simplicity we will identify (O,B_O) with $(G/G_f,A/A_f)$ here so that $O = G\cdot f = G/G_f$.

Now if G_f^e is the identity component of G_f then since G is simply connected the fundamental group of O is naturally isomorphic to the discrete group G_f/G_f^e and hence if $\widehat{(G_f/G_f^e)}$ is the character group of G_f/G_f^e one has a natural isomorphism

(6.4.3) $$H^1(O,\mathbb{C}^*) \cong \widehat{(G_f/G_f^e)} \ .$$

Now assume that f is integral so that G_f' is not empty. Since any two elements in G_f' differ multiplicatively by an element of $\widehat{(G_f/G_f^e)}$ it follows that the group $H^1(O,\mathbb{C}^*)$ operates on G_f'. If $\chi \in G_f'$, $c \in H^1(O,\mathbb{C}^*)$ let $c\cdot\chi \in G_f'$ denote the action. One easily has

Proposition 6.4. *If $f \in \underline{g}_0'$ is integral then G_f' is a principal $H^1(O,\mathbb{C}^*)$ homogeneous space so that if an element $\chi \in G_f'$ is fixed the correspondence $c \longrightarrow c\cdot\chi$ induces a bijection*

(6.4.4) $$H^1(O,\mathbb{C}^*) \longrightarrow G_f' \ .$$

Now by Theorem 6.1 any $\chi \in G_f'$ defines a line bundle sheaf L^χ over $(G/G_f,A/A_f) = (O,B_O)$. We now wish to observe that there is a natural connection ∇^χ in L^χ.

For any $u \in A(G)^*$ recall that $R_u \in \text{End}\, A(G)$ has been defined in §3.6 by the right regular action of (G,A) on itself. In particular $R_x \in \text{Der}\, A(G)$ for any $x \in \underline{g}$. The following lemma does not require the integrality of f.

Lemma 6.4.1. *There exists a unique 1-form $\alpha_f \in \Omega^1(G,A)_0$ such that $\langle R_x|\alpha_f\rangle = \langle f,x\rangle 1_G$ for any $x \in \underline{g}$.*

Indeed if x_i is a basis of $\underline{g}$ then by Proposition 2.12.1 it follows that any $\eta \in \text{Der}\, A(G)$ can be uniquely written $\eta = \sum h_i R_{x_i}$ where $h_i \in A(G)$. One defines α_f by the relation $\langle\eta|\alpha_f\rangle = \sum h_i\langle f,x_i\rangle$. It is clearly unique proving Lemma 6.4.1.

Now if $\eta \in \text{Der}\, A(G)$ we will say that η is right (G_f,A_f) invariant if the commutator (in the graded sense)

(6.4.5) $$[\eta,R_u] = 0 \qquad \text{for any } u \in A_f(G_f)^*$$

as operators on $A(G)$. From the definition of (G_f,A_f) one easily has

(6.4.6) $$\langle\eta|\alpha_f\rangle \in B_O(O)$$

for any such vector field. (Recall as in (3.9.7) we regard $B_O(O)$ as a subalgebra of $A(G)$).

Now, as in §3.9 (see (3.9.17)), let $\tau : (G,A) \longrightarrow (O,B_O)$ be the quotient map. If $\xi \in \mathrm{Der}\, B_O(O)$ and $\eta \in \mathrm{Der}\, A(G)$ we will say that ξ and η are τ-related if

(6.4.7)
$$\xi g = \eta g$$

for any $g \in B_O(O)$. We will say that ξ and η are strongly τ-related if they are τ-related and η is right (G_f,A_f) invariant. For example if $x \in \underline{g}$ and $\xi^x \in \mathrm{Der}\, B_O(O)$ is defined as §5.4 and $L_x \in \mathrm{Der}\, A(G)$ is defined as in (3.6.4) then ξ^x and L_x are strongly τ-related by (3.6.5).

Lemma 6.4.2. *If $\xi \in \mathrm{Der}\, B_O(O)$ there always exists $\eta \in \mathrm{Der}\, A(G)$ which is strongly τ-related to η.*

Proof. If x_i is a basis of $\underline{g}$ then by Proposition 2.12.1 one can always find $h_i \in B_O(O)$ such that $\xi = \sum h_i \xi^{x_i}$. If we put $\eta = \sum h_i L_{x_i}$ then η is strongly related to ξ. Q.E.D.

Now let $\chi \in G_f'$. Recall from (6.1.5) that $L^\chi(O) = A(G,\chi) \subseteq A(G)$. Although the η in Lemma 6.4.2 is far from unique one, however, has

Lemma 6.4.3. *Let $\xi \in \mathrm{Der}\, B_O(O)$ and let $h \in L^\chi(O)$. Then if $\eta \in \mathrm{Der}\, A(G)$ is τ-strongly related to ξ one has*

(6.4.8)
$$(\eta + 2\pi i\langle\eta|\alpha_f\rangle)h \in L^\chi(O)$$

and that furthermore (6.4.8) is independent of the choice of η.

Proof. Let g be given by (6.4.8). One has $g \in L^\chi(O)$ by (6.4.6) since η is right (G_f,A_f) invariant. Now if $\eta_1, \eta_2 \in \mathrm{Der}\, A(G)$ are strongly τ-related to ξ then $\zeta = \eta_1 - \eta_2$ vanishes on $B_O(O)$. It follows (since $\tilde{R}_y$, for $y \in \underline{g}$, vanishes nowhere on G) that if y_j is a basis of $\underline{g}_f$ we can find $g_j \in A(G)$ such that

(6.4.9)
$$\zeta = \sum_i g_j R_{y_j} .$$

But then by definition of $A(G,\chi)$ one has

(6.4.10) $$\zeta h = \sum_j g_j \langle y_j, \hat{\chi}\rangle$$

where $\hat{\chi} \in (A_f)_{\mathbb{C}}(G_f)$ is defined in (6.1.7). But clearly $\langle y_j, \hat{\chi}\rangle = -2\pi i\langle f, y_j\rangle$. However $\langle \zeta|\alpha_f\rangle = \sum g_j\langle f, y_j\rangle$ by definition of α_f. Thus one has $(\zeta + 2\pi i\langle\zeta|\alpha_f\rangle)h = 0$ proving the lemma.

One can now define a connection ∇^{χ} in L^{χ} by the formula

(6.4.11) $$\nabla^{\chi}_{\xi}h = (\eta + 2\pi i\langle\eta|\alpha_f\rangle)h$$

where $h \in L^{\chi}(O)$, $\xi \in \mathrm{Der}\, B_O(O)$, are arbitrary and $\eta \in \mathrm{Der}\, A(G)$ is a vector field which is strongly τ-related to ξ.

The following theorem settles the question as to which coadjoint orbits (O, B_O, ω_O) is $[\omega_O]$ integral. If $[\omega_O]$ is integral we recall that $L_{\omega_O}(O, B_O)$ is a principal $H^1(O, \mathbb{C}^*)$ homogeneous space. The following theorem also determines $L_{\omega_O}(O, B_O)$ in such a case.

Theorem 6.4. Let g be any finite dimensional real graded Lie algebra and let (G,A) be the corresponding simply-connected graded Lie group. Let $f \in g_0'$ and let (O, B_O, ω_O) be the (G,A) orbit defined by f for the coadjoint action. Then the cohomology class $[\omega_O] \in \mathrm{Coh}^2(\Omega_{\mathbb{C}}(O, B_O))$ is integral if and only if f is integral (i.e. G_f' is not empty). Furthermore in such a case one has

(6.4.12) $$\omega_O = \mathrm{curv}\,(L^{\chi}, \nabla^{\chi})$$

where (G_f, A_f) is the isotopy subgroup of (G,A) at f, $\chi \in G_f'$ and $(L^{\chi}, \nabla^{\chi})$ is the line bundle sheaf with connection over (O, B_O) defined by (6.4.11) and Lemmas 6.4.2 and 6.4.3. Finally the map

(6.4.13) $$G_f' \longrightarrow L_{\omega_O}(O, B_O)$$

defined by the correspondence $\chi \longmapsto [(L^{\chi}, \nabla^{\chi})]$ is an isomorphism of principal $H^1(O, \mathbb{C}^*)$ homogeneous spaces.

Proof (Sketched). By Theorem 4.7 one has $[\omega_{\bar{O}}]$ is integral if and only if $[\kappa\omega_O]$ is integral. But by Theorem 5.7.1 in [8] and Remark 6.4, $[\kappa\omega_O]$ is integral if and only if f is integral. This proves the first statement. Now if f is integral, $\chi \in G_f'$ and $\omega = \text{curv}\,(L^{\chi},\nabla^{\chi})$ then one proves $\omega = \omega_0$ by showing that

$$<\xi^x,\xi^y|\omega> = <\xi^x,\xi^y|\omega_0> \tag{6.4.14}$$

for any $x, y \in \underline{g}$. The argument is essentially the same as the argument in the proof of Proposition 5.7.2 in [8] in that one shows both left and right sides of (6.4.14) are equal to $<L_x,L_y|d\alpha_f> \in B_0(O)$.

Now if $[(\tilde{L}^{\chi},\tilde{\nabla}^{\chi})] = \ell \in L_c(O)$ then $\ell \in L_{\kappa\omega}(O)$ by (6.2.6). But using the notation of §5.7 in [8] it follows easily that $\Lambda^{\ell} = \chi_{G_f}$. But clearly $(c\cdot\chi)_{G_f} = c\cdot\chi_{G_f}$ for $c \in H^1(O,\mathbb{C}^*)$. But then by Corollary 1 to Theorem 5.7.1 in [8] one has $[(\tilde{L}^{c\cdot\chi},\tilde{\nabla}^{c\cdot\chi})] = c\cdot\ell$. This, however, implies

$$[(L^{c\cdot\chi},\nabla^{c\cdot\chi})] = c \cdot [(L^{\chi},\nabla^{\chi})] \tag{6.4.15}$$

for any $c \in H^1(O,\mathbb{C}^*)$ by Proposition 6.2.1. The final statement of Theorem 6.4 then follows from Propositions 6.4 and 4.10.2.

6.5. Now let (X,B,ω) be any graded symplectic manifold and let $\underline{g}$ be any finite dimensional real graded Lie algebra.

Assume that one has a Poisson representation

$$\lambda : \underline{g} \longrightarrow B(X) .$$

Assume also that the cohomology class $[\omega]$ is integral so that one has a line bundle sheaf with connection (L,∇) such that $\omega = \text{curv}\,(L,\nabla)$. Thus one has prequantization

$$\nu_L : B(X) \longrightarrow \text{End}\, L(X) .$$

But now if

$$\gamma_{L,\lambda} : \underline{g} \longrightarrow \text{End}\, L(X) \tag{6.5.1}$$

is defined by putting $\gamma_{L,\lambda}(x) = \nu_L(\lambda(x))$ it follows that $\gamma_{L,\lambda}$ is a representation of $\underline{g}$ on $L(X)$ by continuous operators. See Proposition 6.3.1 and Remark 6.2.2.

Remark 6.5. One of the aspects of prequantization ν_L is that $\nu_L(1_X) \neq 0$ so that $L(X)$, as a $B(X)$-module, does not factor through $\underline{a}(X)$, the graded Lie algebra of Hamiltonian vector fields, with respect to the exact sequence $B(X) \longrightarrow \underline{a}(X) \longrightarrow 0$. Thus if $X = U$ in the notation of (5.3.11) and $\underline{g} = P$ is the graded Heisenberg Lie algebra spanned by 1_U and a B-Darboux coordinate system, as in (5.3.11), and λ is the injection map then $\gamma_{L,\lambda}$ is a faithful representation of P and does not factor through the commutative graded Lie algebra $\underline{p}$. Also recalling Remark 5.3.3 when X reduces to a point so that $\underline{a}(X)$ is a simple graded Lie algebra (noted by Kac-Kaplansky) then, with respect to $\gamma_{L,\lambda}$, $L(X)$ is a module for the central extension $\underline{g} = B(X)$ (see (5.3.8)) and not a module for $\underline{a}(X)$.

Now let (G,A) be a simply connected graded Lie group corresponding to $\underline{g}$. We will say that $\gamma_{L,\lambda}$ is integrable if it extends to a smooth representation

(6.5.2) $$\gamma_{L,\lambda} : A(G)^* \longrightarrow \text{End } L(X)$$

of (G,A) on $L(X)$. See §6.1. The argument in Remark 5.4.1 shows that any such extension is unique. The point is that if (X,B,ω) is a Hamiltonian (G,A) space with respect to λ then $\gamma_{L,\lambda}$ is indeed integrable and furthermore $\gamma_{L,\lambda}$ is just an induced representation. We will state and prove the theorem for the case of an orbit of the coadjoint action. The general case follows similarly using the orbit covering theorem, Theorem 5.5.2.

Let $f \in \underline{g}_0'$ and let (O,B_O,ω_O) be the (G,A) orbit defined by f. Thus by Theorem 5.5.1 (O,B_O,ω_O) is a Hamiltonian (G,A) space with respect to the Poisson representation $\lambda_O : \underline{g} \longrightarrow B_O(O)$ defined in (5.5.1). Recall that in the notation of §5.4 $\lambda_O(y) = \varphi^y$ for $y \in \underline{g}$. Now assume f is integral and let $\chi \in G_f'$. Then $[\omega_O]$ is also integral and if (L^χ,∇^χ) is the corresponding line bundle sheaf with connection over (O,B_O) defined in §6.4 one has $\omega_O = \text{curv } (L^\chi,\nabla^\chi)$ by Theorem 6.4. Furthermore, also by Theorem 6.4, (L^χ,∇^χ) is, up to equivalence, the most general line bundle sheaf with connection over (O,B_O) having this property. Now we will write γ_χ for $\gamma_{L,\lambda}$ where $L = L^\chi$ and $\lambda = \lambda_O$. Thus

$$\gamma_\chi : \underline{g} \longrightarrow \mathrm{End}\ L^\chi(O)$$

is the representation of $\underline{g}$ in $B_O(O)$ obtained by the composition of prequantization with the Poisson representation λ_O. Explicitly, one has for any $y \in \underline{g}$, $h \in L^\chi(O)$

$$\gamma_\chi(y)h = (\nabla^\chi_{\xi^y} + 2\pi i\varphi^y)h \tag{6.5.4}$$

using the notation of §5.4 and §6.4.

Theorem 6.5. *Let $\underline{g}$ be any finite dimensional real graded Lie algebra and let (G,A) be the corresponding simply connected graded Lie group. Assume $f \in \underline{g}_0'$ is integral and let $\chi \in G_f'$. Let (O,B_O,ω_O) be the (G,A) orbit defined by f for the coadjoint action so that (O,B_O,ω_O) is a graded symplectic manifold on which (G,A) operates. Let L^χ be the line bundle sheaf over (O,B_O) defined in §6.1 and let γ_χ be the representation of $\underline{g}$ on $L^\chi(O)$ defined above, as the composite of prequantization with the Poisson representation $\underline{g} \longrightarrow B_O(O)$, $x \longrightarrow \varphi^x$ (using the notation of §5.4). Then γ_χ is integrable uniquely defining a smooth representation*

$$\gamma : A(G)^* \longrightarrow \mathrm{End}\ L^\chi(O) \tag{6.5.5}$$

of (G,A) on $L^\chi(O)$. Furthermore γ is an induced representation. In fact one has

$$\gamma = \mathrm{Ind}\ \chi\ . \tag{6.5.6}$$

(*See* §6.1).

Proof. If $h \in L^\chi(O) = A(G,\chi)$ in the notation of §6.1 and $y \in \underline{g}$ then by (6.4.8) one has

$$\nabla^\chi_{\xi^y} h = L_y h + (2\pi i\langle L_y | \alpha_f\rangle)h \tag{6.5.7}$$

since L_y is strongly τ-related to ξ^y. Thus we have only to prove

$$-\langle L_y | \alpha_f\rangle = \varphi^y \tag{6.5.8}$$

where, as in §6.4, we are regarding $B_O(O) \subseteq A(G)$. We will prove (6.5.8) by showing that any element $u \in A(G)^*$ yields the same value when applied to both sides of (6.5.8). Let $u \in A(G)^*$. Now the inclusion $B_O(O) \subseteq A(G)$ came about from the identification of (O,B_O) with $(G/G_f,A/A_f)$. But then

(6.5.9) $$\langle u,\varphi^y\rangle = \langle u\cdot\delta_f,\psi^y\rangle$$

where $\psi^y \in A_{\mathfrak{g}'}(\mathfrak{g}_0')$ is defined in §5.4, $\delta_f \in A_{\mathfrak{g}'}(\mathfrak{g}_0')^*$ is the group-like element corresponding to f, and dot denotes the coadjoint action of (G,A). But $\langle u\cdot\delta_f,\psi^y\rangle = \langle\delta_f,s(u)\cdot\psi^y\rangle$ by (3.9.5) where $s(u)\cdot\psi^y$ is given by the corresponding coaction representation. However one has $s(u)\cdot\psi^y = \psi^{s(u)\cdot y}$ by definition of the coadjoint action where $s(u)\cdot y = \mathrm{Ad}\,(s(u))(y)$. See (3.11.10). Thus $\langle u\cdot\delta_f,\psi^y\rangle = \langle f,s(u)\cdot y\rangle$ and hence

(6.5.10) $$\langle u,\varphi^y\rangle = \langle f,s(u)\cdot y\rangle\ .$$

On the other hand for any $g \in A(G)$ one has $\langle u,g\rangle = (\widetilde{L_{s(u)}g})(e)$ by definition of the left regular representation. Thus

(6.5.11) $$\langle u,\langle L_y|\alpha_f\rangle\rangle = (\widetilde{L_{s(u)}\langle L_y|\alpha_f\rangle})(e).$$

But by Proposition 4.3.5 and (3.6.5) $\theta(L_x)\alpha_f = 0$ for any $x \in \mathfrak{g}$. Also by (3.6.5) α_f is invariant under the isomorphism $(G,A) \longrightarrow (G,A)$ corresponding to L_a for any $a \in G$. Thus by Proposition 4.3.5 one has $L_{s(u)}\langle L_y|\alpha_f\rangle =$ $\langle L_{s(u)\cdot y}|\alpha_f\rangle$. But $(\widetilde{\langle L_{s(u)\cdot y}|\alpha_f\rangle})(e) = (\widetilde{\langle L_{s(u)\cdot y}}|\tilde{\alpha}_f\rangle)(e)$. However $(\widetilde{L_{s(u)\cdot y}})(e) = -(\widetilde{R_{s(u)\cdot y}})(e) = -s(u)\cdot y$. Hence

(6.5.12) $$\langle u,\langle L_y|\alpha_f\rangle\rangle = -\langle f,s(u)\cdot y\rangle$$

by definition of α_f. The theorem then follows from (6.5.10). Q.E.D.

<u>Example</u>. Consider the case where $\mathfrak{g}$ is the graded Heisenberg Lie algebra such that $\dim \mathfrak{g}_0 = 1$. See Remark 5.3.4. (This is the opposite extreme of the usual Heisenberg Lie algebra). Let $n = \dim \mathfrak{g}_1$. One has the identification $\mathfrak{g}_0' = \mathbb{R}$. Each $r \in \mathbb{R}$ is integral and G_r' has only one element χ_r. If $r = 0$ then the representation $\mathrm{ind}\,\chi_r$ given by Theorem 6.5 is 1-dimensional. If $r \neq 0$ then $\mathrm{ind}\,\chi_r$ is 2^n dimensional and the image of $\mathrm{ind}\,\chi_r$ itself is isomorphic to a Clifford algebra $C_r(\mathfrak{g}_1)$ over $\mathfrak{g}_1$. The representation $\mathrm{ind}\,\chi_r$ is a multiple of either 1 or 2 irreducible representations of (G,A) depending on the parity of n. (Of course, since we have not considered polarizations one should not expect $\mathrm{ind}\,\chi_r$ to be irreducible itself). These irreducible

components, over all $r \in \mathbb{R}$, are, of course, irreducible for $\underline{g}$ and run through all irreducible representations (necessarily finite dimensional as one easily shows) of $\underline{g}$ which are unitarizable for G. More generally this last statement is true for any $\underline{g}$ such that $\underline{g}_0$ is central in $\underline{g}$.

References

1. C. Chevalley, The Algebraic Theory of Spinors, Columbia Univ. Press, New York, 1954.
2. L. Corwin, Y. Neeman and S. Sternberg, Graded Lie algebras in mathematics and physics, Review of Modern Physics 47 (1975), 573-603.
3. R. Godement, Théorie des Faisceaux, Hermann, Paris, 1958.
4. R. Heyneman and M. Sweedler, Affine Hopf algebras I, Journal of Algebra 13 (1969), 192-241.
5. V. G. Kac, Lie superalgebras, Uspehi Matem. Nauk. (to appear).
6. I. Kaplansky, Graded Lie algebras I, II, University of Chicago, Preprints, 1975.
7. A. Kirillov, Unitary representations of nilpotent Lie groups, Uspehi. Matem. Nauk, 17 (1962), 57-110.
8. B. Kostant, Quantization and unitary representations, Lectures in Modern Analysis and Applications III, Springer-Verlag, vol. 170 (1970), 87-207.
9. B. Kostant, On the definition of quantization, Géométrie Symplectique et Physique Mathématique, Colloques Internationaux, CNRS 237 (1975)., 187-210.
10. J. Milnor and J. Moore, On the structure of Hopf algebras, Ann. of Math. 81 (1965), 211-264.
11. A. Nijenhuis and R. W. Richardson, Cohomology and deformations in graded Lie algebras, Bull. Amer. Math. Soc. 72, 1-29 (1966).
12. D. Quillen, Rational homotopy theory, Ann. of Math. 90 (1969), 205-295.
13. D. Simms and N. Woodhouse, Lectures on Geometric Quantization, Lecture Notes in Physics, Springer-Verlag, vol. 53 (1976).
14. J. Souriau, Structures des systèmes dynamiques, Dunod, Paris (1970).
15. M. Sweedler, Hopf algebras, Benjamin, New York, 1969.
16. G. Warner, Harmonic Analysis on Semi-Simple Lie Groups I, Springer Verlag, New York, 1972.

GAUGE FIELDS AS QUANTIZED CONNECTION FORMS

Meinhard E. Mayer

Departments of Mathematics and Physics

University of California, Irvine, CA 92717

1. INTRODUCTION AND HISTORICAL SURVEY

A. This is an expanded version of my lecture at the Symposium. The oral presentation was aimed mainly at the mathematicians in the audience, who knew more than I do about principal and associated fibrations, holonomy groups, etc., and to whom I was supposed to explain the physical aspects of gauge theories. Once this purpose was achieved (given the slightly biased way in which I presented the physics) it was probably clear to most mathematicians and general relativists in the audience what comes next: a reformulation of gauge theory in mathematical language (which had been done before) and an attempt to "axiomatize" a quantum theory of gauge fields by introducing the notion of quantized connection form (which seems to be new).

Due to the time limitation of a one-hour lecture the physicists in the audience who did not have a differential-geometric background were shortchanged. Since these notes will have a wider circulation, I have decided to include a section, intended for physicists, which explains the basic differential-geometric concepts used and illustrating them by translating familiar expressions from gauge theory into this new language.

One of the purposes of this set of notes is to convince physicists involved in gauge theories that some (if not all) the difficulties they encounter in "quantizing" such theories are related to the differential geometric nature of the objects it deals with (i. e., to the fact that the gauge potentials and field strengths are respectively a connection form and its curvature 2-form, rather than a vector and a tensor, as usually stated), and that a systematic use of modern differential geometry may help overcome the most difficult of these problems.

In addition, I wish to reemphasize the fact first pointed out by H. Loos [29], that the holonomy group of the gauge fibration may play a more fundamental role in describing the physical symmetries than the gauge group itself, and that it is quite possible that while the gauge group is nonabelian, the "physical holonomy group" may be abelian. We discuss possible mechanisms which may lead to this peculiar form of "symmetry-breakdown", making it plausible that this may be brought about by the requirement that in order to guarantee duality (in the sense used in the algebraic approach to quantum field theory) an internal symmetry group (read internal holonomy group) must, according to a theorem of Doplicher, Haag and Roberts [13], admit only one-dimensional irreducible representations.

B. The organization of these notes is as follows: In Subsection C of this introduction we give a brief survey of the history of gauge invariance together with some references. The list of references is by no means complete; in particular we could not try to list the numerous articles which have appeared in the physics literature in the past five years in connection with the unified gauge theory of weak and electromagnetic interactions, referring the reader to the reviews [1, 49].

Section 2 gives an outline of the basic ideas of gauge theory for the benefit of mathematicians, illustrated on the example of scalar electrodynamics. It attempts to explain the physical terminology and to make it obvious that the principal and associated bundles (albeit trivial) are a natural framework for this theory. This section may be skipped by physicists familiar with the subject.

Section 3 describes in qualitative terms, for the benefit of the physicist reader, the basic concepts of principal and associated fibrations, connections and holonomy groups. This section should be skipped by mathematicians, who may find its loose style repugnant.

Section 4 contains the basically new attempt at setting up a set of Gårding-Wightman [18] type axioms for connections and curvature foms by introducing the concept of quantized differential form (This term has been used by I. E. Segal [38] in a different context and with a different meaning). It makes plausible the use of the Bleuler-Gupta indefinite metric used in quantum electrodynamics and explains its extension to the nonabelian case.

Section 5 contains some arguments for considering the holonomy group of the gauge connection as the physical internal symmetry group of the theory and discusses the relation to duality mentioned above.

Section 6 acknowledges a number of weak points in the discussion and points out some facts which might be useful in quanizing the general theory of relativity, where the base space of the bundle participates more directly in the physics.

As already mentioned, the Bibliography, though extensive, is by no means exhaustive, and I would like to apologize in advance to any author who has been inadvertently overlooked. Time did not allow me to carry out an extensive library search.

Unfortunately, time did also not allow me to discuss the Higgs mechanism which has an obvious geometric nature. I hope to return to this problem in a future publication.

C. Brief Historical Outline. The term Eichinvarianz (gauge invariance) was introduced in 1918 by Hermann Weyl, who investigated it in the context of electromagnetism in 1929 [50]. Although the gauge ambiguity for electromagnetic potentials had been known for some time, one must consider Emmy Noether's paper [32] on invariant variational principles as the precursor of present day gauge theory. Already the early formulations of quantum electrodynamics [15, 34] recognize the difficulties involved in reconciling the gauge invariance of the potentials and other requirements, but it was not until the 1950-s that modern gauge theory was born.

Schwinger [37] extended Weyl's idea and treated the quantized electromagnetic field as a consequence of the "local gauge invariance of the charged matter fields. This leads directly to Maxwell's equations.

Soon thereafter, C. N. Yang and R. L. Mills[51] extended this idea to a field theory with the nonabelian symmetry group SU(2), deriving for the gauge vector fields (which are now noncommutative, since they take values in the Lie algebra of SU(2)) a system of nonlinear equations which bear their name. A rereading of the Yang-Mills paper shows that the "vector fields" are indeed connection coefficients and that the field strengths are the curvature forms of these, related to them by the well-known structure equations. The Yang-Mills equations then become Bianchi identities (cf. Section 2).

The Yang-Mills approach was generalized in 1956 to arbitrary internal symmetry groups by R. Utiyama [46], and independently by the author in his dissertation [30]. It was recognized that a generalization of Gell-Mann's principle of minimal coupling to nonabelian gauge theories requires replacing partial derivatives by "gauge-covariant derivatives" in the field equations; this immediately suggested the geometric interpretation of the gauge vector potentials as connections (Utiyama went further and pointed out that the Christoffel symbols in general relativity could be interpreted as the result of subjecting the Lorentz group to the "gauge principle"; this point of view was further investigated by Thirring [42] and Kibble [25].

Gauge theories underwent a rapid development beween 1957 and 1961 (here is an incomplete list of papers which I remember[2, 6, 19, 25, 33, 35, 42], and was material in the discovery of SU(3) symmetry by Ne'eman and Gell-Mann. A large number of people continued to work on the quantization and solution of gauge theories [29, 16, 14, 11, 47,48], but it was not until 1967 -- 1970 that Weinberg [48], Salam[36] and 't Hooft [43] discovered the unified theory of weak and electromagnetic interaction, which revived the interest in gauge theories. This model was made possible by the discovery by Higgs, Kibble and Brout and Englert [8, 20, 22, 26] of a mechanism of symmetry breaking of gauge fields which allows them to acquire mass and gets rid of some massless

Goldstone bosons (cf. the review [3]).

After 1971 there was a deluge of models for gauge theories of weak and electromagnetic interactions and also for strong interactions. The activity in this area has been stimulated in the last year by the discovery of the new ψ (J) particles and the various "charm" models. For references I refer the reader to the reviews [1, 49] and to recent issues of Phys. Rev. Letters or Physics Letters.

At the same time, relatively few people paid attention to the geometric aspects of gauge theory. Most of the papers deal with classical theories, although they emphasize the goemetric aspect [45, 17, 23, 24].

My interest in the geometric aspects of gauge theories was rekindled by Trautman's lecture at the 1973 Bonn Symposium [45]. It resulted in a few announcements [31] and led to this lecture. Another stimulus came from remembering a discussion I had with Hank Loos in 1968 or 1969, where he tried to convince me of the fundamental role played by the holonomy group of the connection, fact which I fully grasped only during the last two years.

In concluding this review, I would like to mention the definitive treatment of the Bleuler-Gupta indefinite metric quantization of gauge fields which was given by Strocchi and Wightman [40] and where references to earlier work by Strocchi can be found.

ACKNOWLEDGEMENT

It is my pleasant duty to express my gratitude to Konrad Bleuler and the other organizers of this Symposium for their warm hospitality and to the many collaborators who made the success of the Symposium possible.

I think it is appropriate to end this introduction with the following quotation from the work of Bonn's most famous native son:

DER SCHWER GEFASSTE ENTSCHLUSS.

(L. van Beethoven, String quartet No 16 in F major, Op. 135).

2. GAUGE THEORY. A PRIMER FOR MATHEMATICIANS

In this section I shall attempt to introduce mathematicians to the quantum theory of gauge fields (for the classical theory cf., e. g., [45, 17, 23, 24]). I shall start with a simple example: the quantum theory of a "complex scalar field" (charged spin-0 field) which through the principle of local gauge invariance produces its own electromagnetic field. The structure group (gauge group) in this case is $U(1) \cong SO(2)$, i. e., abelian. Generalizing this example on the one hand to arbitrary spin and on the other hand to an arbitrary compact gauge group (e. g., SU(2) for the Yang-Mills theory, or $SU(2) \otimes U(1)$ for the Weinberg-Salam model), we arrive at the set of "axioms" to be described in Section 4.

A quantized spin 0 (scalar) field is a linear mapping $\varphi[f]$ from a space of complex-valued test functions $f \in \mathcal{S}$ ($\mathcal{S}$ is the Schwartz space of infinitely differentiable functions on Minkowski space M_4 which decrease at infinity faster than any power of the Euclidean distance to the origin) to unbounded normal operators on a separable Hilbert space $\mathcal{H}$ satisfying the following requirements (known as the Gårding-Wightman axioms [18]):

i) The operators φ and φ^* are defined on a common dense domain $\mathcal{D}$ which is invariant under $\varphi[f]$.

ii) a. $\varphi[\alpha f + \beta g] = \alpha\varphi[f] + \beta\varphi[g]$, $\alpha, \beta \in C$, $f, g \in \mathcal{S}$ on $\mathcal{D}$.

b. $\varphi^*[\alpha f] = \bar{\alpha}\varphi[\bar{f}]$ (* means adjoint, $\bar{\ }$ means complex conjugate).

c. If $f_\iota \to f$ in $\mathcal{S}$, then $\varphi[f_\iota] \to \varphi[f]$ weakly.

iii) In $\mathcal{D}$ operates a unitary representation of the proper orthochronous Poincaré group $P_+^\uparrow$ (the affine group of M_4 preserving the indefinite quadratic form $x_0^2 - x_1^2 - x_2^2 - x_3^2$; an element of this group is of the form (a, Λ), where Λ is a 4×4 matrix such that $\Lambda^T G \Lambda = G$, G is the metric matrix, and $\Lambda^{00} \geq 0$, $\mathrm{Det}\Lambda = +1$; the composition law of Poincaré transformations is: $(a', \Lambda')(a, \Lambda) = (a' + \Lambda' a, \Lambda'\Lambda))$, $U(a, \Lambda)$ such that

$$U(a, \Lambda)\,\varphi[f]U^{-1}(a, \Lambda) = \varphi[P_{a,\lambda}f], \tag{2.1}$$

where

$$[P_{a,\Lambda}f](\mathbf{x}) = f[\Lambda^{-1}(x - a)]\ ^{\dagger} \tag{2.2}$$

and the domain $\mathcal{D}$ is assumed invariant under the representation U.

iv) There exists a unique vector $\Omega \in \mathcal{D}$ invariant under U and which is cyclic for the algebra generated by the field operators.

v) The spectrum of the self-adjoint generators of the translation subgroup is in the forward lightcone; more precisely, for $U(a, 1) = \exp(iP_\mu a^\mu)$ we have $(\Psi, (P_0^2 - P_1^2 - P_2^2 - P_3^2)\Psi) \geqq 0$ for all $\Psi \in \mathcal{D}$ and $P_\mu \Omega = 0$ (the <u>spectral condition</u>).

vi) If the supports of the test functions f and g are spacelike separated then $[\varphi[f], \varphi[g]] = 0$ (<u>local commutativity</u>).

We shall assume for simplicity that φ is a <u>free field</u> satisfying the Klein-Gordon equation

$$(\Box + m^2)\varphi = 0, \quad \Box = \partial_\mu \partial^\mu \tag{2.3}$$

(interpreted as a distribution p. d. e.).

For our purposes (inventing electromagnetism) it is simpler to replace the second-order equation (2.3) by a system of first-order equations (the so-called Duffin-Kemmer-Petiau equations) by defining the five fields (in the rest of this section, until further notice, we proceed formally, as if the fields were classical, i. e., we do not consider the ordering important) [7, 30]

$$u_0 = (i/m)\partial_0\varphi, \ \ldots \ , u_3 = (i/m)\partial_3\varphi \ , u_4 = \varphi, \tag{2.4}$$

and the six 5×5 matrices (Duffin-Kemmer matrices in a special representation)

$$\beta^0 = \begin{pmatrix} 0 & 0 & 0 & 0 & -1 \\ 0 & 0 & 0 & 0 & 0 \\ 0 & 0 & 0 & 0 & 0 \\ 0 & 0 & 0 & 0 & 0 \\ 1 & 0 & 0 & 0 & 0 \end{pmatrix}, \qquad \beta^1 = \begin{pmatrix} 0 & 0 & 0 & 0 & 0 \\ 0 & 0 & 0 & 0 & -1 \\ 0 & 0 & 0 & 0 & 0 \\ 0 & 0 & 0 & 0 & 0 \\ 0 & -1 & 0 & 0 & 0 \end{pmatrix},$$

---

† In naive field-theory notation: $[P_{a,\Lambda}\varphi](x) = \varphi(\Lambda x + a)$.

$$\beta^2 = \begin{pmatrix} 0&0&0&0&0\\ 0&0&0&0&0\\ 0&0&0&0&-1\\ 0&0&0&0&0\\ 0&0&-1&0&0 \end{pmatrix}, \qquad \beta^3 = \begin{pmatrix} 0&0&0&0&0\\ 0&0&0&0&0\\ 0&0&0&0&0\\ 0&0&0&0&-1\\ 0&0&0&-1&0 \end{pmatrix}, \tag{2.5}$$

$$\beta^4 = \begin{pmatrix} 1&0&0&0&0\\ 0&-1&0&0&0\\ 0&0&-1&0&0\\ 0&0&0&-1&0\\ 0&0&0&0&1 \end{pmatrix}, \qquad M = \begin{pmatrix} m&0&0&0&0\\ 0&m&0&0&0\\ 0&0&m&0&0\\ 0&0&0&m&0\\ 0&0&0&0&m \end{pmatrix}.$$

We also define the "conjugate field" (a row-vector if $u = (u_a)$ is considered a column)

$$\bar{u} = u^*\beta^4. \tag{2.6}$$

Then the equations (2.3) and its adjoint become the Duffin-Kemmer-Petiau equations

$$(i\beta^\mu\partial_\mu - M)u = 0, \quad i\partial_\mu\bar{u}\beta^\mu + \bar{u}M = 0 \tag{2.7}$$

which can formally be thought of as the Euler equations for the Lagrangian density (which is ill defined unless one makes a convention about operator products)

$$L = (i/2)[\bar{u}\beta^\mu\partial_\mu u - (\partial_\mu\bar{u})\beta^\mu u] - \bar{u}Mu = (1/m)\ \varphi^*(\Box + m^2)\varphi.$$

We can subject the u, $\bar{u}$ (or the φ,φ^*) to the gauge transformations of the first kind

$$u \to e^{i\chi}u\ , \quad \bar{u} \to \bar{u}e^{-i\chi} \tag{2.8}$$

where χ is a number (or an operator independent of x). The Lagrangian density L is formally invariant under the transformations (2.8) and therefore, according to Emmy Noether's first theorem [32], the quantity

$$j^\mu = \bar{u}\beta^\mu u \tag{2.9}$$

is "conserved", i. e.,

$$\partial_\mu j^\mu = 0. \tag{2.10}$$

Classically, this implies conservation of the "charge", i. e.,

$$Q = \int j^0 d^3x$$

does not depend on time (this is a consequence of Gauss' theorem, applied to a region in M_4 between two planes of constant time, if the fields fall off rapidly enough at spacelike infinity; for a rigorous discussion cf.[40]). Neglecting for the moment the difficulty involved in defining the operator products, we may consider j^μ as a field operator (the difficulties are the same as in other areas of field theory, and will not be important for the heuristic discussion of this section, which aims at motivating the axioms of Section 4).

Since the transformation (2.8) is "unobservable" (it leaves all the observables, including current and 4-momentum, invariant) nothing prevents us from assuming that the quantity χ in (2.8) depends on the point $x \in M_4$. We shall assume that $\chi(x)$ is a smooth function (preferably belonging to S, although for most of the local arguments it suffices to assume that it is in C^∞). If $\chi(x)$ is an operator (as it must be in a quantized field theory), smoothness and commutation properties have to be defined appropriately; in particular, since in classical gauge theories the function χ is usually taken to be a solution of the wave equation, one can specify commutation relations only on a spacelike hypersurface, since the wave equation propagates the commutator.

Mathematically $\exp(i\chi(x))$ defines a section through the trivial principal bundle $U(1)\times M_4$. If we consider $u(x)$ as an operator-valued functional on rapidly decreasing sections of a trivial line-bundle, the "local gauge transformation"

$$\begin{aligned} u(x) &\to \exp(i\chi(x))u(x) \\ \bar{u}(x) &\to \bar{u}(x)\exp(-i\chi(x)), \end{aligned} \tag{2.11}$$

is an action in the fibers of the line bundle. The expressions (2.11) are in general meaningless. To give them a more precise meaning we use the heuristic representation of $u[f]$ as an operator $u(x)$ "smeared" with the test function $f(x)$, with the symbolic notation:

$$u[\bar{f}] = \int \bar{f}(x)u(x)\, d^4x, \quad \bar{u}[f] = \int \bar{u}(x)f(x)d^4x, \tag{2.12}$$

where f and $\bar{f}$ are classical solutions of the Duffin-Kemmer equations (2.7) (the combinations in (2.12) have been chosen so as to yield single operators, rather than quintuplets of operators). The operators $u[\bar{f}]$ and $\bar{u}[f]$ are then weak solutions (distibutional solutions) of the Duffin-Kemmer equations. The gauge transformations (2.8) or (2.11) can then be defined to act on the test functions:

$$\begin{aligned}(Gu)[\bar{f}] &= u[\bar{f}\exp(-i\chi)] = \exp(iQ\chi)u[\bar{f}]\exp(-iQ\chi),\\ (G\bar{u})[f] &= \bar{u}[\exp(i\chi)f] = \exp(-iQ\chi)\bar{u}[f]\exp(iQ\chi).\end{aligned} \tag{2.13}$$

When χ becomes dependent on the point x, the differentiation law for distributions

$$\partial_\nu u[g] = -u[\partial_\nu g] \tag{2.14}$$

implies that the gauge transformation G does not commute with differentiation and that we have

$$(\partial_\mu G - G\partial_\mu)u[f] = iu[\bar{f}\partial_\mu\chi\exp(-i\chi(x))] \tag{2.15}$$

i. e., the u and $\bar{u}$ no longer satisfy the Duffin-Kemmer equations, but there appears an additional term involving the gradient of χ. One may compensate for the appearance of this term by adding to the gradient operator in (2.7) a 1-form with values in the Lie algebra of the gauge group, to be denoted by $A_\mu(x)$ and which is chosen in such a way that the "gauge-covariant derivative"

$$\nabla_\mu = \partial_\mu - iA_\mu \tag{2.16}$$

does commute with the gauge transformation G (we have chosen the "electric charge" e = 1, for simplicity). It is easy to see that the operator A is defined by (2.15) to be

$$A_\mu u[\bar{f}\exp(-i\chi)] = u[\bar{f}\partial_\mu\chi\exp(-i\chi)] \tag{2.17}$$

and hence, when u, $\bar{u}$ are subjected to the gauge transformation (2.13) the "vector potential" A must undergo the "gauge transformation of the second kind"

$$A_\mu \to A_\mu - \partial_\mu\chi. \tag{2.18}$$

Thus, replacing everywhere, in particular in the Lagrangian L, the partial derivatives by (2.16) and subjecting u, $\bar{u}$ and A to the simultaneous transformations (2.11), invariance is restored. Since A is a connection one-form, its curvature two-form $d \wedge A = F$, in components

$$F_{\mu\nu} = \partial_\mu A_\nu - \partial_\nu A_\mu , \tag{2.19}$$

is gauge-independent, closed, i. e., satisfies the homogeneous Maxwell equation

$$\partial \wedge F = \varepsilon^{\mu\nu\kappa\lambda} \partial_\nu F_{\kappa\lambda} = 0. \tag{2.20}$$

It is natural to include the forms A and F among the dynamical variables (although it will turn out that not all components can be treated as such in a quantized theory); we have to complete the set of equations satisfied by u and $\bar{u}$:

$$\begin{aligned} (i\beta^\mu \nabla_\mu - M)u &= 0 \\ i\nabla_\mu \bar{u}\beta^\mu + \bar{u}M &= 0, \end{aligned} \tag{2.21}$$

with an equation relating A and F to u, $\bar{u}$, which, by analogy to classical electrodynamics had better be in the form

$$\partial_\mu F^{\mu\nu} = j^\nu. \tag{2.22}$$

This is easily achieved by adding to the free-field Lagrangian L a coupling term

$$L_{int} = j^\mu A_\mu \tag{2.23}$$

and a term containing the curvature form (choosing the simplest quadratic invariant in the curvature form)

$$L_{EM} = \tfrac{1}{2} F^{\mu\nu}(\partial_\mu A_\nu - \partial_\nu A_\mu), \tag{2.24}$$

so that variation of the total Lagrangian

$$L + L_{EM} + L_{int} \tag{2.25}$$

leads to the equations (2.19), (2.20), (2.21), (2.22).

All these equations have to be made more precise by specifying the meaning of the operator products (e. g., by means of Wick-ordering) and by introducing the appropriate commutation relations for the gauge fields A, F.

One can construct a fully satisfactory quantum field theory of a free electromagnetic field satisfying the Maxwell equations (2.20) and (2.23) with $j_\mu = 0$, and where the fields F satisfy the commutation relations

$$[F_{\kappa\lambda}(x), F_{\mu\nu}(y)] = i(g_{\lambda\nu}\partial_\kappa\partial_\mu - g_{\lambda\mu}\partial_\kappa\partial_\nu - g_{\kappa\nu}\partial_\lambda\partial_\mu + g_{\kappa\mu}\partial_\lambda\partial_\nu)D(x - y)$$
$$= A_{\kappa\lambda\mu\nu}D(x - y) \qquad (2.26)$$

where $D(x)$ is the Pauli-Jordan invariant distribution $\delta(x^2)$. Then the vacuum expectation value of the product of two fields is

$$(\Omega, F_{\kappa\lambda}(x)F_{\mu\nu}(y)\Omega) = A_{\kappa\lambda\mu\nu}D^{(+)}(x - y), \qquad (2.27)$$

where $D^{(+)}$ is the positive-frequency part of D and $A_{\kappa\lambda\mu\nu}$ is defined by Eq. (2.26). However, if one introduces the potential A as a neutral vector field (transforming as a vector under Poincaré transformations) one runs into the following difficulty. Form the vacuum expectation values

$$K_{\mu\nu}(x - y) = (\Omega, A_\mu(x)A_\nu(y)\Omega) \qquad (2.28)$$

which are tempered distributions, whose Fourier transforms are measures and vanish outside the forward lightcone. Hence

$$K_{\mu\nu} = -g_{\mu\nu}(c_1 + c_1'(-i)D^{(+)}) - c_2(-i)\partial_\mu\partial_\nu D^{(+)}. \qquad (2.29)$$

Positive-definiteness of the Hilbert-space inner product implies

$$K_{00} \geq 0,\ K_{11} \geq 0,\ K_{22} \geq 0,\ K_{33} \geq 0,$$

hence $c_1 = c_1' = 0$, $c_2 \geq 0$, which, using (2.19), (2.27) shows that the left-hand side of (2.27) vanishes, in contradiction with (2.27). In addition, the validity of the Maxwell equation

$$\partial_\mu F^{\mu\nu} = 0 ,$$

together with commutativity of the potentials $A_\mu(x)$, $A_\nu(y)$ for space-like $x - y$ also contradicts (2.27) (the left-hand side must vanish).

A way out of this difficulty has been proposed by Bleuler [5] and Gupta [20], who in addition to the positive definite inner product (,) in Hilbert space have introduced an indefinite continuous sesquilinear form determined by a Hermitean operator η, $\eta^2 = 1$:

$$\{\Psi, \Phi\} = (\Psi, \eta\Phi). \qquad (2.30)$$

Expectation values of physically meaningful quantities are defined by

$$\langle R \rangle = \{\Omega, R\Omega\} = (\Omega, \eta R\Omega) \qquad (2.31)$$

and representations of the Poincaré group are supposed to leave the sesquilinear form invariant. The modified Lorentz condition

$$\partial^{\mu}A_{\mu}^{(-)} = 0, \qquad (2.32)$$

singles out those states in H which are physical. These states span a subspace H' on which the form (2.30) is positive semidefinite. The subspace of H', where

$$\{\Psi, \Psi\} = 0 , \qquad (2.33)$$

will be denoted by H''. State vectors describing physical states belong to the space $H_{phys} = H'/H''$ (completed, if necessary). The operator A_{μ} is then second-quantized by introducing the appropriate Fock space and satisfies a gauge-dependent commutation relation (eq. (2.37) in [40] where more details can be found).

The Maxwell equation (2.22) is valid in a weak sense:

$$\{\Phi, \partial_{\mu}F^{\mu\nu}\Psi\} = 0, \text{ all } \Phi, \Psi \in H'. \qquad (2.34)$$

It is easy to construct examples, showing that adding vectors in the unphysical part of the Hilbert space to a physical vector corresponds to subjecting A to a gauge transformation of the second kind.

The case of nonabelian gauge groups is treated similarly. Let $u^{(a)}$ be a field transforming under a representation (irreducible, unitary) $\rho(G)$ of a semisimple compact (f-parameter) Lie group G, and such that the field equations

$$i\beta^{\mu}\partial_{\mu}u^{(a)} - Mu^{(a)} = 0 \qquad (2.35)$$

are invariant under G (in particular, this means that the matrices β^{μ} commute with the representation $\rho(g)$). Then Noether's theorem implies that there exist f conserved currents (summation convention!):

$$j^{\mu}_{(\lambda)} = i\bar{u}^{(a)}T_{(a)(b)(\lambda)}\beta^{\mu}u^{(b)}, \qquad (2.36)$$

$$(a), (b) = 1, 2, \ldots, \mathrm{Dim}\rho(G),$$

$$(\lambda) = 1, 2 \ldots, f ,$$

where the T are the generators of the group G in the representation $\rho(G)$. In particular, if the group transforms the $u^{(a)}$ (i. e., the $u^{(a)}$ are the basis of the lowest-dimensional faithful representation), the T are the structure constants of the group, as in the paper of Yang and

Mills[51].

Making G into the structure group of a (trivializable) principal fibration over M_4, the representation space of ρ becomes an associated vector bundle, i. e., we have a different representation of the same dimension at each space-time point and the choice of the trivializing section of the principal fibration produces a distinguished section in the associated bundle (choice of gauge).

Repeating the heuristic argument that led in the abelian case to the appearance of the connection form A and the curvature form F we arrive at the following situation.

1. To each representation ρ (generator T_ρ) is associated a connection 1-form U^ρ. The invariance of the field equations (2.35) is preserved if we replace ∂_μ by the gauge-covariant derivatives

$$\nabla^\rho_\mu = \partial_\mu - iT_\rho U^\rho_\mu \tag{2.37}$$

where the U are multilinear in tensor products of representations (just like covariant derivatives of higher order tensors) and under the representation section $\rho(x)$ induced by a section of the principal bundle they transform as

$$U^\rho_\mu \rightarrow \rho^{-1}U^\rho_\mu\rho - \rho^{-1}\partial_\mu\rho , \tag{2.38}$$

typical of connection forms.

The curvature form of U is given by the exterior covariant derivative of U, hence by the structure equation we have:

$$F = DU = d\wedge U - \tfrac{1}{2}U\wedge U \tag{2.39}$$

or, in components,

$$F^\rho_{\mu\nu} = \partial_\mu U^\rho_\nu - \partial_\nu U^\rho_\mu - [U^\rho_\mu , U^\rho_\nu] \tag{2.40}$$

where, in anticipation of noncommutativity, we have replaced the wedge product by the Lie bracket (commutator bracket). Eq. (2.40) is the Yang-Mills-Utiyama definition of the "field strength", and if one forms the quadratic invariant associated to the curvature 2-form F, one is almost unambiguously led to the Lagrangian for the gauge field [29]. Quantization of the nonabelian gauge field will be discussed in Sec. 4.

3. A CRASH COURSE IN FIBER BUNDLES FOR PHYSICISTS

This section attempts the impossible: to explain the differential geometric concepts used in the next sections, to physicists who are at best familiar with the rudiments of Riemannian geometry encountered in general relativity. As was made plausible in Section 2 and will be developed further in Section 4, local gauge theory fits extremely well in to the theory of vector bundles associated to principal fibrations with the gauge group as structure group. I shall try to explain in terms as simple as possible the concepts introduced, and will list the basic definitions and properties of the objects used in this approach. The reader who finds this section difficult to understand should not get discouraged, but should try to fill in the gaps by referring to the many excellent textbooks available: [9, 10, 12, 27, 28, 39, 41], just to name a few.

Most of the definitions and properties are local, and can be best illustrated by recalling how a <u>differential manifold</u> can be viewed as the result of patching together coordinate patches -- called charts.

<u>3.1</u>. A <u>differential manifold</u> is by definition a Hausdorff metrizable topological space X on which is given an equivalence class of <u>atlases</u>. An <u>atlas</u> is a collection of <u>charts</u> and a <u>chart</u> is a triplet $c = (U, \varphi, E)$, where U is an open set of X, E is a Banach space (in most interesting cases, finite dimensional, i. e., $\mathbb{R}^n$) and φ is a bijection of U onto an open set of E. Two charts $c = (U, \varphi, E)$ and $c' = (U', \varphi', E')$ are compatible if $\varphi(U \cap U')$ (and $\varphi'(U \cap U')$) are open in E (respectively E') and the <u>transition diffeomorphisms</u> $\varphi' \circ \varphi^{-1}$ and $\varphi \circ \varphi'^{-1}$ are C^∞-maps. An <u>atlas</u> is a collection of pairwise compatible charts. Two atlases are <u>equivalent</u> (C^∞-equivalent) if their union is again an atlas, and this is the equivalence relation referred to above when we called a manifold a space equipped with an equivalence class of atlases. A chart is nothing else than what physicists usually call a coordinate system and the example of the sphere shows that in order to coordinatize

the surface of the sphere by means of spherical coordinates we need at least two "charts". Thus the "atlas" of the surface of the earth will contain at least two "charts", i. e., maps. Another familiar example is the case of a Lie group G (nowadays defined as a group which is also a differentiable manifold, the group structure being compatible with the manifold structure, i. e., the mapping $(x, y) \to xy^{-1}$ of $G \times G$ into G is a "morphism of manifolds" - a C^{∞}-map. A homogeneous space G/H of a Lie group, where H is a Lie subgroup, is also a manifold. Important notions related to manifolds which we will use without repeating the definitions are: tangent space at a point x, to be denoted by $T_x(X)$, tangent linear map $T_x(f)$, where f is a mapping of manifolds; the dual of the tangent space is the cotangent space $T^*_x(X)$, formed of cotagent vectors or covectors, differentials and vector fields (a warning is necessary, though, vector fields in differential geometry are sections of the tangent bundle, cf. infra, i. e., in coordinates, $\partial/\partial x_i$, rather than the conventional definition of "a vector at each point" to which physicists are used).

3.2. A fibration (or fiber bundle) is a triplet (X, B, p) where the base space B and the bundle space X are manifolds and p is a C^{∞}- surjection of X onto B called the projection satisfying the requirement of local triviality: for each $b \in B$ there exists an open neighborhood U and a manifold F such that $p^{-1}(U)$ can be mapped diffeomorphically onto $U \times F$:

$$\varphi: p^{-1}(U) \to U \times F, \text{ with } p(\varphi^{-1}(b, y)) = b, \quad (3.1)$$

where $b \in U$, $y \in F$. The inverse image $p^{-1}(b)$ is a closed submanifold of X called the fiber at b ($p^{-1}(b) = X_b$). If B is connected all fibers are isomorphic to F -- the typical fiber.

A fibration of the type $(B \times F, B, pr_1)$, where pr_1 means projection on the first factor in the Cartesian product, is called trivial (or a trivial bundle). An isomorphism of a fibration onto a trivial one is called a trivialization. Loosely speaking any fibration is the result of "patching" together local bundle charts which are trivial, giving them a "twist". The most intuitive example of a nontrivial fibration is the Möbius strip, which locally is isomorphic to a Cartesian rectangle, the rectangles being patched together with a twist, which adds up to a change in orientation. The base space of this fibration is isomorphic to a circle and the "typical fiber" is a line segment.

A section of a fibration $\lambda = (X, B, p)$ (also called a cross-section of the fiber bundle) is a mapping $s:B \to X$, such that $p \circ s = Id_B$ (the identity map of B). In a trivial bundle a section is simply a map $s:b \mapsto s(b, y)$ such that $p(s(b, y)) = b$; in other words, a section in a trivial bundle is the graph of a function on the base space with values in the fiber. This is locally true for sections in all fibrations. A fibration does not necessarily admit global sections. One can easily define continuity and differentiability of sections. If nothing else is assumed, sections will be understood to be C^∞. If the typical fiber is a vector space (the fibration is then a vector bundle) sections can be added, multiplied by scalars, etc. One can easily develop an integration theory for sections, as well as a distribution theory: consider C^∞ sections in a vector bundle with compact support, or of rapid decrease at infinity (if X is noncompact). Then vector-valued distributions will be linear functionals on spaces of sections, defined in the same manner as ordinary distributions (cf., e. g., [12], Chapter VXII).

3.3. A principal fibration (principal fiber bundle) is a fibration where all fibers are isomorphic to a Lie group. More precisely, let B be a differential manifold and G a Lie group. A principal fibration of base space B and structure group G is a quadruplet $\lambda = (P, G, B, \pi)$, where P is a manifold on which G acts freely (i. e. without fixed points) on the right (denoted by $(x, g) \mapsto x \cdot g$) and π is a C^∞-surjection of P onto B, subject to the condition of local triviality: for each $b \in B$ there exists an open neighborhood U of b and an isomorphism $f: U \times G \to \pi^{-1}(U)$ such that for $u \in U$ and $g, g' \in G$

$$\pi(f(u, g)) = u \text{ and } f(u, gg') = f(u, g) \cdot g'. \quad (3.2)$$

The triplet (P, B, π) is a fibration and the equivalence relation induced in P by the projection π is the same as the one defined by the group G (the points projected onto the same point b in the base space are all transformed into one another by the structure group, which acts "vertically" in P).

Therefore, if a Lie group G acts freely and properly on a manifold P from the right, the quadruplet $(P, G, P/G, \pi)$, where P/G is the orbit space of G in P and π is the canonical projection of P on P/G, is a principal fibration, and every principal fibration may be thought of as being of this type.

A <u>morphism</u> of principal fibrations $\lambda \to \lambda'$ is a triplet (f, φ, h), where $f: P \to P'$, $h: B \to B'$ are diffeomorphisms (manifold-morphisms), and $\varphi: G \to G'$ is a Lie group homomorphism.

The principal fibration $(B \times G, G, B, pr_1)$ is called <u>trivial</u>, and any morphism of a principal fibration (p. f.) into a trivial p. f. is called a <u>trivialization</u>. If a p. f. $\lambda = (P, G, B, \pi)$ admits a (global) section s, it is trivializable, the trivialization being given by

$$f_s^{-1}(b, g) = s(b)\cdot g, \quad b \in B, \ g \in G, \qquad (3.3)$$

and φ and h are the appropriate identity maps. The question whether a p. f. admits global sections, i. e., is trivializable, is an important topic in differential geometry and leads to the theory of characteristic classes. Most of the p. f. with which we will be dealing are trivializable, the different gauges corresponding to different trivializations (in a physically nontrivial manner).

<u>3.4.</u> A <u>vector bundle associated to a principal fibration</u> (one may consider more general associated fiber bundles, but we shall need only vector bundles) is defined as follows. Let $\lambda = (P, G, B, \pi)$ be a p. f. and let V be a vector space which is the substrate of a (left) representation r(G) of the structure group G:

$$r: (g, v) \mapsto r(g)v, \quad v \in V, \ g \in G. \qquad (3.4)$$

One may think of G acting on the <u>right</u> on the product manifold $P \times V$ by $(x \in P, v \in V)$:

$$(x, v)\cdot g = (x\cdot g, r(g^{-1})v). \qquad (3.5)$$

The orbit space of this action (i, e., the set of pairs taken into one another by the action) is denoted by $P \times^G V = (P \times V)/G$ and is a manifold on which G acts on the right. Let $\pi_V(x, v)$ be the element of B equal to $\pi(x)$ for all $(x, v) \in P \times^G V$. Then $(P \times^G V, B\ \pi_V)$ is a fibration (actually a vector bundle, since locally it is isomorphic to $U \times V$, where U are open subsets of B). A manifold E is called a <u>vector bundle associated to the p. f. λ of fiber-type V</u>, if there is a morphism $\rho: P \times V \to E$ having the following property:

$$\rho(x\cdot g, r(g^{-1})v) = \rho(x, v), \quad x \in P, v \in V, g \in G, \qquad (3.6)$$

(the map ρ is called a <u>framing</u> or <u>frame map</u> of E) and the quotient map $\bar{\rho}: P \times^G V \to E$ is an isomorphims of manifolds. In other words, the quadruplet $(P \times V, G, E, \rho)$ is a principal fibration and the frame map satisfies

$$\rho(x\cdot g, v) = \rho(x, r(g)v), \quad x \in P, g \in G, v \in V.$$

We can define a projection in E by

$$\pi_E(\rho(x, v)) = \pi(x) \quad \text{for } x \in P, v \in V \qquad (3.7)$$

in terms of which the triplet (E, B, π_E) is a fibration. The fiber $E_b = \pi_E^{-1}(b)$ is a closed submanifold of E. The mapping $\theta_x(v) = \rho(x, v)$ is a vector space isomorphism (i. e. a linear invertible map) of V on E_b and this map is equivariant under the action of G.

Finally, let s be a local section of P over $U \subset B$. Then the frame map induces an isomorphism $\psi: (b, v) \to \rho(s(b), v)$ of $U \times V$ onto $\pi_E^{-1}(U)$. Every section σ of the associated bundle $P \times^G V$ over U can be uniquely expressed in the form $\sigma; b \to s(b)\cdot\zeta(b)$, where ζ is a mapping of U into the vector space V. We have $\zeta(b) = \Xi(s(b))$, where $\Xi = \zeta\circ\pi|s(U)$

<u>3.5. Examples.</u> a) Let the group G be $U(1) \cong SO(2)$ (the group of constant phase transformations which is isomorphic to the circle group) and let B be the segment [0, 1] of the real line. Then we can form the trivial principal bundle illustrated in Fig. 1.

The cylinder represents the bundle space P and the line on it

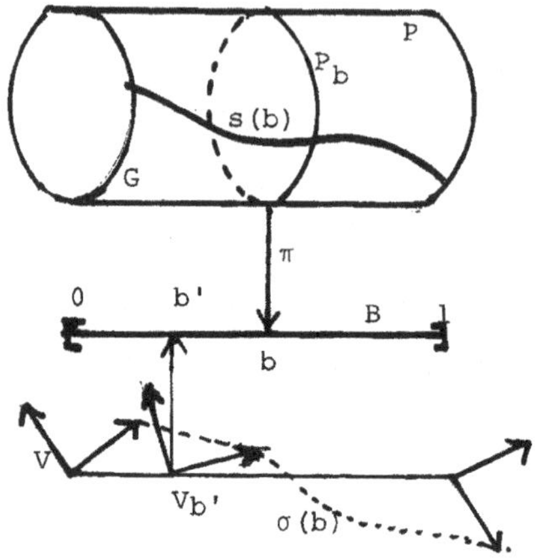

Fig. 1

represents a section s(b). The section shows us "how the group identity" varies along the bundle space, and the identification of s(b) with the base space realizes the trivialization we talked about (roughly speaking, the section tells us which way we have sectioned the cylinder to make it into a rectangle, the cartesian product [0, 1] ×SO(2)). The lower string of "vector spaces" symbolizes a vector bundle (with two dimensional fiber) which is associated to our boundle (in this case the illustration is oversimplified, since we have represented the group as a circle; in a manner of speaking we change the frame along the base space). Had we identified the endpoints 0 and 1 of the base space the section would no longer be continuous (but contiuous sections still exist on the torus obtained in this manner).

b) Let G be SU(2) and B = M_4 be Minkowski space. Then the trivial p. f. (M_4 × SU(2), SU(2), M_4, pr_1) is the principal fibration of the Yang-Mills gauge theory. Any p. f. isomorphic to it will be trivialized by a global section, which we will call a <u>gauge</u>, but every gauge leads to a different trivialization. Let V be the 2-dimensional complex space of the lowest-dimensional faithful representation of SU(2) (the isospinor space); in the manner described above we can form the vector bundle of type V associated to the Yang-Mills bundle. A section of this bundle will be a classical isospinor field (a quantized field will be defined in Section 4 as an operator-valued distribution on infinitely differentiable, rapidly decreasing sections of the vector bundle. A choice of section in the isospinor bundle produces local or global trivializations, which allow us to proceed in the manner familiar to physicists.

3.6. Connections in Principal Fibrations are devices which allow us to compare tangent vectors to fibers above different points in a p. f. In the case of field bundles to be considered in Section 4 (the field bundle is associated to the p. f. with the gauge group as structure group), the connections in associated vector bundles discussed in subsection 3.7, will allow is to define a covariant differentiation and the gauge field.

There are three equivalent ways of defining a connection in a p. f. (also called a principal connection):

i) A connection in the principal fibration (or principal connection (when no danger of confusion, simply: connection)) $\lambda = (P, G, B, \pi)$ is a C^{∞}-family of subspaces H_p in T_p ($p \in P$), the tangent space to the bundle space at p, called horizontal subspaces, such that

$$1) \quad T_p = H_p \oplus V_p,$$

where V_p is called the space of vertical tangent vectors (tangent vectors to the fibers), and

$$2) \quad H_{p \cdot g} = H_p \cdot g,$$

where the right-hand side denotes the right action of the group element g on the tangent vector in H_p, i. e. the subspaces H_p are right-invariant.

This definition is understood very easily in terms of a chart, i. e., restricting the fibration to an open subset U of B, where the bundle is isomorphic to $U \times G$. We can then map the tangent space $T_b(B)$ into $T_{(b,g)}(U \times G)$ in such a way that the latter splits into a horizontal component $H_{(b,\ g)}$ and a vertical component $V_{(b,\ g)}$ supplementary to it. The mapping C which maps $T_b(B) \times TG$ into $T_{(b,\ g)}(U \times G)$ lifts any tangent vector to the base space into a horizontal vector in $T_{(b,\ g)}$ and satisfies the conditions imposed by 1), 2) (cf.[12], Sec. 20.2.2, for details). In particular, $C((b,k),(b,g)) = ((b,k), P(b,s)k)$, for $(b,g) \in U \times G$ and k a vector in $T_b(U)$, where $P(b,g)$ is a homomorphism of the vector space $T_b(U)$ into $T_g(G)$. For later reference, we denote by Q(b) the homomorphism $P(b, e)$, where e is the identity of the group G. Obviously, Q(b) is a homomorphism of the tangent space $T_b(B)$ into the Lie algebra $g = T_e(G)$.

ii) Let $T_p^*(P)$ be the cotangent space of P at p, i. e., the space of all linear forms on T_p. The space V_p^* is the set of all linear forms which vanish on H_p. We can then define a connection as giving in place of H_p the annihilator V_p^*. But the space V_p is tangent to the fiber, which is isomorphic to the Lie group and hence is itself a Lie algebra. The elements of V_p^* are then Lie-algebra valued 1-forms (if $X_1, \ldots, X_n$ are the basis vectors of a Lie algebra g, and $\omega^1, \ldots, \omega^n$ are n one-forms , the Lie-algebra valued form ω is defined by $\omega = \sum X_a \otimes \omega_a$, such that the commutator bracket becomes a sum of tensor products of Lie brackets of the X-s and wedge products of the ω-s). Thus a <u>connection</u> can be defined by giving a Lie-algebra valued one-form ω which on the fiber P_b reduces to $g^{-1}dg$. In an open set U of B around b we have then

$$\omega(p) = g_U^{-1}dg_U + ad(g_U^{-1})\theta_U(b, db) \qquad (3.8)$$

where $b = \pi(p)$, $\theta_U(b, db)$ is a g-valued one form on U (the Q(b) described above) and

$$\omega(p\cdot g) = ad(g^{-1})\ \omega(p). \qquad (3.9)$$

iii). The third definition of a connection is closest to the one familiar to physicists from general relativity, which tells us that a connection coefficient (Christoffel symbol) does not transform "as a tensor", i. e., under the adjoint representation of the structure group but that one has to add the "derivatives of the transformation coefficient"; translated into our language, a coordinate transformation is a change of chart, hence the third definition tells us how a connection behaves under a change of chart. Let U, V be two overlapping open sets in B, such that $b \in U \cap V \neq \emptyset$. In $U \cap V$ we have a transition homomorphism $\psi_{UV} \in G$. Then in $\pi^{-1}(U \cap V)$ the expression (3.8) becomes

$$\theta_U = \psi_{UV}^{-1}d\psi_{UV} + ad(\psi_{UV}^{-1})\theta_V. \qquad (3.10)$$

In coordinates the 1-form θ_U has the expression

$$\theta_U(x,\ dx) = \Gamma^{(U)}_{\mu\nu} x^\mu dx^\nu \tag{3.11}$$

with summation over repeated indices. The coordinate in the Lie algebra has not been specified , therefore only two indices appear on the connection coefficient in place of the familiar three (in fact the usual connection coefficient is in the tangent bundle of the Riemannian manifold, rather than in the pricipal bundle -- which in this case is called the frame bundle; subsection 3.7 treats that case).

The horizontal projection (made possible by the existence of a connection) of the exterior differential of a q-form in P is a q + 1-form called the exterior covariant differential and denoted by D. The exterior covariant differential for the 1-form ω is defined by

$$\Omega = D\omega = d\omega + \tfrac{1}{2}[\omega,\ \omega] \tag{3.12}$$

respectively for θ_U

$$\Theta_U = D\theta_U = d\theta_U + \tfrac{1}{2}[\theta_U,\ \theta_U]. \tag{3.13}$$

The Lie-algebra valued 2-forms defined by these equations are called the curvature forms of the connection and the equations themselves the structure equations (when ω is the canonical 1-form ω= de of the Lie algebra $\Omega = 0$ and the structure equation reduces to the Maurer-Cartan equation $d\omega + \frac{1}{2}[\omega,\ \omega] = 0$).

Under a change of chart Θ_U transforms under the adjoint representation ("as a covariant tensor of rank 2" in the older terminology)

$$\Theta_U = ad(\Psi^{-1}_{UV})\Theta_V. \tag{3.14}$$

(In coordinates the relation between Θ and θ is given by an expression of the form (2.40), i. e. the Yang-Mills-Utiyama relation between field and potential.)

The exterior covariant differential of Ω vanishes, yielding the Bianchi identity (Faraday part of the Maxwell or Yang-Mills equations)

$$D\Omega = d\Omega + [\omega,\ \Omega] = 0 \tag{3.15}$$

which implies

$$D([\omega,\ \omega]) = 0. \tag{3.16}$$

3.7. Linear Connections in Associated Vector Bundles allow us to compare vectors at different points in the base space and thus to define covariant derivatives of sections of the bundle. Since the definition is essentially local, we shall carry out the construction in a chart.

Let $\lambda = (P, G, B, \pi)$ be a p. f., with a principal connection ω. Let E be a vector bundle associated to λ with typical fiber F: $E = \bar{\rho}\cdot(P \times^G F)$. We want to define a connection in E which will be called a linear connection .

Let $P: T(B) \times_B P \to T(P)$ be the mapping defined in the last paragraph on page 21 (here $T(B) \times_B P$ denotes the fibered product, i. e., the subset of the cartesian product which is mapped into the same point of the base space by the two projections -- T(B) being the tangent bundle of B; P was defined in charts on p. 21), mapping which defines the splitting of $T_{(b, g)}(P)$ into the horizontal and vertical subspaces. For each point $b \in P$ and vector $u_b \in E_b$ there exists an element $p_b \in P_b$ and a vector $f \in F$ such that $u_b = \bar{\rho}(p_b, f)$, where $\bar{\rho}$ is the frame map. For a tangent vector $k_b \in T_b(B)$ we define

$$C_b(k_b, u_b) = \bar{\rho}(P_b(k_b, p_b), f) = \tilde{P}(b,g)\cdot f \quad (3.17)$$

where $P(b, g)$ is a homomorphism of the vector space $T_b(B)$ into $T_g(G)$. $C_b(k_b, u_b)$ is independent of the choice of the pair (p_b, f) such that $\bar{\rho}(p_b, f) = u_b$, since any other such pair is of the form $(p_b\cdot g, r(g^{-1})f)$.

If π_E denotes the projection of E on B, we have $\pi_E(\bar{\rho}(p, f)) = \pi(p)$, hence

$$T(\pi_E)C_b(k_b, u_b) = T(\pi)P_b(k_b, p_b) = k_b. \quad (3.18)$$

It is easy to check that the mapping $u \mapsto C_b(k_b, u_b)$ is a linear mapping of E_b into $(T(E))_{k_b}$ since $k_b \mapsto P_b(k_b, p_b)$ is linear from $T_b(M)$ to $T_{p_b}(P)$, and $k_b \mapsto C_b(k_b, u_b)$ is linear from $T_b(M)$ to $T_{u_b}(E)$. In a chart

$$C_b((b,k), (b,f)) = ((b,f), (k,-\Gamma_b(k,f))) \quad (3.19)$$

where

$$\Gamma_b(k, f) = -\rho_*(Q(b)\cdot k))\cdot f \quad (3.20)$$

where ρ_* is the inverse image of ρ in the associated bundle and is a homomorphism of g (the Lie algebra of G) into the Lie algebra $g\ell(F)$ of the typical fiber, and Q(b) is defined on top of p. 22.

In components we have

$$\Gamma_b^\alpha(k, f) = \sum_{\beta,\gamma} \Gamma_{\beta\gamma}^\alpha(b)k^\beta f^\gamma, \tag{3.21}$$

i. e., the connection is given by the connection coefficients in the right-hand side, which under a change of chart transform like the familiar Christoffel symbols of a Riemannian manifold.

<u>3.8. Covariant Derivatives of Mappings and Sections</u> in a vector bundle E of typical fiber V, associated to a principal fibration λ = (P, G, B, π) can be defined as follows. Let M be a differential manifold and let S: M → E be a C^∞-map. Let $z \in M$ and let $h_z \in T_z(M)$ be a tangent vector, approximating a displacememnt in M. The tangent vector $T_z(S)\cdot h_z$ in E has both horizontal and vertical components. We may use the connection C in E to extract from it the horizontal component, and thus to be able to compare vectors tangent to the fiber E_b at the same point (it is useful to consider S as a lifting to E of a map f of M into B). The vector

$$T_z(s)\cdot h_z - C_{\pi_E(S(z))}(T_z(\pi_E\circ S)\cdot h_z, S(z)) \tag{3.22}$$

is in the tangent space $T_{S(z)}(E_{\pi_E(S(z))})$ to the fiber through S(z) of E. The fiber being a vector space there is a canonical linear bijection $\tau_{S(z)}: T_{S(z)}(E_{\pi_E(S(z))}) \to E_{\pi_E(S(z))} = V$ (cf. [12], Sec. 16.5.2) which identifies the tangent to the fiber with the fiber. This finally yields a vector in the fiber denoted by

$$\nabla_{h_z}\cdot S = \tau_{S(z)}(T_z(S)\cdot h_z - C_{\pi_E(S(z))}(T_z(\pi_E\circ S)\cdot h_z, S(z))) \tag{3,23}$$

called the <u>covariant derivative of the map S at z relative to the connection C, in the direction of the vector h_z</u>.

Any (local) section S(b) of the bundle E can be expressed in terms of a (local) section s(b) of the principal fibration as follows:

$$S(b) = s(b)\cdot\zeta(s(b)) \tag{3.24}$$

where ζ is a C^∞-map of a neighborhood of $p_b = s(b)$ into $E_b = \rho V$. The covariant derivative of the section S(b) in the direction h_b is then defined by (3.23), with z replaced by b, h_z replaced by h_b , and owing to the fact that S is a section $T_z(\pi_E \circ S)$ becomes the identity map in $T_b(B)$. $T_b(S)\cdot h_b$ can further be calculated by "differentiating" (3.24) according to the Leibniz formula for tangent mappings. Denoting $k = T_b(s)\cdot h_b \in T_{p_b}(P)$, with $T(\pi)\cdot k = h_b$ and noting that $T_{s(b)}(\zeta).k$ is in $T_{\zeta(s(b))}(E_b)$ and equal to $\tau^{-1}_{\zeta(s(b))}(\theta_k\cdot\zeta)$, where θ_k is the Lie derivative of the function ζ in the direction of the vector k, we obtain (with a slight abuse of notation) the <u>first fundamental formula</u>

$$\nabla_{h_b}\cdot S = \bar{\rho}(\theta_k\cdot\zeta + \tau_\zeta((\omega(s(b)).k)\cdot\zeta(s(b)), \tag{3.25}$$

where ω is the connection one-form (3.8), which replaces the principal connection (the latter being introduced via Eq. (3.17) in place of C).

It is easy to check that ∇ is a derivation, and satisfies rules of the Leibniz type (for scalar functions it reduces to the Lie derivative). Furthermore, one can introduce the curvature associated to the connection C via the covariant derivative (cf. [12], section 17.20).

<u>3.9. Holonomy Groups and the Ambrose-Singer Theorem</u> will play some role in our further discussions. We recall some simple facts about paths and loops in manifolds.

A <u>path</u> in a manifold X is a piecewise smooth mapping of the interval [0, 1] into X, $t \mapsto \gamma(t)$.A <u>loop</u> is a path γ such that $\gamma(0) = \gamma(1)$. The set of loops through a point x ($\gamma(0) = x$) can be endowed with a group structure if we define "multiplication" of loops by going around the first loop followed by going around the second one (with an appropriate change of scale in the parameter t) and the inverse by reversing the sense in which a loop is traversed.

If $\lambda = (P, G, B, \pi)$ is a principal fibration with connection ω (for brevity we denote a connection by its Lie-algebra-valued one-form) and E is an associated vector bundle, then any path $\gamma(t)$ in B can be lifted to a horizontal path z(t) in P (i, e., there is a unique such path which is projected into γ by the bundle projection and whose tangent vectors are all horizontal). Moreover, the path γ and the connection ω determine an isomorphism of the fiber $E_{\gamma(0)}$ onto the fiber $E_{\gamma(1)}$ called <u>parallel displacement</u> of vectors. In particular, if γ is a loop, the two fibers actually coincide, and the horizontal lift induces in P a fiber-isomorphism $[\gamma]:P_b \to P_b$ of the fiber over $b = \gamma(0)$ onto itself, respectively a linear isomorphism of E_b onto itself. The composition law of loops then implies that the set of all such isomorphisms is a group called the <u>holonomy group of the connection at b</u>, and denoted by K_b. If the loops are contractible to zero at b we obtain a subgroup of K_b, called the restricted holonomy group, K_b^0. If b_1 and b_2 are two point in B which can be joined by a path, the holonomy groups at these points are isomorphic; therefore if B is connected it makes sense to speak of <u>the holonomy group</u> of the fibration. If in addition B is paracompact (admits locally finite coverings) then K_b is a Lie subgroup of the structure group G, K_b^0 is a normal subroup and the quotient $K_b/K^0{}_b$ is countable [27, 27a, 41].

The Lie algebra of K_b is called the holonomy algebra of the connection ω. The <u>holonomy theorem of Ambrose and Singer</u> establishes a 1 - 1 correspondence between the holonomy algebra k_b and the curvature two-form Ω: the holonomy algebra k_b is equal to the subspace of g (the Lie algebra of the structure group G) spabbed by all elements $\Omega_u(X,Y)$, where u is an element of the "holonomy bundle" (the set of all points in P which can be joined to a point by a horizontal curve-- it is a fibration with K_b as structure group) and X and Y are horizontal vectors at u. The theorem shows that if $\Omega = 0$, i. e., the bundle is <u>flat</u>, and B is simply connected , the holonomy group is trivial.

For further details on holonomy groups and the Ambrose-Singer theorem we refer the reader to the literature quoted at the beginning of this section.

4. QUANTIZED CONNECTION FORMS

In this section we attempt to arrive inductively at a set of axioms modeled on the Gårding-Wightman axioms for fields[18], allowing us to incorporate connection and curvature forms associated to gauge invariance into quantum field theory. This will hopefully convince us that the new quantized objects, whether they are operator-valued multipliers of operator-valued distributions or a representation of the Lie-algebra valued forms as morphisms of the operator fields, are objects of a different nature than the quantized fields associated to the charged particles, and therefore it should not be too surprising that one is led to a new quantization scheme, involving indefinite metric, and possibly other unphysical states known as Faddeev-Popov ghosts[14].

There are two basic properties of the gauge fields as known in the heuristic approach which we would like to see emerging from the axiomatic treatment, namely the masslessness of the gauge field (i. e., the fact that the spectrum of the generators fo space-time translations for these objects has support on the light cone) and the fact that the Maxwell or Yang-Mills equations satisfied by the connection and curvature forms are to be interpreted as valid for expectation values in physical states, rather than as operator equations.

At the end of the section we sketch an alternative approach, by noting that some of these results can be reformulated in terms of the Wightman distributions (vacuum expectation values of products of field operators) with Ward-like identities expressing the properties of gauge-covariant differentiation. We also make some remarks on what it would take to incorporate these quantized forms into an algebraic approach to quantum field theory.

We first recall the Gårding-Wightman approach to a field which transforms under an irreducible representation D(G) of a compact "internal" symmetry group (i. e., a group whose action on the fields commutes with that of the Poincaré group).

For simplicity the fields φ_α transforming under the representation D will be assumed to be Lorentz scalars (generalization to other fields introduces only notational complications). In naive field theory the unitary representation on the Hilbert space which implements the finite dimensional unitary representation D(G) has the following effect on the field operators $\varphi_\alpha(x)$:

$$U(g)\varphi_\alpha U(g)^{-1} = \sum_\beta D_{\alpha\beta}(g)\varphi_\beta. \tag{4.1}$$

In dealing with a multi-component field as an operator-valued distribution it is convenient to introduce multi-component test functions in the spirit of Schwartz' vector-valued distributions (we have done this in Section 2, when the test function was allowed to become complex, cf. Eq.(2.12)). We therefore consider complex d-dimensional test-function vectors $f_\alpha(x)$, each belonging to the adopted test function space $\mathcal{D}$ or $\mathcal{S}$ and transforming under the d-dimensional representation D(G) of G, and define the operator-valued distribution (the last expression being of heuristic value only):

$$\varphi(f) = \sum_\alpha \varphi_\alpha(\bar{f}_\alpha) = \sum_\alpha \int \varphi_\alpha(x)\bar{f}_\alpha(x)d^4x, \tag{4.2}$$

where the complex conjugate $\bar{f}$ has been introduced for convenience. We then "transpose the action" of D onto $\bar{f}$:

$$\begin{aligned} \sum\int D_{\alpha\beta}\varphi_\beta(x)\bar{f}_\alpha(x)d^4x &= \sum\int \varphi_\beta \overline{D_{\alpha\beta}(g)\bar{f}_\alpha(x)}\, d^4x \\ &= \sum\int \varphi_\beta[\bar{D}_{\alpha\beta}(g)f_\alpha(x)]d^4x \\ & \sum_{\alpha\beta}\int \varphi_\beta[D(g^{-1})]_{\beta\alpha}\bar{f}_\alpha(x)d^4x, \end{aligned} \tag{4.3}$$

where we have used the unitarity of D.

Hence we adopt as a definition for the transformation property of $\varphi(f)$

$$U(g)\varphi(f)U(g)^{-1} = \varphi(D^T(g^{-1})\bar{f}), \tag{4.4}$$

where D^T is the transpose of the matrix D.

Next we replace the functions f by cross sections of a d-dimensional vector bundle associated to the principal fibration $\lambda = (P, G, M, p)$ by means of the representation D^T of the group G. Any such section f can be considered as the "pullback" of a section $s(x)$ of λ via the relation

$$f(x) = s(x)\zeta(s(x)), \qquad (4.5)$$

where ζ is a C^{∞}-map of a neighborhood of $s(x) = p_x \in P_x$ into the fiber E_x of the representation-bundle (or into the vector space of the representation D, to which each fiber of the associated bundle is isomorphic). In particular, the support properties of $f(x)$ will carry over to the mapping ζ (the vanishing of the section $s(x)$ does not make sense unless G is a vector-group). For simplicity, since we will have problems of multiplication, we assume that the test-sections $f(x)$ belong to the space $\mathcal{D}(M_4, E)$, where E denotes the associated vector bundle.

Since the group G acts on the sections $s(x)$ by right multiplication we can represent the action of a section $s'(x)$ of λ on a section $f(x)$ of E by a "point-dependent" representation $D^T(s'^{-1}(x))$:

$$(s'\cdot f)(x) = D^T(s'^{-1}(x))f(x) = s'(x)s(x)D^T(s'^{-1}(x))\ \zeta(s's),$$

where the product and inverse of a section is defined pointwise.

We now postulate the following action of a section $s(x)$ of λ on the operator-valued distribution $\varphi(f)$ (the notation in the left-hand side is symbolic, since we do not know whether there exists a unitary operator $U(s)$ implementing the action; if the principal bundle is trivial, we may assume the section $s(x)$ to deviate from the identity section $e(x)$ only in a small neighborhood of x, and then (4.4) implies (4.6); for nontrivial bundles this point requires more careful consideration, unless we take s to be highly localized):

$$U[s]\varphi(f)U[s]^{-1} = \varphi(s^{-1}(x)\cdot f(x)) = (s\cdot\varphi)(f). \qquad (4.6)$$

Since f is of compact support, the middle term should be well defined for local or global sections s (which as always are assuned infinitely differentiable). Note that in a certain sense $U[s]$ is a functional of the section s, fact which is symbolized by the square bracket.

If in the absence of the gauge group G the field φ satisfied a linear partial differential equation (e. g., the Klein-Gordon or Duffin-Kemmer equations, appropriate for a scalar field), this will no longer be true, since the action of a section (i. e., gauge transformation of the second kind) does not commute with differentiation (i. e., space-time translation). We must therefore replace the derivatives ∂_μ or the exterior differentials of forms, d, by covariant derivatives ∇_μ or the covariant exterior differentials associated to the connection of the principal fibration.

Since it is difficult to see how to define these operations directly for the operator-valued distributions on cross sections of the field bundle, we use the traditional distribution-theoretic approach, "transposing" the operations from test functions (or test cross sections) to the operator-valued linear forms defining the quantized fields. We often use the heuristic representation (4.2), but for economy of space do not write things down in detail here.

Our starting point is the first fundamental formula (3.25), but by a slight abuse of notation we will omit the frame map $\bar{\rho}$ and the tangent mapping τ_ζ. We will also make use in intermediate steps of the calculations of a "moving frame" in E, i. e., we expand a section f(x) in terms of the linearly independent sections

$$e_\alpha(x) = D^T(s^{-1}(x))_{\alpha\beta}e_\beta, \qquad (4.7)$$

where e_β is a fixed basis for the framing representation of the associated vector bundle of test-objects, and s(x) is a (local) section of the principal fibration.

In terms of this moving frame, the connection coefficients can be written as matrix-valued one forms, so that if

$$f(x) = \sum_\alpha c_\alpha e_\alpha(x), \qquad (4.9)$$

the covariant derivative of the section f in the direction x^μ can be written in the form

$$\nabla_\mu f(x) = \sum\{\partial_\mu c_\beta(x)e_\beta(x) + c_\beta(x)T_\mu^{\beta\alpha}e_\alpha(x)\}, \qquad (4.10)$$

where T_μ is the connection matrix in the moving frame and summation is from 1 to d over repeated indices. The $c_\beta(x)$ are C^∞-functions of x with appropriate support or decrease properties.

We now define the covariant derivative of the field $\varphi(f)$ in the usual manner of distribution theory, i. e., by "transposing" the operations to test-sections:

$$(\nabla_\mu \varphi)(f) = \varphi(-\nabla_\mu f) = \sum \int \varphi_\alpha(x)[-\nabla_\mu f_\alpha(x) - T_\mu^{\alpha\beta}(x) f_\beta(x)]d^4x$$

$$:= (\partial_\mu \varphi)(f) + (Y_\mu \cdot \varphi)(f), \qquad (4.11)$$

where the last expression Y_μ is the coordinate expression of the 1-form Y which we call <u>quantized connection form</u> (or quantized Yang-Mills field, hence the notation Y; we will use M for the field 2-form M = DY, where D is the covariant exterior differential). The precise meaning of the form Y requires further investigation. Since it is the action on the operator φ of the Lie-algebra valued form T which acts on the test-sections, and hence belongs to the multiplier-class of the test-section space chosen, it is clear that **Y** and M are not simply operator-valued distributions (or more correctly, operator-valued currents, since they are differential forms), at least not on the same test-section space as used for the field. We have the alternative of interpreting Y (and M) as either an operator-valued multiplier of the distributions describing the charged field, or as an operator-morphism, mapping the operator φ into a new operator $Y \cdot \varphi$. In addition, the "operator" Y is a representation of the Lie algebra g of the structure group of the gauge fibration, hence in addition to the domain problems posed by having Y "act" on the fields, there arise the usual domain problems associated with representations of Lie algebras by unbounded operators, Gårding domains and the appropriate analytic vectors, etc. We postpone a detailed discussion of these finer points to another publication, listing here a conceivable set of postulates for the forms Y, M constructed by analogy with the Gårding-Wightman axioms for the interacting electromagnetic field, as abstracted from perturbation theory by Strocchi and Wightman[40]. A more detailed analysis may show that some of these postulates have to be modified (e. g., in the sense that the forms Y and M cannot be defined as operators, but only as sesquilinear forms).

1. There exists a Hilbert space H with a distinguished sesquilinear form $\langle\cdot,\cdot\rangle$, with respect to which the representation $U(P_+^\uparrow)$ of the Poincaré group and the representation $U(G)$ of the gauge group is unitary.

2. The field φ is defined as an operator-valued distribution on an appropriate space of test-sections of the associated bundle, and the quantized connection forms Y and M = DY, are operator-multipliers in a sense which has to be made more precise. The current-operator is defined as the exterior covariant derivative of the dual $^*M$ of the 2-form M (see below for the correct interpretation of this):

$$D^*M = {}^*J \tag{4.12}$$

and is an operator-valued distribution (or a sesquilinear form) which depends on the "charged" fields. It too is a representation of the Lie algebra of the gauge group.

3. There exists a distinguished subspace $H' \subset H$ on which the sesquilinear form is nonnegative; let H'' denote the subspace where the sesquilinear form vanishes. Then the physical Hilbert space is the quotient

$$H_{phys} = H'/H''. \tag{4.13}$$

4. There is a common dense domain $\mathcal{D} \subset H'$ which is invariant under the representations U, and under the action of observables such as M, J. The Yang-Mills (Maxwell) equations are valid as expectation-value equations in physical states:

$$\langle \Phi, DM\Psi \rangle = 0, \tag{4.14}$$

$$\langle \Phi, [D^*M - {}^*J]\Psi \rangle = 0, \tag{4.15}$$

for all vectors $\Phi, \Psi \in H'$ and Ψ is in the domain of M and J (or, in the alternate interpretation (4.14) and (4.15) are to be interpreted as sesquilinear forms).

5. There exists a unique vector Ω in H', invariant under the translation subgroup of the Poincaré group and cyclic for the fields Y, J, φ.

6. The spectral condition, either in the usual form (the Fourier

transforms of $\langle \Phi, M\Omega \rangle$, $\langle \Phi, {}^*J\Omega \rangle$ $\langle \Phi, \varphi\Omega \rangle$, etc. have support in the closed forward lightcone $\bar{V}_+$) or in the strengthened form (iv') of Strocchi and Wightman [40].

We do not know whether these postulates are compatible and whether there exists a nontrivial model satisfying them, but the perturbation expansions of nonabelian gauge theories [1] and quantum electrodynamics in the various gauges, lead us to believe that they are as plausible as the corresponding axioms for quantum electrodynamics, on which they are modeled. A detailed study of these axioms is in progress. It should be noted that the uniqueness of the vacuum may not hold in the presence of symmetry breakdown. The whole problem of symmetry breakdown and Higgs-Kibble mechanism (which seems to be associated to the fact that the gauge fields are connection forms) requires careful and detailed study.

Before closing this section, I wish to outline an alternate approach, based on the Wightman functions (Wightman distributions) and Ward-Takahashi identities. Let $W(x_1, \ldots, x_n)$ denote the Wightman function associated to the fields φ. We have omitted the "internal" indices, which show that under the representation of the gauge group W transforms as a tensor product of n fields f. Subjecting each field φ in the vacuum expectation value W to the transformation (4.4) one can see how W transforms. Replacing the fixed transformation (4. 4) by a "local" gauge transformation (4.6) we obtain the tensor action of a section s on W.

The gauge-covariant derivative of W with respect to one of its "coordinates" will raise the order of the Wightman function, much in the same way as the insertion of an electromagnetic vertex into a T-ordered product of Dirac fields leads to the appropriate vertex function in perturbation theory (cf., e. g., the derivation of the Ward identity in the books [7, 30]). Ultimately, one obtains (the divergence of) the Wightman function involving the n original fields plus the field Y. Invoking a nuclear theorem one is then led to the operator Y.

5. HOLONOMY GROUPS AS INTERNAL SYMMETRY GROUPS

This section discusses a generalization of a proposal by Loos [29] to consider the "internal holonomy group", i. e., the holonomy group associated to the gauge connection, as the symmetry group of a field theory. The heuristic reasoning behind this is the following. If we transport parallel a frame (a set of linearly independent sections) in the bundle E around a closed loop which is a lift of a spacelike path in the underlying manifold (Minkowski space; if there are no "magnetic monopoles" such loops are always contractible) the frame undergoes a linear transformation. The totality of these linear transformations is a representation of the restricted holonomy group at the point x, which is a subrepresentation of D(G). The holonomy groups at different points of a connected base-space are isomorphic to one another; we can therefore speak about the holonomy group of the principal fibration.

The Ambrose-Singer theorem stated in Section 3 tells us that the holonomy group is closely related to the curvature form of the connection, consequently, for regions with nonvanishing expectation values of the curvature form M the holonomy group is nontrivial.

On the other hand, even if the gauge group of the principal fibration is nonabelian (as seems to be necessary in the gauge theories popular in physics), the holonomy group (which is a Lie subgroup of G) may turn out to be abelian, i. e., its irreducible unitary representations are one-dimensional (what Yang calls a "nonintegrable phase factor [52]). This may turn out to be a very desirable feature, which adds weight to the choice of the holonomy group as the actual physical "internal symmetry group", for the following reasons.

If the holonomy group is abelian, the curvature field M must be Maxwellian, rather than of the Yang-Mills-Utiyama type, in agreement with the experimental fact that no nonabelian vector fields have been observed (there may be other reasons for this, and this alone would not be sufficient reason for adopting the present viewpoint).

Furthermore, it has been proved in the algebraic framework [13] that nonabelian internal symmetries (which, when exact, are tantamount to parastatistical behavior of the fields, like the color degeneracy of quark models) are incompatible with "duality" (more precisely: duality for the field algebra requires that the irreducible representations of the gauge group -- we shall say holonomy group -- be one-dimensional).

Here duality means roughly the following: Let $F(O)$ be the field algebra (von Neumann algebra generated by the unobservable fields) of a double cone O in Minkowski space and let O' denote the causal complement of O, i. e., the set of all points in Minkowski space that are spacelike to O. Then duality means that $[F(O)]' = F(O')$, where the first prime denotes the commutant of the von Neumann algebra. This property was known for free fields from the early work of Araki and has recently been proved to hold for a scalar field theory satisfying the Wightman axioms by Bisognano and Wichmann[4]. It is therefore tempting to conjecture that duality holds (at least in the vacuum superselection sector) for all quantum field theories.

However, this would deprive us of the possibility of having gauge groups which are nonabelian, unless we interpret the Doplicher-Haag-Roberts theorem quoted above to refer to the holonomy group, rather than the gauge group itself.

We are thus led to the tempting conjecture: The duality requirement on the field theory forces the holonomy group of the gauge connection to be abelian, whereas the structure group G itself may remain nonabelian (and observable only through the multiplet structure of the fields; the latter do not have to have the same masses within the same multiplet, since the action of the translation group commutes only with the holonomy group, thus solving an old puzzle). The "gauge fields" are either Maxwellian (as required by the Ambrose-Singer theorem) or they are somehow "confined", so that there are no macroscopic regions with nonvanishing Yang-Mills fields. The details of this picture need still to be filled in.

Finally, some remarks about gauge theories in an algebraic approach to field theory. Here the basic object is a net of C*-algebras $A(\mathcal{O})$ labeled by double cones $\mathcal{O}$ in Minkowski space and containing all the observables of the theory and associated nets of "unobservable" fields $F(\mathcal{O})$ which are von Neumann algebras obtained from the algebras $A(\mathcal{O})$ by the action of localized morphisms. At first it seems that the quasi-local structure of the net $F(\mathcal{O})$ and the gauge bundle concept are incompatible, and that therefore the algebraic framework is not suited for a discussion of gauge fields. For detailed definitions, cf. [13].

On the other hand, various generalizations are open. One of these could be the following. Assume that the quasilocal structure holds only in the tangent space to the base space of the principal and associated bundles, so that in each tangent space we can construct the quasilocal observable and field algebra nets. By using the projection from the tangent bundle and liftings into the tangent spaces at various points of the underlying manifold, one can then construct morphisms between the quasilocal nets associated to tangent spaces, morphisms which play the role of connections, since they allow us to compare quantities at various points. One is naturally led to the concept of curvature (the mismatch in circling an infinitesimal loop) and holonomy group (the automorphism of the quasilocal structure induced by transport along all closed loops starting at a point). These morphisms might play the role of the gauge fields in the algebraic approach.

The same approach might also be useful in attempts to formulate quantum field theory (or statistical mechanics) on a Riemannian manifold with prescribed metric (e. g., statistical mechanics in a strong gravitational field). This topic is currently under investigation by one of my students (M. Kovacich, forthcoming Ph. D. thesis).

6. CONCLUDING REMARKS

In this lecture I have tried to formulate a quantum theory of gauge fields by treating them as "quantized differential forms" (this term was coined by Segal [38] in a somewhat different context; it should be noted that Segal was among the first to advocate a geometric approach to the theory of quantization of nonlinear systems and that many current ideas on quantization can be found in his early papers on nonlinear systems [38]).

There are several shortcomings of this presentation which hopefully will be removed in future, more carefully prepared, publications. However, it should already be clear that the geometric nature of gauge transformations and the fact that the gauge fields are representations of Lie-algebra valued differential forms, has a deeper significance than originally suspected. I am convinced that the peculiarities of gauge fields (such as the need for Bleuler-Gupta quantization or the appearance of Faddeev-Popov ghosts or Higgs bosons) will prove to be a direct consequence of their geometric nature. A detailed analysis of the Higgs mechanism for the suppression of the Goldstone bosons and for the appearance of masses for the gauge particles (cf.,e. g., [3]) will show that the fact responsible for this is the connection nature of the "vector potentials". A rigorous and detailed examination of these problems within the framework of axiomatic field theory is very important and will form the subject of further publications.

Finally, a word about general relativity. Utiyama and others [25, 42, 46] have interpreted the Christoffel symbols as the gauge potentials associated to the gauge bundle with the Lorentz group as structure group. Many attempts at quantization of general relativity (Arnowitt-Deser-Misner, De Witt, Faddeev-Popov, etc.; cf. the book "Gravitation" by C. Misner, K. Thorne and J. A. Wheeler for an exhaustive list of references) were based on treating the Christoffel symbols as operators satisfying canonical commutation relations and additional conditions, at least in the linear approximation. A successful completion of the quantization program outlined in this lecture for the case of Yang-Mills fields, where the gauge fields are interpreted as operator-multipliers representing the connection one-form and its curvature two-form, might yield important techniques which could be applied in attempts to treat the Christoffel symbols and the curvature tensor of general relativity in a similar vein.

The remarks made about the possible algebraic approach to gauge theory at the end of Section 5, if successful within that framework, might open up a door towards an algebraic treatment of field theory in the framework of general relativity: the underlying Riemannian structure of the space-time manifold should emerge as a result of treating the connection and curvature of the latter in the same manner as one treats the connection forms associated to other gauge bundles. It is too optimistic to hope that this simplistic idea will actually solve the problem, but it might provide a useful testing ground which needs to be explored further. As a first attempt in this direction we are investigating the formulation of statistical mechanics (in terms of local algebras of observables) in the tangent bundle of a Riemannian manifold (such as the 3-space in a comoving frame of a prescribed metric). There too the concepts of connection and holonomy seem to play a very important role.

POSTSCRIPT (November 10, 1975).

After this manuscript was concluded, Professor C. N. Yang kindly sent me some preprints of joint work with T. T. Wu [52] which discuss in greater detail the concept of "nonintegrable phase factors". Wu and Yang establish a correspondence between physical and geometrical terminology which is close to that espoused here, but dealing with nonquantized connection forms.

Moreover, the "nonintegrable phase factors" used by Wu and Yang are in fact representations of the holonomy group in the field bundle. In addition to a detailed discussion of electromagnetism and magnetic monopoles (in which case the base space is no longer simply connected and the gauge bundles become nontrivial) they also discuss the phase factors and monopoles of the Yang-Mills theory, based on both SO(3) and SU(2) as structure groups. The interesting conclusion reached by Wu and Yang is that there are infinitely may types of monopoles for the U(1) gauge group (electromagnetism), one type for SU(2) and 2 types for the SO(3) group.

It is remarkable that the number of "types of monopoles" (i. e. types of noncontractible loops in the base space) is identical to the order of the fundamental group of the three structure groups considered (the fundamental group is Z for U(1), Z_2 for SO(3) and trivial for SU(2) cf. [12]). I feel that this cannot be a coincidence and requires further investigation. It is yet another illustration of the interweaving of physics and geometry!

BIBLIOGRAPHY

[1] E. Abers and B. W. Lee, Gauge Theories, Physics Reports 9, No 1, 1 -- 141 (1973).

[2] R. L. Arnowitt and S. I. Fickler, Phys. Rev. 127, 1821 (1962).

[3] J. Bernstein, Spontaneous Symmetry Breaking, Gauge Theories, Higgs Mechanism and All That, Rev. Mod. Phys. 46, 1 (1974)

[4] J. Bisognano and E. Wichmann, J. Math. Phys. 16, 985 (1975).

[5] K. Bleuler, Helv. Phys. Acta 23, 567 (1950).

[6] S. Bludman, Nuovo Cimento 9, 433(1958)

[7] N.N. Bogolyubov and D. V. Shirkov, Vvedenie v teoriyu kvantovannykh poleĭ (Introduction to the Theory of Quantized Fields), Moscow, 1957 [Engl. Transl., Interscience, N.Y., 1959]; 2nd Russian Ed. Moscow, 1973.

[8] R. Brout and F. Englert, Phys. Rev. Lett. 13, 321 (1964).

[9] N. Bourbaki, Variétés différentielles et analytiques, Fascicule de résultats. Hermann, Paris § 1 -- 7, 1967; § 8 -- 15, 1971.

[10] S. S. Chern, Geometry of Characteristic Classes, Proc. 13-th Biennial Seminar, Canadian Math. Congr., 1972 pp. 1--40.

[11] B. S. De Witt, Phys. Rev. 162, 1195, 1239 (1967).

[12] J. Dieudonné, Treatise on Analysis, vol. III, 1972; vol. IV, 1974; Academic Press, N. Y. [French original: Éléments d'analyse, Gauthier Villars, Paris, 1970, 1971]

[13] S. Doplicher, R. Haag and J. E. Roberts, Fields, Observables and Gauge Transformations, I; II, Commun. Math. Phys. 13, 1 -- 23 (1969); 15, 173 -- 200 (1971). Local Observables and Particle Statistics, I, Ibid. 23, 199 -- 230 (1971); S. Doplicher and J. E. Roberts, Fields, Statistics and Nonabelian Gauge Groups, Ibid. 28, 331 -- 348 (1972).

[14] L. D. Faddeev and V. N. Popov, Feynman Diagrams for the Yang-Mills Field, Phys. Lett. 25B, 29 (1967); Kiev Preprint, 1967; L. D. Faddeev, Teor. Mat. Fiz. 1, 3 (1969) [Theoret. Mathem. Phys. 1, 1 (1969]].

[15] E. Fermi, Rev. Mod. Phys. 4, 87 (1932).

[16] R. P. Feynman, Acta Physica Polonica 26, 697 (1963).

[17] P. L. Garcia, Gauge Algebras, Curvature and Symplectic Structure, to appear in J. Differ. Geom.; Reducibility of the Symplectic Structure of Classical Fields with Gauge Symmetry, this volume,p.

[18] L. Gårding and A. S. Wightman, Fields as Operator-Valued Distributions in Relativistic Quantum Field Theory, Arkiv för Fysik 28, 129 -- 184 (1964).

[19] M. Gell-Mann and S. L. Glashow, Ann. Phys. (N. Y.) 15, 437 (1961).

[20] S. N. Gupta, Proc. Phys. Soc. (Lond.) 61A, 68 (1950).

[21] G. S. Guralnik, C. R. Hagen and T. W. B. Kibble, Phys. Rev. Lett. 13, 585 (1964).

[22] P. W. Higgs, Phys. Rev. Lett. 12, 132 (1964); Phys. Rev. 145, 1156 (1966).

[23] H. Kerbrat-Lunc, Ann. Inst. Poincaré 13A, 295 (1970).

[24] R. Kerner, Ann. Inst. Poincaré 9A, 143 (1968).

[25] T. W. B. Kibble, J. Math. Phys. 2, 212 (1961).

[26] T. W. B. Kibble, Phys. Rev. 155, 1554 (1967).

[27] S. Kobayasi and K. Nomizu, Foundations of Differential Geometry, Vols. 1 and 2, Wiley(Interscience), N. Y., 1963 and 1969.

[28] A. Lichnerowicz, Théorie globale des connexions et des groupes d'holonomie, Ed. Cremonese, Roma 1955.

[29] H. G. Loos, Internal Holonomy Groups of Yang-Mills Fields, J. Math. Phys. 8, 2114--2124 (1967); Phys. Rev. 188, 2342 (1969).

[30] M. E. Mayer, Thesis, Unpubl. University of Bucharest, 1956; Preprint, JINR, Dubna, 1958 and Nuovo Cimento 11, 760--770 (1959); also Cîmpuri cuantice și particule elementare (Quantized Fields and Elementary Particles), Ed. Tehnica, Bucharest, 1959.

[31] M. E. Mayer, Fibrations, Connections and Gauge Theories (An Afterthought to the Talk of A. Trautman), Proc. of the International Symposium on New Mathematical Methods in Physics, K. Bleuler and A. Reetz, Eds., Bonn 1973; Talk at Intern. Congress of Mathematicians (Abstract N4) Vancouver, B. C., August 1974; Gauge Fields as Quantized Connection Forms, to be published.

[32] E. Noether, Invariante Variationsprobleme, Nachr. Ges. Göttingen (math.-phys. Klasse) 1918, 235--257.

[33] V. I. Ogievetskii and I. V. Polubarinov, Zh. Eksp. Teor. Fiz. 41, 247 (1961) [Sov. Phys. JETP 14, 179 (1962)].

[34] W. Pauli, Die allgemeinen Prinzipien der Quantenmechanik, Handb. d. Physik, Bd. XXIV, 1. T., 1933 (2nd Ed. Bd. V, 1. T. 1958) Springer-Verlag.

[35] A. Salam and J. C. Ward, Nuovo Cimento 11, 568 (1959).

[36] A. Salam, Nobel Symposium 1968, Almquist & Viksell, Stockholm.

[37] J. Schwinger, Phys. Rev. 82, 914 (1951); 125,1043; 127, 324 (1962); 91, 714 (1953).

[38] I. E. Segal, Proc. Nat. Acad. Sci. USA 41, 1103 (1955); 42, 670 (1956); Quantization of Nonlinear Systems, J. Math. Phys. 1, 468 -- 488 (1960); Quantized Differential Forms, Topology 7, 147 --171 (1968).

[39] S. Sternberg, Lectures on Differential Geometry, Prentice-Hall, Englewood Cliffs, N. J. , 1963.

[40] F. Strocchi and A. S. Wightman, Proof of the Charge Superselection Rule in Local Relativistic Quantum Field Theory, J. Math. Phys. 15, 2198 -- 2224 (1974).

[41] R. Sulanke and P. Wintgen, Differentialgeometrie und Faserbündel, Birkhäuser Verlag, Basel 1972.

[42] W. E. Thirring, Ann. Phys. (N. Y.) 16, 96 (1961).

[43] G. 't Hooft, Nucl. Phys. B33, 173; B35, 167 (1971).

[44] G. 't Hooft and M. Veltman, Nucl. Phys. B44, 189 (1973).

[45] A. Trautman, Infinitesimal Connections in Physics, Proc. Internat. Symposium on New Mathematical Methods in Physics, K. Bleuler and A. Reetz, Eds. Bonn, 1973 and earlier work quoted there.

[46] R. Utiyama, Invariant Theoretical Interpretation of Interaction, Phys. Rev. 101, 1597 (1957).

[47] M. Veltman, Nucl. Phys. B21, 288 (1971).

[48] S. Weinberg, Phys. Rev. Lett. 19, 1264 (1967).

[49] S. Weinberg, Rev. Mod. Phys. 46, 255 (1974) [cf. also Scientific American, 231, 50 (1974)].

[50] H. Weyl, Gravitation und Elektrizität, Stzber. Preuss. Akad. Wiss. 1918, 465 -- 480; Z. Physik 56, 330 (1929).

[51] C. N. Yang and R. L. Mills, Conservation of Isotopic Spin and Isotopic Gauge Invariance, Phys. Rev. 96, 191 (1954).

[52] C. N. Yang, Integral Formalism for Gauge Fields, Phys. Rev. Lett. 33, 445--447 (1974); T. T. Wu and C. N. Yang, Concept of Nonintegrable Phase Factors and Global Formulation of Gauge Fields, Preprint ITP-SB 75/31, Stony Brook, 1975 (Phys. Rev., to be published).

COMPLEX LINE BUNDLES AND THE MAGNETIC FIELD OF A MONOPOLE

Werner H. Greub

Department of Mathematics
University of Toronto

1. Introduction. The motion of an electron in a magnetic field F is described by the Schrödinger equation

$$-\frac{1}{2m}\sum_{\mu=1}^{3}(\partial_\mu + ie\,A_\mu)^2\,\psi = E\cdot\psi . \tag{1}$$

Here m, e and E denote respectively the mass, the charge and the energy of the particle and A is the vector potential of F. Thus A is a differential form of degree 1 satisfying the relation

$$\frac{\partial A_\nu}{\partial x^\mu} - \frac{\partial A_\mu}{\partial x^\nu} = F_{\nu\mu} . \tag{2}$$

It should be observed that the vector potential A will not globally exist in general. In fact, consider the magnetic field of a single magnetic pole. It is given by the 2-form

$$F = \frac{\lambda}{r^3}(x^1dx^2\wedge dx^3 + x^2dx^3\wedge dx^1 + x^3dx^1\wedge dx^2)$$

$$r^2 = \sum_\mu (x^\mu)^2$$

where λ is the pole strength. In this case F is a closed 2-form in the deleted 3-space $\dot{\mathbb{R}}^3$ which can not be derived from a global vector potential as follows from Stoke's theorem, since

$$\int_{S^2} F = 4\pi\lambda \neq 0 .$$

It is the purpose of this contribution to show that equation (1) becomes globally meaningful if we interpret ψ and A in the following way:

(i) ψ is a cross-section in a complex line bundle (rather than a scalar function).

(ii) The functions ieA_μ are the parameters of a linear connection ∇ in this bundle (rather than the components of a 1-form).

Thus we write

$$\nabla_\mu = \frac{\partial}{\partial x^\mu} + ieA_\mu$$

Then the Schroedinger equation reads

$$-\frac{1}{2m}\sum_{\mu=1}^{3} \nabla_\mu{}^2\psi = E\cdot\psi$$

and equation (2) says that the curvature form of the linear connection ∇ is given by

$$R_{\nu\mu} = ie\left(\frac{\partial A_\nu}{\partial x^\mu} - \frac{\partial A_\mu}{\partial x^\nu}\right) = ie\,F_{\nu\mu}$$

This leads to the following mathematical problem:
Given a closed 2-form Φ on a smooth manifold M, when does there exist a complex line bundle ξ over M and a linear connection in ξ such that the corresponding curvature form is Φ ?

2. Complex line bundles. Let $\xi = (E,\pi,M,\mathbb{C})$ be a complex line bundle over a smooth manifold M and let ∇ be a linear connection in ξ. Then the curvature form R of ∇ is a complex valued 2-form on M and the Bianchi identity states that this form is closed. Thus it represents a de Rham cohomology class of M.

A de Rham cohomology class $[\Phi]$ of degree p is called integral, if

$$\int_{z_i} \Phi \in \mathbb{Z}, \quad i = 1\ldots k,$$

where Φ is a closed form representing $[\Phi]$ and $z_i\ldots z_k$ is a basis of $H_p(M,Z)$

Theorem: If R is the curvature form of a linear connection in a complex line bundle over M then

$$\Phi = \frac{1}{2\pi i} R$$

represents an integral class. Conversely, if Φ is a closed 2-form on M representing an integral class, then there is a complex line bundle ξ over M and a linear connection in ξ such that

$$R = 2\pi i\Phi .$$

Moreover, if this condition holds and M is simply connected, then ξ is uniquely determined up to a bundle isomorphism.

For the proof, cf. [1] .

3. <u>The magnetic field of a monopole.</u> Now consider the magnetic field of a monopole, $F = \lambda \cdot \Omega$, where Ω is the closed 2-form in $\dot{\mathbb{R}}^3$ given by

$$\Omega = \frac{1}{r^3}\,(x^1 dx^2 \wedge dx^3 + x^2 dx^3 \wedge dx^1 + x^3 dx^1 \wedge dx^2) \tag{3}$$

Since

$$\int_{S^2} \Omega = 4\pi$$

and since S^2 represents a basis of $H^2(\dot{\mathbb{R}}^3, \mathbb{Z})$, it follows that the 2-form $\frac{1}{4\pi}\Omega$ represents an integral class. Thus $\frac{1}{2\pi}F$ represents an integral class if and only if λ satisfies Dirac's condition

$$\lambda = \frac{n}{2e} \qquad n \in Z \tag{4}$$

Thus, by the theorem, (4) is necessary and sufficient for the existence of a complex line bundle ξ^n over $\dot{\mathbb{R}}^3$ and a linear connection in ξ^n such that

$$R = \frac{in}{2}\Omega \qquad n \in Z$$

It should be noted that the integral $\int_{S^2} F$ represents the <u>magnetic charge.</u> Thus condition (4) says that the magnetic charge is an integer multiple of $\frac{2\pi}{e}$.

4. <u>Construction of the bundles ξ^n</u> . We shall now construct the complex line bundles ξ^n $(n \in \mathbb{Z})$ explicitly. Consider the principal bundle

$$(\dot{Q}\ ,\pi,\ \dot{\mathbb{R}}^3, S^1)$$

where Q denotes the space of quaternions and $\mathbb{R}^3$ is the subspace of Q spanned by the vectors e_1, e_2, e_3 . The bundle projection is given by

$$\pi(x) = x \cdot e_3 \cdot \bar{x} \qquad x \in \dot{Q}\ ,$$

where $\bar{x}$ is the conjugate of x.

Next, let Φ_n denote the representation of S^1 in $\mathbb{C}$ defined by

$$\Phi_n(\alpha)z = \alpha^{-n}z \qquad \alpha\epsilon S^1,\ z\epsilon\mathbb{C}$$

and let ξ^n be the associated vector bundle (cf. [2]).

To obtain a linear connection in ξ^n consider the principal connection in $\dot{Q}$ defined by the 1-form

$$\omega(x,h) = \frac{1}{|x|^2} \langle h, xe_3 \rangle,\ x\epsilon\dot{Q}\ ,\ h\epsilon Q$$

This connection induces via the representation Φ_n a linear connection ξ^n . A straightforward calculation shows that the corresponding curvature form is given by

$$R = \frac{in}{2}\,\Omega$$

where Ω is the 2-form in $\dot{\mathbb{R}}^3$ given by (3).

5. Solution of the Schroedinger equation. Finally we return to the Schroedinger equation

$$-\frac{1}{2m} \sum_{\mu=1}^{3} \nabla_\mu{}^2\psi = E\cdot\psi\ , \qquad (5)$$

where ψ is a cross-section in the complex line bundle ξ^n and ∇ is the linear connection obtained in the last section. In view of the canonical isomorphism between cross-sections in ξ^n and Φ_n - equivariant functions on $\dot{Q}$ equation (5) is equivalent to equation

$$-\frac{1}{2m} \sum_{\mu=1}^{3} \partial_\mu^2 f = E\cdot f\ , \qquad (6)$$

where f is a function on the total space $\dot{Q}$ equivariant under the right action of S^1 . Here ∂_μ denotes the derivative of f in the direction of the horizontal lift of e_μ . Using the ansatz

$$f(x) = g(|x|^2)\cdot h\left(\frac{x}{|x|}\right)\ ,\ x\epsilon\dot{Q}$$

we obtain from (6) an ordinary second order differential equation for f and a partial differential equation for h . These equations can be solved explicitly in terms of Bessel functions and the Wigner coefficients for the irreducible representations of $SU(2)$ respectively.

References

[1] B. Kostant, Quantization and unitary representations, in: Springer lecture notes in mathematics, Heidelberg 1970.

[2] W. Greub et al., Connections, Curvature and Cohomology, Vol. II (Academic Press, N.Y. 1973).

[3] W. Greub and H.R. Petry, Minimal Coupling and complex line bundles, J. Math. Phys. 16, 1347, 1975.

CONCLUSIONS FROM AN EXTENDED GAUGE PRINCIPLE OF DIRAC'S EQUATION

Leopold Halpern*
Dept. of Physics, Florida State University
Tallahassee, Fla. 32306

I. Introduction and Summary

Unified theories of electromagnetism and gravitation are usually understood to involve geometrization. Geometry has however advanced so far that any gauge theory is a geometrized theory of a principal fiber bundle. The first step in an attempt to create a unified theory, usually triggers as a reaction the citation of Weyl's phrase: "What God has separated man should not unite." A manifestation of the divine decision may be seen for example in the necessity to combine different irreducible representations of a general invariance group. The second step (more often not performed) consists in finding an extension of the group which provides a wider view in the man-made universe of continuum mathematics, remedies the evil and hopefully adjusts to the real universe.

The first step suggested here, consists already in an extension of the basic invariance group by the combination of the electromagnetic gauge group and the group of similarity transformations of the Dirac spinor. Weyl's phrase applies nonetheless because the two groups are combined in form of a direct product and eigen values of the generators of the second group can assume half integer values whereas those of the first admit only integers. Bearing in mind however that hardly any combination of groups a priori appears crazier than that of rotations with the Lorentz transformations we perform this unification as a first step. One obtains this way formally electromagnetic and gravitational theory unifiedly derived as a gauge theory - (as good as a metric gravitational theory can be derived from gauge arguments). The unification (or more cautiously: pre-unification) prescribes however already in its simplest form, new Lagrangians which differ from the Einstein Lagrangian $\sqrt{g}R$ by nonlinear terms giving rise to derivatives of the fourth order. The very simplest version is in fact formally equivalent to Weyl's derivation of electromagnetism from gauge invariance (Reference 1) and Utiyama's derivation of general relativity from local Lorentz invariance (Reference 3) with the difference that the simplest Lagrangian which can be constructed consists in its gravitational part only of nonlinear terms. Lack of knowledge on the physical interpretation of such a Lagrangian suggests to extend the invariance one step further to a mass renormalization of the Dirac equation which automatically leads to addition of the Einstein Lagrangian. A rather unnatural combination of the available invariants in this context could even yield the Einstein Lagrangian alone beforehand.

The analogy between electromagnetic gauge transformations and spin rotations becomes already rather apparent in Weyl's pioneer work in which the general two component spinor theory is introduced (Ref. 1). The possibility of using Lorentz-spin rotation to derive general relativity according to Utiyama's method has also been recognized (Ref. 10, 15). The quoted references consider all possible spin representations, whereas we have restricted our considerations to Dirac spinors. We consider spinors as fundamental to physics and geometry. Such a point of view has been emphasized by Penrose (Ref. 7) and the motivation of our work was influenced by work of O. Klein (Ref.) and A. Sakharov (Ref. 4, 18, 9) in which the gravitational theory is obtained from the covariance of Dirac's equation and second quantization of the spinor field, considered as a fundamental field. While our work was in progress a preprint of a paper by Wu and Yang appeared (Ref. 17, sect. 9) in which it is suggested to extend the phase factors from complex numbers to the largest division algebra, the quaternions. This is a point of view which we originally considered, yet abandoned, because the spinors (even the two component spinors to which one may restrict oneself when considering only proper Lorentz rotations) form themselves a non-associative algebra with zero divisers. (Ref. 19); one can hardly gain from a division algebra of their phase factors. We deal here with the group of nonsingular matrices as a generalization of the phase. We also want to draw the most general conclusions from Dirac's equation and therefore do not make restriction to two component spinors. Our similarity transformations are only formally equivalent to Lorentz transformations of spinors in the simplest cases. We later hope to obtain more physical results from the remaining transformations which formally go beyond local Lorentz invariance (and in fact partly beyond any law physically realized in extended Lorentz frames).

Our principle of unification suggests also a generalization of Dirac's magnetic monopole to the gravitational variables. Much of the ambiguity of choice for such a generalization manifested in J. Dowkers and Y. Dowkers, and in J. Dowkers and Roches interesting work on this subject (Ref. 10, 15) is here eliminated. Results obtained by Wu and Yang (Ref. 17) in a formally similar situation allow us to deny preliminarily the physical nature of such a generalization for the simplest analog of the electromagnetic gauge group. The monopole constructed explicitely by Dowker and Roche (Ref. 15) is based on the gauge group of the linear approximation to general relativity and we have doubts on the physical validity of an extension of the argument to the invariance group of general relativity in this context. The results quoted here with our method should apply to the complete theory. The general invariance group considered will pose more problems and promises more results. A paper in which the group of electromagnetic gauge transformations and the group of spin rotations are truly unified within a larger invariance group is hoped to appear soon.

II Outline of the Formalism of the covariant Dirac Equation

We deal with entities which are four dimensional spin matrices and have a definite character w.r.t. coordinate transformations. They are called correspondingly scalar, vector- and tensor operators. The vector operators γ^μ fulfill the modified Dirac algebra:

$$\{\gamma^\mu, \gamma^\lambda\} = 2g^{\mu\lambda}(x) \tag{1}$$

There exists a vector operator Γ'_μ so that the extended covariant derivative (e.c.d.) of the γ^λ vanishes:

$$\gamma^\mu_{|\lambda} \equiv \gamma^\mu{}_{;\lambda} + [\gamma^\mu, \Gamma'_\lambda] = 0 \qquad \gamma^\mu{}_{;\lambda} \equiv \gamma^\mu{}_{,\lambda} + \Gamma^\mu_{\lambda\alpha}\gamma^\alpha \tag{2}$$

This entity is determined up to its trace by eq. (2). Γ'_ν is expressible in terms of the γ-matrices and their covariant derivatives as follows:

$$\Gamma'_\nu = -\tfrac{1}{8}\,\{\tfrac{1}{3}\mathrm{Tr}(s_{\alpha\rho}\gamma^\rho{}_{;\nu})\gamma^\alpha + \mathrm{Tr}(\gamma_\beta\gamma_{\alpha;\nu})s^{\alpha\beta} +$$

$$+(g^{-1})\mathrm{Tr}(\gamma_5\gamma_{\alpha;\nu})\gamma_5\gamma^\alpha - \tfrac{1}{4}(g)^{-1}(\gamma_5\gamma_\rho\gamma^\rho_{;\nu})\gamma_5\} + A_\nu 1 \tag{2a}$$

where $g \equiv -\det(g_{\mu\nu})$ and A_ν is an arbitrary vector field. We shall later consider more general e.c.d. with vector operators $\Gamma\mu$ (prime ommitted) for which $\gamma^\mu_{|\lambda} \neq 0$. The Dirac spinor Ψ transforms as scalar w.r.t. coordinate transformations. One needs a further scalar operator α, the hermitizing matrix with the properties:

$$\alpha^+ = \alpha \ , \quad -\gamma^{\mu\,+} = \alpha\gamma^\mu\alpha^{-1} \tag{3}$$

The e.c.d. $\alpha_{|\lambda} \equiv \alpha_{;\lambda} + \alpha\Gamma_\lambda + \Gamma^+_\lambda\alpha$ (4)

is supposed to vanish for $\Gamma_\lambda = \Gamma'_\lambda$, for which also $\gamma^\mu_{|'\lambda}$ vanishes. α is determined by equs. (3) up to a real factor. One defines: $\bar\Psi \equiv \Psi^+\alpha$. It is useful to introduce also: $\xi \equiv (g)^{1/4}\Psi$ and $\bar\xi \equiv \xi^+\alpha \equiv (g)^{1/4}\bar\Psi$.

Similarity transformations with nonsingular matrix S transform the above entities as follows:

$$\Psi \to S^{-1}\Psi\ ,\ \gamma^{\mu} \to S^{-1}\gamma^{\mu}S\ ,\ \alpha \to S^{+}\alpha S\ ,\ \Gamma_{\mu} \to S^{-1}\Gamma_{\mu}S - S^{-1}S_{,\mu} \tag{5}$$

We use also $\gamma_{\alpha} = \gamma^{\mu}g_{\mu\alpha}$. The Leibniz rule applies to products for the e.c.d. and $g_{\mu\alpha;\lambda} \equiv g_{\mu\alpha,\lambda} - g_{\zeta\alpha}\Gamma^{\zeta}_{\mu\lambda} - g_{\mu\zeta}\Gamma^{\zeta}_{\mu\alpha} \equiv 0$

The field: $\Phi_{\mu\nu} = \Gamma_{\nu,\mu} - \Gamma_{\mu,\nu} - [\Gamma_{\mu},\Gamma_{\nu}]$ (6)

transforms homogenously: $\Phi_{\mu\nu} \to S^{-1}\Phi\mu\nu S$ according to the adjoint re-resentation. If the e.c.d. of the γ^{μ} and of α vanish, one finds the relation: $\Phi'_{\lambda\mu} = \tfrac{1}{4}R_{\mu\lambda\alpha\beta}\ s^{\alpha\beta} + (A_{\mu,\lambda} - A_{\lambda,\mu}).\mathbf{1}ie$ (7)

where $s^{\alpha\beta} = \tfrac{1}{2}[\gamma^{\alpha},\gamma^{\beta}]$, $A_{\mu}(x)$ is arbitr. real vectorfield, e = real constant, i = imaginary unit.

We note the relations:

$$R = R^{\mu\lambda}{}_{\lambda\mu} = -\tfrac{1}{2}\ \mathrm{Tr}(s^{\mu\lambda}\Phi'\mu\lambda)\ ,\ R_{\mu\lambda\alpha\gamma} = \tfrac{1}{2}\ \mathrm{Tr}\ (s_{\mu\lambda}\Phi'_{\alpha\gamma}) \tag{8}$$

Therefore:

$$\mathrm{Tr}(\Phi'_{\mu\nu}\Phi'^{\mu\nu}) = -\tfrac{1}{2}\ R_{\mu\lambda\alpha\gamma}R^{\mu\lambda\alpha\gamma} - 4\ e^{2}\ f_{\mu\lambda}f^{\mu\lambda} \tag{9}$$

with $f_{\mu\lambda} = A_{\lambda,\mu} - A_{\mu,\lambda}$

The e.c.d. of a spinor is defined as:

$$\Psi_{|\mu} \equiv \Psi_{;\mu} - \Gamma_{\mu}\Psi \quad \text{and} \quad \bar{\Psi}_{|\mu} \equiv \bar{\Psi}_{;\mu} + \bar{\Psi}\Gamma_{\mu} \tag{10}$$

We write the Dirac equation for $\Gamma_{\mu} = \Gamma'_{\mu}$:

$$\gamma^{\mu}\Psi_{|'\mu} + \mu\Psi = 0\ ,\quad \bar{\Psi}_{|'\mu}\gamma^{\mu} - \mu\Psi = 0 \qquad \mu = \frac{mc}{\hbar} \tag{11}$$

The spinor Ψ in (11) may be replaced by ξ because the covariant derivative of g vanishes. The action of the Dirac field is then:

$$\mathcal{L}_{D} = \tfrac{1}{2}(\bar{\xi}\gamma^{\alpha}\xi_{|\alpha} - \bar{\xi}_{|\alpha}\gamma^{\alpha}\xi) + \mu\bar{\xi}\xi \tag{11a}$$

Γ'_{μ} can be transformed away at any point by suitable coordinate transformations and S-transformations. The most general form for Γ_{μ} is thus: $\Gamma_{\mu} = \Gamma'_{\mu} + M_{\mu}$ where M_{μ} is a vector operator which transforms according to the adjoint representation $M_{\mu} \to S^{-1}M_{\mu}S$.

The e.c.d. become:

$$\gamma^{\mu}_{|\lambda} \equiv \gamma^{\mu}_{;\lambda} + [\gamma^{\mu},\Gamma_{\lambda}] = [\gamma^{\mu},M_{\lambda}] \tag{2'}$$

and $\alpha_{|\lambda} \equiv \alpha_{;\lambda} + \alpha\Gamma_{\lambda} + \Gamma^{+}_{\lambda}\alpha = \alpha M_{\lambda} + M^{+}_{\lambda}\alpha$ (3')

The hermitian Lagrangian would assume in this general case an additional term: $\tfrac{1}{2}\xi^{+}(M^{+}_{\mu}\alpha + \alpha M_{\mu})\ \gamma^{\mu}\xi$ (11a')

and the Dirac equation becomes:

$$\gamma^{\mu}\psi_{|\mu} + \tfrac{1}{2}[\gamma^{\mu}_{,}M\mu]\psi + \tfrac{1}{2}(M^{+}_{\mu}\alpha + \alpha M\mu)\gamma^{\mu}\psi - \mu\psi = 0$$

respect. (11')

$$\bar{\psi}_{|\mu}\gamma^{\mu} + \tfrac{1}{2}\bar{\psi}[\gamma^{\mu}_{,}M_{\mu}] - \tfrac{1}{2}\ \psi^{+}(\alpha M_{\mu} + M^{+}_{\mu}\alpha)\gamma^{\mu} + \bar{\psi}\mu = 0$$

We shall deal in general only with Γ_{μ} for which $\alpha_{|\mu} = 0$

III The Group of Similarity Transformations and its Gauge Field

The Lagrangian $\mathcal{L}_D$ is not only invariant w.r.t. general coordinate transformations but also w.r.t. similarity transformations with non-singular matrix fields S. (see eq. (II.5)) The electromagnetic gauge transformations are special transformations of this type with $S=e^{i\chi}.\mathbf{1}$ (χ real); they form an invariant subgroup. Γ_μ is altered by a term: $-i\chi_{,\mu} = -ieA_\mu$ by such a transformation. A_μ is the real electromagnetic gauge potential and e in our units where $\hbar = c = \mathbf{1}$ is a dimensionless, empirically determined constant. Our point of view in the present work is, that this procedure should be extended from the electromagnetic gauge group to the full group of similarity transformations, not only for the potentials but even for their fields. Formally the first step is almost performed already in the generally covariant Dirac equation. We can, for example, define the generalized vector potentials in terms of the vector operator Γ_μ as follows:

$$e\mathcal{A}_\mu = \Gamma_\mu \tag{1}$$

The $\mathcal{A}_\mu$ are here 4x4 matrices. We have chosen the same constant e for all the potentials. The electromagnetic potential is then:

$$A_\mu = \frac{i}{8}\,\mathrm{Tr}(\mathcal{A}_\mu + \mathcal{A}_\mu^+) \tag{2}$$

The constant is however not necessarily the same for all potentials; GL(4c) is expressible as the direct product of SL(4c) and the group of complex multiples of the unit matrix to which the electromagnetic gauge group belongs. We shall see that our assumption of a universal e suggests a fundamental unit of length.

The generalized potentials should give rise to fields which transform homogenously w.r.t. similarity transformations. These fields are according to eq. (II.6) and eq. (1):

$$F_{\mu\nu} = \mathcal{A}_{\nu,\mu} - \mathcal{A}_{\mu,\nu} - e[\mathcal{A}_\mu, \mathcal{A}_\nu] \tag{3}$$

The number of potentials is in general larger than the number of independent fields which can be formed out of the potentials. The conditions restrict for example the Γ'_μ such that the $F_{\mu\nu}$ according to eq. (II.7) are formed only out of linear superpositions of the six matrices $s^{\mu\nu}$ and multiples of the unit matrix. The $s^{\mu\nu}$ in a local inertial frame are the generators of Lorentz transformations of the spinors. We need here not even consider the local Lorentz frames and our principal fiber bundle with the groups of similarity transformations is not the same as that of Lorentz frames but in this particular case there is a formal resemblance of our method to a special case of Utiyamas method to obtain the gravitational theory from the invariance w.r.t. local Lorentz transformations. (Ref. 3)

Our basic assumption in this work is, that electromagnetic gauge transformations and the other similarity transformations belong together in one gauge group. The potentials and Yang-Mills fields are also united in eqs. (1 and 3). We give now independent degrees of freedom to the gauge fields $F_{\mu\nu}$ and form a Lagrangian out of them of which we require that it manifests the same unification. The Maxwell Lagrangian:

$$\mathcal{L}_M = -\frac{\sqrt{g}}{4} f_{\mu\nu}f^{\mu\nu} \tag{4}$$

suggests thus:

$$\mathcal{L}_G = \frac{\sqrt{g}}{32} \mathrm{Tr}\{F_{\mu\nu}F^{\mu\nu} + \mathrm{h.c.}\} \tag{5}$$

eq. (I 9) shows us that if the e.c.d. of γ^μ and α vanish this is equal to:

$$\mathcal{L}_G = -\frac{\sqrt{g}}{32e^2} R_{\mu\lambda\alpha\beta}R^{\mu\lambda\alpha\beta} - \tfrac{1}{4}\sqrt{g}\, f_{\mu\lambda}f^{\mu\lambda} \tag{5'}$$

$\mathcal{L}_G$ is formed out of the A_μ and their derivatives and out of the metric tensor $g_{\mu\nu}$. The $g_{\mu\nu}$ are however expressible in terms of the γ^μ and the Γ_μ are in general not independent of the γ^μ. If one chooses Γ'_μ, the conditions $\gamma^\mu_{|\nu} = \alpha_{|\nu} = 0$ allow to express the traceless part of Γ'_μ by the γ^ν according to eq. (II.2a). One can thus vary the Lagrangian in this case w.r.t. the γ^μ and the electromagnetic potentials. The γ^μ must be varied such that eq. (II.1) remains true:

$$\delta\gamma^\mu = \varepsilon^\mu{}_\nu\gamma^\nu \qquad\qquad \delta g^{\mu\nu} = \varepsilon^{\mu\nu} + \varepsilon^{\nu\mu} \tag{6}$$

$\varepsilon^{\mu\nu}$ is a tensor field, the antisymmetric part of which gives rise to similarity transformation of the γ^μ which are not of physical significance. One obtains: (Ref. 5)

$$\delta\Gamma'_\nu = \tfrac{1}{2}\varepsilon_{\nu\rho;\sigma}s^{\rho\sigma} + ie\delta A_\nu \tag{6'}$$

When varying $\mathcal{L}_G$ (eq. 5) w.r.t. $\varepsilon^{\mu\nu}$ one can consider its Form in (eq. 5') and the fact that the variation: (Ref. 21)

$$\delta\{\sqrt{g}\,(R^2 - 4R_{\mu\nu}R^{\mu\nu} + R_{\mu\nu\rho\sigma}R^{\mu\nu\rho\sigma})\} = 0 \tag{7}$$

one can replace this way all the $R_{\mu\nu\rho\sigma}$ by $R_{\mu\nu}$, R and obtains:

$$-\frac{\sqrt{g}}{8e^2}\{4R^{\mu\nu;\sigma}{}_\sigma - 4R^{\mu\sigma;\mu}{}_\sigma + 8R^{\mu\sigma}R_{\nu\sigma} - 2g^{\mu\nu}R_{\alpha\beta}R^{\alpha\beta} + 2R;^{\mu\nu} + 2RR^{\mu\nu} - \tfrac{1}{2}g^{\mu\nu}R^2\} - \sqrt{g}\{f^\mu{}_\beta f^{\nu\beta} - \tfrac{1}{4}g^{\mu\nu}f_{\alpha\beta}f^{\alpha\beta}\} \tag{8}$$

The term in the second paranthesis is the symmetric energy tensor of the electromagnetic field. The term in the first paranthesis is a symmetric tensor which contains fourth derivatives of the $g_{\mu\nu}$.

Variation of $\mathcal{L}_D$ (eq. II.11a) yields the symmetric energy tensor density of the Dirac field:

$$\sqrt{g}\; T_{\mu\nu} = \tfrac{1}{4}(\bar{\xi}\gamma_\mu \xi_{|\nu}' - \bar{\xi}_{|\nu}'\gamma_\mu \xi + \bar{\xi}\gamma_\nu \xi_{|\mu}' - \bar{\xi}_{|\mu}'\gamma_\nu \xi) \tag{9}$$

One recognizes that the gravitational field equations in the vacuum resulting from (eq. 8) admit $R^{\mu\nu} = 0$ as solution and therefore all the solutions of Einstein's vacuum field equations. The presence of matter leads however to ambiguities. (Ref. 12). To overcome such difficulties, one has of course the possibility to add the simplest invariant density (see eq. I.8)

$$\mathcal{L}_G' = - \frac{e}{2\kappa} \sqrt{g}\; \mathrm{Tr}(s^{\mu\nu} F_{\mu\nu}) = \frac{1}{\kappa} \sqrt{g} R \quad \text{(if e.c.d. of } \gamma^\mu \text{ vanishes)} \tag{10}$$

to the Lagrangian. $\kappa = \frac{8\pi G}{c^4}$ is Einstein's constant and G Newton's gravitational constant. $\mathcal{L}_G'$ is the simplest invariant that can be formed with $F_{\mu\nu}$ and it does not contain the electromagnetic field; it disturbs nevertheless the symmetry which we introduced due to the occurance of the constant κ. This constant has in our units where $\hbar = c = 1$ the dimension of a length squared. We can thus restore the symmetry by choosing a preferred unit of length such that for example $e^2 = 2\kappa$. (Ref. 2). The fundamental length in this case is $\approx 10^{-34}$cm. The Lagrangian $\mathcal{L}_G$ (eq. 5) shows the symmetry with the electromagnetic field but it is not the only possible Lagrangian which gives rise to Maxwell's equations and gravitational field equations of the fourth order. One still has the choice of:

$$\mathcal{L}_G'' = \sqrt{g}\; \mathrm{Tr}(F_{\mu\nu} F^\mu{}_\beta s^{\nu\beta}) \tag{11}$$

or

$$\mathcal{L}_G''' = \sqrt{g}\mathrm{Tr}(F_{\mu\nu} s^{\mu\nu} F_{\alpha\beta} s^{\alpha\beta}) \tag{12}$$

if the $\gamma^\mu{}_{|\nu} = 0$, $\alpha_{|\mu} = 0$ then

$$\mathcal{L}_G'' = \frac{\sqrt{g}}{e^2} \{R_{\mu\nu} R^{\mu\nu} - \tfrac{1}{4} R_{\mu\nu\rho\sigma} R^{\mu\nu\rho\sigma}\} \tag{11'}$$

and $$\mathcal{L}_G''' = \frac{\sqrt{g}}{e^2} R^2 + 8 \sqrt{g}\; f_{\mu\nu} f^{\mu\nu} \tag{12'}$$

The Lagrangian: $\sqrt{g}\mathrm{Tr}(F_{\mu\nu} s^{\mu\alpha} s^{\nu\beta} F_{\alpha\beta})$ is equivalent to $\tfrac{1}{2}\mathcal{L}_G'''$.

One sees that due to the relation (7) one can always choose a linear superposition of the given Lagrangians such that the Einstein-Maxwell-Dirac Theory results without fourth order derivative terms of the gik. Fourth order Lagrangians alone lead to difficulties when matter is present. The same needs however not to be true when also the Einstein Lagrangian $\mathcal{L}_G'$ is present. Indeed, quantum considerations demand an admixture of fourth order Lagrangians of the order of magnitude as a sum of $\mathcal{L}'_G + \mathcal{L}_G$ would yield. (Ref. 22) The Einstein Lagrangian has then by far the largest coefficient.

We shall not extend here our considerations to the case where the e.c.d. of γ^μ, α do not vanish and new fields are introduced by the M_μ (eq. II 2'). Laurent (Ref. 5) has worked out one case without electromagnetic fields and obtains a theory that differs somewhat from general relativity. Novello (Ref. 11) considered such degrees of freedom for the weak interactions. We would like still to remark that a M_μ of the simplest form:

$$M_\mu = m\gamma_\mu \qquad (m = \text{constant}) \tag{13}$$

which only gives rise to a mass renormalization in the Dirac-Lagrangian $\mathcal{L}_D$ (see eq. II. 11a and eq. II.11') will yield antomatically an admixture of the Einstein Lagrangian when exclusively the Lagrangian $\mathcal{L}_G$ (eq. 5) with such generalized $F_{\mu\nu}$ is introduced. One obtains:

$$\sqrt{g}\{-\tfrac{1}{32e^2}R_{\mu\lambda\alpha\beta}R^{\mu\lambda\alpha\beta} - \tfrac{1}{4}f_{\mu\lambda}f^{\mu\lambda} + \tfrac{m^2}{8}R\} \tag{5a}$$

if $m = \mu$ the electron mass, the coefficient is of course very small compared to κ^{-1} and we don't know at present whether the solutions yield physical results. The unrenormalized mass may however be chosen arbitrarily; one sees that even in this simple case of a modified theory one can work with the Lagrangian $\mathcal{L}_G$ alone which exhibits our symmetry considerations most beautifully.

IV On the Gravitational Analog of Magnetic Monopoles

The gauge fields $F_{\mu\nu}$ eq. (III.3) transform as the adjoint representation of the group of similarity transformations and their e.c.d. is consequently:

$$F_{\mu\nu|\rho} = F_{\mu\nu;\rho} + e\,[F_{\mu\nu}, \Gamma_\rho] \tag{1}$$

the identity:

$$F_{\mu\nu|\rho} + F_{\rho\mu|\nu} + F_{\nu\rho|\mu} \equiv 0 \tag{2}$$

is a consequence of the covariance. Applying our guidelines we consider a monopole of the field dual to $F_{\mu\nu}$: $F^{*\mu\nu} = \tfrac{1}{2}\varepsilon^{\mu\nu\rho\sigma}F_{\rho\sigma}$ which according to eq. (2) should fulfill:

$$F^{*\mu\nu}{}_{|\nu} \equiv 0 \tag{2a}$$

We restrict our considerations to the case $\gamma^{\mu}{}_{|\nu} = \alpha_{|\nu} = 0$ so that eq. (2) implies also the Bianchi identities for the Riemann tensor:

$$R_{\mu\nu\alpha\beta;\rho} + R_{\mu\nu\rho\alpha;\beta} + R_{\mu\nu\beta\rho;\alpha} \equiv 0 \qquad (2b)$$

the generalization of the magnetic monopole, which we consider can thus not leave the metric unmodified.

Let us return to the potentials A_{μ}. The potentials due to a monopole can in general not be given singularity free in all space. A way to describe the field of the monopole is to give nonsingular potentials in several, partially overlapping regions such that in points common to two such regions the potentials are related by a gauge transformation.

Wu and Yang (Ref. 17) have examined such gauges for some non-abelian groups as SU_2 and SO_3.

We have shown in section III that if the e.c.d. of $\gamma^{\mu}{}_{|\alpha}$ vanish, the remaining invariance group is the direct product of the electromagnetic gauge group and SL2C of which SU_2 is a compact subgroup. Some of the main results of section III (apart from those including the rest mass) could thus have been conveniently obtained using the two component spinor formalism. (Ref. 18) The results of Wu and Yang for SU_2 can thus be applied to the present case:

1. The generators of the spinor representation have half integer eigen values, those of the adjoint representation which transforms the potentials have integer eigen values; as a result, the Dirac quantization condition becomes here: $\frac{eg}{\hbar c}$ = integer instead of $\frac{2eg}{\hbar c}$ = integer for the electromagnetic case. (g=monopole charge).

2. Equation (2a) in the non-abelian case is not a conservation law and consequently the monopole charge is not an invariant of the group operations; in case of the singly connected group SU_2 all gauges are of one type and can be transformed into each other by global gauge transformations so that there remains essentially only one independent source free (integer = zero) solution for the potential for which the different regions where potentials were defined can be fused into one region.

Wu and Yang considered the SU_2 group of the Yang-Mills field so that their group operations give rise to modified charge distributions --a real physical modification which alters the quantum number of the monopole. Our group operations are similarity transformations (we do not even consider them as rotations of fictitious local frames). The monopole charge can therefore hardly be a physical concept in our case and the same is probably true for the present gravitational analog of a monopole as a whole. Caution is however required in drawing such a conclusion in particular because global gauge types are in our case related to the topology of space time and only a subgroup of our invariance group has been considered. A subsequent investigation is hoped to reveal the features in the more general case.

Literature References

1. H. Weyl, Zeit. f. Physik 56, p. 330 (1929).
2. O. Klein, Ark. f. Math., Astron. o. Fysik Vol 34A, Nr. 1, pp. 1-17.
3. R. Utiyama, Phys. Rev. Vol. 101, pp. 1537-1607 (1956).
4. O. Klein, Nuclear Physics B21, p. 253 (1970).
5. B. Laurent, Ark. f. Fysik, Vol. 16, nr. 25, p. 263 (1959).
6. A. Trautmann, Theory of Gravitation in The Physicist's Conception of Nature (ed. J. Mehra), Reidel Dordrecht, Boston 1973.
7. R. Penrose: Structure of Spacetime in Batelle Rencontrec (ed. C. M. De Witt & J. A. Wheeler), New York (1968); p. 121.
8. C.N. Yang & R.L. Mills, Phys. Rev. Vol. 96, Nr. 1, p. 191 (1954).
9. A.D. Saharov, Abstracts 5th International Conference on Gravitation and Theory of Relativity, Publishing House of Tbilisi Univ. (1968).
10. J.S. Dowker & Y.P. Dowker, Proc. Phys. Soc. 87, pp. 65-78 (1966).
11. M. Novello, Phys. Rev. D, Vol. 8, Nr. 8; pp. 2398-2400 (1973).
12. E. Pechlaner & R. Sexl, Comun. math. Phys. 2, pp. 165-175.
13. P.A.M. Dirac, Proc. Roy. Soc. (London) A133, p. 60 (1931).
14. P.A.M. Dirac, Pys. Rev. Vol. 74, Nr. 7, pp. 817-830 (1948).
15. J.S. Dowker & J.A. Roche, Proc. Phys. Soc. 92, pp. 1-8 (1967) and literature quoted there.
16. C.N. Yang, Phys. Rev. Letters 33, p. 445 (1974).
17. T.T. Wu & C.N. Yang: Concept of Nonintegrable Phase Factors and Global Formulation of Gauge Fields. Preprint ITP-SB 75/31.
18. A.D. Sakharov, Doklady Akad. Nauk. SSSR 177, p.70-71 (1967).
19. A. Sommerfeld, Atombau und Spektrallinien, Chapter IV, F. Vieweg Braunschweig (1951).
20. L. Infeld & B.L. v.d. Waerden, Sitzber. preuss. Akad. Wiss. Physik math. Klasse, p. 380 (1933).
21. C. De Witt, Les Houches Course 1963 Lecture of B. De Witt, Wiley N.Y. (1963).
22. L. Halpern, Arkiv f. Fysik, Vol. 34, nr. 43, p. 539 (1967).

---

*Acknowledgements

This research was supported in part by U.S.E.R.D.A. under grant number AT-(40-1)-3509.

I thank Prof. P.A.M. Dirac and Prof. J. Lannutti for the possibility to work in their Institute.

REDUCIBILITY OF THE SYMPLECTIC STRUCTURE OF CLASSICAL FIELDS WITH GAUGE-SYMMETRY

Pedro L. García

Departamento de Matemáticas
Universidad de Salamanca, Espana

Introduction

Every "classical field" has an associated (formal) symplectic structure $(\mathcal{V}, \omega_2, A)$, where $\mathcal{V}$ is the "solution manifold" of the field equations, ω_2 is a "symplectic metric" on $\mathcal{V}$ canonically defined by the lagrangian, and A is a "Poisson algebra" of functions of $\mathcal{V}$ to which belong all interesting dynamical quantities (energy, linear and angular moments etc.) [3]. In certain particular cases [1], [7], $\mathcal{V}$ can be endowed with a (infinite-dimensional) differentiable structure such that ω_2 is a closed 2-form on $\mathcal{V}$ and A is a subring of the ring of differentiable functions on $\mathcal{V}$. The case when ω_2 is irreducible is the best for a good dynamical theory. Nevertheless, there exist fundamental examples (electromagnetic field, Yang-Mills fields etc.) where ω_2 is not irreducible. In such a situation, it would be desirable that a differentiable proyection $p: \mathcal{V} \longrightarrow \bar{\mathcal{V}}$ could be found relative to which ω_2 be projected on an irreducible metric $\bar{\omega}_2$ on $\bar{\mathcal{V}}$, and A on a subring $\bar{A}$ of the ring of differentiable functions on $\bar{\mathcal{V}}$. Thus $(\bar{\mathcal{V}}, \bar{\omega}_2, \bar{A})$ will be the appropiate symplectic structure in order to develop the dynamic theory of the field under study. But, the problem of endowing with an adequate differentiable structure already being dificult, how can we hope to find this hypothetical projection. Faced with this, we could start with the simplest (and in a certain sense, previous) question of characterising those tangent vectors in the radical of ω_2 which are zero on A. On the other hand, this would be precisely the case for vectos tangent to the fibres of $p:\mathcal{V} \longrightarrow \bar{\mathcal{V}}$. In this paper we want to start the study of this question for "classical fields with gauge-symmetry", including, among others, the electromagnetic field and all versions of Yang-Mills fields. More precisely, <u>we try to see that in the free case, the Lie algebra defined by the gauge-symmetry of the field belongs to the radical of its symplectic metric, and is zero on the ring of its dynamical quantities</u>. We start with the simple case of the electromagnetic field and then we pass to the general case of non-abelian gauge-symmetry. The question remains open of the type of modifications to be made in the results obtained when we pass to the "interaction case". We think that this last question can be easily attacked from our viewpoint after the geometrical formulation established for it in [6].

1. Some notions and results of Calculus of Variations

We collect here some facts from the Geometrical Theory of the Calculus of Variations developped in [3], upon which all our treatment will rest.

A "classical field" is given by a variational problem geometrically defined by a fiber bundle $\pi:E \longrightarrow V$ (configuration space) on a manifold V with a volume element η (space-time) and a real differentiable function $\mathcal{L}$ (lagrangian) defined on the fiber bundle $\bar{E}$ of the 1-jets of local sections of π (state space). To a such structure one can associate two differential forms θ and Θ , the <u>structure 1-form</u> of the 1-jet fiber bundle $\bar{E}$ and the <u>Poincaré-Cartân form</u> corresponding to the lagrangian $\mathcal{L}$, from them the geometrical theory of the variational problem can be intrinsically developped [2] [3] [5]. θ is an 1-form on $\bar{E}$ valued in the induced vector bundle $p^*T^V(E)$, where $T^V(E)$ is the vertical tangent bundle of E and p is the canonical projection of $\bar{E}$ on E. Its local expression with respect to a system of natural local coordinates $(x_i z_j p_{ij})$ on $\bar{E}$ is:

$$\theta = \sum_j (dz_j - \sum_i p_{ij} dx_i) \cdot p^* \frac{\partial}{\partial z_j}$$

In terms of this 1-form, one can characterise in a very operative way the 1-jet extension $\bar{s}$ of a section $s: V \longrightarrow E$ and the 1-jet extension $\bar{D}$ of a vector field D on E. In particular, this last notion admits the following, very useful, variant: a vector field D_s on E, defined along a section $s: V \longrightarrow E$, being given, will be called 1-jet extension of D_s to the unique vector field $\bar{D}_{\bar{s}}$ on $\bar{E}$, defined along $\bar{s}$, such that $p(\bar{D}_{\bar{s}}) = D_s$ and $L_{\bar{D}_{\bar{s}}} \theta \big|_{\bar{s}} = 0$. The Poincaré-Cartān form Θ corresponding to the lagrangian $\mathcal{L}$ is the n-form on $\bar{E}$, n=dim V, whose expression is $\Theta = \theta \wedge \Omega - \mathcal{L}\eta$, where Ω is an (n-1)-form on $\bar{E}$ valued in the dual vector bundle of $p^*T^V(E)$, whose local expression is:

$$\Omega = \sum_{i,j} (-1)^i \frac{\partial \mathcal{L}}{\partial p_{ij}} dx_1 \wedge \ldots \wedge \widehat{dx_i} \wedge \ldots \wedge dx_n \cdot p^* dz_j$$

and where de exterior product $\theta \wedge \Omega$ is taken with respect to the bilineal product of the duality notion.

The extremals of the variational problem are now characterised as follows:

Theorem 1.1 (Cartân)

A section $s:V \longrightarrow E$ is extremal if and only if its 1-jet extension $\bar{s}$ varifies:

$$iXd\,\Theta \Big|_{\bar{s}} = 0$$

for every vector field X on $\bar{E}$.

The set of extremals will be called the solution manifold of the given classical field, and we shall write it $\mathcal{V}$. We shall think of $\mathcal{V}$ as some sort of "infinite dimensional differentiable manifold", an which we give the following notion of "tangent space".

Definition 1.1

A Jacobi field along an extremal s is a vertical vector field D_s on E, defined along s, such that its 1-jet extension $\bar{D}_{\bar{s}}$ verifies:

$$iXL_{\bar{D}_{\bar{s}}}\, d\Theta \Big|_{\bar{s}} = 0$$

for every vector field X on $\bar{E}$. The set of all such fields is a vector space, which we shall call the tangent space to the solution manifold $\mathcal{V}$ at s, and we will write $T_s(\mathcal{V})$.

There are two basic notions of Calculus of Variations which admit inmediate interpretations in terms of the solution manifold: infinitesimal symmetries and Noether invariants.

An infinitesimal symmetry of the given variational problem is a π-projectable vector field D on E such that its 1-jet extension $\bar{D}$ verifies $L_{\bar{D}}(\mathcal{L}\eta)=0$. We have now the following result:

Theorem 1.2 (Noether).

The restriction to the 1-jet extension $\bar{s}$ of an extremal s of the (n-1)-form $i\bar{D}\Theta$ corresponding to an infinitesimal symmetry D, is a closed form.

Now, if one takes the cohomology class in $H^{n-1}(V, \mathbb{R})$ defined by $i\bar{D}\Theta\big|_{\bar{s}}$, one can interpret these (n-1)-forms $i\bar{D}\Theta$ (called Noether invariants), as functions on the solution manifold $\mathcal{V}$ valued in $H^{n-1}(V, \mathbb{R})$. This sugests us to take this cohomology space as "scalars" in place of real numbers. In particular, if $V=\mathbb{R}$, we get $H^0(V, \mathbb{R})=\mathbb{R}$ and we are in the ordinary case.

On the other hand, if D_s is the vertical component, with respect to an extremal s, of an infinitesimal symmetry D, one can prove that D_s is a Jacobi vector field along s. Thus the mapping $s\in\mathcal{V}\longrightarrow D_s$ induced by an infinitesimal symmetry D defines a vector field on the solution manifold.

Definition 1.2

The alternating m-linear mappings from $T_s(\mathcal{V})\times\overset{m}{\dots}\times T_s(\mathcal{V})$ into $H^{n-1}(V, \mathbb{R})$ will be called m-forms on $\mathcal{V}$.

The following are two remarkable examples of this concept: if X_s is a Jacobi field along an extremal s and $i\bar{D}\Theta$ is a Noether invariant, then $i\bar{X}_s d(i\bar{D}\Theta)\big|_{\bar{s}}$ is closed, thus allowing us to define a linear mapping from $T_s(\mathcal{V})$ into $H^{n-1}(V, \mathbb{R})$, which we shall call the differential of $i\bar{D}\Theta$ at s, and we will write $d(i\bar{D}\Theta)_s$. As it is usual the scalar $d(i\bar{D}\Theta)_s(X_s)$ will be called derivative of $i\bar{D}\Theta$ with respect X_s and will be writed $X_s(i\bar{D}\Theta)$. On the other hand, let D_s and D'_s be two Jacobi fields along the extremal s. One can prove, analogously, that $i\bar{D}'_{\bar{s}} i\bar{D}_{\bar{s}} d\Theta \big|_{\bar{s}}$ is closed. Then we can define a hemisymetric bilinear mapping:

$$T_s(\mathcal{V}) \times T_s(\mathcal{V}) \xrightarrow{(\omega_2)_s} H_{n-1}(V, \mathbb{R})$$

which generalizes in a natural way the ordinary sumplectic metric of Analytical Dynamics.

In terms of these new concepts, an infinitesimal symmetry is related with its corresponding Noether invariant as follows:

<u>Theorem 1.3</u>

Let D be an infinitesimal symmetry and $i\bar{D}\Theta$ its corresponding Noether invariant. An extremal s being given, if $D_s \in T_s(\mathcal{V})$ is the vertical component of D with respect to s, one has:

$$iD_s \cdot (\omega_2)_s = -d(i\bar{D}\Theta)_s$$

Finally, the set of infinitesimal symmetries has a Lie algebra structure with respect to the ordinary operations for vector fields. This structure can be carried to the corresponding set A of Noether invariants. This is the <u>Poisson algebra</u> of the given classical field. Its relation with the symplectic structure is as follow:

<u>Theorem 1.4</u>

Let D, D' be two infinitesimal symmetries with their corresponding Noether invariants $i\bar{D}\Theta$, $i\bar{D}'\Theta$. Then, to every point s of the solution manifold $\mathcal{V}$, one has:

$$(\omega_2)_s(D_s, D'_s) = \{i\bar{D}\Theta,\ i\bar{D}\Theta\}(s)$$

where the second member is the value taken at s by the Poisson bracket of the functions $i\bar{D}\Theta$ and $i\bar{D}'\Theta$.

2. The case of free electromagnetic field

In this case, V is an orientable riemannian manifold with metric tensor g, of arbitrary signature, and with riemannian volume element η, E is the cotangent bundle $T^*(V)$ of V, and $\mathcal{L}$ is the function defined on the 1-jet fiber bundle of $T^*(V)$ by the formule:

$$\mathcal{L}(\bar{\omega}_x) = -\frac{1}{4}\cdot g_x(d\omega, d\omega)$$

where $\bar{\omega}_x$ is the 1-jet at x of a local 1-form ω on V.

If f is a differentiable function on V, the infinitesimal generator D_f of the one-parameter group τ_t of automorphisms of $T^*(V)$ given by:

$$\tau_t\omega_x = \omega_x + t(df)_x$$

defines an infinitesimal symmetry of this variational problem. It is called a gauge-symmetry of the free electromagnetic field. The set of all them is an (abelian) real Lie algebra of vector fields on $T^*(V)$, which locally coincides with the ideal of vertical infinitesimal symmetries of the field.

Theorem 2.1

If $(\mathcal{V}, \omega_2)$ is the symplectic solution manifold of the free electromagnetic field and D is a gauge-symmetry of the said field, then the tangent vector $D_s \in T_s(\mathcal{V})$ defined by D at a point $s \in \mathcal{V}$, satisfies:

$$iD_s \cdot (\omega_2)_s = 0$$

i.e. the gauge-symmetries belong to the radical of the symplectic metric.

<u>Sketch of the proof</u>

We try to see that, if X_s is an arbitrary Jacobi field along an extremal s, then the (n-1)-form $i\bar{X}_{\bar{s}} i\bar{D} d\Theta \big|_{\bar{s}}$ is exact. A calculus proves that this (n-1)-form has the expression:

$$df \cdot (iF^2\eta)$$

where f is the function on V defining the gauge-symmetry D, F^2 is a certain hemisymmetric tensor on V depending of X_s, i means contraction of one index of F^2 with the volume element η, and $\cdot$ means contraction of the other index of F^2 with df. By operating with the riemannian connection of V one has:

$$df \cdot (iF^2\eta) = d(ifF^2\eta) - i(f\,\mathrm{div}F^2)\eta$$

Last, we prove that $\mathrm{div}F^2 = 0$, for X_s is a Jacobi field .

With respect to the other point that we have mentioned in the introduction, we have the following:

Let $\mathcal{D}$ be the idealizator of the Lie algebra $\{D_f\}$ of gauge-symmetries in the Lie algebra of all infinitesimal symmetries. One has the extension of real Lie algebras:

$$0 \longrightarrow \{D_f\} \longrightarrow \mathcal{D} \xrightarrow{\ \pi\ } \bar{\mathcal{D}} \longrightarrow 0$$

where $\bar{\mathcal{D}}$ is the image of $\mathcal{D}$ by the canonical projection of $T^*(V)$ on V. The elements of $\bar{\mathcal{D}}$ are usually called <u>external infinitesimal symmetries</u>. In particular, when V is the Minkowski space, the Lie algebra of the Poincaré group belongs to $\bar{\mathcal{D}}$, and the corresponding Noether invariants are the interesting dynamical quantities (energy, linear and angular moments etc.).

Using the fact that $\{D_f\}$ is a ideal of $\mathcal{D}$, one can prove, in a way similar to the previous theorem, the following result:

<u>Theorem 2.2</u>

If $i\bar{D}\Theta$ is the function on $\mathcal{V}$ defined by the Noether invariant corresponding to an infinitesimal symmetry $D\in\mathcal{D}$, and X is an arbitrary gauge-symmetry, then, for every point $s\in\mathcal{V}$, one has:

$$X_s(i\bar{D}\Theta) = 0$$

If, in particular, D is a gauge-symmetry, then the function $i\bar{D}\Theta$ is identically zero.

3. <u>The general case</u>

The notion of "free gauge-invariant field" has been introduced, in a very simple and general way, in [4] as follows.

Let $p:P \longrightarrow V$ be a principal bundle with structural group G with Lie algebra $\mathcal{G}$, let L(P) be its adjoint bundle, i.e. the vector bundle associated to P by the adjoint representation of G on $\mathcal{G}$, and let {f} be the algebra of real differentiable functions on V.

Definition 3.1

The Lie $\{f\}$-algebra Γ of global sections of the adjoint bundle $L(P)$ will be called the gauge algebra of the principal bundle P.

If $G=U(1)$, then $L(P)=V\times\mathbb{R}$ and then Γ is the abelian Lie algebra $\{f\}$. This is the gauge algebra in the electromagnetic field theory.

In [4] is proved that connections on the principal bundle P can be canonically identified with global sections of the affine bundle $\pi:E\longrightarrow V$ corresponding to the vector bundle $Hom(T(V), L(P))$. E is called the fibre bundle of connections of the principal bundle P. If $G=U(1)$, then E is the affine bundle corresponding to the cotangent bundle $T^*(V)$, on which, we have developped the variational theory of the free electromagnetic field. As in this case, every element γ of the gauge algebra Γ defines a one-parameter group τ_t of vertical automorphisms of E. If D_γ is the infinitesimal generator of τ_t, one can prove that the mapping $\gamma\in\Gamma\longrightarrow D_\gamma$ is a homomorphism from the real Lie algebra Γ in the real Lie algebra of vector fields on E. Thus one gets a real Lie algebra $\{D_\gamma\}$, $\gamma\in\Gamma$, of (vertical) vector fields on E, which generalizes in an obvious way the abelian Lie algebra $\{D_f\}$ of gauge-symmetries of free electromagnetic field.

Definition 3.2

A free gauge-invariant field is defined by a variational problem whose configuration space is the affine bundle $\pi:E\longrightarrow V$ of connections of a given principal bundle, and whose lagrangian $\mathcal{L}$ is invariant by the above real Lie algebra $\{D_\gamma\}$. We shall call the D_γ, gauge-symmetries of the field.

A basic theorem proved by Utiyama, of which we have given in [4] a geometrical version, characterizes this type of variational problems in terms of the notion of "curvature of a connection". In particular, this allows us to obtain a good idea of how ample is the class of variational problems included in the above definition.

Now, let us suppose, as in the case of the free electromagnetic field, that V is an orientable riemannian manifold with metric tensor g of arbitrary signature. By taking the riemannian volume element η to build up the variational theory, one can prove that Theorems 2.1 and 2.2 established in the preceding paragraph can be generalized, without any change, to the present situation.

We shall end by giving a brief idea of how to procede in this case. If, for instance, we wanted to prove the Theorem 2.1, it will be a matter of seeing, as in the case of electromagnetic field, that if X_s is an arbitrary Jacobi field along an extremal s and D is a gauge-symmetry, then the (n-1)-form $i\bar{X}_{\bar{s}} i\bar{D} d\Theta|_{\bar{s}}$ is exact. As it is well known [3], in order to make differential calculus on $\bar{E}$ valued in $p^*T^V(E)$ one must take a connection (with vertical torsion zero) on $T^V(E)$, which, on the other hand, can be a completely arbitrary one. As, in our case, $T^V(E)$ is the induced vector bundle π^* Hom(T(V), L(P)), the riemannian connection on V and a connection on $\pi^*L(P)$ introduced in [4], allow us to define a special connection on $T^V(E)$, relative to which we can calculate the above (n-1)-form. Proceding of this way one gets:

$$i\bar{X}_{\bar{s}} i\bar{D} d\Theta|_{\bar{s}} = \text{exact } (n-1)\text{-form} + \omega_{n-1}|_{\bar{s}}$$

where ω_{n-1} is a (n-1)-form on $\bar{E}$, defined along $\bar{s}$, depending on X_s and D. Now, applying Utiyama's Theorem, one has $\omega_{n-1}|_{\bar{s}}=0$, for X_s is a Jacobi field.

References

1. P. Chernoff and J. Marsden. Properties of Infinite Dimensional Hamiltonian Systems, Lecture Notes in Mathematics. Springer-Verlag (1974)

2. P. Dedecker. Calculus of Variations, formes differentielles et champs geodesiques, Colloque International de Geometrie Differentielle, Strasbourg (1953).

3. P. García. The Poincaré-Cartán Invariant in the Calculus of Variations, Symposia Mathematica, vol. 14 (1974).

4. P. García. Gauge Algebras, Curvature and Symplectic Structure (to appear).

5. H. Goldschmidt and S. Sternberg. The Hamilton-Cartán Formalism in the Calculus of Variations, Ann. Inst. Fourier, 23, (1973).

6. A. Pérez-Rendón. A Minimal Interaction Principle for Classical Fields, Symposia Mathematica, Vol. 14 (1974).

7. I. Segal. La varieté des solutions d'une équation hyperbolique non-linéare d'ordre 2, Collège de France, París (1964).

NEW GEOMETRICAL DYNAMICS

André Lichnerowicz
Collège de France, Paris

My purpose in this lecture is to give a general survey on the geometrical dynamics, according to different recent works. We consider here essentially the frame of the classical analytical dynamics, from a global point of view, for systems which have a finite number of degrees of freedom. Using Banach manifolds, problems and a part of results can be extended to continuum media and to fields.

One of the motivations of these works is the following : if we understand truly classical analytical dynamics, we have a chance to understand more easily quantum dynamics and to obtain new invariant tools. It appears that problems and results are interesting and are absolutely not trivial, from the point of view of the differential geometry, even for classical dynamics. In this lecture, the proofs are generally absent.

I - POISSON MANIFOLDS.

1 - Notion of Poisson manifold.

a) Let W be a differentiable, connected, paracompact manifold of dimension m and class C^∞. All the considered elements are supposed C^∞. We denote $\{x^A\}$ $(A, B, \ldots = 1, \ldots, m)$ a local chart of W of domain U and we set $N = C^\infty(W ; R)$. A i-tensor is, by definition, a skewsymmetrical contravariant tensor of order i

For such tensors, Schouten and Nijenhuis have introduced an important tool, the Schouten bracket ; if A(resp. B) is a i-tensor (resp. j-tensor), $[A, B]$ is a $(i+j-1)$-tensor defined in the following way : for each closed $(i+j-1)$-form β, we have :

$$i([A,B])\beta = (-1)^{ij+j} i(A)\, di(B)\beta + (-1)^{i} i(\beta) di(A)\beta \tag{1-1}$$

where i is the interior product. For $i = 1$, $[A, B] = \mathcal{L}(A)\, B$, where $\mathcal{L}$ is the Lie derivative we have :

(1-2) $$[A,B] = (-1)^{ij}[B,A]$$

Moreover, if C is a k-tensor, we have the pseudo "Jacobi identity" [7]

(1-3) $$S(-1)^{ij}[[B,C],A] = 0$$

where S is the summation after circular permutation. An elementary calculus gives for components of $[A, B]$ on the domain of an arbitrary local chart :

(1-4) $$[A,B]^{k_2\ldots k_{i+j}} = \frac{1}{(i-1)!j!}\varepsilon^{k_2\ldots k_{i+j}}_{I_2\ldots I_i\ldots J_j} A^{RI_2\ldots I_i}\partial_R B^{J_1\ldots J_j} + \frac{(-1)^i}{i!(j-1)!}\varepsilon^{k_2\ldots\ldots k_{i+j}}_{I_1\ldots I_i J_2\ldots J_j} B^{RJ_2\ldots J_j}\partial_R A^{I_1\ldots I_i}$$

where $\partial_R = \partial/_{\partial x^R}$ and where ε is the skewsymmetrical Kronecker tensor-indicator.

b) Consider on W a <u>2-tensor G</u> of rank 2n everywhere ($G^n \neq 0$ everywhere, $G^{n+1}=0$); G is called of codimension $h = m-2n$. We denote a, b, c indexes taking h values, p, q, r ... indexes taking 2n values. Introduce on the space $N=C^\infty(W ; R)$ the braket $\{\ ,\ \}_G$ (generalized Poisson braket) defined by :

(1-5) $$\{u,v\}_G = i(G)(du \wedge dv) \qquad (u,v \in N)$$

If $u, v, w \in N$, we have :

$$S\{\{u,v\}_G, w\}_G = \frac{1}{2}\, i([G,G])(du\wedge dv\wedge dw)$$

we set the following definition :

<u>Definition</u> - <u>A structure of Poisson manifold is defined on a manifold W of dimension m by a 2-tensor G of constant rank $2n$ (with $h = m-2n$) such that</u>

(1-6) $$[G,G] = 0$$

For a Poisson manifold (W, G), (1-5) defines on N a structure of Lie algebra (dynamical Lie algebra)

c) A symplectic structure is generally defined on a manifold W of dimension $2n$ by a closed 2-form F of rank $2n$. We denote $\mu : TW \to T^*W$ the isomorphism of bundles defined by $\mu(X) = -i(X)F$; this isomorphism can be extended in a natural way to the tensor bundles. Let G be the 2-tensor $\mu^{-1}(F)$ of rank $2n$; the Poisson bracket of (W, F) is defined by (1-5) and G satisfies (1-6). A symplectic structure is nothing other as a Poisson structure (W, G) of codimension $0(h=o)$. Moreover if A is a i-tensor, we have :

$$\mu([G,A]) = d\mu(A) \tag{1-7}$$

It is well known that there exist on a symplectic manifold (W, G) atlas of canonical charts $\{x^p\} = \{x^\alpha, x^{\bar\alpha}\}$ ($\alpha = 1, \dots n$; $\bar\alpha = \alpha+n$) ; for such a chart, G admits only non vanishing components

$$G^{\alpha\bar\alpha} = -G^{\bar\alpha\alpha} = 1$$

d) Let (W, G) be a Poisson manifold of codimension $h \neq 0$; G defines on W by $G^{AB}\omega_B = 0$ a pfaffian system which is integrable and so a foliation of W of codimension h. The tensor G determines on each connected component of a leaf a structure of symplectic manifold. I have proved

Proposition - A Poisson manifold (W, G) admits a foliation of codimension h by symplectic manifolds. There exist on (W, G) atlas of so-called canonical chart $\{x^a, x^p\} = \{x^a, x^\alpha, x^{\bar\alpha}\}$ such that G has only as non vanishing components :

$$G^{\alpha\bar\alpha} = -G^{\bar\alpha\alpha} = 1 \tag{1-8}$$

In particular $G^{Ba} = 0$ and all the components of G are constant

2 - G-cohomology of a Poisson manifold

If A is a i-tensor, $[G, A]$ is a (i+1)-tensor. According to (1-3), the operator $\partial : A \to -[G, A]$ is a cohomology operator ($\partial^2 = 0$) satisfying

$$\partial(A \wedge B) = \partial A \wedge B + (-1)^i A \wedge \partial B \tag{2-1}$$

and also

(2-2) $$\partial[A,B] = -[\partial A,B] - (-1)^{i}[A,\partial B]$$

The cohomology defined by ∂ on the exterior algebra (cf(2-1)) of the skewsymmetrical contravariant tensors is called the G- cohomology of the Poisson manifold. The Schouten bracket induces (cf (2-2))a bracket on the classes of G-cohomology.

Consider a contractile domain U of W. Let A be a i-tensor on U which is a cocycle($\partial A = 0$) ; A is called <u>constant for the foliation</u> if for each canonical chart of domain U , A is of type (i, o), with components constant along the leafs. We have on U

1°) If $i > h$, each i-tensor A which is a cocycle is a coboundary (local triviality)

2°) If $i \leqslant h$, each i-tensor A which is a cocycle is the sum of a coboundary and of a constant tensor for the foliation

We denote $H^{i}(W; G)$ the ith G- cohomology space of (W, G). These spaces can be inter-preted in terms of a cohomology of the dynamical Lie algebra of the Poisson manifold. For a symplectic manifold $H^{i}(W; G) \simeq H^{i}(W; R)$ (de Rham cohomology)

II - CONTACT MANIFOLD AND INVARIANT HAMILTONIAN FORMALISM

3 - Exact symplectic manifolds

a) Let (W,F) be a <u>symplectic manifold</u> of dimension m = (2n+2). A <u>symplectic infinitesimal transformation</u> (s.i.t) - or <u>locally hamiltonian</u> infinitesimal trans formation - is defined by a vector field X such that $\mathcal{L}(X)\ F = 0$, that is such that the 1-form $\mu(X)$ is <u>closed</u>. Let L be the Lie algebra of the s.i.t ; for $X, Y \in L$

(3-1) $$\mu([X,Y]) = di(G)(\mu(X)\wedge\mu(Y))$$

Let L^* be the image by μ^{-1} of the space of the exact 1-forms. We have $[L,L] \subset L^*$ and L^* is the Lie algebra of the globally hamiltonian vector fields. I have proved that $[L, L] = L^*$ and that dim. $L/L^* = b_1(W)$, where $b_1(W)$ is the first Betti number of the homology of W with compact supports.

b) Suppose now that F is exact (W being necessarily non compact) : we choose a 1-form ω of W such that $F = d\omega$ and we consider the structure (W, ω) (with $(d\omega)^{n+1} \neq 0$ everywhere). Introduce the vector field Z defined by :

$$i(Z)d\omega = \omega$$

that is $Z = -\mu^{-1}(\omega)$. We have :

$$(3\text{-}2) \qquad i(Z)\omega = 0$$

and so

$$(3\text{-}3) \qquad \mathcal{L}(Z)\omega = \omega \qquad \mathcal{L}(Z)d\omega = d\omega$$

We denote L_ω the Lie algebra of the i.t preserving ω ; L_ω is the subalgebra of L defined by the elements X such that $[Z, X] = 0$; X belongs to L_ω if and only if

$$\mu(X) = du \qquad u = i(X)\omega \in N$$

Therefore $L\omega \subset L^*$ and we have :

$$(3\text{-}4) \qquad \mathcal{L}(Z)u = u$$

Let N_ω be the subspace of N defined by the solutions of (3-4). We have :

Proposition - The Lie algebra $L\omega$ is isomorphic to the Lie algebra defined on $N\omega$ by the Poisson bracket, for the isomorphism

$$\begin{cases} \tau : X \in L_\omega \to u = i(X)\omega \in N_\omega \\ \tau^{-1} : u \in N_\omega \to X = \mu^{-1}(du) \in L_\omega \end{cases}$$

c) We suppose in the following part that ω is $\neq 0$ everywhere. Let R be the equivalence relation defined on W by the integral curves of Z, π the projection of W on the quotient space $\hat{W} = W/R$ we suppose

Assumption (A) : $\hat{W}$ admits a structure of differentiable manifold such that π is a submersion

$\hat{W}$ is orientable ; the corresponding elements are denoted with a ^ . Each element X of L_ω is projectable by π ; we set $\hat{X} = \pi_*(X)$.

Let x be a point of W. If $\{x^K\}$ $(K = 1,\ldots,n,\bar{0},\bar{1},\ldots,\bar{n})$ is a chart of $\hat{W}$ of domain $\hat{U}$ containing $\pi(x)$, introduce a chart $\{x^o = z_U, x^K\}$ of W of domain U containing x such that on U

$$Z^o = 1 \qquad\qquad Z^K = 0$$

If $w_U = e^{z_U}$, we have :

$$\mathcal{L}(Z|_U)w_U = w_U \qquad\qquad w_U > 0$$

It follows that, on U, the 1-form $\omega|_U/w_U$ satisfies :

$$i(Z|_U)(\omega|_U/w_U) = 0 \qquad\qquad \mathcal{L}(Z|_U)(\omega|_U/w_U) = 0$$

there exists then on $\pi(U)$ a 1-form $\hat{\omega}_U$ such that

(3-5) $$\omega|_U = w_U \pi^* \hat{\omega}_U \qquad\qquad w_U > 0$$

with

$$\hat{\omega}_U \wedge (d\hat{\omega}_U)^n \neq 0$$

The $\hat{\omega}_U$'s define on $\hat{W}$ a contact structure. According to a theorem of Gray and Strong this contact structure can be defined by a global contact 1-form $\hat{\omega}$: there is on $\hat{W}$ a 1-form $\hat{\omega}$ such that :

$$\hat{\omega}_U = f_U \hat{\omega}|_{\pi(U)} \qquad\qquad f_U > 0$$

Let $w > 0$ be the function element of N such that

$$\omega = w\pi^* \hat{\omega}$$

It follows from (3-3) that w belongs to N_ω.

d) We denote $\hat{L}$ the Lie algebra of the infinitesimal automorphisms of the contact structure $(\hat{W}, \hat{\omega})$, that is the Lie algebra of the vector fields $\hat{X}$ such that $(\mathcal{L}(\hat{X}) + \hat{a})\hat{\omega} = 0$, where $\hat{a}$ is a function. If $\hat{N} = C^\infty(\hat{W}; R)$, we know that $\hat{L}$ is isomorphic to the Lie algebra defined on $\hat{W}$ by the Jacobi bracket corresponding to $\hat{\omega}$, for the isomorphism :

$$\sigma : \hat{X} \in \hat{L} \rightarrow \hat{U} = i(\hat{X})\hat{\omega} \in \hat{N}$$

We can prove

Theorem - Under the assumption (A), the Lie algebra L_ω of the infinitesimal automorphisms of (W, ω) is isomorphic by $\pi_* : X \rightarrow \hat{X}$ to the Lie algebra $\hat{L}$ of the infinitesimal automorphisms of the contact manifold $(\hat{W}, \hat{\omega})$.

The Lie algebra N_ω (which is isomorphic to L_ω) is isomorphic by : $\hat{u} \in \hat{N} \rightarrow u = w\ \pi^* \hat{u}$ to the Lie algebra $\hat{N}$ (which is isomorphic to $\hat{L}$) defined by the Jacobi bracket.

4 - Hamiltonian formalism [6]

a) We consider a classical dynamical system defined by means of its space-time M of configurations, where M is a differentiable manifold of dimension $(n+1)$. We denote $\{q^i\} = \{q^o, q^\alpha$ $(i = 0, 1, \ldots, n ; \alpha = 1, \ldots n)$ a local chart of M of domain V ; q^o corresponds eventually to the time. Let $T^* M$ be the cotangent bundle of M, $\psi : T^* M \rightarrow M$ the canonical projection. We denote p a point of $T^* M$; on $\psi^{-1}(v)$ we have a local chart $\{p_i , q^i\}$; $T^* M$ admits a canonical 1-form ω (the Liouville form) such that $F = d\omega$ defines $T^* M$ as an exact symplectic manifold. We have locally :

$$\omega|_{\psi^{-1}(v)} = p_i dq^i = p_o dq^o + p_\alpha dq^\alpha \qquad (4\text{-}1)$$

Consider the manifold W defined by $T^* M$ without the null section. The main vector field Z of the structure (W, ω) is given locally by :

$$Z|_{\psi^{-1}(v)} = \sum_i p_i \frac{\partial}{\partial p_i} \qquad (4\text{-}2)$$

Z is the generator of the one-parameter group of the homotheties of W . The corresponding quotient manifold $\hat{W}$, that is the bundle of the directions of covectors of M , is by definition the manifold of the states of the considered dynamical system. We know that ω defines on $\hat{W}$ a contact structure $\hat{\omega} = 0$. A solution of the problem of motion is an integral curve of a vector field $\hat{X} \in \hat{L}$

The vector field $\hat{X}$ is the projection on $\hat{W}$ of a determined vector field $X \in L_\omega$ of W. An integral curve of $\hat{X}$ is the projection on $\hat{W}$ of convenient integral curves of X. An element X of L_ω is given by $\mu^{-1}(d\mathcal{H})$, where $\mathcal{H}$ (generalized homogeneous hamiltonian) is an element of $N\omega$, that is a function of W satisfying

$$\mathcal{L}(Z)\mathcal{H} = \mathcal{H}$$

so that $\mathcal{H}$ is homogeneous of degree 1 in p. It follows that if $c(s)$ is an integral curve of X on W, we have locally the differential system :

$$(4\text{-}3)\quad \frac{dq^\alpha}{ds}(c(s)) = \frac{\partial\mathcal{H}}{\partial p_\alpha}(c(s)), \quad \frac{dp_\alpha}{ds}(c(s)) = -\frac{\partial\mathcal{H}}{\partial q^\alpha}(c(s)), \quad \frac{dq^o}{ds}(c(s)) = \frac{\partial\mathcal{H}}{\partial p_o}(c(s)), \quad \frac{dp_o}{ds}(c(s)) = -\frac{\partial\mathcal{H}}{\partial q^o}(c(s))$$

Globally, along $c(s)$, we have for each variable :

$$(4\text{-}4)\qquad \frac{dv}{ds} = \{\mathcal{H}, v\}$$

$\mathcal{H}$ is a first integral of (4-3) or (4-4). If e is a given constant, we call <u>pseudoenergy hypersurface</u> a regular connected component Σ_e of dimension $(2n+1)$ of $\mathcal{H}^{-1}(e)$; other hypersurfaces are deduced from Σ_e by homothety. The 2-form F induces on Σ_e a closed 2-form $\tilde{F}$ of rank $2n$; X is tangent to Σ_e and defines on Σ_e a vector field $\check{X}$ such that $i(\check{X})\tilde{F} = 0$. Conversely each vector field of Σ_e for which $\tilde{F}$ vanishes is proportional to $\check{X}$.

b) All the following considerations are local. Let p ($\mathcal{H}(p) = e$) be a point of Σ_e. Suppose $\frac{\partial\mathcal{H}}{\partial p_o}(p) \neq 0$. The relation

$$(4\text{-}5)\qquad \mathcal{H}(p_o, p_\alpha, q^i) = e$$

which defines locally Σ_e, can be solved in p_o, on a neighborhood U of p, according to

$$(4\text{-}6)\qquad p_o + H_U(p_\alpha, q^i) = 0$$

If $c(s)$ is an integral curve of X on Σ_e, we can substitute on U for the parameter s the parameter $t = q^o$. We obtain a curve $t \to \gamma(t)$ which is solution of the differential system :

$$(4\text{-}7)\quad \frac{dq^\alpha}{dt}(\gamma(t)) = \frac{\partial H_U}{\partial p_\alpha}(\gamma(t)) \,,\quad \frac{dp_\alpha}{dt}(\gamma(t)) = -\frac{\partial H_U}{\partial q^\alpha}(\gamma(t)) \,,\quad \frac{dH_U}{dt}(\gamma(t)) = \frac{\partial H_U}{\partial t}(\gamma(t))$$

(4-7) is the usual system of Hamilton. The form $\tilde{F}$ can be written on $\Sigma_e \cap U$:

$$\tilde{F}|_U = -dH_U \wedge dt + dp_\alpha \wedge dq^\alpha \;;$$

we exply thus the classical results of Poincaré-Cartan. If we substitute $k\mathcal{H}$ to $\mathcal{H}$ ($k = c^{te} \neq 0$) (4-7) is preserved, s is changed by a constant factor. For $\mathcal{H} = 1$, s is nothing other as the usual lagrangian action.

Conversely, suppose H_U given on U . We can consider (4-6) as the equation of Σ_1 in U . The equation of Σ_e deduced of Σ_1 by homothety is then :

$$\frac{p_o}{e} + H_U\left(\frac{p_\alpha}{e}, q^i\right) = 0$$

This relation can be solved in e and gives locally

$$\mathcal{H}(p_o, p_\alpha, q^i) = e$$

where $\mathcal{H}$ is necessary homogeneous of degree 1 in p and is defined in a domain of W generated by the homotheties.

We see that each usual system of Hamilton (4-7) can be translated in a differential system (4-4), with a generalized homogeneous hamiltonian $\mathcal{H}$ and the usual action as parameter ; (4-4) can be considered as the general invariant form of the equations of motion and we have obtained an absolute hamiltonian formalism indépendant from any choice of chart (or reference system) in the space-time of the configurations.

The formalism corresponding to the contact structure is absolutely unpleasant for effective calculus.

III - CANONICAL MANIFOLDS AND CANONICAL TRANSFORMATIONS

5 - Canonical manifolds. [4]

In the previous study, the time has been considered as a simple coordinate. We change now the point of view

a) Let $(\tilde{W}, \tilde{G})$ be a symplectic manifold of dimension (2n+2) and we introduce a differentiable map $t : x \in \tilde{W} \to \mathbb{R}$ of rank 1 every where, which is the global time projection. We associate to t the hamiltonian vector field $\tilde{P} \in \tilde{L}^*$ defined by $\tilde{\mu}^{-1}(dt)$. We have :

$$(5\text{-}1) \qquad \mathcal{L}(\tilde{P})\tilde{G} = 0 \qquad\qquad \mathcal{L}(\tilde{P})t = 0$$

Let S be the equivalence relation defined on $\tilde{W}$ by the integral curves of $\tilde{P}$, p the projection of $\tilde{W}$ on the quotient space $W = \tilde{W}/S$. We suppose that W admits a differentiable structure of dimension (2n+1) such that p is a submersion. The 2-tensor $\tilde{G}$ is projectable on W by p and $G = p_* \tilde{G}$ has the rank $2n$ and satisfies

$$(5\text{-}2) \qquad [G, G] = 0$$

Similarly t is the image of a function t on W such that :

$$(5\text{-}3) \qquad [G, t] = 0$$

b) More generally a canonical manifold (W, G, t) is a Poisson manifold of codimension 1, such that the foliation is defined by the global equation $t = \text{const}$.

Let (Σ_τ, F_τ) be a connected leaf corresponding to $t = \tau$, which is a symplectic manifold. In general, there is not on W a closed 2-form F inducing F_τ on Σ_τ for each τ. We have

Proposition - A canonical manifold (W, G, t) admits a 2- form F if and only if it is possible to deduce (W, G, t) by quotient of a symplectic manifold $(\tilde{W}, \tilde{G}, t)$, where $\tilde{W}$ is the product $W \times \mathbb{R}$

6 - Canonical transformations of $(\tilde{W}, \tilde{F}, t)$

a) Let (W, G, t) be a canonical manifold, I_t the ideal of the exterior algebra of the forms ψ such that $dt \wedge \psi = 0$. A closed form ψ of I_t is called I_t-exact if there exists $\chi \in I_t$ such that $\psi = d\chi$. We obtain on I_t an I_t-cohomology with is locally trivial.

b) Let $(\tilde{W}, \tilde{F}, t)$ and $(\tilde{W}', \tilde{F}', t)$ two diffeomorphic symplectic manifolds of dimension (2n+2), with time projections such that the corresponding $\tilde{P}$ end $\tilde{P}'$ define submersions. A diffeomorphism $\tilde{\rho} : \tilde{W} \to \tilde{W}'$ is a canonical transformation of the first manifold onto the second manifold if $\tilde{\rho}$ satisfies the three following conditions

1) $\tilde{\rho}$ preserves t, let $\tilde{\rho}^* t = t$

2) the image of $\tilde{P}$ is $\tilde{P}'$, let $\tilde{\rho}_* (\tilde{P}) = \tilde{P}'$

It follows that $\tilde{\rho}$ defines a diffeomorphism $\rho : W \to W'$ preserving t

3) the image by ρ of G is G', let $\rho_* (G) = G'$

ρ induces a symplectomorphism of (Σ_τ, F_τ) on (Σ'_τ, F'_τ). It is easy to see that there exists a closed 2-form γ on W, element of I_t such that :

$$\tilde{\rho}^*(\tilde{F}') - \tilde{F} = p^*\gamma$$

If γ is I_t-exact, we have $\gamma = dv \wedge dt$. Similar considerations are valid for canonical infinitesimal transformations of $(\tilde{W}, \tilde{F}, t)$

7 - Canonical transformations of (W, G, t)

a) A canonical (resp. infinitesimal) transformation ρ (resp. X) of the canonical manifold (W, G, t) is an automorphism (resp. an infinitesimal automorphism) of the manifold ; G and t are preserved. We denote L^* the Lie algebra of the canonical i. t., L^* the ideal of L defined by the vectors X such that $X = [G, u]$, where $u \in N$; L/L^* is abelian.

b) Consider a canonical manifold (W, G, t) admitting a 2-form F and let ρ be a canonical transformation. There is a closed 2-form $\gamma \in I_t$ such that

$$\rho^*F - F = \gamma$$

If γ is I_t-exact, we have $\gamma = dv \wedge dt$.

c) Let $(\hat{W}, \hat{F})$ a symplectic manifold of dimension 2n. We denote $\hat{W}$ the manifold $W \times \mathbb{R}$, π the projection $W \to \hat{W}$, t the projection $W \to \mathbb{R}$. The 2-tensor $\hat{G}$ of $\hat{W}$ defines on W a structure of canonical manifold (W, G, t) admitting the 2-form $\pi^* \hat{F}$. If ρ is a globally canonical transformation, we have

$$\rho^*(\pi^*\hat{F}) - \pi^*\hat{F} = dv \wedge dt$$

It is the relatively trivial case studied for example by Abraham - Marsden (time independant constraints)

The notion of canonical manifold appears also if we introduce a regular hamiltonian $\tilde{H}$ on a symplectic manifold $(\tilde{W}, \tilde{G})$ and will study the Lie algebra of the first integrals of the corresponding differential system.

IV - LIE ALGEBRAS FOR A POISSON MANIFOLD

8 - The Lie algebras L_G, L, L^* , L^c and their derivations. [9]

a) Consider a Poisson manifold (W, G) of dimension m, codimension $h(m-h = 2n)$. We introduce the ring $\mathcal{A}$ of the functions $a \in N$ such that

$$[G,a] = 0 \tag{8-1}$$

A Poisson infinitesimal transformation is an infinitesimal automorphism of (W, G) defined by a vector field X preserving G . We denote as L_G the Lie algebra of the Poisson i.t ; L_G is an $\mathcal{A}$-modulus.

Consider now the subspace L of L_G defined by the elements X of L_G tangent to the foliation ; L^* is the subspace of L defined by the vector fiels X such that $X = [G, u]$, where $u \in N$; L and L^* are $\mathcal{A}$-modulus and are also ideals of L_G ; the quotient algebra L/L^* is abelian.

b) We say that X defines a conformal Poisson infinitesimal transformation if X preserves G up to a scalar factor. We have

$$\mathcal{L}(X)G = a_X G \tag{8-2}$$

where necessarily $a_X \in \mathcal{A}$. We denote as L^c the Lie algebra of the conformal Poisson i. t. ; L and L^* are ideals of L^c, but L_G is not an ideal of L^c. The notations L, L^* corresponding to a canonical manifold or a symplectic manifold are consistent with our present notations.

A derivation of a Lie algebra L is an endomorphism D of L such that for any $Y, Z \in L$

(8-3) $$D[Y,Z] = [DY,Z] + [Y,DZ]$$

Concerning the derivations of the Lie algebras L, L^*, L^c, L_G, I have proved the following theorem

Theorem - Let (W, G) be a Poisson manifold.

1°) Each derivation of L (resp. L^*) is given by $Y \to [X, Y]$, where $X \in L^c$

2°) Each derivation of the Lie algebra L^c is an inner derivation

3°) Each derivation of L_G is given by $Y \in L_G \to [X, Y] \in L_G$, where X belongs to the normalizer $\mathcal{N}(L_G ; L^c)$ of L_G in L^c.

c) I have proved that each local derivation of the dynamical Lie algebra N is given by $\mathcal{D} = \mathcal{L}(X) + a_X$, where $X \in L^c$. But N can admit non local derivations ; precise results can be obtained.

All these results are valid for symplectic manifolds and canonical manifolds. If (W, G) is a non compact symplectic manifold, all the derivation of N are local. If (W, G) is a compact symplectic manifold, each derivation $\mathcal{D}$ of N is given by the formula

$$\mathcal{D}_u = \mathcal{L}(X)\, u + \lambda \int_W u\, \eta \qquad (\lambda \in R)$$

where $X \in L$ and where η is the symplectic volum element.

9 - 1- differentiable cohomology of the Lie algebra N

a) The Chevalley-Eilenberg cohomology of the Lie algebra N, for exemple, is defined in the following way : a i-cochain C of N is an alternate i-linear map of N^i in N, the 0-cochain being identified with the elements of N. The coboundary of the i-cochain C is the (i+1)-cochain ∂C defined by :

(9-1) $$\partial C(u_0,\dots,u_i) = \frac{1}{i!}\, \varepsilon^{\lambda_0\dots\lambda_i}_{0\dots i} \{u_{\lambda_0}, C(u_{\lambda_1},\dots u_{\lambda_i})\}$$
$$- \frac{1}{2(i-1)!}\, \varepsilon^{\lambda_0\dots\lambda_i}_{0\dots i}\, C(\{u_{\lambda_0},u_{\lambda_1}\},u_{\lambda_2},\dots,u_{\lambda_i})$$

where $u_\lambda \in N$. The space of the 1-cocycles of N is the space of the derivations of N, the space of the exact 1-cocycles is the space of the inner derivations of N.

A i-cochain C is local if, for each $u_1 \in N$ such that $u_1|_U = 0$ on a domain U, we have $C(u_1,\dots u_i)|_U = 0$. If C is local, ∂C is local.

b) A i-cochain C is called 1-differentiable if it is defined by means of first order differential operators on the elements of N. Each 1-differentiable 1-cochain C of N admits a decomposition $C = A + B$, where A (resp. B) is defined by a i-tensor (resp (i-1)-tensor so that locally

$$A(u_1,\dots u_i)|_U = A^{k_1\dots k_i}\,\partial_{k_1}u_1\dots\partial_{k_i}u_i$$

$$B(u_1,\dots u_i)|_U = \frac{1}{(i-1)!}\,\varepsilon^{\lambda_1\dots\lambda_i}_{1\dots i}\,B^{k_2\dots k_i}\,u_{\lambda_1}\,\partial_{k_2}u_{\lambda_2}\dots\partial_{k_i}u_{\lambda_i}$$

An 1-differentiable 1-cochain such that the part of type B vanishes is called pure. We have the following theorem which gives an interpretation of the G-cohomology

Theorem - 1°) The coboundary of a 1-differentiable pure 1-cochain can be defined from the i-tensor A by the (i+1)-tensor $\partial A = -[G, A]$

2°) The ith cohomology space $H^i_{(p)}(N)$ of N for the pure 1-differentiable cohomology is isomorphic to the ith cohomology space $H^i(W; G)$ of the Poisson manifold.

c) Let $C = (A,B)$ a 1-differentiable cochain on N. Straight forward calculus give

$$\partial(A,B) = (-[G,A] + G\wedge B, [G,B]) \tag{9-2}$$

Moreover, we can introduce the exterior product

$$(A,B)\wedge(A',B') = (A\wedge A', B\wedge A' + (-1)^i A\wedge B') \tag{9-3}$$

Let $e(G)$ be the operator on the tensors defined by the exterior product by G ; $e(G)$ acts on the G-cohomology classes of the manifold (W, G) we have proved

Theorem - The ith 1- differentiable cohomology space $H^i(W)$ of the Lie algebra N of the Poisson manifold (W, G) is isomorphic to the space :

where $P^{i-1}(W; G)$ is the kernel of the operator $e(G) : H^{i-1}(W; G) \to H^{i+1}(W; G)$ and where $Q^i(W; G)$ is the image by $e(G)$ of $H^{i-2}(W; G)$. The exterior product (9-3) induces on the set of the 1-differentiable cohomology classes a structure of cohomology algebra.

If G is exact, $H^i(N)$ is isomorphic to $H^{i-1}(W; G) \oplus H^i(W; G)$. Similar results can be obtained for the others Lie algebras.

V - DEFORMATIONS OF THE LIE ALGEBRA N FOR THE SYMPLECTIC CASE

10 - Formal 1-differentiable deformations of N . [3]

a) Let (W, G) be a symplectic manifold of dimension 2n. We denote as $E(N; \lambda)$ the space of the formal functions in λ with coefficients in N. Consider an alternate bilinear map $N \times N \to E(N;\lambda)$ which gives a formal series in λ

(10-1) $$[u,v]_\lambda = \{u,v\} + \sum_{r=1}^{\infty} \lambda^r C_r(u,v)$$

where the $C_r(u, v)$ are 2 cochaines on N which can be extended to $E(N;\lambda)$ in a natural way; (10-1) defines a formal deformation of the Lie algebra N if we have formally the Jacobi identity

(10-2) $$S\left[[u,v]_\lambda, w\right]_\lambda = 0$$

According to Gerstenhaber, (10-2) can be translated by :

(10-3) $$\partial C_t = E_t \qquad (t = 1,2,\ldots)$$

where

(10-4) $$E_t(u,v,w) = \sum_{\substack{r+s=t \\ r,s\geq 1}} SC_s(C_r(u,v),w)$$

If (10-3) satisfied for $t = 1,\ldots, q-1$, we have $\partial E_q = 0$ and E_q is a 3-cocycle of N . The class defined by E_q is the obstruction for the order q to the construction of a formal deformation of N .

We say that

(10-5) $$[u,v]_\lambda = \{u,v\} + \lambda C(u,v)$$

defines an infinitesimal deformation of N if the Jacobi identity is satisfied up to the order 2 that is if C is a 2-cocycle of N .

A formal (resp infinitesimal) deformation of N is 1-differentiable if the C_r (resp. C) are supposed 1-differentiable. This restriction gives a consistent frame for the deformations, according to the following lemma

Lemma - If C, C' are two 1-differentiable 2-cochain on N, the 3-cochain D defined by

$$2D(u,v,w) = SC(C'(u,v).w) + SC'(C(u,v),w)$$

is 1-differentiable.

It follows that in this context, E_t is a 1-differentiable 3-cocycle on N.

b) Consider a formal series in λ

(10-6)
$$T_\lambda = Id + \sum_{s=1}^{\infty} \lambda^s T_s$$

where T_s is a differential operator of order s on N ; T_λ acts naturally on $E(N; \lambda)$. We say that (10-1) is a trivial formal deformation if there exists (10-6) such that the identity

(10-7)
$$T_\lambda [u,v]_\lambda - \{T_\lambda u, T_\lambda v\} = 0$$

is formally satisfied. I have proved the consistence of this definition. Similar definition for a trivial infinitesimal deformation ; C is then exact. We can prove

Proposition - The space of the 1-differentiable infinitesimal deformations of N, modulo the trivial deformations is isomorphic to $H^2(N) \simeq P^1(W; G) \oplus H^2(W; R) / Q^2(W; G)$

11 - Inessential deformations.

a) Consider a formal series in λ :

(11-1)
$$G_\lambda = G + \sum_{r=1}^{\infty} \lambda^r G_r$$

where the G_r are 2-tensors such that the identity

(11-2)
$$[G_\lambda, G_\lambda] = 0$$

is formally satisfied ; $\{u, v\}_{G_\lambda} = i(G_\lambda)(du \wedge dv)$ defines a 1-differentiable formal deformation of N which is deduced from a formal deformation of the geometrical structure. A 1-differentiable formal deformation is called inessential if there exist G_λ and T_λ such that

$$T_\lambda [u,v]_\lambda - \{T_\lambda u, T_\lambda v\}_{G_\lambda} = 0 \tag{11-3}$$

Similar definition for an inessential infinitesimal deformation. We have

Theorem - The space of the 1-differentiable infinitesimal deformations of N, modulo the inessential deformations, is isomorphic to $P^1(W; G)$.

b) Suppose the main 2-form F exact. We have :

$$H^2(N) \simeq H^1(W;R) \oplus H^2(W;R) \qquad H^3(N) \simeq H^2(W;R) \oplus H^3(W;R)$$

We can prove

Theorem - Let (W, F) a symplectic manifold such that F is exact. If $b_1(W) \neq 0$, $b_2(W) = b_3(W) = 0$ (Betti numbers), the dynamical Lie algebra N admits essential, 1-differentiable, formal deformations and in particular non trivial deformations

We note that inessential deformations (but non trivial deformations) can admit non trivial dynamical or physical interpretations.

c) We have proved

Proposition - Let (W, F) be an exact symplectic manifold ($F = d\omega$, $\mu(Z) = \omega$) and let β be a non exact closed 1-form such that $i(Z)\beta = \text{const.}$ If $B = \mu^{-1}(\beta)$

$$[u,v]_\lambda = \{u,v\} + \lambda C(u,v)$$

where

$$C = (Z \wedge B, B)$$

defines a rigourous deformation of N, which is essential.

We have such a situation if $W = T^* M$, where $b_1(M)$ is $\neq 0$

We remark, in conclusion, that Poisson manifolds of arbitrary codimension give a natural geometrical frame for the classical Dirac brackets ; the corresponding theory is very simple, but too long for this talk ([2] and [9]).

R E F E R E N C E S

1 A. Avez, A. Lichnerowicz and A. Diaz-Miranda J. of Diff. Geom. 9 (1974), 1-40

2 P.A. Dirac Canad J. of Math 2 (1950), 129-148 ; Lectures on quantum Mechanics Yeshiva Univ. New York 1964

3 M. Flato, A. Lichnerowicz and D. Sternheimer C.R. Acad Sci. Paris A 279 (1974), 877-881. Déformations 1-differentiables des algèbres de Lie attachées à une variété symplectique ou de contact Compos. Math. 1975 (to appear)

4 M. Flato, A. Lichnerowicz and D. Sternheimer Algèbres de Lie attachées à une variété canonique J. Math. mures et appl. (1975) (to appear)

5 A. Lichnerowicz C.R. Acad. Sc. Paris A 277 (1973), 215-220 ; J. de Math. pures et appl. 53 (1974), 459-484

6 A. Lichnerowicz C. R. Acad. Sc. Paris A 280 (1975), 37-40 ; variétés symplectiques, variétés canoniques et systèmes dynamiques oct. 1974 ; vol in memory of E.T. Davies (to appear) ; C.R. Acad. Sc. Paris A 280 (1975), 1217-1220

7 A. Nijenhuis Indag. Math. 17 (1955), 390-403

8 M. Gerstenhaber Ann. of Math. 79 (1964), 59-103

9 A. Lichnerowicz C.R. Acad. Sc. Paris A 280 (1975), 523-527 ; variétés de Poisson ; J. of Diff. Geom. (to appear)

1o I. Segal Symplectic structures and the quantization problem for wave equations. Conf. on symplectic geometry, Ist. di Alta Matematica, Rome January 1973

ON THE GENERALIZATION OF SYMPLECTIC GEOMETRY TO MULTIPLE INTEGRALS IN THE CALCULUS OF VARIATIONS [+)]

by

Paul Dedecker

Université Catholique de Louvain

1. Introduction

Before going into symplectic geometry, we want first to insist briefly on fundamental peculiarities of multidimensional Calculus of Variations, especially on the fact that the local properties of such a problem are imbricated into the global geometry of Grassmann manifolds. The core of the situation lies in the correct definitions of the following three concepts: 1° Legendre transformation, 2° regularity, 3° phase space. Too often the classical analyses present inadequately these concepts, they hide the interference of the global into the local and they use invariantness with respect to an inadapted pseudogroup. All this with the result that important problems (f.i. one related to Electromagnetism) appear inaccurately as irregulars, owing to a too narrow approach. It turns out, moreover, that our presentation raises natural problems of equivalence (local and global), classification (local and global) and stability.

The generalization to multiple integrals of the symplectic structure arising in classical Mechanics is by no means trivial. We believe that it lies in a sheaf of spectral sequences producing most interesting invariants, of both local and global character, reflecting the geometry of the structure 11 , [12] . These invariants belong to homology theory but they are substantially more complicated than their usual analogues in the algebraic topology of a fibre bundle. It is worthwile to mention that these invariants are a special case of other ones reflecting interesting properties of systems of partial differential equations, more generally, of exterior differential systems.

2. Sketch of the generalized symplectic structure.

To a "regular" variational problem of dimension p over a manifold $\mathcal{V}$ of dimension $n = p + q$, is associated a phase space $\mathcal{E}$ provided with a differential form Ω of degree p . In the classical case (namely for $p = 1$) $\mathcal{E}$ is as usual and Ω is the Pfaffian

+) A table of contents is placed on page 456

form $\Sigma p_i dq^i - Hdt$ of the relative integral invariant of H. Poincaré--E. Cartan. In general, $\mathcal{E}$ is something like a "fibred manifold" with base a "contact manifold" $\tilde{\mathcal{V}}^p$ of $\mathcal{V}$, while the "fibre-map" $\mathcal{E} \to \tilde{\mathcal{V}}^p$ generalizes the Legendre transformation and is an isomorphism only for $p = 1$ or $p + 1 = n$. Specifically $\tilde{\mathcal{V}}^p = \tilde{\mathcal{V}}$ is the manifold of p-dimensional contact elements (of some differential order k) of V. The variational structure defines a filtration of the algebra of differential forms over the phase manifold $\mathcal{E}$: it appears that Ω is of filtration zero, while $d\Omega$ is of filtration two. In the classical case $(p = 1)$, the extremals form a symplectic manifold M whose r-chains correspond exactly to the elements of the groups $E^1_{u,o}$, $E^1_{u,1}$ ... in the homology spectral sequence; moreover, the symplectic structure is precisely defined by $d\Omega$. We can consider a point of M as an equivalence class of points in $\mathcal{E}$, namely as an equivalence class of initial conditions for the problem of extremals. For p arbitrary, an initial condition for the problem of extremals is more complicated and one can think of it as a closed submanifold $\mathcal{W} = \mathcal{W}_{p-1}$ of dimension p-1 in $\mathcal{E}$. Considering as "equivalent" two such (p-1)-manifolds $\mathcal{W}$ if they lie in the same extremal, yields the idea of the "pseudo-manifold" M of their equivalence classes. It is out of question to discuss here whether the quotient object M is a "manifold" in some appropriate sense and, no doubt, its "local structure" cannot be completely tame by essence. However, one has good reasons to approach M by studying its "moral" groups of r-chains, namely $E^1_{u,p-1}$, $E^1_{u,p}$, ... In particular the classical "symplectic structure" generalizes in the fact that $d\Omega$ induces elements Γ_i in the terms $E_2^{i,p-i+1}$, $i = 2,3,\dots,p+1$. About them, see section 15, in particular the proof of theorem 15.2.

In the classical case $(p = 1)$, only the values $i = 1,2$ are significant and this cocycle Γ_1 is locally the coboundary of a cochain Θ in $E_1^{1,p-1}$ or, equivalently, Γ_2 vanishes in $E_2^{2,p-1}$. In other words $d\Omega$ of filtration two is locally the differential of a form Θ of filtration one, so that, locally again, there exists a (p-1)-form Σ such that

$$\Omega = d\Sigma + \Theta$$

An identical formula holds for abritrary p provided $\Gamma_2, \Gamma_3, \dots \Gamma_{p+2}$ vanish, which is the case for at least a wide class of problems. This probably opens the way to an important generalization of the theory of waves, in particular of Geometrical Optics and, perhaps, of Quantum Mechanics. The problem, moreover, seems related to relatively old work of E. Vessiot [35] .

To clarify intuitively the above-mentioned filtration, let us say that, in the phase manifold $\mathcal{E}$, a smooth r-dimensional family of extremals (those are submanifolds of dimension p) spans a (p+r)-manifold $X_{r,p}$ and that a (p+r)-differential form vanishing on each such $X_{r,p}$, is said to be of filtration $\geq$ r+1 (and of complementary degree $\leq$ p+1). Dually $X_{r,p}$ is of "filtration" $\leq$ r, yielding the significant evaluation of a form of filtration s over a smooth cochain or manifold of the same or higher filtration.

3. Bibliographical and epistemological remarks.

An effort has been made recently by H. Goldschmidt and S. Sternberg [19] to give an extension to multiple integrals of the classical Hamiltonian formalism valid for simple integrals as analysed and described by Elie Cartan [3] . The subject is not new and was already treated by C. Caratheodory [6] in 1929 and Th. de Donder [14] and Hermann Weyl [36] in 1935. However, there exists a great gap between Caratheodory's and de Donder-Weyl's interpretations. The latter, contrary to Caratheodory's, does not have the "invariantive" properties one should expect, with respect to "geometric" transformations, i.e. those mixing time and space coordinates as was precisely achieved by E. Cartan in the case of simple integrals. This lead Th. Lepage to describe a large family of theories containing both those of Caratheodory and de Donder-Weyl as special cases, through the so-called "congruences of Lepage" [27] which were later interpreted in terms of fibre bundle theory [8] , [10], [12] and were also shown to carry a flavor of "Lagrange multipliers".

Surprisingly, many authors dealing recently with multiple integrals, among them Goldschmidt-Sternberg (loc. cit.) and Palais-Smale [32] follow de Donder-Weyl's line and base their study on a narrow "invariantive" point of view, slipping over fundamental aspects of E. Cartans's contribution. Indeed, they start with a fibred manifold $\xi: \mathcal{V} \longrightarrow \mathcal{T}$ in some vague sense and one could even consider a foliation; here we shall assume that $\mathcal{V}$ and $\mathcal{T}$ are smooth manifolds of dimensions n = p+q and p respectively and that ξ is a smooth map of rank p everywhere. Then they consider the space $J^k(\xi)$ of k-jets of local sections $s: \mathcal{T} \longrightarrow \mathcal{V}$ (k-jet or: jet of differential order k). In dynamical language, $\mathcal{T}$ is a multidimensional "time" manifold (or a manifold of times).

Considering a "Lagrange function", i.e. a sufficiently smooth function L on $J^k(\xi)$, they consider the problem of extremizing the integral

$$I(s) = \int_D L(\tilde{s})\omega \qquad (3.1)$$

where: (i) D is a compact domain of $\mathcal{T}$ (that is a submanifold with boundary of the same dimension p as $\mathcal{T}$), (ii) $\tilde{s}: D \longrightarrow J^k(\xi)$ is the k-flow (or canonical lifting) of a section $s: D \longrightarrow \mathcal{V}$ and (iii) ω is a volume element or a p-form on $\mathcal{T}$, vanishing nowhere.

In local coordinates, the description is as follows. Let (t^α), $1 \le \alpha \le p$, be local coordinates in $\mathcal{T}$. We understand that there exist local coordinates (x^r), $1 \le r \le n$, in $\mathcal{V}$ such that ξ is locally defined by

$$t^1 = x^1,\ t^2 = x^2, \ldots,\ t^p = x^p$$

Such a local coordinate system (x^r) in $\mathcal{V}$ will be said <u>compatible</u> with the fibration. A local section $s: \mathcal{T} \longrightarrow \mathcal{V}$ is then defined locally by equations

$$x^i = f^i(t^1, \ldots, t^p), \qquad p+1 \le i \le p+q = n .$$

At a point $t = (t^\alpha) \in \mathcal{T}$ the value $\tilde{s}(t)$ of the k-flow of s is determined by the values $x^i = f^i(t)$ and the derivatives

$$x^i_\alpha = \frac{\partial f^i}{\partial t^\alpha},\ x^i_{\alpha\beta} = \frac{\partial^2 f^i}{\partial t^\alpha \partial t^\beta}, \ldots,\ x^i_{(\alpha)} = \frac{\partial^k f}{\partial t^{(\alpha)}}$$

where $(\alpha) = (\alpha_1, \ldots, \alpha_k)$ is a succession of k integers from 1 to p. Thus $J^k(\xi)$ is a manifold with local coordinates

$$t^\alpha, x^i, x^i_\alpha, x^i_{\alpha\beta}, \ldots, x^i_{(\alpha)}$$

(with obvious symmetry relations in the multi-indices).

The pseudogroup governing the situation just described is defined by local diffeomorphisms of the type

$$\bar{t}^\alpha = F^\alpha(t^1, \ldots, t^p), \quad \bar{x}^i = G^i(t^1, \ldots, t^p, x^{p+1}, \ldots, x^n) \quad \text{or}$$

$$\bar{t}^\alpha = F^\alpha(t^\beta), \quad \bar{x}^i = G^i(t^\beta, x^j) \qquad (3.2)$$

No doubt, many physical and geometrical problems fit within such a structure, but many others don't. In the classical case of Mechanics, when $p = 1$ and $\mathcal{T}$ is a time line, that kind of structure applies only if time is an invariant concept, a situation which is against the "Principle of Relativity", when applicable. The main feature of the preceding theory is that $J^k(\xi)$ is fibred over $\mathcal{V}$ with fibre a Euclidean space.

However, this fibre is also an open set in the kth-order Grassmann manifold $G^k_{p,q}$ of p-dimensional contact elements of differential order k in $\mathbb{R}^{p+q}$. The latter is a definitely non topologically trivial manifold which must be present in the study of any variational problem on $\mathcal{V}$ dealing with p-dimensional submanifolds without transversality condition to the fibres of a fibration $\mathcal{V} \longrightarrow \mathcal{C}$, or to the leaves of some foliation. One important such problem is the p-dimensional Plateau problem: there $\mathcal{V}$ is an n-dimensional Riemann manifold and one looks for the p-dimensional minimal submanifolds with prescribed boundary. More general problems of this type, namely without transversality condition, are sometimes improperly called "problems in parametric form"; they are fundamental for a fully comprehensive generalization of E. Cartan's work. Such problems are governed by the general pseudogroup of transformations mixing the "time" coordinates t^α, with the "space" coordinates x^i, that is the pseudogroup of transformations of the form:

$$\bar{t}^\alpha = F^\alpha(t^\beta, x^j), \quad \bar{x}^i = G^i(t^\beta, x^j) \qquad (3.3)$$

We proceed to discuss briefly the situation for $k = 1$.

4. Definition of a first order variational problem of degree p.

Let $\mathcal{V} = \mathcal{V}_n$ be an n-dimensional smooth (e.g. C^∞) manifold. We assume $p<n$ and we put $q = n-p$. By a p-dimensional contact element of $\mathcal{V}$ at a point $x \in \mathcal{V}$, we mean a p-dimensional linear subspace $X = X_p$ in the tangent space $T_x(\mathcal{V})$. We denote by $\tilde{\mathcal{V}} = \tilde{\mathcal{V}}^p = \tilde{\mathcal{V}}^p_n$ the space of all these contact elements with obvious projection $\pi: \tilde{\mathcal{V}} \to \mathcal{V}$. This is a fibred manifold with fibre $G_{p,q}$, the Grassmann manifold of p-dimensional linear subspaces of $\mathbb{R}^n$.

Let $\mathcal{A}$ be the complete atlas defining the smooth structure of $\mathcal{V}$, consisting of maps, or charts, $\psi: U_\psi \to V_\psi$ of source U_ψ, an open part of $\mathcal{V}$, to V_ψ, an open part of $\mathbb{R}^n$. This defines local coordinates $x^r = x^r_\psi$ in U_ψ and also the p-form

$$\omega = \omega_\psi = dx^1_\psi \wedge \ldots \wedge dx^p_\psi \quad \text{in} \quad U_\psi \qquad (4.1)$$

The contact elements at the points $x \in U_\psi$ form an open subset of $\tilde{\mathcal{V}}$

which we denote by $\tilde{U}_\psi = \pi^{-1}(U_\psi)$ and which is isomorphic to $U_\psi \times G_{p,q}$. It contains the open subset $U^\#_\psi$ of those X_p's in $\tilde{U}_\psi$ which do not annihilate ω_ψ, a subset isomorphic to $U_\psi \times \mathbb{R}^{p\cdot q}$. If we denote by (x^i_α) a $p\times q$ matrix, an element $X \in U^\#_\psi$ at a point $x \in U_\psi$ is spanned by the tangent vectors a at x with ψ-components

$$a^\alpha_\psi = \lambda^\alpha , \qquad a^i_\psi = \Sigma\, x^i_\alpha \lambda^\alpha ,$$

where the λ^α's are parameters. Here, we use indexes $\alpha,\beta,\ldots$, $i,j,\ldots$, $r,s,\ldots$ with the following ranges:

$$1\leq \alpha,\beta,\ldots\leq p, \quad p+1 \leq r,s,\ldots\leq p+q = n \tag{4.2}$$

$$1\leq r,\ s,\ldots\leq n$$

The correspondence $(x^i_\alpha) \mapsto X$ yields a local chart

$$\psi^\#:\ U^\#_\psi \rightarrow V_\psi \times \mathbb{R}^{p\cdot q}$$

if we put

$$\psi^\#(X) = (x^r, x^i_\alpha) = (x^r_\psi, x^i_{\alpha,\psi}) \quad \text{where} \quad (x^r_\psi) = \psi\circ\pi(X).$$

The family $\mathcal{A}^\# = \{\psi^\#:\ U^\#_\psi \rightarrow V^\#_\psi \subset \mathbb{R}^n \times \mathbb{R}^{p\cdot q}\}$ is a (non complete) atlas defining $\tilde{\mathcal{V}}$ as a smooth manifold and the local representation of the projection π with respect to the pair $(\psi^\#, \psi)$ is obviously the map

$$(x^r, x^i_\alpha) \mapsto (x^r)$$

Let $\mathcal{W}$ be a p-dimensional smooth manifold with boundary and let $f: \mathcal{W} \rightarrow \mathcal{V}$ be an immersion, namely a smooth map of rank p everywhere. At each point $t \in \mathcal{W}$ the tangent space $T_t(\mathcal{W})$ induces an $X = \tilde{f}(t)$ in $T_{f(t)}(\mathcal{V})$. We thus have another immersion $\tilde{f}: \mathcal{W} \rightarrow \tilde{\mathcal{V}}$ making commutative the triangle

$$\begin{array}{ccc} & & \tilde{\mathcal{V}} \\ & \overset{\tilde{f}}{\nearrow} & \downarrow \pi \\ \mathcal{W} & \xrightarrow[f]{} & \mathcal{V} . \end{array}$$

For every $t \in \mathcal{W}$, $\tilde{f}(t)$ lies in some $U^\#_\psi$. One can then find local coordinates (t^α) in a neighbourhood W of t in $\mathcal{W}$, such that the local expression of f be of the form

$$x^1 = t^1, \ldots, x^p = t^p , \qquad x^i = f^i(t^1, \ldots, t^p);$$

the local expression of $\tilde{f}$ is then:

$$x^\alpha = t^\alpha, \quad x^i = f^i(t^\alpha), \quad x^i_\alpha = \frac{\partial f^i}{\partial t^\alpha} .$$

In the Calculus of Variations, one is interested in computing an integral over the pair $(\mathcal{W}, f)$ the value of which does not depend only on the immersion f, but also on its orientation around each point $f(t)$. This means that this integral involves a differential form not belonging properly to the space $\mathcal{V}$, but to the space $\tilde{\mathcal{V}}$. Locally it will be written as

$$I(f) = \int L\left(t^\alpha, f^i(t^\alpha), \frac{\partial f^i}{\partial t^\alpha}\right) dt^1 \wedge \ldots \wedge dt^p. \tag{4.3}$$

The function L is called a <u>Lagrangian function</u>; it depends on the local chart ψ and is of the type

$$L = L(t^\alpha, x^i, x^i_\alpha) = L_\psi(t^\alpha, x^i, x^i_\alpha)$$

For instance, if $\mathcal{V}$ is just $\mathbb{R}^n$ provided with the usual Euclidean metric, the p-dimensional area of $f(\mathcal{W})$ involves a function L which is the square root of the sum

$$1 + \Sigma (x^i_\alpha)^2 + \sum_{\alpha<\beta}^{i<j} \begin{vmatrix} x^i_\alpha & x^j_\alpha \\ x^i_\beta & x^j_\beta \end{vmatrix}^2 + \sum_{\alpha<\beta<\gamma}^{i<j<k} \begin{vmatrix} x^i_\alpha & x^j_\alpha & x^k_\alpha \\ x^i_\beta & x^j_\beta & x^k_\beta \\ x^i_\gamma & x^j_\gamma & x^k_\gamma \end{vmatrix}^2 + \ldots$$

One sees, however, that the differential form L

$$L(x^\alpha, x^i, x^i_\alpha)\, dx^1 \wedge \ldots \wedge dx^p \tag{4.4}$$

is not an arbitrary differential form on $\tilde{\mathcal{V}}$ since it does involve only the differentials of the local coordinates in U_ψ, not the differentials dx^i_α along the fibres of $\pi: \tilde{\mathcal{V}} \to \mathcal{V}$. Such a differential form over $\tilde{\mathcal{V}}$ (more generally over a fibred manifold) is called <u>semi-basic</u>. It is not essentially monomial as in (4.4) but could be an expression

$$\sum_{r_1 < \ldots < r_p} A_{r_1 \ldots r_p}(x^\alpha, x^i, x^i_\alpha) dx^{r_1} \wedge \ldots \wedge dx^{r_p} \tag{4.5}$$

whose integral over $f(\mathcal{W})$ would always be reducible to the form (4.3). However, a general expression of the type (4.5) happens to be most important for many reasons. One is the invariantness with respect to the pseudogroup of transformations (3.3). Indeed a monomial like (4.4) is not invariant and will transform into the general expression (4.5). But we want to observe that a transformation of type (3.2) transforms a monomial of type (4.4) into one of the same type.

It is interesting to see that any immersion $\tilde{f}$ is an integral of the Pfaffian semi-basic equations

$$\omega^i \equiv dx^i - \Sigma x^i_\alpha dx^\alpha = 0 , \tag{4.6}$$

a highly not completely integrable system.

Clearly, the two systems

$$dx^1,\dots,dx^p,dx^{p+1},\dots dx^n ,$$

$$dx^1,\dots,dx^p,\ \omega^{p+1},\dots,\ \omega^n ,$$

are bases for the (free) module of local semi-basic forms. Owing to their importance, we describe the corresponding bases for the (free) module of semi-basic forms of degree p . Starting with the monomial (4.1), we may eliminate one factor dx^α and replace it by dx^i , producing a form

$$\theta^i_\alpha = dx^1\wedge\dots\wedge dx^{\alpha-1}\wedge dx^i\wedge dx^{\alpha+1}\wedge\dots\wedge dx^p .$$

Similarly, we can eliminate two factors dx^α,dx^β and replace them by dx^i,dx^j , obtaining a form $\theta^{ij}_{\alpha\beta}$ .More generally we define forms $\theta^{ijk}_{\alpha\beta\gamma}$, etc... . Instead of replacing the dx^α's by the dx^i's , we can replace them by the ω^i's obtaining forms which we shall denote ω^i_α, $\omega^{ij}_{\alpha\beta}$, $\omega^{ijk}_{\alpha\beta\gamma}$,...For example if $\alpha<\beta$ we have

$$\omega^{ij}_{\alpha\beta} = dx^1\wedge\dots\wedge\omega^i\wedge\dots\wedge\omega^j\wedge\dots\wedge dx^p .$$

The corresponding bases for the module of semi-basic p-forms are then the following:

$$\omega,\theta^i_\alpha,\theta^{ij}_{\alpha\beta},\theta^{ijk}_{\alpha\beta\gamma},\dots \text{ and} \tag{4.7}$$

$$\omega,\omega^i_\alpha,\ \omega^{ij}_{\alpha\beta},\ \omega^{ijk}_{\alpha\beta\gamma},\dots \tag{4.8}$$

A differential form (4.4), after a coordinate transformation (3.3) becomes of one of the following types:

$$A\omega \quad + \Sigma A_i^\alpha \theta_\alpha^i + \sum_{\alpha<\beta} A_{ij}^{\alpha\beta}\theta_{\alpha\beta}^{ij} + \dots \text{ or}$$

$$M\omega \quad + \Sigma M_i^\alpha \omega_\alpha^i + \sum_{\alpha<\beta} M_{ij}^{\alpha\beta}\omega_{\alpha\beta}^{ij} + \dots \qquad (4.9)$$

where the coefficients depend on all coordinates. The first type is just a new presentation of (4.5). The interest of the second expression is that the coefficient M is the new lagrangian function: indeed, integrating over a solution $\tilde{f}$ of the system (4.6), only the first term $M\omega$ is not zero.

We define a <u>multiplicity</u> (more precisely a p-<u>multiplicity</u>) to be a smooth map $F: \mathcal{W} \to \tilde{\mathcal{V}}$ such that: (i) $\mathcal{W}$ is a manifold of dimension p and (ii) F is of the form $\tilde{f}$ for some $f: \mathcal{W} \to \mathcal{V}$. The last condition clearly means that $F = \tilde{f}$ for $f = \pi \circ F$. Let us now consider a smooth map $F: \mathcal{W} \to \tilde{\mathcal{V}}$ for which $\mathcal{W}$ is p-dimensional and for which $f = \pi \circ F$ is an immersion (that is, of rank p at each point $t \in \mathcal{W}$); it is easy to check that the three following conditions are equivalent:

(i) F <u>is a multiplicity</u>;

(ii) F <u>is an integral manifold of the Pfaffian system</u> (4.6);

(iii) F <u>is an integral manifold of the exterior differential system</u>:

$$\omega_\alpha^i = \omega_{\alpha\beta}^{ij} = \omega_{\alpha\beta\gamma}^{ijk} = \dots = 0 . \qquad (4.10)$$

To be more explicit, both the systems (4.6) and (4.10) are defined in the open sets $U_\psi^\#$ (where ψ runs over the atlas $\mathcal{A}$) and we mean that, for each $\psi \in \mathcal{A}$ and for each open part W_ψ of $\mathcal{W}$ such that $F(W_\psi) \subset U_\psi^\#$, the restriction of F to W_ψ annihilates every form of the corresponding system.

In each $U_\psi^\#$ (more generally, in every open set U of $\mathcal{V}$) we have the algebra $\mathcal{B}(U_\psi^\#)$ (resp. the algebra $\mathcal{B}(U)$) of semi-basic forms; this is an algebra over the ring of arbitrary smooth functions. In it, the forms $\omega_{\alpha,\psi}^i$ generate an ideal $\mathcal{I}(U_\psi^\#)$ and we shall denote by $\mathcal{I}^p(U_\psi^\#)$ the module of p-forms in this ideal. [Notations like $\mathcal{I}_\psi(U_\psi^\#)$ and $\mathcal{I}_\psi^p(U_\psi^\#)$ would be more precise.] We also have the algebra (over the same ring) $\mathcal{G}(U_\psi^\#)$ of all differential forms in $U_\psi^\#$ and, in it, the ideal $\mathcal{J}(U_\psi^\#)$ of forms generated by the ω_ψ^i's and their differentials

$$d\omega^i = -\sum dx_\alpha^i \wedge dx^\alpha .$$

The latter is a differential ideal (that is: stable by exterior differentiation) and we shall denote by $\mathcal{J}^p(U_\psi^\#)$ the module of p-forms into it.

If U is an open subset of $U_\psi^\#$, we shall denote by $\mathcal{I}_\psi(U)$ the ideal of $\mathcal{B}(U)$ generated by the restrictions of the forms ω_ψ^i to U . We define similarly $\mathcal{I}_\psi^p(U)$, $\mathcal{J}_\psi(U)$, $\mathcal{J}_\psi^p(U)$. It is clear that, in case $U_\psi = U_\psi \cap U_\varphi$ is non empty, on has

$$\mathcal{I}_\psi(U_{\psi\varphi}^\#) = \mathcal{I}_\varphi(U_{\varphi\psi}^\#) \quad , \text{etc....}$$

<u>Definition 4.1.</u> Let $\tilde{\mathcal{O}}$ be an open set in $\tilde{\mathcal{V}}$. A <u>p-dimensional variational problem</u> of differentiable order one on $\mathcal{V}$ is defined by a family $\mathcal{L} = \{L_\psi\}_{\psi\in\mathfrak{A}}$ of smooth functions

$$L_\psi : \tilde{\mathcal{O}} \cap U_\psi^\# = O_\psi^\# \longrightarrow R$$

such that, for every non empty intersection $O_{\psi\varphi}^\# = O_\psi^\# \cap O_\varphi^\#$, it is the case that, in this intersection, the difference

$$L_\psi dx_\psi^1 \wedge \dots \wedge dx_\psi^p - L_\varphi dx_\varphi^1 \wedge \dots \wedge dx_\varphi^p$$

belongs to $\mathcal{I}_\psi^p(O_{\psi\varphi}^\#) = \mathcal{I}_\varphi^p(O_{\psi\varphi}^\#)$; in other words, this difference has to be a linear combination of the forms ω_α^i, $\omega_{\alpha\beta}^{ij}$, $\omega_{\alpha\beta\gamma}^{ijk}$,... corresponding to either the chart ψ or the chart φ . One may also ask, additionally, that $\pi(\tilde{\mathcal{O}}) = \mathcal{V}$, otherwise the problem would be one over the manifold $\pi(\tilde{\mathcal{O}})$. It is, however, essential to consider the possibility of a problem limited to an open set $\tilde{\mathcal{O}}$ strictly smaller than $\tilde{\mathcal{V}}$.[A classical example is the <u>Dirichlet</u> problem corresponding to $p = 2$, $q = 1$, $\mathcal{V} = \mathbb{R}^3$ and the integral

$$I = \int (p^2+q^2)\,dx \wedge dy$$

where x,y,z are coordinates in $\mathbb{R}^3$ and $z = z(x,y)$, $p = \partial z/\partial x$, $q = \partial z/\partial y$. There, $\mathbb{R}^3$ can be considered as one U_ψ and the problem cannot be extended out of $\tilde{\mathcal{O}} = U_\psi^\# \approx \mathbb{R}^3 \times \mathbb{R}^2$, with coordinates x,y,z,p,q .]

A general problem of the type just described will be denoted $\mathcal{P}_1 = \mathrm{Var}_1(\mathcal{V}, \tilde{\mathcal{O}}, \mathcal{L})$.

Of course, these definitions can be presented with greater

elegance in sheaf language. We have the sheaf $\mathcal{B}$ of germs of semi-basic forms on $\tilde{\mathcal{V}}$ and a subsheaf of ideals $\mathcal{J}$ induced by the ω^i_ψ's . In degree p , these sheafs have, as components, sheafs of modules $\mathcal{B}^p$ and $\mathcal{J}^p$. Clearly the local forms

$$L_\psi dx^1_\psi \wedge \ldots \wedge dx^p_\psi$$

associated to the family $\mathcal{L}$ of definition 4.1 define a section $\sigma_{\mathcal{L}}$ over $\tilde{\mathcal{O}}$ of the quotient sheaf $\mathcal{B}^p/\mathcal{J}^p$;

$$\sigma_{\mathcal{L}} : \tilde{\mathcal{O}} \longrightarrow \mathcal{B}^p/\mathcal{J}^p ;$$

reciprocally, a section σ over $\tilde{\mathcal{O}}$ of the same sheaf defines uniquely a family $\mathcal{L}_\sigma$. We can thus speak of a variational problem $\mathrm{Var}_1(\mathcal{V},\tilde{\mathcal{O}},\sigma)$ and one has

$$\mathrm{Var}_1(\mathcal{V},\tilde{\mathcal{O}},\mathcal{L}) = \mathrm{Var}_1(\mathcal{V},\tilde{\mathcal{O}},\sigma_{\mathcal{L}})$$

Correspondingly, we shall denote by $\mathcal{G}$ the sheaf of germs of all differential forms on $\tilde{\mathcal{V}}$ and $\mathcal{J}$ the subsheaf of ideals corresponding to the local modules $\mathcal{J}(U^\#_\psi)$. In it, we have the sheaf $\mathcal{J}^p$ of modules, consisting of the germs of degree p.

It is clear that for an immersion $f: \mathcal{W} \to \mathcal{V}$ as above and such that $\tilde{f}(\mathcal{W})$ lies in the open set $\tilde{\mathcal{O}}$, one can compute, in an obvious sense, the integral

$$I_{\mathcal{L}}(f) = \int_{\tilde{f}} \sigma_{\mathcal{L}} . \qquad (4.11)$$

Considering then "variations of f with fixed boundary", we are interested in the immersions f for which the "first variation " $\delta I_{\mathcal{L}}$ of the integral (4.9) vanishes.

Definition 4.2. The function $I_{\mathcal{L}}(-)$ will be called the action of the problem $\mathrm{Var}_1(\mathcal{V},\tilde{\mathcal{O}},\mathcal{L})$. The immersions f on which the first variation $\delta I_{\mathcal{L}}$ vanishes are called the extremals of the first order problem $\mathrm{Var}_1(\mathcal{V},\tilde{\mathcal{O}},\mathcal{L})$. [We are not interested in finding proper maxima and minima.]

5. Definition of a zero order variational problem of degree p .

We now consider a smooth N-dimensional manifold $\mathcal{H}$ together with a differential form θ of degree p over $\mathcal{H}$. Again, if $f: \mathcal{W} \to \mathcal{H}$ is a smooth map originating from a smooth compact manifold with boundary of dimension p , one can compute the integral

$$I_\Theta(f) = \int_f \Theta \ . \qquad (5.1)$$

Definition 5.1. We call zero-order variational problem over $\mathcal{V}$, associated to the p-form Θ the problem of finding those immersions f which annihilate the "first variation" δI_Θ of (5.1) for "variations" of f fixed over the boundary $\partial \mathcal{W}$ of $\mathcal{W}$. This problem is denoted $\mathrm{Var}_0(\mathcal{V}, \Theta)$ and an immersion on which δI_Θ vanishes is called an extremal of this zero-order problem.

We shall see the importance of looking for extremals of a restricted type. Suppose, for example that we are given, in addition, a smooth "projection" $\pi : \mathcal{H} \to \mathcal{Y}$ where $\mathcal{Y}$ is another smooth manifold. We will then be interested in the extremals $f : \mathcal{W} \to \mathcal{H}$ subject to the condition that $\pi \circ f : \mathcal{W} \to \mathcal{Y}$ be an immersion, that is, be of rank $p = \dim(\mathcal{W})$ everywhere. In other words, we are looking for the extremals lying in a "generic position" with respect to $\mathcal{Y}$; equivalently, which are transversal to the fibres of π. This defines a new variational problem of differential order zero and dimension p which we denote $\mathrm{Var}_0(\mathcal{H}, \Theta, \pi)$.

It turns out that the core of the theory of a first order variational problem is to make it equivalent (in an appropriate sense) to a zero-order problem. We shall see that this involves a condition of regularity; that the equivalence is possible for a regular problem and that the regularity condition makes it possible to define a phase space with canonical coordinates which is the natural frame into which generalize naturally the classical Hamilton-Poincaré- E.Cartan formalism of classical Mechanics, applicable to the simple integral problems.

Our results were already presented in [8], [10] and [12] and are significantly more general than those more recent of [19]. It is important to mention here that our theory is invariant with respect to the general pseudogroup of transformations (3.3), while the theory of Goldschmidt and Sternberg is invariant only under the subpseudogroup of transformations of type (3.2). These authors introduce a regularity condition based on a so-called Legendre transformation which they derive from a kind of universal condition (lemma 3.1 of [19]). However, neither this transformation nor this condition have a fully invariant meaning. Their Legendre transformation is invariant only through (3.2) and was already used in [14] and [36], while in [6] Caratheodory had another one (see remark 8.6).

6. Notion of regularity.

The integral (4.3) can be written, locally in $U^{\#}_{\psi}$, as

$$I(f) = \int_{\tilde{f}} L_{\psi}(x^{\alpha},x^{i},x^{i}_{\alpha})dx^{1}_{\psi}\wedge\ldots\wedge dx^{p}_{\psi} = \int_{\tilde{f}} L_{\psi}\omega_{\psi}$$

As the integration domain is a solution of the equations (4.6) or (4.8), the value of the integral is not altered if we add to the differential form $L_{\psi}\omega_{\psi}$ a linear combination of the first members in (4.8), which yields the form $\bar{\Omega}_{\psi}\varepsilon\ \mathcal{B}^{p}(O^{\#}_{\psi})$:

$$\bar{\Omega}_{\psi} = L_{\psi}\omega + \sum \Lambda^{\alpha}_{i}\ \omega^{i}_{\alpha} + \sum \Lambda^{\alpha\beta}_{ij}\ \omega^{ij}_{\alpha\beta} + \sum \Lambda^{\alpha\beta\lambda}_{ijk}\ \omega^{ijk}_{\alpha\beta\lambda} + \ldots; \tag{6.1}$$

the sums are restricted to the combinations of indices satisfying

$$\alpha<\beta<\ldots \quad \text{and} \quad i<j<\ldots \tag{6.2}$$

As we see, the parameters Λ are in number

$$N = N_{p,q} = \binom{p}{1}\cdot\binom{q}{1} + \binom{p}{2}\cdot\binom{q}{2} + \ldots + \binom{p}{k}\cdot\binom{q}{k},\ k = \min(p,q)$$
$$= \binom{p+q}{p}-1 = \binom{n}{p}-1$$

They can be taken as arbitrary functions defined in $O^{\#}_{\psi}$, but we want to consider them as coordinate functions in $\mathbb{R}^{N}$, so that (6.1) represents a differential form over the product $O^{\#}_{\psi}\times\mathbb{R}^{N}$. It is then easy to glue together these products over the intersections $O^{\#}_{\psi\varphi} = O^{\#}_{\psi}\cap O^{\#}_{\varphi}$, so as to obtain a unique bundle manifold

$$\bar{\chi}: \bar{\mathcal{E}} \longrightarrow \tilde{\mathcal{O}}$$

of fibre $\mathbb{R}^{N}$ and structure group the affine group $A(N,\mathbb{R})$, provided with a unique differential form $\bar{\Omega}$ of restriction $\bar{\Omega}_{\psi}$ over $O^{\#}_{\psi}\times\mathbb{R}^{N}$.

In the problem $\mathrm{Var}_1(\mathcal{V},\tilde{\mathcal{O}},\mathcal{L})$, we look for extremals $\tilde{f}: \mathcal{W}\to\tilde{\mathcal{V}}$ subject to the conditions (4.8). The parameters Λ thus appear as Lagrange multipliers and it is reasonable to compare our initial first order problem $\mathrm{Var}_1(\mathcal{V},\tilde{\mathcal{O}},\mathcal{L})$ to the zero-order problem $\mathrm{Var}_0(\bar{\mathcal{E}},\bar{\Omega},\pi\circ\bar{\chi})$. This yields the following:

<u>Theorem 6.1. For every extremal</u> $\bar{f}: \mathcal{W}\to\bar{\mathcal{E}}$ <u>of the problem</u> $\mathrm{Var}_0(\bar{\mathcal{E}},\bar{\Omega},\pi\circ\bar{\chi})$, <u>the following properties hold:</u>

(i) the image $\bar{f}(\mathcal{W}) \cap U_\psi^{\#} \times \mathbb{R}^N$ lies in the submanifold of $\bar{\mathcal{E}}$ defined by the local equations

$$\Lambda_i^\alpha = \frac{\partial L_\psi}{\partial x_\alpha^i} ; \tag{6.3}$$

(ii) the map $\bar{f}$ is an integral manifold of the sheaf $\mathcal{J}^p$; equivalently $\bar{f}$ satisfies locally the equations (4.10); or its projection $\bar{\chi}\circ\bar{f}: \mathcal{W} \to \tilde{\mathcal{V}}$ is the canonical lifting $\tilde{f}: \mathcal{W} \to \tilde{\mathcal{V}}$ of the immersion $f = \pi\circ\bar{\chi}\circ\bar{f}: \mathcal{W} \to \mathcal{V}$;

(iii) this last immersion f is an extremal of the initial problem $\mathrm{Var}_1(\mathcal{V}, \tilde{\mathcal{O}}, \mathcal{L})$.

Conversely, let $f: \mathcal{W} \to \mathcal{V}$ be an extremal of the first order problem $\mathrm{Var}_1(\mathcal{V}, \tilde{\mathcal{O}}, \mathcal{L})$; then any smooth map $\bar{f}: \mathcal{W} \to \bar{\mathcal{E}}$ which lifts the contact mapping $\tilde{f}: \mathcal{W} \to \tilde{\mathcal{V}}$ of f (that is any smooth map $\bar{f}$ such that $\bar{\chi}\circ\bar{f} = \tilde{f}$) and satisfies (i), is an extremal of the zero-order problem $\mathrm{Var}_0(\bar{\mathcal{E}}, \bar{\Omega}, \pi\circ\bar{\chi})$.

We are led to consider the sub-bundle $\tilde{\mathcal{E}}$ of $\bar{\mathcal{E}}$ defined locally by the local equations (6.3). This is a new bundle over $\tilde{\mathcal{O}}$, with fibre $\mathbb{R}^M$ of dimension $M = M_{p,q} = N - p\cdot q$ and structure group the affine group $A(M, \mathbb{R})$. The projection $\tilde{\chi}: \tilde{\mathcal{E}} \to \tilde{\mathcal{O}}$ is the restriction of $\bar{\chi}$ and we have over $\tilde{\mathcal{E}}$ the global differential form $\tilde{\Omega}$, deduced from $\bar{\Omega}$, of which the local expression reads

$$\tilde{\Omega}_\psi = L_\psi\ \omega + \sum \frac{\partial L_\psi}{\partial x_\alpha^i}\ \omega_\alpha^i + \sum \Lambda_{ij}^{\alpha\beta}\ \omega_{\alpha\beta}^{ij} + \sum \Lambda_{ijk}^{\alpha\beta\gamma}\ \omega_{\alpha\beta\gamma}^{ijk} + \dots \tag{6.4}$$

with summations restricted by the (6.2) combinations.

The problem $\bar{\mathcal{P}}_0 = \mathrm{Var}_0(\bar{\mathcal{E}}, \bar{\Omega}, \pi\circ\bar{\chi})$ induces over $\tilde{\mathcal{E}}$ a new problem $\tilde{\mathcal{P}}_0 = \mathrm{Var}_0(\tilde{\mathcal{E}}, \tilde{\Omega}, \pi\circ\tilde{\chi})$ and, clearly, an extremal of the first one lying in $\tilde{\mathcal{E}}$ is by necessity an extremal of the second problem. The converse needs not be true at all. It is thus a remarkable fact that the converse holds true provided some light condition is satisfied.

This condition is precisely what we call the condition of regularity as it coincides with the usual one for simple integrals(i.e. $p = 1$). This condition is also the key to the Legendre transformation, to canonical coordinates and to phase space.

The classical procedure to determine the extremals of the new problem on $\tilde{\mathcal{E}}$, is to compute $d\tilde{\Omega}$ and to form the exterior differential system obtained by equating to zero all "derivatives" of $d\tilde{\Omega}$ with respect to all differentials of the local coordinates. The derivatives with respect to the $d\Lambda$'s yield the "higher order" equations

$$\omega^{ij}_{\alpha\beta} = \omega^{ijk}_{\alpha\beta\gamma} = \dots = 0 \tag{6.5}$$

and it is important to look now for the derivatives with respect to the dx^i_α's . Those taken modulo the preceding terms give exactly the equations

$$\frac{\partial^2 M}{\partial x^i_\alpha \partial x^i_\beta} \omega^j_\beta = M^{\alpha\beta}_{ij} \omega^j_\beta = 0 \qquad \text{mod } (6.4) \tag{6.6}$$

where $M = M_\psi$ is the locally defined function

$$M = M_\psi = L - \sum^{i<j}_{\alpha<\beta} \Lambda^{\alpha\beta}_{ij}(x^i_\alpha x^j_\beta - x^j_\alpha x^i_\beta) \tag{6.6.a}$$

Deriving (6.5) as we did will be referred to as the "mod (6.4) trick": see remark 8.2 for the implication. From this we deduce that, if the matrix

$$\left(\frac{\partial^2 M}{\partial x^i_\alpha \partial x^j_\beta}\right) = (M^{\alpha\beta}_{ij}) \tag{6.7}$$

is regular, then equations (6.4) and (6.5) yield the total system (4.10), so that an extremal in $\tilde{\mathcal{E}}$ with regular projection into $\mathcal{V}$ will also have to satisfy the equations (4.6) (since it cannot annihilate the "volume element" $\omega = dx^1 \wedge \dots \wedge dx^p$).

It is clear also from the preceding discussion that the equations

$$\mathrm{Det}(M^{\alpha\beta}_{ij}) = 0 \tag{6.8}$$

defined locally into each $U^{\#}_\psi \times \mathbb{R}^N \cap \tilde{\mathcal{E}}$ have an invariant meaning, i.e. define the same subset over non empty intersections $U^{\#}_{\psi\varphi}$. Thus these equations define a unique global subset $\tilde{S}$ os .

<u>Definition 6.2</u>. We define this subset S of $\tilde{\mathcal{E}}$ as the <u>irregular locus</u> and we denote by $\mathcal{E}$ its complement $\tilde{\mathcal{E}} - S$.

This is an open part of $\tilde{\mathcal{E}}$ over which $\tilde{\Omega}$ induces a p-form which we shall denote Ω . If we denote by $\chi : \mathcal{E} \to \tilde{\mathcal{O}}$ the restriction of $\tilde{\chi} : \tilde{\mathcal{E}} \to \tilde{\mathcal{O}}$, we have a new variational problem $\mathcal{P}_0 = \mathrm{Var}_0(\mathcal{E}, \Omega, \pi \circ \chi)$. The latter is just the restriction of $\tilde{\mathcal{P}}_0$ over an open subset $\mathcal{E}$ of $\tilde{\mathcal{E}}$ and, therefore, the extremals of $\mathcal{P}_0$

are exactly those of $\tilde{\mathcal{P}}_o$ lying in $\mathcal{E}$.

Definition 6.3. A point $X \in \tilde{\mathcal{O}}$ is called a regular contact element of the initial variational problem $\mathcal{P}_1$ = Var($\mathcal{V}$, $\tilde{\mathcal{O}}$, $\mathcal{L}$) of order one if there exist points $\tilde{x} \in \mathcal{E}$ over it (i.e. such that $\chi(\tilde{X}) = X$) . The problem $\mathcal{P}_1$ is called regular if every $X \in \tilde{\mathcal{O}}$ is regular.

Of course, an element $X \in U^{\#}_{\psi}$ is regular if the matrix

$$\left(\frac{\partial^2 L}{\partial x^i_\alpha \partial x^j_\beta}\right) \tag{6.9}$$

is regular, but this is a sufficient condition which is necessary only in cases $p = 1$ or $q = 1$. There are indeed fundamental cases in which (6.6) is regular for appropriate choice of the extra variables $\Lambda^{\alpha\beta}_{ij}$ while (6.9), corresponding to the vanishing of these parameters, is not. The example of the electromagnetic field equations is given below.

The following result generalizes one due E. Cartan [1, chapter II, no 16] and was first announced in [6] .

Theorem 6.4. Every extremal $F: \mathcal{W} \to \mathcal{E}$ of the problem $\mathcal{P}_o$ = $\mathrm{Var}_o(\mathcal{E}, \Omega, \pi\circ\chi)$, satisfies the local systems (4.6) and its projection $f = \pi\circ\chi\circ F: \mathcal{W} \to \mathcal{V}$ is an extremal of the original first order problem $\mathcal{P}_1$ = $\mathrm{Var}_1(\mathcal{V}, \tilde{\mathcal{O}}, \mathcal{L})$. Conversely, for every extremal $f: \mathcal{W} \to \mathcal{V}$ of this first order problem $\mathcal{P}_1$, consider its canonical flow or lifting $\tilde{f}: \mathcal{W} \to \tilde{\mathcal{O}} \subset \tilde{\mathcal{V}}$; then, every smooth immersion $F: \mathcal{W} \to \mathcal{E}$ of projection $\chi\circ F = \tilde{f}$ into $\tilde{\mathcal{O}}$ (in other words: every lifting F of $\tilde{f}$) is an extremal of the zero-order problem $\mathcal{P}_o$.

Roughly speaking, a regular first order problem $\mathcal{P}_1$ is "equivalent" to the zero-order problem $\mathcal{P}_o$.

Remark 6.5. Our method compared to the multiplier rule. We have mentioned that the parameters Λ in (6.1) and (6.4) behave like "Lagrange multipliers". However, and as far as I know, the classical way of introducing them consists in replacing the first order Lagrangian

$$L\left(x^\alpha, x^i, \frac{\partial x^i}{\partial x^\alpha}\right)$$

over the manifold $\mathcal{V}$ by another one, of the same order,

$$L'(x^\alpha, x^i, \frac{\partial x^i}{\partial x^\alpha}, x^i_\alpha, \lambda^\alpha_i) = L(x^\alpha, x^i, x^i_\alpha) + \sum \lambda^\alpha_i (x^i_\alpha - \frac{\partial x^i}{\partial x^\alpha}) \qquad (6.10)$$

over $\mathcal{V} \times \mathbb{R}^{p \cdot q}$, that is over the product of $\mathcal{V}$ by the "space of multipliers". In other words, the classical method starts with a first order problem $\mathcal{P}_1$ and passes to a second problem $\mathcal{P}'_1$ of the same order. The originality of our method which follows the mentioned result of E. Cartan is to pass from a problem $\mathcal{P}_1$ of order one to successive problems $\overline{\mathcal{P}}_0$, $\widetilde{\mathcal{P}}_0$ and $\mathcal{P}_0$ of order zero. The "multiplier rule" applied to (6.10) yields both

$$\lambda^\alpha_i = \frac{\partial L}{\partial x^i_\alpha} \quad \text{and} \quad x^i_\alpha = \frac{\partial x^i}{\partial x^\alpha}$$

but is essentially governed by the pseudogroup (3.2) and is clearly of more limited scope.

7. Regularity and electro-magnetic field MAXWELL equations.

The usual regularity condition presented in [14] , [19] , [36] is the regularity of (6.9); our condition is substantially more general. Both conditions coincide in the case of simple integrals (that is for $p = 1$) where it reduces to the classical regularity condition of Mechanics:

$$\mathrm{Det}\left(\frac{\partial^2 L}{\partial \dot{q}^i \partial \dot{q}^j}\right) \neq 0 \ ;$$

both conditions do also coincide for problems involving one single "unknown function", that is for $q = 1$. Moreover, it should also be observed that, for $1<p<n-1$, that is for p and $q > 1$, regularity of (6.9) is not invariant under a pseudogroup of type (3.3) but only under transformations of type (3.2).

Regularity will also appear as the key to the generalization of the Hamilton-Cartan formalism to multiple integrals of the Calculus of variations. This means that the good definition of this concept is fundamental, even in the case of physical problems whose internal geometric structure is governed by the reduced pseudogroup (3.2).

One important and very suggestive example is the one giving the 4-vector-potential of electromagnetic field and the Maxwell equations. Here we have four independent variables x^i and four unknown functions $y^i = y^i(x^j)$, the components of the 4-potential $(i,j = 1,2,3,4)$. Denoting by g_{ij} the components of the Lorentz metric tensor, we form the Lagrangian

$$L = \sum g_{ii} g_{jj} \left(\frac{\partial y^i}{\partial x^j} - \frac{\partial y^j}{\partial x^i}\right)^2 \qquad (7.1)$$

yielding the Maxwell equations [16] , [26] . In this special case, the matrix (6.9) is highly irregular while (6.7) is regular for infimitesimal values of the parameters.

A miniature of the phenomenon corresponds to $p = q = 2$, two independent variables x^1, x^2, two unknown functions $y^3(x^1,x^2)$, $y^4(x^1,x^2)$ and the Lagrangian

$$L = \frac{1}{2} \left(\frac{\partial y^3}{\partial x^2} - \frac{\partial y^4}{\partial x^1}\right)^2 . \qquad (7.2)$$

In that case there is one single parameter $u = \Lambda^{12}_{34}$ and the matrix (6.7) is

$$\begin{pmatrix} 0 & 0 & 0 & u \\ 0 & 1 & -1-u & 0 \\ 0 & -1-u & 1 & 0 \\ u & 0 & 0 & 0 \end{pmatrix}$$

which is regular for $u \neq 0,-2$, while (6.7), that is

$$\begin{pmatrix} 0 & 0 & 0 & 0 \\ 0 & 1 & -1 & 0 \\ 0 & -1 & 1 & 0 \\ 0 & 0 & 0 & 0 \end{pmatrix} ,$$

is suggestively irregular.

8. LEGENDRE transformation, canonical coordinates, phase space.

The local expression of $\tilde{\Omega}$ on $\tilde{\mathcal{E}}$ or of Ω on $\mathcal{E}$ is given by (6.4) if we use the base (4.8) of the free module of semi-basic p-forms. Passing to the base (4.7) yields an expression

$$\tilde{\Omega} \text{ or } \Omega = P^{\alpha}_{i} \theta^{i}_{\alpha} + P^{\alpha\beta}_{ij} \theta^{ij}_{\alpha\beta} + P^{\alpha\beta\gamma}_{ijk} \theta^{ijk}_{\alpha\beta\gamma} + \dots - H\,\omega \qquad (8.1)$$

with summations omitted and restricted by inequalities (6.2). One sees that the new coefficients P and H are defined by

$$\left.\begin{aligned}
p_i^{\alpha} &= \frac{\partial L}{\partial x_\alpha^i} - \textstyle\sum_\beta^j \Lambda_{ij}^{\alpha\beta} x_\beta^j + \sum_{\beta<\gamma}^{j,k} \Lambda_{ijk}^{\alpha\beta\gamma} x_\beta^j x_\gamma^k - \dots ,\\
p_{ij}^{\alpha\beta} &= \Lambda_{ij}^{\alpha\beta} - \textstyle\sum_\gamma^k \Lambda_{ijk}^{\alpha\beta\gamma} x_\gamma^k + \dots ,\\
p_{ijk}^{\alpha\beta\gamma} &= \Lambda_{ijk}^{\alpha\beta\gamma} - \textstyle\sum_\gamma^l \Lambda_{ijkl}^{\alpha\beta\gamma\delta} x_\delta^l + \dots ,\\
&\dots\dots
\end{aligned}\right\} \tag{8.2}$$

$$H = -L + \sum \frac{\partial L}{\partial x_\alpha^i} x_\alpha^i - \sum_{\alpha<\beta}^{i,j} \Lambda_{ij}^{\alpha\beta} x_\alpha^i x_\beta^j + \dots \quad . \tag{8.3}$$

Note that these formulas involve functions $\Lambda_{ij}^{\alpha\beta}$, $\Lambda_{ijk}^{\alpha\beta\gamma}$,... indexed by combinations no more restricted inequalities (6.2); these functions are defined by two conditions: (i) of coinciding with those defined as local coordinates on $\tilde{\mathcal{E}}$ and subject to these inequalities; (ii) of providing an antisymmetric symbol in both the greek and latin indices. In other words the functions $\Lambda_{ij}^{\alpha\beta}$, $\Lambda_{ijk}^{\alpha\beta\gamma}$,... for $\alpha<\beta<\dots$ and $i<j<\dots$ are part of the local coordinates over $\tilde{\mathcal{E}}$ and, for $i<j<k$, $\alpha<\beta<\gamma$, we suppose that

$$\Lambda_{ij}^{\alpha\beta} = -\Lambda_{ji}^{\alpha\beta} = -\Lambda_{ij}^{\beta\alpha} = \Lambda_{ji}^{\beta\alpha} ,$$

$$\Lambda_{ijk}^{\alpha\beta\gamma} = \Lambda_{jki}^{\alpha\beta\gamma} = \Lambda_{kij}^{\alpha\beta\gamma} = -\Lambda_{jik}^{\alpha\beta\gamma} = \dots$$

$$= \Lambda_{ijk}^{\beta\gamma\alpha} = -\Lambda_{ijk}^{\beta\alpha\gamma} = \dots,$$

etc.

Proposition 8.1. We consider a point $\tilde{X}\in\tilde{\mathcal{E}}$ with local coordinates $(x^\alpha, x^i, x_\alpha^i, \Lambda_{ij}^{\alpha\beta}, \Lambda_{ijk}^{\alpha\beta\gamma},\dots)$ with respect to a local chart ψ in the atlas $\mathcal{A}$. The equations (8.2) define a transformation

$$(x_\alpha^i, \Lambda_{ij}^{\alpha\beta}, \Lambda_{ijk}^{\alpha\beta\gamma},\dots) \longrightarrow (p_i^\alpha, p_{ij}^{\alpha\beta}, p_{ijk}^{\alpha\beta\gamma},\dots) \tag{8.4}$$

whose jacobian is different from zero at the necessary and sufficient condition that $\tilde{X}$ be outside the irregular locus S of $\tilde{\mathcal{E}}$

Remark 8.2. Both the proof of this statement and that of formulas (8.2,3) require some attention. It was kind of surprising that in theorem 6.1 only the "first order" parameters Λ_i^α were involved, while the "higher" ones $\Lambda_{ij}^{\alpha\beta}, \Lambda_{ijk}^{\alpha\beta\gamma}$ remained arbitrary. Similarly, it was curious that the duly motivated condition of "regularity" of matrix (6.7), essential in the definition of the irregular locus and in theorem 6.4, touches only the "second order" parameters $\Lambda_{ij}^{\alpha\beta}$ and not the higher ones as results from (6.6a). This result was obtained by the "mod (6.4) trick" and, not using it would yield a regularity condition where the higher parameters would seemingly be involved. At first sight, these higher parameters are also present in the jacobian determinant of transformation (8.2) and it requires some pain to eliminate them there. At this time, I know no trick paralleling the alluded "mod (6.4) trick" so as to make this independence more immediate or obvious.

Proposition 8.1 shows that transformation (8.4) is a first reasonable candidate for the Legendre transformation as it coincides with the classical one for $p = 1$. We also see that, around every point $\tilde{X} \varepsilon \mathcal{E}$ (that is around every point of $\tilde{\mathcal{E}}$ outside the irregular locus), there are local coordinates $(x^\alpha, x^i, P_i^\alpha, P_{ij}^{\alpha\beta}, \ldots)$ such that the fundamental p-form Ω has the local expression (8.1) with H a function of these variables.

Definition 8.2. Such coordinates around $\tilde{X}$ are called canonical and the function H of these variables is called the corresponding Hamiltonian function. The coordinates $P_i^\alpha, P_{ij}^{\alpha\beta}, \ldots$ are called the (local) momenta. The domain of canonical coordinates is called a phase neighboorhood and the manifold $\tilde{\mathcal{E}}$ is called the (global) phase space of the variational problem $\mathcal{P}_1$: see def. 6.2.

In the classical case $p = 1$, the Hamiltonian function has the form $H = H(t, q^i, p_i)$ and may be an arbitrary function of these variables subject to the condition that the matrix

$$\left(\frac{\partial^2 H}{\partial p_i \partial p_j}\right)$$

be regular. Indeed, in this case (8.4) and its inverse read

$$p_i = \frac{\partial L}{\partial \dot{q}^i}, \quad \dot{q}^i = \frac{\partial H}{\partial p_i},$$

$$L = L(t, q^i, \dot{q}^i), \; H = \frac{\partial L}{\partial \dot{q}^i} \dot{q}^i - L \; .$$

We could say that, in the classical case $p=1$, any "generic" function H of the canonical variables is the Hamiltonian of a regular variational problem. This, however, is no longer the case for arbitrary p as expressed by the coming interesting theorem 8.3.

To its effect it will be useful to present a new form of the momenta, writing them as

$$(P_{r_1 \ldots r_p})$$

where the indices are subject to the following conditions:

(i) $$1 \le r_\alpha \le p+q = n ,$$

(ii) $$r_1 < r_2 < \ldots < r_p ,$$

(iii) $$(r_1, r_2, \ldots, r_p) \neq (1,2,\ldots,p) .$$

The fundamental p-form Ω can then be written as

$$\Omega = \sum P_{r_1 \ldots r_p} dx^{r_1} \wedge \ldots \wedge dx^{r_p} - H\, dx^1 \wedge \ldots \wedge dx^p$$

with $H = H(x^\alpha, x^i, P_{r_1 \ldots r_p})$. [One could write $H = - P_{1 \ldots p}$.]

<u>Theorem 8.3. Every Hamiltonian function H of a regular variational problem satisfies the system of polynomial partial differential equations expressing the fact that the equations</u>

$$a^{1,2,\ldots,p} = 1 , \qquad a^{r_1 \ldots r_p} = \frac{\partial H}{\partial P_{r_1 \ldots r_p}} \tag{8.5}$$

<u>define a simple p-vector.</u>

In other words, around any point $\tilde{X}$ of the phase space, the projection $\chi : \mathcal{E} \to \tilde{\mathcal{O}}$ is locally defined by

$$\chi : (x^\alpha, x^i, P^\alpha_i, P^{\alpha\beta}_{ij}, \ldots) \longrightarrow (x^\alpha, x^i, x^i_\alpha)$$

where

$$x^i_\alpha = \frac{\partial H}{\partial P^\alpha_i} \tag{8.6}$$

and the further partial derivatives of H with respect to the other momenta satisfy polynomial equations of the type

$$\frac{\partial H}{\partial P^{\alpha\beta\gamma}_{ijk}} = \mathrm{Det}\begin{pmatrix} \frac{\partial H}{\partial P^{\alpha}_{i}} & \frac{\partial H}{\partial P^{\alpha}_{j}} & \frac{\partial H}{\partial P^{\alpha}_{k}} \\ \frac{\partial H}{\partial P^{\beta}_{i}} & \frac{\partial H}{\partial P^{\beta}_{j}} & \frac{\partial H}{\partial P^{\beta}_{k}} \\ \frac{\partial H}{\partial P^{\gamma}_{i}} & \frac{\partial H}{\partial P^{\gamma}_{j}} & \frac{\partial H}{\partial P^{\gamma}_{k}} \end{pmatrix} . \tag{8.7}$$

This has several consequences. One is that the "undetermination" of the fundamental form Ω due to the higher parameters $\Lambda^{\alpha\beta}_{ij},\ldots$ is counterbalanced by the Hamiltonian function being subdued by a remarkable system of partial differential equations. Another is that equations (8.6) offer a second reasonable candidate to the Legendre transformation. It also (as (8.2))reduces to the classical one for $p = 1$ and it has, furthermore, the other interpretation which follows.

At every point x of $\mathcal{V}$ we have the space $\Lambda^p T_x(\mathcal{V})$, namely the space of all p-vectors at x, or the degree p component of the exterior algebra of the tangent space $T_x(\mathcal{V})$. $\Lambda^p T_x(\mathcal{V})$ is a vector space of dimension $\binom{n}{p}$ and, corresponding to a basis in $T_x(\mathcal{V})$, a p-vector $\xi\varepsilon\Lambda^p T_x(\mathcal{V})$ has $N+1 = \binom{n}{p}$ real components

$$\xi^{r_1\ldots r_p} , \quad 1 \leq r_1<\ldots<r_p < p+q = n . \tag{8.8}$$

Among them, the simple p-vectors form a Grassmann cone $\Gamma_{p,q}$ defined by well-known polynomial equations in the $\xi^{r_1\ldots r_p}$'s.
The hyperplane at infinity in $\Lambda^p T_x(\mathcal{V})$ is a real projective space of dimension N intersecting $\Gamma_{p,q}$ in a Grassmann manifold $G_{p,q}(x)$ which we may canonically indentify to the fibre $\tilde{\mathcal{V}}_x$ at $x\varepsilon\mathcal{V}$ in $\tilde{\mathcal{V}} = \tilde{\mathcal{V}}^p$. For a chart ψ in the atlas $\mathcal{A}$, a contact element $X\varepsilon\ U^{\#}_{\psi}$ with coordinates $(x^\alpha, x^i, x^i_\alpha)$ corresponds to the point at infinity of $\Gamma_{p,q}$ along the simple p-vector whose components $\xi^{r_1\ldots r_p}$ are the minors of degree p corresponding to the columns $r_1,\ldots,r_p$ in the matrix

$$\begin{Vmatrix} 1 & 0 & & 0 & x_1^{p+1} & x_1^{p+2} & \dots & x_1^n \\ 0 & 1 & & 0 & x_2^{p+1} & x_2^{p+2} & & x_2^n \\ & & \ddots & \vdots & & & & \\ 0 & 0 & \dots & 1 & x_p^{p+1} & x_p^{p+2} & & x_p^n \end{Vmatrix} .$$

In view of this, equations (8.6) which are the local expression of the projection $\chi:\mathcal{E} \longrightarrow \mathcal{V}$, can be interpreted as sending the point of momenta $P_{r_1 \dots r_p}$ to the point at infinity of $\Gamma_{p,q}$ along the simple p-vector defined by

$$\xi^{1 \dots p} = 1, \quad \xi^{r_1 \dots r_p} = \frac{\partial H}{\partial P_{r_1 \dots r_p}} .$$

The map $\tilde{\chi}:\tilde{\mathcal{E}} \longrightarrow \tilde{\sigma}$ corresponds to a bona fide fibre bundle in the sense of Ehresmann or of Steenrod, with fibre $\mathbb{R}^{N'}$ of dimension

$$N' = N'_{p,q} = N - p \cdot q = \binom{n}{p} - p \cdot q - 1$$

and structure group $A(N', \mathbb{R})$, the real affine group. This dimension is exactly the number of higher paramenters $\Lambda^{\alpha\beta}_{ij}$, $\Lambda^{\alpha\beta\gamma}_{ijk}$ .. which were first introduced by Lepage [27] as arbitrary functions over $U^{\#}_{\psi}$. He characterized the form (6.4) as the most general semi-basic one satisfying the conditions

$$\Omega \equiv L \cdot \omega \ (\mathrm{mod}\ \omega^i) \ , \ d\Omega \equiv 0 \ (\mathrm{mod}\ \omega^i)$$

which we refer to as Lepage congruences. The globalization and bundle organization was done later by Dedecker and that is why A. Liesen [29] called $(\tilde{\mathcal{E}}, \tilde{\chi}, \tilde{\sigma})$ the Lepage-Dedecker bundle of problem $\mathcal{P}_1$. Certainly both the global semi-basic form $\tilde{\Omega}$ over $\tilde{\mathcal{E}}$ and its restriction Ω to $\mathcal{E}$ deserve the name of Cartan's fundamental p-form of the problem.

It is clear that the pairs $(\tilde{\chi}, \tilde{\Omega})$ and (χ, Ω) [or more precisely the systems $(\tilde{\mathcal{E}}, \tilde{\chi}, \tilde{\sigma}; \tilde{\Omega})$ and $(\mathcal{E}, \chi, \tilde{\sigma}; \Omega)$] are canonically associated to a first order variational problem $\mathcal{P}_1$; and so are the zero order problems $\tilde{\mathcal{P}}_0$, $\tilde{\mathcal{P}}_0$ and $\mathcal{P}_0$. In an old language, they would have been called "invariants" of the problem, which seems to be the sense of the word "invariant" in the title of communication

[16] by Pedro Luis García. In modern terminology, I would rather call them covariants of problem $\mathcal{P}_1$. More precisely, I do not see any practical use of defining the categories of first order and of zero order variational problems, but this can be done formally as "local categories" (or as "situses" in the sense of A. Grothendieck) with relatively few morphisms, so that these "covariants" become good standing functors and even functors of a very specific, differential geometric type.

Definition 8.4. The transformation (8.2) or (8.4) corresponds to a local chart $\psi \in \mathcal{A}$ and will be called the associated (local) horizontal Legendre transformation. Similarly (8.6) depends also on ψ and will be called the associated (local) vertical Legendre transformation. As (8.6) is just the local expression of $\chi : \mathcal{E} \longrightarrow \tilde{\mathcal{O}}$, it is perfectly reasonable to define χ as the (global) vertical Legendre transformation. It is also possible to define a (global) horizontal Legendre transformation as a global map of bundles over $\mathcal{V}$ (not over $\tilde{\mathcal{O}}$) which happen to be an isomorphism outside the irregular locus. Time and space are missing to explicit this idea here which shall be discussed elsewhere [13].

The vertical and horizontal Legendre transformations coincide for p or q = 1. We believe that both are compatible with the general idea of passing from point coordinates to tangential coordinates as discussed for example in Courant - Hilbert [7]. But the bifurcation for p and q>1 (i.e. for $p \cdot q > 1$) is due to the fact that, in such situation, the "hodograph" of an immersion $\mathbb{R}^p \longrightarrow \mathbb{R}^{p+q}$ is a map

$$\mathbb{R}^p \longrightarrow \Gamma_{p,q} \subset \mathbb{R}^{N+1} \quad , \quad N+1 = \binom{p+q}{p}$$

so that, at points of this hodograph, there are infinitely many tangent hyperplanes, making the "dual hodograph" a hypersurface in the dual of $\mathbb{R}^{N+1}$.

Remark 8.5. This indetermination had already puzzled Caratheodory [6] who eliminated the superfluous parameters Λ in (6.4) by choosing the form

$$\Omega = L\left(dx^1 + \frac{1}{L}\frac{\partial L}{\partial x_1^i}\,\omega^i\right) \wedge \left(dx^2 + \frac{1}{L}\frac{\partial L}{\partial x_2^i}\,\omega^i\right) \wedge \dots \wedge \left(dx^p + \frac{1}{L}\frac{\partial L}{\partial x_p^i}\,\omega^i\right),$$

that is by imposing that the fundamental Cartan's p-form be monomial. He produced this way a most elegant generalization of the classical theory, generalization which has the great advantage (over that of de Donder-Weyl and continuators) of being invariant under the general pseudogroup of transformations (3.3). It is, however, clear from the above discussion that his choice does not yield the most general notion of regularity.

The notion of regularity corresponding to a choice like Caratheodory's, that is to a section $\lambda : \tilde{\mathcal{O}} \longrightarrow \tilde{\mathcal{E}}$, is not in general the restriction of our regularity.

Remark 8.6. Let us mention here that our condition of regularity is quite different from another one which should rather be called ellipticity. See E. Bombieri, Proceedings of the International Congress of Mathematicians, Vancouver 1974, vol.1, p. 54.

9. Equivalence, classification, stability.

In the classical case $p = 1$, two regular problems of the same dimension n are locally equivalent around two points of the phase space. Indeed the phase space $\mathcal{E}$ has $2n-1 = 2q+1$ dimensions in this case and Cartan's fundamental form Ω is the most general Pfaffian form in this number of variables; in other words, around a point $X \in \mathcal{E}$, one can choose coordinates (s, α_i, β^i) such that

$$\Omega = ds + \sum \alpha_i d\beta^i .$$

For p arbitrary, consider two problems with phase spaces $\mathcal{E}_1$, $\mathcal{E}_2$ and fundamental forms Ω_1, Ω_2. For $\tilde{X}_i \in \mathcal{E}_i$, we shall say that the problems are locally equivalent at these points if there exist neighbourhoods U_i of $\tilde{X}_i$ in $\mathcal{E}_i$ and a diffeomorphism $f : U_1 \longrightarrow U_2$ compatible with Ω_1 and Ω_2, that is transporting the restriction of Ω_1 in U_1 to that of Ω_2 in U_2; then f is also called a local isomorphism at $\tilde{X}_1$ or at $(\tilde{X}_1, \tilde{X}_2)$. Suppose that the problems are defined on two manifolds $\mathcal{V}_i$ of the same dimension; we can say that they are locally equivalent around two points $X_i \in \mathcal{V}_i$ if there exists points $\tilde{X}_i \in \mathcal{E}_i$ above them and satisfying the last condition. Local equivalence is thus a problem of local classification of exterior p-forms which if far from trivial and eventually unsolved. Without having a formal proof, we believe that the following are examples of non locally equivalent regular variational problems. Consider in $\mathbb{R}^4$, with coordinates x, y, z, Z, two problems for surfaces $z = z(x,y)$, $Z = Z(x,y)$ and

$$p = \frac{\partial z}{\partial x} , \ q = \frac{\partial z}{\partial q} , \ P = \frac{\partial Z}{\partial x} , \ Q = \frac{\partial Z}{\partial y} :$$

(A) the Plateau problem governed by the area integral

$$\int \sqrt{1+p^2+q^2+P^2+Q^2+(pQ-Pq)^2} \ dx \wedge dy \ ; \tag{9.1}$$

(B) the miniaturized electromagnetism defined by (7.2) or

$$\int \frac{1}{2} (q-P)^2 dx \wedge dy \ . \tag{9.2}$$

The latter yields the equations

$$\frac{\partial}{\partial x} \left(\frac{\partial z}{\partial y} - \frac{\partial Z}{\partial x}\right) = \frac{\partial}{\partial y} \left(\frac{\partial z}{\partial y} - \frac{\partial Z}{\partial x}\right) = 0$$

whose general integral can be written

$$\begin{cases} z = \Phi(x,y) \\ Z = \kappa x + \int \frac{\partial \Phi}{\partial y} dx \end{cases} \quad \text{or} \quad \begin{cases} z = -\kappa y + \int \frac{\partial \Psi}{\partial x} dy \\ Z = \Psi(x,y) \end{cases}$$

where Φ (or Ψ) is an arbitrary function and κ an arbitrary constant. We hope that a combination of this together with the theory of minimal surfaces can provide a formal proof of non local equivalence. These problems can also be compared with the one defined by

$$\int \frac{1}{2} (p^2+q^2+P^2+Q^2) \ dx \wedge dy \tag{9.3}$$

whose extremals are pairs (z,Z) of harmonic functions.

Clearly, if local equivalence is not a trivial question, so is a fortiori the problem of global equivalence. We thus have problems of both local and global classification. The global one for $p = 1$ is related to the classification problem of dynamical systems. It is also possible to consider, over a manifold $\mathcal{V}$, more precisely over an open set $\tilde{\mathcal{O}} \subset \tilde{\mathcal{V}}$, a family of variational problems depending on parameters. Then arise problems of global and local stability, that is the question whether one problem in the family remains equivalent to itself either globally or locally around a point $X \in \tilde{\mathcal{O}}$ when the parameters undergo a small change. More generally, the stability problem arises considering the totality of variational problems over $\tilde{\mathcal{O}}$, endowed with an appropriate topology: for one such problem $(\mathcal{P}_1)_0$, we shall say that it is stable around a point $X \in \tilde{\mathcal{O}}$ if any other problem $\mathcal{P}_1$ in a sufficiently small neighbourhood is locally equivalent to it around X.

For $p = 1$, every regular variational is stable in any reasonable smooth topology. I would like to conjecture that the Plateau problem and the Dirichlet problem are similarly stable but that the miniaturized problem (7.2) or the general electromatism problem (7.1) are not.

These conjectures seem to be interesting open questions and the problem is also related to a remark at top of page 229 in [19]. However, the study mentioned there should be extended to our larger phase space. See also section 12 below and [13].

10. Generalized canonical equations.

Let $\mathcal{P}_0 = (\mathcal{X}, \Theta, \pi)$ a zero-order variational problem of degree p as in §5. One easily finds the equations of the extremals: take the differential $d\Theta$ in some coordinate neighbourhood $U \subset \mathcal{X}$ with local coordinates $x^1,\dots,x^N$ and form the equations

$$\frac{\partial(d\Theta)}{\partial(dx^s)} = 0 , \qquad s = 1,\dots,N \tag{10.1}$$

They consist in equating to zero N differential forms of degree p and constitute what we have called the first associated system of $d\Theta$ [12]. The usual associated system of $d\Theta$, as defined by E. Cartan, is obtained by taking the derivatives of order p, instead of order one (so that the "first associated system" coincides with the "associated system" in case $p = 1$). Let ξ_s be the field of tangent vectors in U whose components ξ_s^t are zero if $t \neq s$ and 1 for $t = s$. Then each equation (10.1) reads exactly

$$i(\xi_s)d\Theta = 0 \qquad \text{or} \qquad \xi_s \lrcorner d\Theta = 0 \tag{10.2}$$

where $i(\ )$ or $\lrcorner$ indicate the interior product. These equations, confined to coordinate neighbourhoods can be globalized either by taking

$$i(\xi)d\Theta = 0 \qquad \text{or} \qquad \xi \lrcorner d\Theta = 0 \tag{10.3}$$

where ξ is an arbitrary field of tangent vectors to $\mathcal{X}$, at least in the C^r-case, $1 \leq r \leq \infty$, either considering the sheaves generated by the first members in (10.1) or (10.2); the latter globalization applies also in the analytic case. (Indeed (10.3) imply (10.1) only in case a C^r-Urysohn's lemma is valid).

The extremals of $\mathcal{P}_0$ are thus the smooth maps $f: \mathcal{W} \to \mathcal{X}$ ($\dim \mathcal{W} = p$) solutions of one of the above systems and such that $\pi \circ f: \mathcal{W} \to \mathcal{Y}$ be immersions. Say that f is a solution of, for example (10.3), means that the inverse image by f of $i(\xi)d\Theta$, that is the p-form

$$f^*[i(\xi)d\Theta]$$

which is a p-form over $\mathcal{W}$, is identically zero.

We thus see that the extremals of a <u>zero-order</u> problem $\mathcal{P}_o$ are given by a system of exterior differential equations in the sense of E. <u>Cartan</u>- E. <u>Kähler</u> (for which we refer to [5], [9], [12]) with an additional regularity condition with respect to the projection $\pi : \mathcal{X} \to \mathcal{Y}$. So are also characterized the extremals of a regular <u>first-order</u> variational problem $\mathcal{P}_1$ over the phase space $\mathcal{E}$. For $p = 1$, the latter coincides with $\tilde{\mathcal{V}}$ (or $\tilde{\mathcal{O}}$) and this exterior system is the Pfaffian system of the classical canonical equations

$$dq^i - \frac{\partial H}{\partial p_i} dt = 0 , \qquad dp_i + \frac{\partial H}{\partial q^i} dt = 0 , \tag{10.4}$$

together with the energy equation

$$\frac{\partial H}{\partial q^i} dq^i + \frac{\partial H}{\partial p_i} dp_i = 0 \quad \text{or} \quad dH - \frac{\partial H}{\partial t} dt = 0 . \tag{10.5}$$

This is a completely integrable and thus involutive Pfaffian system.

The situation is substantially more complicated in the general case when $p > 1$. For example, for $p = 2$, the generalized canonical equations are

$$dP^1_i \wedge dx^2 + dx^1 \wedge dP^2_i + \sum dP_{ij} \wedge dx^j + \frac{\partial H}{\partial x^i} dx^1 \wedge dx^2 = 0 , \tag{10.6-i}$$

$$dx^i \wedge dx^2 - \frac{\partial H}{\partial P^1_i} dx^1 \wedge dx^2 = 0 , \quad dx^1 \wedge dx^i - \frac{\partial H}{\partial P^2_i} dx^1 \wedge dx^2 = 0 , \tag{10.6-ii}$$

$$dx^i \wedge dx^j - \frac{\partial H}{\partial P_{ij}} dx^1 \wedge dx^2 = 0 , \tag{10.6-iii}$$

together with <u>two</u> <u>energy equations</u> (or one 2-<u>vector-energy equation</u>):

$$\begin{cases} dP^2_i \wedge dx^i - \left(dH - \frac{\partial H}{\partial x^1} dx^1\right) \wedge dx^2 = 0 , \\ dP^1_i \wedge dx^i - \left(dH - \frac{\partial H}{\partial x^2} dx^2\right) \wedge dx^1 = 0 . \end{cases} \tag{10.7}$$

One complication is that <u>this system is not complete</u>: in other words, the left members generate an ideal $\mathcal{i}$ in the algebra of local differential forms but there are forms not belonging to $\mathcal{i}$ and vanishing on every solution (or on every extremal). If we denote by $\bar{\mathcal{i}}$ the ideal of all forms vanishing on all extremals, we have a strict inclusion $\mathcal{i} \subset \bar{\mathcal{i}}$. In particular, if $\theta \in \mathcal{i}$ it is in general the case that $d\theta \notin \mathcal{i}$, while $d\theta \in \bar{\mathcal{i}}$. Moreover, $\mathcal{i}$ contains no form of degree $<p$, while we have seen that $\bar{\mathcal{i}}$ contains the Pfaffian forms

$$\omega^i = dx^i - \sum \frac{\partial H}{\partial P_i^\alpha} dx^\alpha . \tag{10.8}$$

For $p = 1$, $\mathcal{i} = \bar{\mathcal{i}}$ and it is the case that (10.5) is an algebraic consequence of (10.4), i.e. the left members of (10.4) generate an ideal coinciding with $\mathcal{i}$ and, thus, containing the first member of (10.5). For $p > 1$, let $\mathcal{j}$ be the ideal of all forms vanishing on the totality of maps $f: \mathcal{W} \to \mathcal{E}$ satisfying the two conditions: 1° be solution of (10.6-i) + (10.6-ii) and 2° have a regular projection $\pi\circ\chi\circ f: \mathcal{W} \to \mathcal{V}$. Then $\omega^i \in \mathcal{j} = \bar{\mathcal{i}}$ and, in particular, the left-members of (10.7) belong to $\mathcal{j}$. More precisely, the energy equations are algebraic consequences of (10.6-i) completed by $\omega^i = 0$. Also, by the property of type (8.7), namely

$$\frac{\partial H}{\partial P_{ij}} = \begin{vmatrix} \dfrac{\partial H}{\partial P_i^1} & \dfrac{\partial H}{\partial P_j^1} \\ \dfrac{\partial H}{\partial P_i^2} & \dfrac{\partial H}{\partial P_j^2} \end{vmatrix}$$

the left member of (10.6-iii) is just $\omega^i \wedge \omega^j$ and, thus, belongs to $\mathcal{j}$.

For further details about the structure of the generalized canonical equations (10.6) we refer the reader to [12] and forthcoming paper [13] discussing this matter in more detail.

The incompleteness of the system of generalized canonical equations brings some difficulties which require a careful study still to be done. The question was already mentioned in [10], [12].

When we have a homogeneous ideal $\mathcal{i}$ in the graded differential algebra $\mathcal{G} = \{\mathcal{G}^n\}$ of exterior differential forms over a manifold $\mathcal{X}$, it yields an interesting decreasing filtration of $\mathcal{G}$ together with the corresponding spectral sequence. For these notions we refer to [28], [12], [34]. The filtration is defined by

$$\mathcal{G} = A^{0} \supset \ldots \supset A^{u} \supset A^{u+1} \quad \ldots \tag{10.9}$$

where we define A^u as the u-th power of $\mathcal{i}$. In general, this filtration is not stable by exterior differentiation, that is one may have

$$dA^u \not\subset A^u$$

and we will have to consider terms $E_r^{u,v}$ with $r < 0$ in the spectral sequence, a phenomenon already provided for in the original presentation by Leray [21] but unfortunately rejected as pathological by all distinguished algebraic topologists who discussed this matter. Suppose for example that $\mathcal{i}$ is generated by linearly independent Pfaffian forms: one has then $dA^u \subset A^{u-1}$ and the terms $E_{-1}^{u,v}$ are to be taken into consideration. In fact, one interpretation of the theorem of Frobenius is that the corresponding system of Pfaffian equations is completely integrable if and only if in (10.11) below with $r = -1$, all arrows δ, δ' are zero, so that we can write

$$\ldots = E_{-2}^{u,v} = E_{-1}^{u,v} = E_{0}^{u,v} .$$

In other words, the system is completely integrable if and only if the only relevant terms E_r occur for $r \geq 0$.

Recall that $C_r^{u,v}$ consists of all forms θ of degree $u + v$ satisfying $\theta \in A^u$ and $d\theta \in A^{u+r}$:

$$C_r^{u,v} = \{\theta \in \mathcal{G}^{u+v} \mid \theta \in A^u,\ d\theta \in A^{u+r}\}.$$

We then have the "cohomology quotients"

$$E_r^{u,v} = C_r^{u,v} / (C_{r-1}^{u+1,v-1} + B_{r-1}^{u,v}) \tag{10.10}$$

where $B_{r-1}^{u,v} = dC_{r-1}^{u-r+1,v+r-2}$. The exterior differentiation d induces

$d\colon C_r^{u,v} \longrightarrow C_r^{u+r,v-r+1}$ and, thus yields diagrams

$$\ldots \longrightarrow E_r^{u-r,v+r-1} \xrightarrow{\delta'} E_r^{u,v} \xrightarrow{\delta} E_r^{u+r,v-r+1} \longrightarrow \ldots \tag{10.11}$$

such that $\delta \circ \delta' = 0$ and one shows that the defect of exactness is

$$\mathrm{Ker}(\delta')/\mathrm{Im}(\delta) \approx E_{r+1}^{u,v} . \qquad (10.12)$$

This is one of the fundamental facts of spectral sequence theory. We have enlarged it in [11], [12] obtaining what we have called the spectral diagram, which is important to develop a theory of integral invariants (a) applicable to multiple integrals of Calculus of variations and (b) generalizing the classical results of H. Poincaré and E. Cartan.

11. The classical integral invariants.

The reason for considering the cohomology quotients (10.9) in algebraic topology and motivation for their generalization deserve some comment. This question is not very clear and I do not know any clarification, even in the fundamental papers of algebraic topologists.

The motivation for considering a $C_r^{u,v}$ seems to be in approximation theory. A true cocycle is an element $x \in A^{u,v}$ (namely a differential form x of degree $u+v$ belonging to A^p) which belongs to $C_\infty^{u,v}$ so that an element $x \in C_r^{u,v}$ is a cocycle mod. A^{u+r} . Without being worried by degree $u+v$, one is lead to define

$$\cdot C_r^u = \{x \in A^u \mid dx \in A^{u+r}\}, \quad \cdot B_r^u = d\cdot C_r^{u-r} \qquad (11.1)$$

Now $A^u \supset A^{u+1}$ so that the quotient A^u/A^{u+1} gives an idea of those elements in A^u which are really of filtration u , that is which do not belong to A^{u+1} . Indeed the filtration $f(x)$ is by definition the maximum integer u such that $x \in A^u$. Then, the image of $\cdot C_r^u \subset A^u$ in A^u/A^{u+1} is exactly $\cdot C_r^u/\cdot C_{r-1}^{u+1}$. But in algebraic topology, people look for differential operators and d sends $\cdot C_r^u$ to $\cdot C_r^{u+r}$ but not $\cdot C_r^u/\cdot C_{r-1}^{u+1}$ to $\cdot C_r^{u+r}/\cdot C_{r-1}^{u+r+1}$ since $\cdot C_{r-1}^{u+1}$ does not go into $\cdot C_{r-1}^{u+r+1}$ but in $\cdot B_{r-1}^{u+r}$ which is not contained in the latter. We however see that $\cdot C_r^u/\cdot C_{r-1}^{u+1}$ goes to

$$\cdot C_r^{u+r}/(\cdot C_{r-1}^{u+r+1} + \cdot B_{r-1}^{u+r}) .$$

We thus have

$$\cdot C^u_r / \cdot C^{u+1}_{r-1} \longrightarrow \cdot C^{u+r}_r / (\cdot C^{u+r+1}_{r-1} + \cdot B^{u+r}_{r-1})$$

and a fortiori

$$\cdot E^u_r = \cdot C^u_r / (\cdot C^{u+1}_{r-1} + \cdot B^{u}_{r-1}) \longrightarrow \cdot E^{u+r}_r = \cdot C^{u+r}_r / (\cdot C^{u+r+1}_{r-1} + \cdot B^{u+r}_{-1}).$$

This yields cochain complexes

$$\ldots \longrightarrow \cdot E^{u-r}_r \longrightarrow \cdot E^u_r \longrightarrow \cdot E^{u+r}_r \longrightarrow \ldots$$

decomposing into (10.11) when degrees are taken into account.

This argument has a dual aspect at chain (or homology) level. One then considers the chain group A of cubical singular homology (see [12], [34]) of class C^k and an increasing filtration

$$\{0\} = A_{-1} \subset A_0 \subset A_1 \subset \ldots \subset A_u \subset A_{u+1} \subset \ldots \tag{11.1}$$

such that $A = \bigcup A_u$. The <u>filtration</u> $f(x)$ of a chain x is now, by definition, the least integer u such that $x \in A_u$. One defines

$$.C^r_u = \{x \in A_u \mid \partial x \in A_{u-r}\} ,$$

$$.B^r_u = \partial .C^r_{u+r} ,$$

$$.E^r_u = .C^r_u / (.C^{r-1}_{u-1} + .B^{r-1}_u) .$$

The boundary operator ∂ then yields chain complexes

$$\ldots \longrightarrow .E^r_{u+r} \xrightarrow{\partial} .E^r_u \xrightarrow{\partial'} .E^r_{u-r} \longrightarrow \ldots$$

whose homology at $.E^r_u$ is exactly

$$\mathrm{Ker}(\partial')/\mathrm{Im}(\partial) \approx .E^{r+1}_u$$

One usually assumes that the A_u's are homogeneous, namely are direct sums of components $A_{u,v}$ of degrees $s = u+v$

$$A_u = \textstyle\sum_v A_{u,v} .$$

Then $.C^r_u = \sum C^r_{u,v}$, where $C^r_{u,v} = .C^r_u \cap A_{u,v}$, yields

$$.E_u^r = \sum_v E_{u,v}^r \ , \ E_{u,v}^r = C_{u,v}^r/(C_{u-1,v+1}^{r-1} + B_{u,v}^{r-1}) \ ,$$

where $B_{u,v}^r = \partial C_{u+r,v-r+1}^r$, and complexes

$$\cdots \longrightarrow E_{u+r,v-r+1}^r \longrightarrow E_{u,v}^r \longrightarrow E_{u-r,v+r-1}^r \longrightarrow \cdots$$

To produce a suggestive example, consider a smooth fibre bundle $\xi : M \longrightarrow B$ where M and B are smooth manifolds and ξ is a smooth onto map. For each point $y \in M$ of projection $x = \xi(y) \in B$ we can choose local coordinates

$$y^1,\ldots y^m \text{ around } y \ , \ x^1,\ldots,x^b \text{ around } x$$

such that ξ sends the point $y = (y^1,\ldots,y^b,y^{b+1},\ldots,y^m)$ to the point

$$x = \xi(y) = (y^1,\ldots,y^b)$$

Then the fibres are defined by the completely integrable Pfaffian system

$$dy^1 = dy^2 = \ldots = dy^b = 0 \ . \tag{11.2}$$

A system of local coordinates $(y^1,\ldots,y^m)$ in M such that there exists a corresponding one $(x^1,\ldots,x^b)$ in B is said to be <u>admissible</u> or <u>compatible</u> with the fibration; the first coordinates $y^1,\ldots,y^b$ are called "B-<u>coordinates</u>" in M .

A differential form ω on M is said to belong to A^u if and only if, its local expression with respect to every such admissible local coordinates in M is of degree <u>at least</u> u in the differentials $dy^1,\ldots,dy^b$ (with coefficients arbitrary functions of all variables $y^1,\ldots,y^b,\ldots,y^m$) . If $A^{u,v}$ denotes the module of forms of degree $u+v$ in A^u , one has

$$\mathcal{G}^{u+v} = A^{0,u+v} \supset \ \ldots \ \supset A^{u,v} \supset A^{u+1,v} \supset \ldots$$

$$A^{u+v+1,-1} = \{0\} \ , \ A^{0,n} = A^{-1,n+1} = A^{-2,n+2} = \ldots$$

This gives rise to (10.9) if one puts

$$A^u = \sum_v A^{u,v} \ .$$

Moreover, $A^{u,v}$ consists of the forms $\mathcal{G}^{u+v}$ belonging to the u-th power of the ideal i of forms locally generated by the left members of (11.2), that is the differentials of the "B-coordinates".

Similarly, a smooth cubical simplex $\psi: I^{u+v} \longrightarrow M$ (here I denotes the closed interval $[0,1]$) is said to belong to $A_{u,v}$ if for fixed $\tau = (t_0^1,\ldots,t_0^u) \in I^u$, the induced map

$$\psi_\tau : I^u \longrightarrow M \ , \ \psi_\tau(s^1,\ldots,s^v) = \psi(t_0^1,\ldots,t_0^u,t^{u+1},\ldots t^{u+v}),$$

$$t^{u+1} = s^1,\ldots,t^{u+v} = s^v \ ,$$

has values in a single fibre (or satisfies all local equations (11.2)). This generates the free groups

$$\{0\} = A_{-1,u+v+1} \subset A_{0,u+v} \subset A_{1,u+v-1} \subset \ldots \subset A_{u+v,0} \ ,$$

$$A_{-1,n+1} = A_{-2,n+2} = \ldots = \{0\} \ , \ A_{n,0} = A_{n+1,-1} = A_{n+2,-1} = \ldots$$

One then gets (11.1) putting

$$A_u = \sum A_{u,v} \ .$$

This example produces the classical cohomology (in the sense of de Rham cohomology) and homology (in the sense of Serre [25]) spectral sequences of a fibre bundle.

But, as we shall show, it also yields the theory of integral invariants of H. Poincaré and, especially, E. Cartan.

Indeed, we can consider a closed compact submanifold $\Gamma = \Gamma_u$ of dimension u in M (or a compact submanifold without boundary) which we shall assume to be "transverse" to the fibres at each point.(N.B.: This transversality condition helps to produce a more "pure" image, but it is, however, inessential, provided it is fulfilled at some point $y \in \Gamma$). The condition at a point $y \in \Gamma$ means that the tangent space $T_y(F_y)$ to the fibre F_y through y and the tangent space $T_y(\Gamma)$, both contained in $T_y(M)$, intersect only at the origin. (Existence of such a situation also implies that $u = \dim(\Gamma) \leq b = \dim(B)$.) In the language of H. Poincaré and E. Cartan, we can imagine that we "move" each point $y \in \Gamma$ along a fibre to a point y' , so that Γ is globally moved to a new similar submanifold Γ'. During this "movement" the points of Γ generate a new submanifold U of dimension $u+1$ in M, with boundary

$$\partial U = \Gamma' - \Gamma . \tag{11.3}$$

The question is then to find a global differential form Θ over M such that, for every such situation, one has

$$\int_\Gamma \Theta = \int_{\Gamma'} \Theta . \tag{11.4}$$

Definition 11.1. Following E. Cartan's terminology, a form Θ of this type is called a relative integral invariant of the local systems (11.2) or, better, of the fibre bundle in consideration.

This is, at least, the offical-historical notion. I want to ask: why should we restrict ourselves by integrating only over u-manifolds Γ without boundary? Equivalently over u-cycles? Algebraically, the important thing is the vanishing of the integral of Θ over ∂U, or of $d\Theta$ over U. The reference to cycles is certainly due to the inspiration by circulation in hydrodynamics, the conservation theorems of circulation and of vorticity. We, however, could refer to any u-dimensional compact submanifold $\Gamma = \Gamma_u$ with boundary, imposing that we restrict ourselves to "movements" leaving that boundary fixed. Historically, this was impossible in the version of Poincaré and contemporary hydrodynamicians who wanted to integrate on domains consisting of simultaneous points and follow these domains along with the physical movement. Such restriction is no longer necessary after the generalization by E. Cartan considering cycles consisting of non simultaneous points, thus making it possible to drop the restriction of integrating over cycles.

Definition 11.2. We shall say that a differential form Θ enjoys the extended relative integral invariant property, or is an extended relative integral invariant, if the formula (11.4) holds for any "movement" from Γ to Γ' leaving the boundary fixed.

Clearly, the usual and the extended relative integral invariant properties are equivalent.

Definition 11.3. If formula (11.4) holds for every compact u-manifold $\Gamma = \Gamma_u$ with boundary and every "movement along the fibres", without fixity restriction on the boundary, then one says that Θ is an absolute integral invariant of the local systems (11.2) or of the fibre bundle $\xi : M \longrightarrow B$.

In this latter kind of movements Γ still generates a $(u+1)$-submanifold U (with corners, not only with boundary) but (11.3) is no-longer valid: one has to consider the u-submanifold with boundary V generated by $\partial\Gamma$ in the movement and the formula becomes

$$\Gamma' - \Gamma = \partial U + V . \tag{11.5}$$

One easily derives the following statements 11.4 through 11.8.

Proposition 11.4. For every u-form θ, the following properties are equivalent:

(i) θ is a relative integral invariant of the bundle;

(ii) for every $(u+1)$-submanifold U, compact with boundary, occuring in (11.3), one has $\int_U d\theta = 0$;

(iii) for every $(u+1)$-chain $c \in A_{u,1}$ one has $\int_c d\theta = 0$;

(iv) $d\theta$ belongs to A^{u+1} or, equivalently, is of filtration $u+1$;

(v) around every point $y \in M$, $d\theta$ is locally the inverse image under ξ of a form defined around $x = \xi(y)$ in B .

This last property means that, in an admissible coordinate neighbourhood $(y^1,\ldots,y^b,\ldots,y^m)$, the local expression of $d\theta$ depends only on the first "B-coordinates" $y^1,\ldots,y^b$ and their differentials. When the fibres are not connected, this does not imply that $d\theta$ be globally the inverse image of a global form on B .

The reader will observe that the equivalence between (iii) and (iv) is just the special case for $r = u+1$, $s = 0$ of the following:

Proposition 11.5. $A^{r,s}$ coincides with the module of differential forms of M orthogonal to $A_{r-1,s+1}$, that is whose integral vanishes on any chain of $A_{r-1,s+1}$.

Proposition 11.6. For every u-form θ , the following properties are equivalent:

(i) θ is an absolute integral invariant;

(ii) for every V and U occuring in (11.5), one has the relations

$$\int_V \theta = \int_U d\theta = 0 ;$$

(iii) θ is an element of $C_1^{u,o}$;

(iv) θ is locally, around every point $y \in M$, the inverse image of a form defined around $x = \xi(y)$ in B .

Moreover, if the fibres are connected, this implies that θ is globally the inverse image of a form of the base.

Proposition 11.7. For $u = 0$, the relative and absolute integral invariants coincide and are nothing else than the first integrals (intégrales premières) of system (11.2).

Proposition 11.8. A differential form θ is a relative integral invariant if and only if its differential $d\theta$ is an absolute integral invariant.

12. Spectral diagram and generalized integral invariants.

Now, in the historical version of the relative integral invariant, we can use, instead of Γ, an arbitrary cycle c of dimension u, that is an element of $C^{\infty}_{u,o} = C^{u+1}_{u,o}$ modulo the boundary of a chain d (playing the role of U) belonging to $A_{u,1}$. This, therefore, introduces the quotient

$$E_{(u)} = C^{u+1}_{u,o}/B^{o}_{u,o}. \tag{12.1}$$

We also see that the integrand θ may be of filtration zero (that is $\theta \in A^{o,u}$); in particular its integral over a cube $\psi \in A_{u-1,1}$ does not need to vanish as shows the example of the linear relative integral invariant of dynamics, in which case the fibres coincide with the trajectories. This justifies the absence of any term like $C^{u}_{u-1,1}$ in the denominator of $E_{(u)}$. This quotient is thus essentially different of any one in the spectral sequence.

Besides $\theta \in A^{o,u}$, one must also have $d\theta \in A^{u+1,o}$, so that θ is a relative integral invariant if and only if $\theta \in C^{o,u}_{u+1}$. Still within the historical presentation, we integrate over cycles, so that we can add to θ any form whose integral vanishes over cycles, e.g. an arbitrary coboundary; in other words θ is defined modulo

$$B^{o,u}_{o} = dA^{o,u-1}.$$

This, therefore, introduces the quotient

$$F^{(u)} = C^{o,u}_{u+1}/B^{o,u}_{o} \tag{12.2}$$

in duality with $E_{(u)}$ in (12.1).

Now, if we use the generalized notion of extended relative integral invariant introduced above, we have to replace the quotients (12.1), (12.2) respectively by

$$G_{(u)} = A_{u,o}/B^{o}_{u,o} = C^{1}_{u,o}/B^{o}_{u,o} \qquad (12.3)$$

$$H^{(u)} = C^{o,u}_{u+1} = C^{o,u}_{u+1}/B^{u+1,-1}_{o} \qquad (12.4)$$

None of the expressions 12.1 through 12.4 appear in the spectral sequence.

Passing to the case of an absolute integral invariant, formula (11.5) means that Γ and Γ' correspond to one single element in the quotient

$$E^{1}_{u,o} = C^{1}_{u,o}/(C^{o}_{u-1,1} + B^{o}_{u,o})$$

which, this time, belongs to the spectral sequence. Certainly, a form $\Theta \in C^{u,o}_{1}$ vanishes over the denominator and $C^{u,o}_{1}$ is essentially the dual term

$$E^{u,o}_{1} = C^{u,o}_{1}/(C^{u+1,-1}_{o} + B^{u,o}_{o})$$

of the spectral sequence, as the denominator $C^{u+1,-1}_{o} + B^{u,o}_{o}$ reduces to zero.

More generally, consider a homology quotient $E^{1}_{u,v}$. We see that a differential form Θ of degree $u+v$, with an integral constant over the classes of $E^{1}_{u,v}$, that is with an integral vanishing over $C^{o}_{u+1,v-1} + B^{o}_{u,v}$ may be considered as a generalized <u>absolute integral invariant</u>. This is certainly the case of a form $\Theta \in C^{u,v}_{1}$ of which it is appropriate to consider the image in $E^{u,v}_{1}$. As is known from the algebraic topology of fibre bundles, an element of $E^{u,v}_{1}$ corresponds to a u-cocyle of the base B with values in the local system of the v-cohomology of the fibres. Similarly, $E^{1}_{u,v}$ corresponds to the group of u-chains of B with coefficients in the v-homology of the fibres. Let us consider a form $\Xi \in C^{o,v}_{1}$, or its image in $E^{o,v}_{1}$: it defines in each fibre a v-dimensional cohomology class induced by Ξ in the ambient space. Therefore, we can consider Ξ as defining a "differentiable function" over B with values in the v-cohomology of the fibres. If $v = o$ and the fibres are connected, this is just a first integral of the system (11.2), or a smooth real valued function over B .

So as to incorporate the relative integral invariants in a similar setting, it is convenient to consider more general homology quotients than those of the spectral sequence, of which (12.1) through (12.4) be special cases. These new quotients have been introduced in

[11] and [12] ; they fit into a large frame, generalizing the spectral sequence, which we have called the spectral diagram.

At this point a change of notation is appropriate and we shall put, for $n = u + u'$:

$$Q^w_{u,u'} = {}_nQ^w_u = \{x \in A_{u,u'} , \partial x \in A_w\} =$$

$$= C^{u-w}_{u,u'} = {}_nC^{u-w}_u ,$$

$$P^w_{u,u'} = {}_nP^w_u = \{x \in A_{u,u'} , x = \partial y , y \in A_w\} =$$

$$= B^{w-u}_{u,u'} = {}_nB^{w-u}_u = \partial({}_{n+1}Q^u_w) ,$$

$$Q^{u,u'}_w = {}^nQ^u_w = \{x \in A^{u,u'} , dx \in A^w\} =$$

$$= C^{u,u'}_{w-u} = {}^nC^u_{w-u} ,$$

$$P^{u,u'}_w = {}^nP^u_w = \{x \in A^{u,u'} , x = dy , y \in A^w\} =$$

$$= B^{u,u'}_{u-w} = {}^nB^u_{u-w} = d({}^{n-1}Q^w_u) .$$

One sees that, for $s > t$, one has

$${}_nQ^u_t \subset {}_nQ^u_s \quad \text{and} \quad {}^nQ^s_r \subset {}^nQ^t_r ,$$

making it possible to consider the expressions

$${}_nE(r,s,t,u) = {}_nQ^u_s/({}_nQ^u_t + {}_nP^r_s) , \qquad t < s ,$$

$${}^nE(r,s,t,u) = {}^nQ^{t+1}_{r+1}/({}^nQ^{s+1}_{r+1} + {}^nP^{t+1}_{u+1}) , \qquad s > t .$$

The fact that A^{u+1} vanishes over A_u implies that the last two expressions are put in duality by integrating a form $\theta \in A^{t+1,t'}$ over a chain $c \in A_{s,s'}$. Indeed, if θ belongs to the numerator of ${}^nE$, it vanishes over the denominator of ${}_nE$; respectively, if θ belongs to the denominator of ${}^nE$, it vanishes over the numerator of ${}_nE$. We also observe that

$$E^r_{u,u'} = {}_nE(u+r-1,u,u-1,u-r), \qquad u+u' = n , \tag{12.5}$$

$$E^{u,u'}_r = {}^nE(u+r-1,u,u-1,u-r), \qquad u+u' = n , \tag{12.6}$$

and that the general duality between ${}_nE$ and ${}^nE$ induces, in particular, the classical duality between these last two terms of the spectral sequence.

Referring to formulas (12.1) through (12.4), we find that

$$E_{(u)} = {}_uE(u,u,-1,-1) \ , \tag{12.7}$$

$$G_{(u)} = {}_uE(u,u,-1,u-1) \, , \tag{12.8}$$

$$F^{(u)} = {}^uE(u,u,-1,-1) \ , \tag{12.9}$$

$$H^{(u)} = {}^uE(u,u,-1,u-1) \, . \tag{12.10}$$

We shall denote by $\mathcal{M}$ the set of quadruples $\mu = (r,s,t,u)$ such that $s > t$ and by $\mathcal{N}$ the set of all pairs $(r,s), \ldots$. We have maps

$$\alpha : \mathcal{M} \to \mathcal{N} \ , \ \beta : \mathcal{M} \to \mathcal{N}$$

such that

$$\alpha(\mu) = \alpha(r,s,t,u) = (r,s) \ , \ \beta(\mu) = \beta(r,s,t,u) = (t,u) \, .$$

We may consider $\mu = (r,s,t,u)$ as a *link* from (r,s) to (t,u) and write

$$\mu : (r,s) \longrightarrow (t,u) \ .$$

If $\nu = (t,u,v,w)$ is another element of $\mathcal{M}$ such that $\beta(\mu) = \alpha(\mu)$, we say that μ and ν are *adjacent* and the pair (μ,ν) is called an *arrow* from μ to ν , what we write

$$\begin{array}{ccccc} & & (t,u) & & \\ & \overset{\mu}{\nearrow} & \Longrightarrow & \overset{\nu}{\searrow} & \\ (r,s) & & & & (v,w) \end{array} \qquad \text{or} \qquad \mu \Longrightarrow \nu \ .$$

Proposition 12.1. For every adjacent pair (μ,ν) , *there are homomorphisms*

$$d_{\nu,\mu} : {}_nE(\mu) \longrightarrow {}_{n-1}E(\nu) \ ,$$

$$d^{\mu,\nu} : {}^nE(\mu) \longleftarrow {}^{n-1}E(\nu) \ ,$$

induced by the boundary and coboundary operators. Moreover, for a diagram

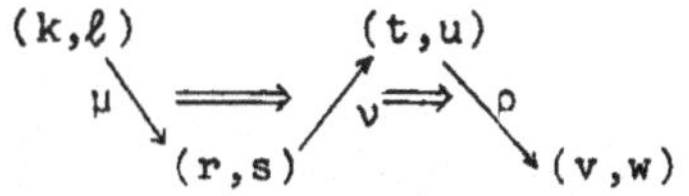

(i.e. for two consecutive arrows), then, in the corresponding diagrams

$$ {}_{n+1}E(\mu) \xrightarrow{d_{\nu,\mu}} {}_{n}E(\nu) \xrightarrow{d_{\rho,\nu}} {}_{n-1}E(\rho) \ , $$

$$ {}^{n+1}E(\mu) \xleftarrow{d^{\mu,\nu}} {}^{n}E(\nu) \xleftarrow{d^{\nu,\rho}} {}^{n-1}E(\rho) \ , $$

the composite arrows vanish:

$$ d_{\rho,\nu} \circ d_{\nu,\mu} = 0 \ , \qquad d^{\mu,\nu} \circ d^{\nu,\rho} = 0 \ . $$

Finally, the defects of exactness are given by

$$ \mathrm{Ker}(d_{\rho,\nu})/\mathrm{Im}(d_{\nu,\mu}) = {}_{n}E(\sigma) \ , $$

$$ \mathrm{Ker}(d^{\mu,\nu})/\mathrm{Im}(d^{\nu,\rho}) = {}^{n}E(\sigma) \ , $$

where $\sigma = (\ell,s,t,v) \in \mathcal{M}$ may be called the derived element of the pair of consecutive arrows (μ,ν) and (ν,ρ) .

We refer the reader to [11], [12] for the proof.

Definition 12.2. For every $\nu = (r,s,t,u) \in \mathcal{M}$, a differential form $\Theta \in {}^{n}Q^{t+1}_{r+1} = {}^{n}C^{t+1}_{r-t}$, or its class in ${}^{n}E(\nu)$ will be called an integral invariant of type (ν) and degree n for the system (11.2) or for the bundle $\xi : M \longrightarrow B$.

From (12.5) we obtain, for $r = 1$,

$$ E_{1}^{u+1,0} = {}^{u+1}E(u+1,u+1,u,u). $$

Together with (12.9), (12.10), this yields the three following elements of $\mathcal{M}$

$$ \mu = (u,u,-1,-1) \ , \ \nu = (u,u,-1,u-1) \ , \ \rho = (u+1,u+1,u,u) $$

and the diagrams (in which $v = u+1$, $w = u-1$)

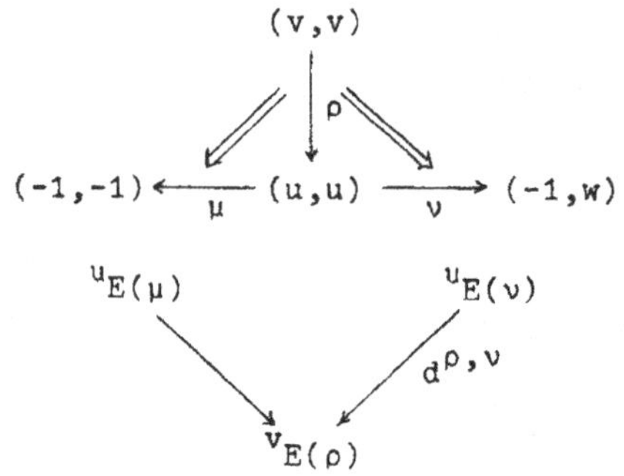

The latter explains how one passes, either from a classical relative integral invariant of degree u in $^uE(\mu)$, or from what we have called an extended relative integral invariant in $^uE(\nu)$, to an absolute integral invariant of degree $u+1$ in $^vE(\rho)$.

We see that, in this more general setting, the notion of "absolute" or "relative" integral invariant becomes rather irrelevant. We, however, see that for a diagram

$$? \overset{x}{\Longleftarrow} \mu \overset{y}{\Longleftarrow} ? \quad ,$$

if an integral invariant in $^nE(\mu)$ lies in the image of $d^x : ? \longrightarrow {}^nE(\mu)$, it is always destroyed by $d^y : {}^nE(\mu) \longrightarrow ?$ The relevant questions, therefore, are the following:

if an integral invariant is destroyed by some d^y , does there exist an arrow x which brings it in the image of d^x ? Or is it in this image for a given x ?

13. Applications to multiple integrals of the Calculus of Variations.

To a regular first-order variational problem $\mathcal{P}_1$ +) of degree p , we have anociated a p-form Ω over a phase space $\mathcal{E}$ and we have seen that the extremals are given by a system of exterior differential equations, that is by an ideal $\mathfrak{i}$ in the algebra of differential forms. But we have mentioned that this ideal is not complete, so that we also have to consider its completion $\bar{\mathfrak{i}}$. Moreover, all this is a local problem so that we rather have to consider the equations as given by sheaves of ideals (which we still denote by $\mathfrak{i}$ and $\bar{\mathfrak{i}}$) in the sheaf of exterior differential forms. This means that all covariants which we shall produce are also sheaf-objects: for example there will be a sheaf of spectral sequences and a sheaf of spectral diagrams in an obvious sense, namely in the sense of contravariant functors defined on the category of open sets of

+) $\mathcal{P}_1 = \mathrm{Var}_1(\mathcal{V}, \tilde{\mathcal{O}}, \mathcal{L})$, see definition 4.1.

$\mathcal{E}$ whose morphisms are the inclusions of a small open set into a larger one.

However, we shall avoid the sheaf language for the sake of simplicity and the careful reader will easily restaure the full situation out of our simplification.

The differential forms of degree n over $\mathcal{E}$ (at this unique place, we mention that this means "over any open set U in $\mathcal{E}$ ") belonging to the u-th power of $\bar{\mathcal{I}}$ form a module $A^{u,v}$, $v = n-u$, yielding a situation absolutely parallel to that associated to a fibre bundle in section 11 above. In particular, we have

$$\ldots = A^{-1,n+1} = A^{0,n} = \mathcal{G}^n \supset A^{1,n-1} \supset A^{2,n-2} \supset \ldots \supset A^{u,v} \supset \ldots$$

$$A^{n+1,-1} = A^{n+2,-2} = \ldots = \{0\} .$$

Here, $\mathcal{G}^n$ denotes the module of n-forms over $\mathcal{E}$.

There is a theorem to the effect of which, there exist p-forms θ^i such that

$$d\Omega = \sum \omega^i \wedge \theta^i$$

where the Pfaffian forms ω^i are those defined by (10.8). This assertion is nothing but the <u>second Lepage congruence</u>, for which we refer to [10], [12] and [27] , and is readily checked. It means that $d\Omega$ belongs to the ideal generated by the ω^i's . But the forms θ^i clearly belong to $\mathcal{I}$ (they are essentially the first members of (10.6-iii)) and we see that $d\Omega$ is of filtration at least two, that is, lies in the second power of the completed ideal $\bar{\mathcal{I}}$. On the other hand, Ω is of filtration zero, otherwise the integral of Ω over any piece of extremal, namely the "Hamiltonian action" over that piece, would always be zero. One thus has

$$\Omega \in C_2^{0,p} = Q_2^{0,p} \tag{13.1}$$

and, for any $i > 0$,

$$d\Omega \in C_i^{2,p-1} = Q_{2+i}^{2,p-1}. \tag{13.2}$$

We also have to dualize the situation, filtering the graded group of differentiable chains. To that effect, we say that a differentiable n-dimensional cube $(n = u+v)$

$$\psi : I^{u+v} \longrightarrow \mathcal{E} , \qquad I = [0,1] ,$$

belongs to $A_{u,v}$ if for any $t_o = (t_o^1,\dots,t_o^u) \in I^u$, the v-cube

$$\psi_{t_o} : I^v \longrightarrow \mathcal{E} \quad , \quad \psi_{t_o}(s^1,\dots,s^v) = \psi(t_o^1,\dots,t_o^p,s^1,\dots,s^q)$$

lies entirely in an integral submanifold of $\bar{i}$. It is well-known that a submanifold of an integral manifold is also an integral manifold and, therefore, we have

$$\{0\} = A_{-1,n+1} \subset A_{0,n} \subset A_{1,n-1} \subset \cdots \subset A_{u,v} \subset \cdots \subset A_{n,0} = A_{n+1,-1} = \cdots$$

Moreover, $A_{n,o}$ is the group of all n-chains, while $A_{o,p}$ is the group of p-chains lying in an extremal.

All this can be generalized to an arbitrary system of exterior differential equations (namely, to any ideal i of $\mathcal{G}$, the algebra of differential forms) and we recall that an <u>integral manifold</u> of i is a submanifold over which vanishes the restriction of any form of i. Of course, the more simple and practicable is the system, the more handy is the situation. No doubt the geometric object consisting of a differentiable fibre bundle as discussed in section 11, or that defined by a simple integral of the Calculus of Variations are more simple and handy than one characterized by a multiple integral whose extremals are the solutions of a system of partial differential equations. However, the case of a multiple integral in C.V. is completely natural and very important. The generalized canonical equations form a very nice example of exterior differential equations and they constitute an involutive system for the dimension p in the sense of E. Cartan [4], a question which we want to discuss in [13].

The complexity of a two-dimensional problem is described by the equations (10.6-iii), (10.8) which have to be completed by $d\omega^i = 0$. For $q = 1$, if we consider the Dirichlet and Plateau integrals corresponding to

$$L = \tfrac{1}{2}(p^2+q^2) \quad \text{and} \quad L = \sqrt{1+p^2+q^2} \tag{13.3}$$

we obtain a five-dimensional phase space $\mathcal{E}$ with local coordinates x,y,z,p,q and the equations

$$dz - pdx - qdy = 0 \;, \qquad dp\wedge dx + dq\wedge dy = 0 \tag{13.4}$$

$$d\theta = 0 \quad \text{for} \quad \theta = Pdy - Qdx \tag{13.5}$$

For the first integrand one must take $P = p$, $Q = q$ and, for the second,

$$P = p\cdot(1+p^2+q^2)^{-\frac{1}{2}} \qquad Q = q\cdot(1+p^2+q^2)^{-\frac{1}{2}}$$

In the case of a simple regular integral the canonical equations form a completely integrable Pfaffian system and a Cauchy datum is just a point of the phase space, through which passes one single trajectory. The fact that, in the general case $(p > 1)$, the canonical equations form an involutive system is essential to understand the generalization of a Cauchy datum, that is of a point is the phase space.

I want to show that, in the general case, such a datum is an integral (p-1)-cycle of the phase space.

We shall limit ourselves to an analytic situation, the only one for which existence theorems are available. Also, so as to have unicity (see remark 13.1 below), we shall limit ourselves to an analytic section $\lambda:\tilde{U} \to \mathcal{E}$ over some open set $\tilde{U}$ of $\tilde{\mathcal{O}}$, avoiding the irregular locus; in other words, the parameters $\Lambda^{\alpha\beta}_{ij}, \Lambda^{\alpha\beta\gamma}_{ijk}, \ldots$ are replaced by functions defined in $\tilde{U} \subset \tilde{\mathcal{O}} \subset \tilde{\mathcal{V}}$. We then denote by $\mathcal{E}_\lambda = \lambda(\tilde{U})$ the subspace of $\mathcal{E}$ defined by λ and we consider the restriction of the canonical equations (we mean of the complete ideal $\bar{\imath}$) to $\mathcal{E}_\lambda$. It is then the case that:

(a_1) through every point of $\mathcal{E}_\lambda$ passes an integral curve $\mathcal{W}_1$;

(a_2) through every integral curve $\mathcal{W}_1$, in general position, passes an integral surface $\mathcal{W}_2$;

.....

(a_p) through every integral manifold $\mathcal{W}_{p-1}$, in general position, passes an integral manifold $\mathcal{W}_p$, that is an extremal manifold, the germ of which along $\mathcal{W}_{p-1}$ is unique.

In these statements, all manifolds $\mathcal{W}_i$ are analytic submanifolds of $\mathcal{E}_\lambda$. In particular, suppose given in $\mathcal{E}_\lambda$ an analytic integral manifold $\mathcal{W}_{p-1}$ (of dimension p-1), compact and without boundary, in generic position at everyone of its points. Then through it and inside $\mathcal{E}_\lambda$ passes exactly one germ μ_p of analytic extremal manifold. Moreover, in the space of germs of analytic submanifolds (without singularity), the connected component of this germ constitutes a unique (maximal) analytic submanifold $\mathcal{W}_p$ which is extremal. It has no reason to be compact and, in general, will be topologically wild. For example, considering the first integral (13.3), as local expression of the global Plateau Problem, it is known that the topological type of a minimal surface $\mathcal{W}_2$ may be arbitrarily complicated.

Remark 13.1. We may also study the solutions of the canonical equations in all of $\mathcal{E}$, without restricting ourselves to a section λ . In that case, statements similar to $(a_1)\ldots(a_p)$ hold, except for unicity in (a_p) . The lack of unicity requires a little more care if one wishes to extend the preceding analysis. The lack of unicity, however, is tamed by the following remark: if $\mathcal{W}_p$ is an integral manifold through $\mathcal{W}_{p-1}$, any other one is obtained moving the points of $\mathcal{W}_p$ not on $\mathcal{W}_{p-1}$ along the fibres of the projection $\pi: \mathcal{E} \longrightarrow \tilde{\mathcal{O}}$.

The important thing is that an analytic integral manifold $\mathcal{W}_{p-1}$, as above, compact and without boundary, constitutes a Cauchy datum (or "initial condition") for the generalized canonical equations.It therefore yields a generalization of a "point" of the phase space for a simple integral and, similarly, the extremal manifold $\mathcal{W}_p$ through $\mathcal{W}_{p-1}$ generalizes the "trajectory" through a "point". Clearly $\mathcal{W}_{p-1}$ defines a cycle of dimension p-1 in $\mathcal{W}_p$ and its homology class in $H_{p-1}(\mathcal{W}_p, Z)$ or $H_{p-1}(\mathcal{W}_p, \mathbb{R})$. This homology class is also a generalization of the "trajectory" through the "point" $\mathcal{W}_{p-1}$. The novelty is that, in general, $H_{p-1}(\mathcal{W}_p)$ has more than one "generator" $\mathcal{W}_{p-1}$ and that another such generator $\mathcal{W}'_{p-1}$ defines the same extremal $\mathcal{W}_p$.

Remark 13.2. First integrals. Considering the examples of problems (13.3), we have $p = 2$, $p-1 = 1$ and, in this case, $\mathcal{E} \equiv \mathcal{E}_\lambda$. An extremal $\mathcal{W} = \mathcal{W}_2$ may have a rich one-dimensional homology and is thus determined by many 1-cycles $\gamma = \mathcal{W}_1$. Equation (13.5) shows that the Pfaffian form θ of $\mathcal{E}$ induces a 1-cocycle on each extremal $\mathcal{W}$, so that

$$F_\theta(\gamma) = \int_\gamma \theta \qquad (13.6)$$

is constant when γ moves along its "trajectory-homology class" in $\mathcal{W}_1$. The integral (13.6) or the Pfaffian form θ thus generalizes a "first integral" of a system of ordinary differential equations.

For $p \geqslant 2$, in general, the generalized first integrals will be differential forms θ of degree p-1 such that $d\theta \in \bar{\mathcal{I}}$, namely the forms of $C_1^{0,p-1}$. However, as we have to integrate these forms, over integral (p-1)-cycles,we have to take them modulo $C_0^{1,p-1} + B_0^{0,p-1}$ so that the real first integrals are the elements of

$$E_1^{o,p-1} = C_1^{o,p-1}/(C_o^{1,p-2} + B_o^{o,p-1}) .$$

It is known that, in the case of a smooth fibre bundle, as in section 11, this term represents the (smooth) "functions of the base with values in the cohomology of dimension (p-1) of the fibre".

For a general regular variational problem $\mathcal{P}_1$, among the forms of the phase space $\mathcal{E}$ belonging to $C_1^{o,p-1}$ are those giving rise to the Poisson and Noether algebras described in [18] and [19]. I like to conjecture that the only "interesting" first integrals (i.e. the forms θ producing non zero elements in the $H^{p-1}(\mathcal{W})$'s for $\mathcal{W}$ an extremal) are exactly those (p-1)-forms of $\mathcal{E}$ for which there exists a field X of tangent vectors such that

$$i(X)d\Omega = d\theta$$

(Hamiltonian vector field).

Note that $i(X)d\Omega$ belongs to $\dot{\imath}$ (by virtue of equations (10.3)), so that the question is whether $d\theta \in \bar{\imath}$ is equivalent to $d\theta \in \dot{\imath}$ for the "interesting" forms θ , i.e. those not in the denominator of $E_1^{o,p-1}$. If the conjecture holds true, then $E_1^{o,p-1}$ carries a structure of Lie algebra (as in the case for p = 1) extending to the Poisson algebra of [19] and the Noether algebra of [18]. Otherwise, this is true only for a part of $E_1^{o,p-1}$.

14. Fundamental integral invariants of a multiple integral.

We consider an extremal $\mathcal{W}_p$ in the phase space $\mathcal{E}$, together with two homologous (p-1)-cycles W_{p-1}^o and W_{p-1}^1 . There thus exists a p-chain $U_p = U_{o,p}$ in M_p such that

$$W_{p-1}^1 - W_{p-1}^o = \partial U_p .$$

Suppose now that all these data depend cyclically on a real parameter t , so that

$$\mathcal{W}_p = \mathcal{W}_p(t) , W_{p-1}^i = W_{p-1}^i(t) , U_p = U_p(t) , i = o,1 , \qquad (14.1)$$

and assume the period of these functions to be 1. Then, when t varies on a small interval, W_{p-1}^i generates a chain of dimension p and filtration 1 belonging to $C_{1,p-1}^1$; similarly, U_p generates a (p+1)-chain of the same filtration, belonging to $C_{1,p}^o$. However, when t runs from 0 to 1 (over a period) W_{p-1}^i and U_p generate respectively a p-cycle $\Gamma^i = \Gamma_{1,p-1}^i$ and a (p+1)-chain $U_{1,p}$, both of filtration 1 fitting in the formula

$$\Gamma^1_{1,p-1} - \Gamma^0_{1,p-1} = \partial U_{1,p}$$

One has, moreover, $\Gamma^i_{1,p-1} \in C^2_{1,p-1}$, $U_{1,p} \in C^0_{1,p}$, so that $\Gamma^0_{1,p-1}$ and $\Gamma^1_{1,p-1}$ represent the same element in ${}_pE(1,1,-1,-1)$.

The direct part of the relative integral invariant of Poincaré-E. Cartan generalizes into the following:

Theorem 14.1. For each such situation, one has

$$\int_{\Gamma^0} \Omega = \int_{\Gamma^1} \Omega \qquad (14.2)$$

in other words, according to definition 12.2, the fundamental form Ω constitutes an integral invariant of degree p and type $\mu = (1,1,-1,-1)$.

Indeed, the difference is equal to

$$\int_{\Gamma^1-\Gamma^0} \Omega = \int_{\partial U_{1,p}} \Omega = \int_{U_{1,p}} d\Omega = 0$$

since $d\Omega$ of filtration 2, vanishes on any chain of filtration one.

There are various converses, one of which is:

Theorem 14.2. Consider a foliation of the phase space $\mathcal{E}$ with p-dimensional leaves (equivalently a completely integrable Pfaffian system whose solutions are those leaves) such that, for the corresponding filtration and for two cycles $\Gamma^0, \Gamma^1 \in C^p_{1,p-1}$ equivalent modulo $B^0_{1,p-1}$, always (14.2) holds. In this case, all leaves of the foliation are extremals.

In other words:

Theorem 14.3. Every foliation of the phase space $\mathcal{E}$ for which the fundamental form Ω is an integral invariant of degree p and type $\mu = (1,1,-1,-1)$, is an extremal foliation, that is a foliation by extremal manifolds.

Proof. The conditions of these theorems imply that, for any field X of tangent vectors, the leaves satisfy the equation $i(X)d\Omega = 0$. See (10.3). We also refer to [8] and [12].

It should be possible to define a certain type of exterior differential system such that the following better statement hold:

Conjecture 14.4. Let $\mathcal{I}$ be an involutive exterior differential system, of some appropriate type, over the phase space $\mathcal{E}$, such that the fundamental form Ω yield an integral invariant of degree p and type $\mu = (1,1,-1,-1)$. Then all its "general" integral p-dimensional manifolds are extremals.

A little more generally, formula (14.1) and these theorems remain valid if we assume to integrate, not only over cycles, but over p-chains $\Gamma^0, \Gamma^1 \in C^1_{1,p-1}$ equivalent modulo $B^0_{1,p-1}$. In other words, formula (13.4) still holds over the quotient

$$C^1_{1,p-1}/B^0_{1,p-1} = {}_pE(1,1,-1,0)$$

generalizing (12.8). In the preceding theorems we can thus say, equivalently, that Ω is an integral invariant of type $\nu = (1,1,-1,0)$. This generalizes what we have called in the classical case $p=1$ the "extended relative integral invariant" property (definition 11.2).

Comparing (11.4) and (14.2), we insist on the analogy between the cases $p=1$ and p arbitrary: Γ (resp. Γ^0) is a cycle generated by <u>points</u> of $\mathcal{E}$ (resp. by <u>integral</u> (p-1)-<u>cycles</u> of $\mathcal{E}$) and Γ' (resp. Γ^1) is obtained from Γ (resp. Γ^0) by moving these points (resp. cycles) along their trajectory (resp. in their homology class in $H_{p-1}(\mathcal{W}_p)$).

We now pass to the generalization of the <u>absolute integral invariant</u> of <u>Poincaré</u>- E. <u>Cartan</u>.

To that effect, assume that instead of (14.1) the data $\mathcal{W}_p$, W^i_{p-1}, $U_p = U_{o,p}$ depend on two real parameters u,v running over a domain D of $\mathbb{R}^2$ with boundary ∂D; for example D may be the standard square $I\times I$, $I = [0,1]$:

$$\mathcal{W}_p = \mathcal{W}_p(u,v)\ ,\ W^i_{p-1} = W^i_{p-1}(u,v)\quad ,\ U_{o,p} = U_{o,p}(u,v)\ ,\ i = 0,1\ ,$$
$$0 \leqslant u \leqslant 1\ ,\ 0 \leqslant v \leqslant 1\ . \tag{14.3}$$

Then, W^i_{p-1} generates an element $\Delta^i = \Delta^i_{2,p-1} \in C^1_{2,p-1}$ and $U_{o,p}$ and element $U_{2,p} \in C^0_{2,p}$. Moreover, one has to consider the "lateral part" $V_{1,p} \in C^0_{1,p}$ of $U_{2,p}$ corresponding to pairs $(u,v) \in \partial D$, yielding the formula

$$\Delta^1_{2,p-1} - \Delta^0_{2,p-1} = \partial U_{2,p} + V_{1,p}$$

analogous to (11.5). Thus Δ^0 and Δ^1 represent the same element of

$$E^1_{2,p-1} = C^1_{2,p-1}/(C^0_{1,p} + B^0_{2,p-1}) = {}_{p+1}E(2,2,1,1)$$

which is (12.5) for $u = 2$, $r = 1$, $n = p+1$. The announced generalization is then:

Theorem 14.4. For each such situation, one has

$$\int_{\Delta^0} d\Omega = \int_{\Delta^1} d\Omega$$

in other words, the differential $d\Omega$ of the fundamental form Ω constitutes an integral invariant of degree $p+1$ and type $\rho = (2,2,1,1)$.

The proof and the converse are left to the reader.

Clearly $d\Omega \in C_1^{2,p-1} = Q_3^{2,p-1}$ and, therefore, has a class in $E_1^{2,p-1} = {}_{p+1}E(2,2,1,1)$. The thing is, however, much more precise since $d\Omega \in C_\infty^{2,p-1}$.

To generalize completely the absolute integral invariants of Mechanics, one has to consider differential forms θ of degree $n = u + v$, $v \geq p - 1$, with a constant integral over the classes of

$$E^1_{u,v} = {}_nE(u,u,u-1,u-1)$$

and this is the case if $\theta \in C_1^{u,v}$. Similarly, forms $\theta \in C_{u+1}^{o,n}$ have a constant integral over the classes of

$$C^{u+1}_{u,q} / B^o_{u,q} = {}_nE(u,u,-1,-1)$$

or of

$$C^1_{u,q} / B^o_{u,q} = {}_nE(u,u,-1,u-1)$$

which, except for the degree, coincide with (12.7) and (12.8). These forms thus generalize the relative integral invariants or what we have called the extended relative integral invariants of Mechanics. A finer analysis requires a bifiltration for which we refer to [12], n° 28, page. 82.

Definition 14.5. We shall say that a differential form $\theta \in C_1^{u,v}$ producing an element in ${}^nE(u,u,u-1,u-1) = E_1^{u,v}$, $u + v = n$, yields an absolute integral invariant of bidegree (u,v) . More generally an element of ${}^nE(p,q,r,s)$, $n = q+q'$, will be said to be an integral invariant of bidegree (q,q') .

Remark 14.6. First integrals and integral invariants. Examples. Clearly, the first integrals considered in remark 13.2 are very good examples of absolute integral invariants of bidegree $(o,p-1)$: that is integral invariants of type (0,0,-1,-1) and degree $p-1$. For example, a holomorphic function $f(z) = u(x,y) + iv(x,y)$ of $z = x + iy$ in a domain $D \subset \mathbb{R}^2$ represents a surface of dimension two in the real 4-dimensional space of variables x,y,u,v. These functions yield the general integral of the Cauchy-Riemann equations and the Cauchy integral

$$\oint f(z)dz$$

is actually a (non trivial) first integral or an integral invariant of bidegree (0,1) according to our definitions. The same is true (for appropriate bidegree) of the generalized Cauchy integral and related formulas in the theory of analytic functions of several complex variables. Similarly, a harmonic function $u = u(x,y)$ in some domain $D \subset \mathbb{R}^2$ represents a surface in $\mathbb{R}^4$ if we introduce the variables $x,y,p = \partial u/\partial x$, $q = \partial u/\partial y$. Then the integral

$$\oint (pdy - qdx)$$

yields a non trivial integral invariant of bidegree (0,1). For example if $D = \mathbb{R}^2 - \{0\}$ and $u = \log\sqrt{x^2+y^2}$, it yields

$$\oint \frac{xdy - ydx}{x^2 + y^2} = \oint d\theta$$

in polar coordinates $x = r\cos\theta$, $y = r\sin\theta$. In a similar frame, harmonic functions of three variables, $u = u(x,y,z)$, yield in seven dimensions: $x,y,z,u,p = \partial u/\partial x$, $q = \partial u/\partial y$, $r = \partial u/\partial z$ the non trivial integral invariant

$$\oiint p\, dy\wedge dz + q\, dz\wedge dx + r\, dx\wedge dy$$

This is nothing but the famous theorem of Gauss concerning the flux of a force derived from a Newtonian potential across a closed surface containing masses in its interior.

There are, of course, many more examples in Geometry and Mathematical Physics.

A first integral, or integral invariant of bidegree (o,v) refers to an "atomic" property, while the integral invariants of bidegree (u,v) refer to objects of u dimensions, as were the object of study of the classical integral invariants of bidegree (u,o). One, therefore, sees that our presentation combines and extends ideas of H. Poincaré and E. Cartan, together with "atomic" properties expressed by the Cauchy integral, the theorem of Gauss, etc... This also includes the classical de Rham cohomology of a manifold; indeed the latter corresponds to integral invariants of bidegree (o,n) when the ideal i reduces to zero.

15. The symplectic structure of a multiple integral.

Advertisement. This section does not want to present formal rigorous results and a lot more work is necessary to obtain a satisfactory theory. We, however, believe that the following ideas may lead to a fruitful generalization of symplectic geometry.

In the case of a simple regular integral of the Calculus of Variations, the canonical equations (10.4) define a foliation of dimension one of the phase space $\mathcal{E}$ which appears, locally, as the product of a 2n-dimensional manifold $M = M_{2n}$ by the real line (the leaf). Local coordinates x^i, y^i, $1 \leq i \leq n$, in M are also (some) local coordinates in $\mathcal{E}$ through the local projection $\mathcal{E} \longrightarrow M$ and the canonical equations form a system equivalent to

$$dx^i = dy^i = 0.$$

The fact that $d\Omega$ defines an absolute integral invariant of degree two, means that $d\Omega$ is the inverse image of a (necessarily closed) 2-form Θ over M_{2n}, while regularity implies that Θ is of maximum rank. This 2-form defines precisely the symplectic structure of the (at least local) manifold M_{2n}, that is the manifold of extremals.

It is an important problem to generalize this situation to multiple integrals. Some people have introduced symplectic structures over manifolds of infinitely many dimensions, but I want to follow ideas more parallel to the approaches suggested by P.L. Garcia and A. Pérez-Rendón [18] and J. Kijowski [23]. These authors consider the "manifold" $\mathcal{V}$ of extremals in the phase space $\mathcal{E}$, limiting themselves eventually (in the case of Garcia and Pérez-Rendón) to the restricted phase space $\mathcal{E}_\lambda$ corresponding to the de Donder-Weyl section λ (that is the section defined by annihilation of all parameters $\Lambda^{\alpha\beta}_{ij}, \Lambda^{\alpha\beta\gamma}_{ijk}, \ldots$). My opinion, however, is that one has to consider, instead, a "manifold" M whose "points" are classes of those objects $\mathcal{W}_{p-1}$ which we have called Cauchy data in section 13. We remind that $\mathcal{W}_{p-1}$ is a submanifold of $\mathcal{E}$, analytic, compact, without boundary, of dimension p-1, integral of the (generalized) canonical equations and in generic position at everyone of its points. Each such $\mathcal{W}_{p-1}$, considered as a cycle, has a class w in $E^1_{0,p-1} = {}_{p-1}E(0,0,-1,-1)$ and we want to consider the totality M of all such classes. It is out of question to discuss here whether this M is a "manifold" in some abstract sense (Banach or more general), a problem which, I believe, has a positive answer, and about which I refer the interested reader to [1]. About the structure of M, let us mention that it is a "simili-manifold" in the sense of [12], p. 204: one can indeed define concepts of differentiable maps of a differentiable manifold into M or from M to a differentiable manifold. In particular, M has differentiable chains of dimension u which are exactly the elements of $E^1_{u,p-1} = {}_nE(u,u,u-1,u-1)$, $n = u+p-1$: Those chains have been described in geometric-heuristic terms in section 14. Dually, a differentiable function $F = F_\tau : M \longrightarrow \mathbb{R}$ is defined by a first integral $\tau \in C^{0,p-1}_1$; the value of F_τ at the class $w \in M$ of $\mathcal{W}_{p-1}$ is defined by the period of τ around $\mathcal{W}_{p-1}$ that is by

$$F_\tau(w) = \oint_{\mathcal{W}_{p-1}} \tau$$

More generally, the differential forms $\alpha \in C^{u,p-1}_1$ or the corresponding absolute integral invariant $[\alpha] \in E^{u,p-1}_1$ of bidegree $(u,p-1)$ can be integrated over the u-chains of M, that is over the elements of $E^1_{u,p-1}$: they thus provide differential forms of degree u over M, an idea which we shall clarify more explicitely in a moment.

Remark 15.1. As we have mentioned in section 13, one extremal E (i.e. an element of the "manifold" $\mathcal{V}$ of Garcia and Pérez-Rendón or of the $\mathcal{H}$ of Kijowski) can be determined by two non homologous cycles $\mathcal{W}_{p-1}$, $\mathcal{W}'_{p-1}$ defining distinct elements w, w' in M. For these, a first integral τ will in general have distinct periods, so that the function F_τ will have distinct values at the points w and w' induced by $\mathcal{W}_{p-1}$ and $\mathcal{W}'_{p-1}$. In other words, a first integral τ does not induce a function on $\mathcal{V}$ (or $\mathcal{H}$). It, however, yields a well defined function

$$F_\tau : M \longrightarrow \mathbb{R}.$$

As he uses $\mathcal{H}$, instead of our M, Kijowski tries to circumvent the difficulty thanks the introduction of a class $\mathcal{C}$ of "admissible initial surfaces" (a.i.s.) subject to a number of axioms ([23], p. 111 and 121), one of which (axiom 6, p. 121) implies that two a.i.s. in the same extremal E are homotopic in E (and therefore, in case they are cycles, homologous in E). His treatment also requires the awkward condition that the restriction of the support of a first integral to an a.i.s. be compact. The latter condition also appears at one point of [19], page 229, line 25.

Returning to the "u-forms" of M, we find that, for $u = 2$, the (p+1)-form $d\Omega$ over the phase space provides a fundamental "2-form" Ξ over M and I believe that the pair (M,Ξ) is a good candidate for the generalized symplectic structure attached to a regular multiple integral problem $\mathcal{P}_1$ of the Calculus of Variations. The situation, however, is deeply different from the classical one. For $p>1$, the object M is a functional space whose geometry is most intimately related to the structure of which it derives, namely the phase space $\mathcal{E}$ and the canonical equations or, equivalently, the pair $(\mathcal{E},\Omega)$. An intrinsic approach of this structure seems much more difficult than in the case of classical symplectic geometry. At the present stage, I do not see any way of defining nor the u-chains, nor the u-forms of M, without constructions deeply rooted into the geometry of the phase space $\mathcal{E}$. We have here a perhaps difficult but most natural situation and it seems more fruitful to study it carefully, rather that going into more easy but formal generalizations. Anyway, the geometry of M seems to show a great deal of rigidity: on the one hand it refers to a rather large set M but, on the other hand, its differential geometric objects seem to be few in number, subject to very hard restrictions.

Let $w \in M$ be represented by a cycle $\mathcal{W}_{p-1} \in C^1_{o,p-1}$, so that $w \in E^1_{o,p-1}$. Suppose that we have a one-parameter family $W_{p-1}(t)$, $t \in R$, such that $\mathcal{W}_{p-1} = W_{p-1}(0)$: this induces a "curve" in M through w. The family can also be viewed as a map

$$W_{p-1}: \mathcal{W}_{p-1} \times \mathbb{R} \longrightarrow \mathcal{E} \tag{15.1}$$

and the 1-jet of the curve at $t = 0$, or the 1-jet of the preceding map at $\mathcal{W}_{p-1}\times\{0\}$, yields the concept of a "tangent vector" $\tilde{w}$ of M at the point w. The 1-jet of (15.1) can be visualized as a smooth (or analytic) function $\tilde{\mathcal{W}}_{p-1}$ assigning a tangent vector of $\mathcal{E}$ at every point of $\mathcal{W}_{p-1}$. In other words, we have the commutative diagram

$$\begin{array}{ccc} & & T(\mathcal{E}) \\ & \overset{\tilde{\mathcal{W}}_{p-1}}{\nearrow} & \downarrow \beta \\ \mathcal{W}_{p-1} & \xrightarrow[i]{} & \mathcal{E} \end{array}$$

where i is the inclusion and β, the projection of $T(\mathcal{E})$ (the tangent space of $\mathcal{E}$). We can also consider a tangent vector $\tilde{w}$ to M at w as an appropriate class of liftings $\tilde{\mathcal{W}}_{p-1}$ of $\mathcal{W}_{p-1}$ to $T(\mathcal{E})$.

More precisely, the complete ideal $\mathcal{j} = \bar{\mathcal{i}}$ over $\mathcal{E}$ yields over $T(\mathcal{E})$ and ideal $\tilde{\mathcal{j}}$ generated by $\alpha^*(\mathcal{j})$ plus the equations of variations or perturbational equations (équations aux variations) of $\mathcal{j}$. [Those characterize when a slight deviation $\mathcal{W} + \delta\mathcal{W}$ from an integral manifold $\mathcal{W}$ of $\mathcal{j}$ still is an integral manifold of $\mathcal{j}$; supposing this deviation depending on a parameter t, one would like to write $\delta\mathcal{W} = \tilde{\mathcal{W}}\cdot\delta t$.] Corresponding to $\tilde{\mathcal{j}}$ we have a filtration of the chain group of $T(\mathcal{E})$ producing homology quotients $\tilde{E}^r_{u,v}$, $.\tilde{E}(p,q,r,s)$, endowed with projections <u>onto</u> induced by α

$$\tilde{E}^r_{u,v} \longrightarrow E^r_{u,v}\ ,\quad .\tilde{E}(p,q,r,s) \longrightarrow .E(p,q,r,s)$$

[The "onto" property is a consequence of the involutive character of $\bar{\mathcal{i}}$.]

Then, a "tangent vector" $\tilde{w}$ to M at w is an element of $\tilde{E}^1_{o,p-1}$ over $w \in E^1_{o,p-1}$.

We denote by $T_w(M)$ the vector space over $\mathbb{R}$ of all these tangent vectors to M at w. Then a "u-form" α of M, that is an element $\alpha \in E_1^{u,p-1}$, or a representative $\alpha \in C_1^{u,p-1}$ induces as follows an alternate function

$$\bar{\alpha}: T_w(M)\times\ldots\times T_w(M) \longrightarrow \mathbb{R}\ .$$

If $\tilde{w}_1,\ldots,\tilde{w}_u$ are u tangent vectors at w, let them be represented by cycles $\tilde{\mathcal{W}}_1,\ldots,\tilde{\mathcal{W}}_u$ over a representative $\mathcal{W}$ of w. Those produce tangent vectors $X_1,\ldots,X_u$ of $\mathcal{E}$ at every point $x \in \mathcal{W}$. We then form the inner product $i(X_1\wedge\ldots\wedge X_u)\alpha$ which induces a (p-1)-form β over $\mathcal{W}$ of which we take the period

$$\bar{\alpha}(\tilde{w}_1,\ldots,\tilde{w}_u) = \int_{\mathcal{W}} \beta\ .$$

The differentials ∂_1 and d_1 of the spectral sequence make the u-chains of M into a complex

$$\cdots \longrightarrow E^1_{u+1,p-1} \xrightarrow{\partial_1} E^1_{u,p-1} \xrightarrow{\partial_1} E^1_{u-1,p-1} \longrightarrow \cdots \qquad (15.1)$$

and the u-forms into a co-complex

$$\cdots \longleftarrow E_1^{u+1,p-1} \xleftarrow{d_1} E_1^{u,p-1} \xleftarrow{d_1} E_1^{u-1,p-1} \longleftarrow \cdots \qquad (15.2)$$

Moreover, the duality between the various terms (by integration of a "u-form" over a "u-chain") yields an obvious Stokes formula.

As mentioned earlier, we have such complexes and co-complexes for every open set U of $\mathcal{E}$ and (15.2) corresponds thus to a sheaf of co-complexes and (15.1) to a co-sheaf of complexes (namely, the $E^1_{u,p-1}$'s are covariant functors of U, while the $E_1^{u,p-1}$'s are contravariant). Therefore, an important question is the local triviality (i.e. local exactness) of (15.1) and (15.2) around a point of $\mathcal{E}$ (or around a point $w \in M$ if this can make sense). Considering $\alpha \in E_1^{u,p-1}$ such that $d_1\alpha$ is zero in $E_1^{u+1,p-1}$, is it the case that $\alpha = d_1\beta$ for some $\beta \in E_1^{u-1,p-1}$? Or, α being definend around a point $X \in \mathcal{E}$, is this the case when α is restricted to a sufficiently small neighbourhood U of X ?

A very important example is given by the form $d\Omega$ which belongs to every $C_r^{2,p-1}$, $r \geq 1$, and, therefore, induces a class $[d\Omega]_r^{2,p-1} = \Xi_r^{2,p-1}$ in each $E_r^{2,p-1}$. However, $\Xi_r^{2,p-1} = 0$ for $r \geq 3$. [A form $\alpha \in C_r^{u,v}$ is often denoted $\alpha_r^{u,v}$ and its class in $E_r^{u,v}$ can be written $[\alpha]$ or $[\alpha]_r^{u,v}$]. The question whether $\Xi_1^{2,p-1} = \Gamma_1 \in E_1^{2,p-1}$ is a d_1-coboundary amounts to look for a form $\Theta^1 = \Theta_1^{1,p-1}$ such that

$$d\Theta^1 = d\Omega - \Psi^3 \text{, with } \Psi^3 = \Psi_0^{3,p-2}, \qquad (15.3.1)$$

which is equivalent to

$$d_1[\Theta^1] = [d\Omega] = \Gamma_1$$

or to $\Gamma_2 = 0$ if $\Gamma_2 = \Xi_2^{2,p-1} \in E_2^{2,p-1}$ is induced by $d\Omega$. But $d\Psi^3 = 0$, so that we have a d_1-cocycle $[\Psi^3]$ in $E_1^{3,p-2}$; this is a d_1-coboundary if and only if there exists a form $\Theta^2 = \Theta_1^{2,p-2}$ such that

$$d\Theta^2 = \Psi^3 - \Psi^4 \text{ , with } \Psi^4 = \Psi_0^{4,p-3} . \qquad (15.3.2)$$

But $d\Psi^4 = 0$, so that we have a d_1-cocycle $[\Psi^4]$ in $E_1^{4,p-3}$ and we look for a form $\Theta^3 \in C_1^{3,p-3}$ such that

$$d\Theta^3 = \Psi^4 - \Psi^5 \qquad (15.3.3)$$

with $\Psi^5 \in A^{5,p-4}$, $d\Psi^5 = 0$

And so on until we eventually cannot find a new form Θ^{i+1} ; we then say that we have an obstruction. If we have reached a pair $(\Theta^{p-1}, \Psi^{p+1})$ such that

$$d\Theta^{p-1} = \Psi^p - \Psi^{p+1}$$

with $\Psi^{p+1} \in A^{p+1,0}$, $d\Psi^{p+1} = 0$

we observe that $C_0^{p+2,-1} = \{0\}$ and the next step will be to find $\Theta^p \in C_1^{p,0}$ such that

$$d\Theta^p = \Psi^{p+1} \qquad (15.3.p)$$

For $p = 1$, there is only one step consisting in finding a form $\Theta \in C_1^{1,p-1}$ such that

$$d\Theta = d\Omega \qquad (15.4)$$

and we know that this is always possible (locally). In general we have to perform at most p steps and, in case no obstruction is encountered, we get the same formula putting

$$\Theta = \Theta^1 + \ldots + \Theta^p \in C_1^{1,p-1} .$$

In these conditions, we have the most interesting fact that there exists locally a (p-1)-form Σ (of filtration zero) such that

$$\Omega = d\Sigma + \Theta \text{ , } \qquad \Theta \in A^{1,p-1} . \qquad (15.5)$$

Then, for every extremal chain $c \in A^{0,p}$, the action integral (4.3) or (4.11) is given by the integral of Σ along the boundary:

$$I(c) = \int_c \Omega = \int_{\partial c} \Sigma \text{ , } \qquad c \in A^{0,p} . \qquad (15.6)$$

For $p = 1$, as we have mentioned, there is no obstruction, so that formulas (15.5) and (15.6) are valid. Σ is then a real function over the phase space which is essentially a complete integral of the Hamilton-Jacobi partial differential equation

$$\frac{\partial s}{\partial t} + H\left(t, q^i, \frac{\partial s}{\partial q^i}\right) = 0$$

when Σ is expressed in terms of appropriate local coordinates (t, q^i, α^i) over $\mathcal{E}$. As one knows, these complete integrals are the key to the equivalence between classical Mechanics and Geometrical Optics, producing the solutions of the canonical equations (10.4) by the theory of waves and envelopes.

Formulas like (15.5) and (15.6) seem to be dramatically absent from the literature on multiple integrals and I know only one instance of it, as an isolated case, namely for minimal surfaces of 3-space represented in a phase space of five dimensions. To that effect, I refer to the old but not to forget book of W. Blaschke, Vorlesungen über Differentialgeometrie, Kap. 6 (Extreme bei Flächen), § 94, Eine Formel von Schwarz für die Oberfläche einer Minimalfläche (Grundlehren d. Math. Wiss., bd. I, J. Springer, zweite Auflage, 1924, p. 173). See also the fünfte Auflage by Blaschke-Leichtweiss, Elementare Differentialgeometrie, Springer Verlag, 1973, p. 338. However, such formulas hold in much more general cases. For example I can prove that they exist in the case of a problem $\mathcal{P}_1$, the lagrangian of which does not depend on the space coordinates x^α, x^i; that is when the $L(x^\alpha, x^i, x^i_\alpha)$ of (4.3) depends only on the x^i_α's. As we shall see in a moment, the local validity of formulas (15.5), (15.6) is dominated by the cohomology of (15.2), that is by the "local" cohomology of the "generalized symplectic manifold" $M = (M, \Xi)$.

Theorem 15.2. *The (local) triviality of the cohomology of* (15.2) *that is the (local) vanishing of* $E_2^{u,p-1}$ *for* $u > 0$, *is sufficient for the existence of forms* $\Theta \in A^{1,p-1}$, $\Sigma \in A^{0,p-1}$ *for which formulas* (15.5) *and* (15.6) *are valid.*

Proof. We have given a step by step construction of the form $\Theta = \Theta^1 + \ldots + \Theta^p$. If Γ_2 is the image of $d\Omega$ in $E_2^{2,p-1}$, we see that the existence of the pair (Θ^1, Ψ^3) for which (15.3.1) holds is equivalent to $\Gamma_2 = 0$. Then, let Γ_3 be the class of Ψ^3 in $E_3^{3,p-2}$: we see that the existence of the pair (Θ^2, Ψ^4) for which

(15.3.2) be valid, is equivalent to $\Gamma_3 = 0$. If this is the case we can define $\Gamma_4 \in E_4^{4,p-3}$, etc... If $\Gamma_2,\ldots,\Gamma_{k-1}$ vanish, let $\Gamma_k \in E_k^{k,p-k+1}$ be the class of ψ^k ; then the existence of a pair $(\otimes^{k-1}, \psi^{k+1})$ such that (13.3.k-1) be valid is equivalent to $\Gamma_k = 0$. The proof is then clear.

As one sees, the situation is largely dominated by the diagrams (15.1) and (15.2).Those can be interpreted formally as the (cosheaf of) chain complexes and the (sheaf of) cochain complexes of the "manifold" M . Indeed,if this "manifold" is a delicate object, its singular complex is a well defined semi-simplicial set, together with its complex of differential forms. Those safe objects are deeply related to our generalized theory of integral invariants and we hope that this shows how the spectral sequence and the spectral diagram offer powerful tools toward a description of the complicated structure consisting of the totality of extremal manifolds in the phase space $\mathcal{E}$ and also of their projections in the manifold $\tilde{\mathcal{V}}$ of contact elements.

I want to mention here that our argument will make extremely unhappy the classical professionals of spectral sequences. It is indeed impossible (or it seems impossible) to compute the $E_2^{u,v}$ terms of our spectral sequence in terms of expressions they are familiar with. They indeed want a formula of the type

$$E_2^{u,v} = H^u(M,H^v(N))$$

when the only candidate for M is our mysterious, hardly existing "manifold" and where $H^v(N)$ would be a ghost-local coefficient system consisting in the v-cohomology of the highly unstable extremal N through a variable point $w \in M$.

I, however, want to suggest that the perfectly well defined $E_2^{u,v}$ should be a way of approaching the ghosts of a difficult M and of the uncatchable and fleeing N .

All this requires a deeper analysis which we wish to develop in forthcoming papers. We believe that the "invariants" Γ_i would become clearer after the computation of the spectral sequence (at least its E_2 term) of the complete ideal defined by the equations (4.6) of the multiplicities would be achieved. The latter spectral sequence should show a great deal of triviality, we believe.

Let us finally mention that formula (15.5), when valid, means that Ω , considered as an element of $C_1^{0,p}$ induces $[\Omega]_1 = 0$ in

$E_1^{o,p}$. The fact that Ω belongs indeed to $C_2^{o,p} \subset C_1^{o,p}$ means that, anyway, $[\Omega]_1$ is a d_1-cocycle; it cannot be a d_1-boundary as $E_1^{-1,p+1} = 0$. However, (15.5) means that the d_o-cycle $[\Omega]_o \in E_o^{o,p}$ is a d_o-boundary (and, therefore, as we have just mentioned, that $[\Omega]_1 = 0$).

APPENDIX. Recently Kijowski introduced a general notion of "p-phase space" of which our pairs ($\mathcal{E}$, Ω) are very special cases; see [23]. This calls for the following remark as his definition seems to be too general to yield practical results. In particular, the "canonical equations" associated to a general p-phase space in the sense of Kijowski have little chance to produce an "involutive system", even when completed. Let us indeed recall that involutiveness is related to the fact that our "Hamilton function" satisfies a very special system of partial differential equations of the first order; without such a condition, the totality of extremals would be a very meager one. Some restriction seems, therefore, necessary on that general notion. We have seen that involutiveness of the generalized canonical equations is essential.

BIBLIOGRAPHY

[1] Ph. Antoine, *Etude de la structure de certains espaces fonctionnels; applications.* Thèse, Université de Paris-Sud (Orsay), 1972.

[2] G.A. Bliss, *Lectures on the Calculus of Variations,* Chicago University Press, 1946.

[3] E. Cartan, *Leçons sur les invariants intégraux.* Paris, Hermann, 1921.

[4] ———, *Les systèmes différentiels extérieurs et leurs applications géométriques.* Paris, Hermann, A.S.I. 1945.

[5] H. Cartan, *Les systèmes d'équations extérieures.* Notes minéographiées, Séminaire Julia, Fac. des Sciences de Paris, 1937.

[6] C. Caratheodory, *Uber die Variationsrechnung bei mehrfachen Integralen.* Acta Szeged, 4 (1929), 193-216.

[7] R. Courant and D. Hilbert, *Methods of Mathematical Physics, I, II.* Interscience Publ., New York 1953, 1962.

[8] P. Dedecker, *Sur les intégrales multiples du Calcul des Variations.* Comptes rendus du IIIe congrès Nat. des Sci., Bruxelles, 1950.

[9] ———, *Les systèmes d'équations extérieures. Equations différentielles extérieures et Calcul des Variations.* Séminaire de Topologie de Strasbourg, 1951 (deux articles).

[10] ——————, *Calcul des variations, formes différentielles et champs géodésiques.* Colloques internat. du C.N.R.S., Strasbourg, 1953.

[11] ——————, *Systèmes différentiels extérieurs, invariants intégraux et suites spectrales.* Convegno internazionale di Geometria Differenziale, Venezia, Padova, Bologna e Pisa, 1953.

[12] ——————, *Calcul des Variations et topologie algébrique.* Mém. Soc. Roy. Sci. Liège, XIX (1957), 1-216.

[13] ——————, à paraître, Accademia Nazionale dei Lincei, Roma, Contributi del Centro Interdisciplinare di Science Matematiche e. l. Applicazioni.

[14] Th. de Donder, *Théorie invariantive du Calcul des Variations.* Paris, Gauthier-Villars, 1935.

[15] P. Funk, *Variationsrechnung und ihre Anwendung in Physik und Technik.* Springer Verl., Grundlehren Math. Wiss. Bd. 94, 1962.

[16] P.L. García, *Geometría simpléctica en la teoría clásica de campos.* Coll. Math. 19 (1968).

[17] ——————, *The Poincaré-Cartan invariant in the Calculus of Variations.* Symposia Mathematica, vol. 14, Istituto Nazionale di Alta Matematica, Roma, 1974.

[18] P.L. García and A. Pérez-Rendón, *Symplectic approach to the theory of quantized fields.* I, Comm. Math. Phys. 13 (1969), 22-44; II, Archiv for Rat. Mechanics and Analysis 43 (1971), 101-124.

[19] H. Goldschmidt and S. Sternberg, *The Hamilton-Cartan formalism in the Calculus of Variations.* Ann. Inst. Fourier, 23 (1973), 203-267.

[20] J. Hadamard, *Leçons sur la propagation des ondes et des équations de l'hydrodynamique.* Paris, Hermann, 1903.

[21] R. Hermann, *E. Cartan's Geometric theory of partial differential equations.* Advances in Math. 1 (1965), 265-317.

[22] ——————, *A differential geometric formalism for multiple integrals in C.V.* A.M.S. Summer Institute in Global Analysis, Berkeley, Ca., July 1-26, 1968. See chapt. V in : Lie Algebras and Quantum Mechanics, Benjamin Lect. Notes.

[23] J. Kijowski, *A finite-dimensional canonical formalism in the classical field theory.* Commun. Math. Phys., 30 (1973), 99-128.

[24] J. Kijowski and W. Szczyrba, *A canonical structure for classical field theories.* Commun. Math. Phys. 46 (1976), 183-206.

[25] R. Klötzler, *Mehrdimensionale Variationsrechnung.* Birkhäuser Verl., Basel und Stuttgart, 1970.

[26] C. Lanczos, *The Variational principles of Mechanics.* University of Toronto Press, 1949.

[27] Th. Lepage, *Champs stationnaires, champs géodésiques et formes intégrables,* I, II. Bull. Acad. Roy. Belg., Classe des Sciences, 28 (1942), 73-92, 247-268.

[28] J. Leray, *L'anneau spectral et l'anneau filtré d'homologie d'un espace localement compact et d'une application continue.* J. Math. Pures et App. (9) 29 (1950), 1-139.

[29] A. Liesen, *Feldtheorie in der Variationsrechnung mehrfacher Integralen*, I, II. Math. Ann. 171 (1967), 194-218, 273-292.

[30] C.B. Morrey, *Multiple integrals in the Calculus of Variations*. Springer Verl., Grundlehren Math. Wiss. Bd. 130, 1966.

[31] M. Morse, *The Calculus of Variations in the large*. Amer. Math. Soc. Colloquium Pub. vol. 14, 1934.

[32] R. Palais and S. Smale, *A generalized Morse theory*. Bull. Amer. Math. Soc., 70 (1964), 165-172.

[33] H. Poincaré, *Leçons sur les méthodes nouvelles de la Mécanique céleste*. Paris, Gauthier Villars, 1892-1899.

[34] J.P. Serre, *Homologie singulière des espaces fibrés; applications*. Ann. of Math. 54 (1951), 425-505.

[35] E. Vessiot, *Sur la théorie des multiplicités et le Calcul des Variations*. Bull. Soc. Math. de France, 40 (1912), 68-139.

[36] H. Weyl, *Geodesic fields*. Ann. of Math., 36 (1935), 607-629.

Contents

A SYMPLECTIC FORMULATION OF PARTICLE DYNAMICS

W.M. Tulczyjew

Department of Mathematics and Statistics

The University of Calgary

Calgary, Alberta, Canada, T2N 1N4

A symplectic formulation of particle dynamics different from the standard formulation based on time evolution concepts is presented. This presentation is intended to serve as an introduction to a symplectic formulation of field dynamics.

1. INTRODUCTION

Let T be a 1-dimensional manifold diffeomorphic to the real line R and identified with R by the choice of a standard chart. The manifold T is interpreted as the *time manifold* and the standard chart corresponds to a standard clock.

Let Q be the *configuration manifold* of a mechanical system. The cotangent bundle T^*Q is denoted by P and called the *phase manifold* of the system. There is a canonical 1-form ϑ on P and the manifold P together with the 2-form $\omega = d\vartheta$ define a symplectic manifold (P,ω).

We denote by φ the canonical projection of $Q\times T$ onto T and by ψ the projection of $P\times T$ onto T. The fibration $(Q\times T,T,\varphi)$ is called the *configuration fibration* and the fibration $(P\times T,T,\psi)$ is called the *phase fibration* of the system.

Histories of the mechanical system are smooth sections of the phase fibration satisfying a first order differential equation. A symplectic theory of such differential equations is developped in subsequent sections.

2. TANGENT VECTORS, FORMS AND DERIVATIONS

Let M be a C^∞-manifold and let TM denote the tangent bundle of M. Elements of TM are equivalence classes of differentiable mappings of R into M with two mappings $\gamma:t \mapsto \gamma(t)$ and $\gamma':t \mapsto \gamma'(t)$ equivalent if $(f\circ\gamma)(0) = (f\circ\gamma')(0)$ and $D(f\circ\gamma)(0) = D(f\circ\gamma')(0)$ for each differentiable function f on M. The equivalence class of γ is denoted by $[\gamma]$ and γ is called an *integral curve* of the vector $v = [\gamma]$. It is clear that integral curves need to be defined only in neighbourhoods of 0 in R. The tangent bundle projection τ_M is the mapping $\tau_M:[\gamma] \mapsto \gamma(0)$. For each differentiable mapping φ of a differential manifold M into a manifold M' there is the differentiable mapping $T\varphi:TM \to TM':[\gamma] \mapsto [\varphi\circ\gamma]$.

The diagram

$$\begin{array}{ccc} TM & \xrightarrow{T\varphi} & TM' \\ \downarrow{\scriptstyle \tau_M} & & \downarrow{\scriptstyle \tau_{M'}} \\ M & \xrightarrow{\varphi} & M' \end{array}$$

is commutative and if $\psi:M'\longrightarrow M''$ is a second differentiable mapping then $T(\psi\circ\varphi) = T\psi\circ T\varphi$.

Let $\gamma:\mathrm{R}^2\longrightarrow M:(s,t)\longmapsto\gamma(s,t)$ be a differentiable mapping. For each $s\in\mathrm{R}$ let γ_s be the mapping $\gamma_s:\mathrm{R}\longrightarrow M:t\longmapsto\gamma(s,t)$. The mapping $\gamma':\mathrm{R}\longrightarrow TM:s\longmapsto[\gamma_s]$ is an integral curve of a vector $w=[\gamma']$ in TTM.

PROPOSITION 2.1. *For each $w\in TTM$ there is a mapping $\gamma:\mathrm{R}^2\longrightarrow M$ such that $w=[\gamma']$.*

Proof: Let $(\tau_M,\eta):TM\longrightarrow M\times\mathrm{R}^n$ be a (local) trivialization of the tangent bundle. The inverse mapping $\xi=(\tau_M,\eta)^{-1}:M\times\mathrm{R}^n\longrightarrow TM$ is trivially completed to a vector field $X=(\xi,0):M\times\mathrm{R}\longrightarrow TM\times T\mathrm{R}^n$. Let $\varphi:\mathrm{R}\times M\times\mathrm{R}^n\longrightarrow M$ be the nontrivial component of the (local) flow $\Phi=(\varphi,pr_3):\mathrm{R}\times M\times\mathrm{R}^n\longrightarrow M\times\mathrm{R}^n$ of X where pr_3 is the projection of $\mathrm{R}\times M\times\mathrm{R}^n$ onto R^n. If $\delta:\mathrm{R}\longrightarrow TM$ is an integral curve of $w\in TTM$ then $\gamma=\varphi\circ(1_{\mathrm{R}}\times(\tau_M,\eta)\circ\delta):\mathrm{R}^2\longrightarrow M$ is a mapping such that $w=[\gamma']$. The mapping γ is in general defined only in a neighbourhood of $(0,0)$ in R^2.

Q. E. D.

Proposition 2.1 makes it possible to represent elements of TTM by differentiable mappings of R^2 into M. In terms of this representation the mapping $\mathrm{R}\longrightarrow M:t\longmapsto\gamma(0,t)$ is an integral curve of $\tau_{TM}(w)$ and the mapping $\mathrm{R}\longrightarrow M:s\longmapsto\gamma(s,0)$ is an integral curve of $T\tau_M(w)$.

We define the *natural involution* $\sigma_M:TTM\longrightarrow TTM$ by $\sigma_M([\gamma'])=[\tilde{\gamma}']$ where $\tilde{\gamma}$ is the mapping $\tilde{\gamma}:\mathrm{R}^2\longrightarrow M:(s,t)\longmapsto\gamma(t,s)$. Using local coordinates it is easy to show that σ_M is a diffeomorphism. Relations $\sigma_M\circ\sigma_M=1_{TTM}$, $\tau_{TM}\circ\sigma_M=T\tau_M$ and $T\tau_M\circ\sigma_M=\tau_{TM}$ hold.

For each manifold M let Φ_M denote the exterior algebra of differential forms on M. If μ is a q-form and $v_1,\ldots,v_q$ are vectors in TM such that $\tau_M(v_1)=\ldots=\tau_M(v_q)$ then $\langle v_1\wedge\ldots\wedge v_q,\mu\rangle$ denotes the *evaluation* of μ on the exterior product $v_1\wedge\ldots\wedge v_q$. If μ is a q-form on M' and $\varphi:M\longrightarrow M'$ a differentiable mapping then $\varphi^*\mu$ is a q-form on M such that $\langle v_1\wedge\ldots\wedge v_q,\varphi^*\mu\rangle=\langle T\varphi(v_1)\wedge\ldots\wedge T\varphi(v_q),\mu\rangle$.

DEFINITION 2.1. A linear mapping $a:\Phi_M\longrightarrow\Phi_{TM}:\mu\longmapsto a\mu$ is called a *derivation of degree r from Φ_M to Φ_{TM}* if

$$\textit{degree }(a\mu)=\textit{degree }\mu+r$$

and

$$a(\mu\wedge\nu)=a\mu\wedge\tau_M^*\nu+(-1)^{qr}\tau_M^*\mu\wedge a\nu$$

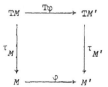

where q = *degree* μ.

We define a derivation i_T of degree -1 from Φ_M to Φ_{TM} by $\langle w_1\wedge\ldots\wedge w_q, i_T\mu\rangle = \langle v\wedge u_1\wedge\ldots\wedge u_q,\mu\rangle$, where μ is a q-form, $v = \tau_{TM}(w_1) = \ldots = \tau_{TM}(w_q)$ and $u_1 = T\tau_M(w_1)$, $\ldots, u_q = T\tau_M(w_q)$. If f is a 0-form in Φ_M (a function on M) then $i_T f = 0$. A derivation d_T of degree 0 from Φ_M to Φ_{TM} is defined by $d_T = di_T + i_T d$ where d is the exterior differential. The relation $d_T d = dd_T$ holds.

An important property of derivations is that a derivation is completely determined by its action on functions and differentials of functions.

Let $w_1, \ldots, w_q$ be elements of TTM such that $\tau_{TM}(w_1) = \ldots = \tau_{TM}(w_q)$. The construction used in the proof of Proposition 2.1 leads to mappings $\gamma_1, \ldots, \gamma_q$ of R^2 into M such that $w_1 = [\gamma_1'], \ldots, w_q = [\gamma_q']$ and $\gamma_1(0,t) = \ldots = \gamma_q(0,t)$. The mappings $\tilde{\gamma}_1', \ldots, \tilde{\gamma}_q'$ of R into TM are integral curves of vectors $\sigma_M(w_1), \ldots, \sigma_M(w_q)$ such that $\tau_M\circ\tilde{\gamma}_1' = \ldots = \tau_M\circ\tilde{\gamma}_q'$.

Let μ be a q-form on M. Let $w_1, \ldots, w_q$ be arbitrary vectors in TTM such that $\tau_{TM}(w_1) = \ldots = \tau_{TM}(w_q)$ and let $\varkappa_1, \ldots, \varkappa_q$ be integral curves of vectors $\sigma_M(w_1), \ldots, \sigma_M(w_q)$ such that $\tau_M\circ\varkappa_1 = \ldots = \tau_M\circ\varkappa_q$. We denote by $\langle\varkappa_1\wedge\ldots\wedge\varkappa_q,\mu\rangle$ the function $\langle\varkappa_1\wedge\ldots\wedge\varkappa_q,\mu\rangle(t) = \langle\varkappa_1(t)\wedge\ldots\wedge\varkappa_q(t),\mu\rangle$. It can be easily verified that the formula $\langle w_1\wedge\ldots\wedge w_q, a\mu\rangle = D\langle\varkappa_1\wedge\ldots\wedge\varkappa_q,\mu\rangle(0)$ defines a derivation a of degree 0 from Φ_M to Φ_{TM}.

PROPOSITION 2.2. *The derivation a defined above is identical with* d_T.

Proof: If f is a function on M then $af = d_T f$ since for each $v \in$ TM we have $(af)(v) = D(f\circ\gamma)(0) = (i_T df)(v) = (d_T f)(v)$. Also $adf = d_T df$ since for each vector $w \in$ TTM we have $\langle w, adf\rangle = D\langle\tilde{\gamma}', df\rangle(0) = D_1D_2(f\circ\gamma)(0,0) = D\langle\gamma', df\rangle(0) = \langle w, d_T df\rangle$, where $\gamma: R^2 \to M$ is a mapping such that $w = [\gamma']$ and D_1 and D_2 denote partial derivatives wth respect to the first and the second argument respectively. Hence $a = d_T$.

Q. E. D.

If functions (x^i), $0 \le i \le n$ are local coordinates of M then functions $(x^i, \dot{x}^j) = (x^i\circ\tau_M, d_T x^j)$, $0 \le i,j \le n$ are local coordinates of TM said to be *associated* with coordinates (x^i).

3. LAGRANGIAN SUBMANIFOLDS AND GENERATING FUNCTIONS [2]

Let P be a manifold and ω a 2-form on P. The form ω is called a *symplectic form* if $d\omega = 0$ and if $\langle u\wedge v,\omega\rangle = 0$ for each $u \in$ TM if and only if $v = 0$. If ω is a symplectic form then (P,ω) is called a *symplectic manifold.*

DEFINITION 3.1. Let P be a manifold of dimension $2n$ and ω a symplectic form on P. A submanifold N of P such that $\omega|N = 0$ and $\dim N = n$ is called a *Lagrangian submanifold* of (P,ω).

The canonical 1-form ϑ_M on the cotangent bundle T*M of a manifold M is defined by $\langle v,\vartheta_M\rangle = \langle T\pi_M(v), \tau_{T^*M}(v)\rangle$ where v is a vector in TT*M and π_M:T*$M \to M$ is the cotangent bundle projection. The canonical 2-form $\omega_M = d\vartheta_M$ is known to be a symplectic form. Hence (T^*M,ω_M) is a symplectic manifold.

Let F be a differentiable function on a manifold M. The 1-form dF is a section $dF: M \to T^*M$ of the cotangent fibration. The image N of dF is a submanifold of T^*M, the mapping $\rho = \pi_M|N : N \longrightarrow M$ is a diffeomorphism and $\vartheta_M|N = \rho^*dF$. Hence $\omega_M|N = 0$ and N is a Lagrangian submanifold of (T^*M, ω_M).

The above construction of Lagrangian submanifolds is generalized in the following proposition.

PROPOSITION 3.1. *Let K be a submanifold of M and F a function on K. The set* $N = \{p \in T^*M;\ \pi_M(p) \in K$ and $\langle u, p\rangle = \langle u, dF\rangle$ for each $u \in TK \subset TM$ such that $\tau_M(u) = \pi_M(p)\}$ *is a Lagrangian submanifold of* (T^*M, ω_M).

Proof: Using local coordinates it is easily shown that N is a submanifold of T^*M of dimension equal to *dim M*. The submanifold K is the image of N by π_M. Let $\rho : N \to K$ be the mapping defined by the commutative diagram

$$\begin{array}{ccc} N & \longrightarrow & T^*M \\ {\scriptstyle \rho}\downarrow & & \downarrow{\scriptstyle \pi_M} \\ K & \longrightarrow & M \end{array}$$

where the horizontal arrows are canonical injections. Then

$$\begin{aligned} \langle v, \rho^*dF\rangle &= \langle T\rho(v), dF\rangle \\ &= \langle T\rho(v), \tau_{T^*M}(v)\rangle \\ &= \langle v, \vartheta_M\rangle \end{aligned}$$

for each vector $v \in TN \subset TT^*M$. Hence $\vartheta_M|N = \rho^*dF$, $\omega_M|N = 0$ and N is a Lagrangian submanifold of (T^*M, ω_M).

Q. E. D.

DEFINITION 3.2. The function F appearing in Proposition 3.1 is called a *generating function* of the Langrangian submanifold N.

The class of Lagrangian submanifolds generated by generating functions is wide enough to include most Lagrangian submanifolds appearing in mathematical physics.

In many applications it is convenient to consider symplectic manifolds which are not directly cotangent bundles but are isomorphic to cotangent bundles.

DEFINITION 3.3. A *special symplectic manifold* is a quintuple $(P, M, \pi, \vartheta, \alpha)$ where (P, M, π) is a differential fibration, ϑ is a 1-form on P and $\alpha : P \longrightarrow T^*M$ is a diffeomorphism such that $\pi = \pi_M \circ \alpha$ and $\vartheta = \alpha^*\vartheta_M$.

If $(P, M, \pi, \vartheta, \alpha)$ is a special symplectic manifold, K a submanifold of M and F a function on K then the set $N = \{p \in P;\ \pi(p) \in K$ and $\langle u, \vartheta\rangle = \langle T\pi(u), dF\rangle$ for each $u \in TP$ such that $\tau_P(u) = p$ and $T\pi(u) \in TK \subset TM\}$ is a Lagrangian submanifold of $(P, d\vartheta)$ said to be generated by F.

4. SYMPLECTIC FORMULATION OF DYNAMICS

Let P be the phase manifold of a mechanical system and let d_T be the derivation from Φ_P to Φ_{TP} defined in Section 2.

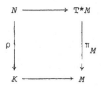

PROPOSITION 4.1. *The manifold* P *together with the form* $\rho = d_T\omega$ *define a symplectic manifold* (TP,ρ).

Proof: Let $\beta : TP \to T^*P$ be the diffeomorphism defined by $\langle u,\beta(v)\rangle = \langle v\wedge u,\omega\rangle$ where u and v are vectors in TP such that $\tau_P(u) = \tau_P(v)$. For each $w \in TTP$ we have

$$\begin{aligned}\langle w,\beta^*\vartheta_P\rangle &= \langle T\beta(w),\vartheta_P\rangle \\ &= \langle T\pi_P(T\beta(w)),\tau_{T^*P}(T\beta(w))\rangle \\ &= \langle T\tau_P(w),\beta(\tau_{TP}(w))\rangle \\ &= \langle \tau_{TP}(w)\wedge T\tau_P(w),\omega\rangle \\ &= \langle w,i_T\omega\rangle.\end{aligned}$$

Hence $i_T\omega = \beta^*\vartheta_P$ and $\rho = d_T\omega = d\beta^*\vartheta_P = \beta^*\omega_P$. It follows that (TP,ρ) is a symplectic manifold isomorphic to (T^*P,ω_P).

Q. E. D.

Since the time manifold T is identified with R each section γ of the phase fibration is a differentiable curve $\gamma : R \to P\times T$ such that $\psi\circ\gamma = 1_R$. Conequently jets of sections are vectors in $T(P\times T)$ and are completely determined by their projections into TP.

THE RECIPROCITY PRINCIPLE OF PARTICLE DYNAMICS. *Projections into* TP *of jets of histories of a mechanical system form a submanifold* D *of* $TP\times T$. *Projections of jets of histories at a time* $t \in T$ *form a Lagrangian submanifold* D_t *of* (TP,ρ).

In standard symplectic formulations of dynamics it is assumed that D_t is a locally Hamiltonian vector field and D is a time dependent locally Hamiltonian vector field [1]. This is always the case in nonrelativistic dynamics. Our Principle is more general. It includes the possibility of a symplectic formulation of relativistic dynamics [3]. More significantly it allows a more uniform treatment of Hamiltonian and Lagrangian dynamics and provides a better basis for generalizations to field dynamics.

5. LAGRANGIAN DYNAMICS

Let v be a vector in TP and w a vector in TTQ such that $\tau_{TQ}(w) = T\pi_Q(v)$. Let further $\gamma : R \to TQ$ and $\varkappa : R \to P$ be integral curves of $\tilde{w} = \sigma_Q(w)$ and v respectively such that $\tau_Q\circ\gamma = \pi_Q\circ\varkappa$. Then $\langle w,\alpha(v)\rangle = D\langle\gamma,\varkappa\rangle(0)$ defines a mapping $\alpha : TP \to T^*TQ$. It can be easily verified using local coordinates that α is a diffeomorphism. Let λ denote the 1-form $d_T\vartheta$.

PROPOSITION 5.1. *The system* $(TP,TQ,T\pi_Q,\lambda,\alpha)$ *is a special symplectic manifold.*

Proof: $\pi_{TQ}\circ\alpha = T\pi_Q$ follows directly from the construction of α. For each $z \in TTP$ we have

$$\begin{aligned}\langle z,\alpha^*\vartheta_{TQ}\rangle &= \langle T\alpha(z),\vartheta_{TQ}\rangle \\ &= \langle T\pi_{TQ}(T\alpha(z)),\tau_{T^*TQ}(T\alpha(z))\rangle \\ &= \langle TT\pi_Q(z),\alpha(\tau_{TP}(z))\rangle.\end{aligned}$$

Let $\zeta : R \to TP$ be an integral curve of $\tilde{z} = \sigma_P(z)$. Then $\gamma = T\pi_Q\circ\zeta$ and $\varkappa = \tau_P\circ\zeta$ are integral curves of $\tilde{w} = \sigma_Q(w) = \sigma_Q(\mathrm{T}\mathrm{T}\pi_Q(z)) = \mathrm{T}\mathrm{T}\pi_Q(\tilde{z})$ and $v = \tau_{\mathrm{T}P}(z)$ respectively and $\tau_Q \circ \gamma = \pi_Q \circ \varkappa$. Hence

$$\begin{aligned} \langle z, \alpha^*\vartheta_{\mathrm{T}Q}\rangle &= \langle w, \alpha(v)\rangle \\ &= \mathrm{D}\,\langle \gamma, \varkappa\rangle(0) \\ &= \mathrm{D}\,\langle \mathrm{T}\pi_Q \circ \zeta, \tau_P \circ \zeta\rangle(0) \\ &= \mathrm{D}\,\langle \zeta, \vartheta\rangle(0) \\ &= \langle z, d_T\vartheta\rangle. \end{aligned}$$

It follows that $\lambda = d_T\vartheta = \alpha^*\vartheta_{\mathrm{T}Q}$.

Q. E. D.

Since $d\lambda = \rho$ submanifolds D_t may be generated by generating functions L_t defined on submanifolds $J_t = \mathrm{T}\pi_Q(D_t) \subset \mathrm{T}Q$. Assuming that this is the case we define a function L on $J = \{(v,t) \in \mathrm{T}Q\times\mathrm{T};\ v \in J_t\}$ by $L(v,t) = L_t(v)$.

DEFINITION 5.1. The function L constructed above is called a *Lagrangian* of the mechanical system.

6. HAMILTONIAN DYNAMICS

Let $\beta:\mathrm{T}P \rightarrow \mathrm{T}^*P$ be the diffeomorphism introduced in the proof of Proposition 4.1 and let χ denote the 1-form $i_T\omega$.

PROPOSITION 6.1. *The system* $(\mathrm{T}P, P, \tau_P, \chi, \beta)$ *is a special symplectic manifold.*

Proof: $\pi_P \circ \beta = \tau_P$ is obvious and the proof of $\chi = \beta^*\omega_P$ is contained in the proof of Proposition 4.1.

Q. E. D.

Since $d\chi = \rho$ submanifolds D_t may be generated by generating functions F_t defined on submanifolds of P. We assume that this is the case and that functions $H_t = -F_t$ are defined on P.

DEFINITION 6.1. The function H defined on $P\times T$ by $H(p,t) = H_t(p)$ is called a *Hamiltonian* of the mechanical system.

7. LOCAL EXPRESSIONS

Let (q^i), $0 \le i \le n$ be local coordinates of Q and $(q^i, \dot{q}^j)$, $0 \le i,j \le n$ the associated local coordinates of $\mathrm{T}Q$. There are local coordinates (q^i, p_j), $0 \le i,j \le n$ of P such that $\langle v,y\rangle = \Sigma_i p_i(y)\dot{q}^i(v)$ where $v \in \mathrm{T}$ and $y \in P$ satisfy $\tau_Q(v) = \pi_Q(y)$. There are finally associated local coordinates $(q^i, p_j, \dot{q}^k, \dot{p}_l)$, $0 \le i,j,k,l \le n$ of $\mathrm{T}P$.

We give local expressions of forms ϑ, ω, λ, χ and ρ:

$$\vartheta = \sum_i p_i dq^i,$$

$$\omega = \sum_i dp_i \wedge dq^i,$$

$$\lambda = \sum_i (\dot{p}_i dq^i + p_i d\dot{q}^i),$$

$$\chi = \sum_i (\dot{p}_i dq^i - \dot{q}^i dp_i)$$

and

$$\rho = \sum_i (d\dot{p}_i \wedge dq^i + dp_i \wedge d\dot{q}^i).$$

If dynamics is described by a Lagrangian L defined on $TQ\times T$ and a Hamiltonian H then equations

$$\sum_i (\dot{p}_i dq^i + p_i d\dot{q}^i) = dL_t(q^i, \dot{q}^j)$$

and

$$\sum_i (\dot{p}_i dq^i - \dot{q}^i dp_i) = -dH_t(q^i, p_j)$$

equivalent to the familiar systems

$$\dot{p}_i = \frac{\partial L}{\partial q^i}, \quad p_i = \frac{\partial L}{\partial \dot{q}^i}$$

and

$$\dot{p}_i = -\frac{\partial H}{\partial q^i}, \quad \dot{q}^i = \frac{\partial H}{\partial p_i}$$

are satisfied on submanifolds D_t.

REFERENCES

[1] Abraham, R. and Marsden, J.E., *Foundations of mechanics*, Benjamin, New York, 1967

[2] Śniatycki, J. and Tulczyjew, W.M., *Generating forms of Lagrangian submanifolds*, Indiana Univ. Math. J., vol. 22, pp. 267-275 (1972)

[3] Tulczyjew, W.M., *Hamiltonian systems, Lagrangian systems and the Legendre transformation*, Symposia Mathematica, vol. 15, pp. 247-258 (1974)

A SYMPLECTIC FORMULATION OF FIELD DYNAMICS

W.M. Tulczyjew

Department of Mathematics and Statistics

The University of Calgary

Calgary, Alberta, Canada, T2N 1N4

A formulation of classical field dynamics in terms of Lagrangian submanifolds of specially constructed symplectic manifolds is presented. This presentation follows closely the formulation of particle dynamics given in the preceding note. Hamiltonian description of field dynamics is not included.

1. INTRODUCTION

Let M be a differential manifold of dimension n. For simplicity we assume that M is oriented. This assumption makes it possible to use differential forms on M in place of vector densities required for correct description of physical field theories.

Let (X,M,ξ) be a differential fibration. For each $p \in M$ there is a fibration (Y_p,X_p,π_p) with base $X_p = \xi^{-1}(p)$. Each fibre $Y_x = \pi_p^{-1}(x)$ is the set of all linear mappings of the tangent space T_xX_p into the vector space $\overset{n-1}{\wedge}T_p^*M$. There is a 1-form ϑ_p on Y_p with values in $\overset{n-1}{\wedge}T_p^*M$ defined by

$$\langle v,\vartheta_p\rangle = \langle T\pi_p(v),\tau_{Y_p}(v)\rangle = \tau_{Y_p}(v)(T\pi_p(v))$$

for each $v \in TY_p$. The 2-form $d\vartheta_p$ with values in $\overset{n-1}{\wedge}T_p^*M$ is denoted by ω_p. Manifolds Y_p are used as fibres of a fibration (Y,M,η). We denote by π the mapping $\pi:Y \to X: y \to \pi_{\eta(y)}(y)$.

The fibration (X,M,ξ) is interpreted as the *configuration fibration* of a physical field. The fibration (Y,M,η) is the *phase fibration*. States of the field are smooth sections of the phase fibration satisfying a first order partial differential equation. In subsequent sections we formulate a symplectic theory of such differential equations.

2. JETS, TANGENT VECTORS AND FORMS

Let (Z,M,ζ) be a differential fibration. For each $p \in M$ let J_p^1Z denote the manifold of jets of sections of the fibration with source p. Elements of the tangent bundle TJ_p^1Z are equivalence classes of mappings of R into J_p^1Z.

PROPOSITION 2.1. *For each vector* $w \in TJ_p^1Z$ *there is a mapping* $\chi:\mathrm{R}\times M \to Z: (s,r) \mapsto \chi(s,r)$ *such that* $\zeta(\chi(s,r)) = r$ *for each* $r \in M$ *and each* $s \in \mathrm{R}$ *and that* w *is the equivalence class of the mapping* $\gamma : \mathrm{R} \to J^1_p Z : s \mapsto \gamma(s)$, *where* $\gamma(s)$ *is the jet of the section* $\chi_s : M \to Z : r \mapsto \chi(s,r)$.

Proof: Let $(x^\varkappa)$, $0 \le \varkappa \le n$ be local coordinates of M such that $x^\varkappa(p) = 0$. Let $(x^\varkappa, z^i)$, $0 \le \varkappa \le n$, $0 \le i \le m$ be local coordinates of Z, $(z^i, z^j_\varkappa)$, $0 \le \varkappa \le n$, $0 \le i,j \le m$ the associated local coordinates of $J^1_p Z$ and $(z^i, z^j_\varkappa, \dot{z}^k, \dot{z}^l_\lambda)$, $0 \le \varkappa,\lambda \le n$, $0 \le i,j,k,l \le m$ the associated coordinates of $TJ^1_p Z$. For each $w \in TJ^1_p Z$ a mapping χ with the required properties is defined by its local expression

$$(s,x^\varkappa) \mapsto (x^\varkappa, z^i(w) + \dot{z}^i(w)s + \sum_{\lambda=1}^{n} (z^i_\lambda(w)x^\lambda + \dot{z}^i_\lambda(w)sx^\lambda)).$$

Q. E. D.

A mapping χ with properties stated in Proposition 2.1 will be called a *representative* of w.

For each $p \in M$ let μ_p be a q-form on Z_p with values in $\overset{k}{\wedge}T^*M$. Let $w^1, \ldots, w^q$ be vectors in $TJ^1_p Z$ with representatives $\chi^1, \ldots, \chi^q$ such that $\chi^1(0,r) = \ldots = \chi^q(0,r)$ for each $r \in M$. For each $r \in M$ let $u^1_r, \ldots, u^q_r \in TZ_r$ be equivalence classes of mappings $\chi^1_r : \mathrm{R} \to Z_r : s \mapsto \chi^1(s,r), \ldots, \chi^q_r : \mathrm{R} \to Z_r : s \mapsto \chi^q(s,r)$. Let finally $\langle u^1 \wedge \ldots \wedge u^q, \mu\rangle$ denote the k-form on M defined by $\langle u^1 \wedge \ldots \wedge u^q, \mu\rangle : M \to \overset{k}{\wedge}T^*M : r \mapsto \langle u^1_r \wedge \ldots \wedge u^q_r, \mu_r\rangle$. The formula $\langle w^1 \wedge \ldots \wedge w^q, d_M \mu_p\rangle = d\langle u^1 \wedge \ldots \wedge u^q, \mu\rangle(p)$ defines a q-form $d_M \mu_p$ on $J^1_p Z$ with values in $\overset{k+1}{\wedge}T^*_p M$. Obvious differentiability conditions must be satisfied for the above construction to be successful.

The relation $dd_M \mu_p = d_M d\mu_p$ holds.

3. GENERALIZED SYMPLECTIC MANIFOLDS, LAGRANGIAN SUBMANIFOLDS AND GENERATING FUNCTIONS

Let P be a manifold and ω a 2-form on P with values in a vector space V of dimension 1. We say that ω is a *symplectic form* if $d\omega = 0$ and if $\langle u \wedge v, \omega\rangle = 0$ for each $u \in TP$ if and only if $v = 0$. If ω is a symplectic form then (P,ω) is called a *symplectic manifold*.

DEFINITION 3.1. Let P be a manifold of dimension $2n$ and ω a symplectic form on P with values in a vector space V of dimension 1. A submanifold N of P such that $\omega|N = 0$ and $\dim N = n$ is called a *Lagrangian submanifold* of (P,ω).

Let X be a differential manifold and V a vector space of dimension 1. We denote by $T^*(X,V)$ the bundle of linear mappings of fibres of the tangent bundle TX into V. We denote by $\pi_{(X,V)}$ the bundle projection and by $\vartheta_{(X,V)}$ the 1-form on $T^*(X,V)$ with values in V defined by $\langle v, \vartheta_{(X,V)}\rangle = \langle T\pi_{(X,V)}(v), \tau_{T^*(X,V)}(v)\rangle$ for each $v \in TT^*(X,V)$. The bundle $T^*(X,V)$ is isomorphic to T^*X. Each choice of a basis in V determines an isomorphism. It is therefore clear that $\omega_{(X,V)} = d\vartheta_{(X,V)}$ is a symplectic form. The theory of Lagrangian submanifolds of $(T^*(X,V), \omega_{(X,V)})$ generated by generating functions is analogous to the corresponding theory of Lagrangian submanifolds of (T^*X, ω_X). If F is a mapping of X into V then the image of $dF : X \to T^*(X,V)$ is a Lagrangian submanifold of $(T^*(X,V), \omega_{(X,V)})$ said to be generated by F.

4. SYMPLECTIC FORMULATION OF FIELD DYNAMICS

Let (Y,M,η) be the phase fibration of a physical field. For each $p \in M$ we have the 2-form $d_M\omega_p$ on J^1_pY with values in the vector space $V_p = \overset{n}{\wedge}T^*_pM$ of dimension 1. The form $d_M\omega_p$ is not in general a symplectic form although $dd_M\omega_p = d_Md\omega_p = 0$. We assume that there is a fibration (Y',M,η'), a symplectic form ρ_p on each fibre Y'_p and a surjective submersion $\sigma:J^1Y \to Y'$. Each fibre J^1_pY is mapped by σ onto the corresponding fibre Y'_p and $d_M\omega_p = \sigma^*_p\rho_p$ where σ_p is the mapping $\sigma_p:J^1_pY \to Y'_p:v \mapsto \sigma(v)$.

THE RECIPROCITY PRINCIPLE OF FIELD DYNAMICS. Jets of states of a physical field form a submanifold D of J^1Y such that for each $p \in M$ the manifold $D_p = \sigma_p(D\cap J^1_pY)$ is a Lagrangian submanifold of (Y'_p,ρ_p).

5. LAGRANGIAN DYNAMICS

Let (X,M,ξ) be the configuration fibration and (Y,M,η) the phase fibration of a physical field. Let w be a vector in TJ^1_pX and v a jet in J^1_pY such that

$$\tau_{J^1_pX}(w) = J^1_p\pi(v)$$

where $J^1_p\pi(v)$ denotes the jet of $\pi\circ\gamma$ if v is the jet of the section $\gamma:M \to Y$. Let further $\chi:R\times M \to X:(s,r) \mapsto \chi(s,r)$ be a representative of w such that $\chi(0,r) = \pi(\gamma(r))$ for each $r \in M$, for each $r \in M$ let $u_r \in TX_r$ denote the equivalence class of $\chi_r:R \to X_r: s \mapsto \chi(s,r)$ and let $<u,\gamma>$ be the n-1-form on M defined by $<u,\gamma> :M \to \overset{n-1}{\wedge}T^*M:r \mapsto <u_r,\gamma(r)>$. The formula $<w,\sigma_p(v)> = d<u,\gamma>(p)$ defines a mapping $\sigma_p:J^1_pY \to Y'_p = T^*(J^1_pX,V_p)$, $V_p = \overset{n}{\wedge}T^*_pM$. It can be easily verified using local coordinates that σ_p is a surjective submersion. The relation

$$\pi_{(J^1_pX,V_p)}\circ\sigma_p = J^1_p\pi$$

follows directly from the definition of σ_p. Let π'_p and ϑ'_p denote

$$\pi_{(J^1_pX,V_p)} \text{ and } \vartheta_{(J^1_pX,V_p)}$$

respectively.

PROPOSITION 5.1. *For each $p \in M$ we have $\sigma^*_p\vartheta'_p = d_M\vartheta_p$.*

Proof: For each $z \in TJ^1_pY$ we have

$$\begin{aligned}<z,\sigma^*_p\vartheta'_p> &= <T\sigma_p(z),\vartheta'_p>\\ &= <T\pi'_p(T\sigma_p(z)),\tau_{Y'_p}(T\sigma_p(z))>\\ &= <TJ^1_p\pi(z),\sigma_p(\tau_{J^1_pY}(z))>.\end{aligned}$$

Let $\zeta:R\times M \to Y$ be a representative of z. Then $\chi = \pi\circ\zeta$ is a representative of $w = TJ^1_p\pi(z) \in TJ^1_pX$,

$$v = \tau_{J^1_pY}(z) \in J^1_pY$$

is the jet of $\gamma = \zeta_0:M \to Y:r \mapsto \zeta(0,r)$ and $\chi(0,r) = \pi(\gamma(r))$ for each $r \in M$. For each $r \in M$ let u_r denote the equivalence class of $\chi_r:R \to X_r:s \mapsto \chi(s,r)$ and let y_r denote the equivalence class of $\zeta_r:R \to Y_r:s \mapsto \zeta(s,r)$.

Then

$$u_r = T\pi_r(y_r), \quad \gamma(r) = \tau_{Y_r}(y_r)$$

and

$$\langle u_r, \gamma(r)\rangle = \langle T\pi_r(y_r), \tau_{Y_r}(y_r)\rangle$$
$$= \langle y_r, \vartheta_r\rangle.$$

Hence

$$\langle z, \sigma_p^* \vartheta'_p\rangle = \langle w, \sigma_p(v)\rangle$$
$$= d\langle u, \gamma\rangle(p)$$
$$= d\langle y, \vartheta\rangle(p)$$
$$= \langle z, d_M \vartheta_p\rangle.$$

It follows that $\sigma_p^* \vartheta'_p = d_M \vartheta_p$.

Q. E. D.

Let (Y', M, η') be the fibration with fibres $Y'_p = T^*(J^1_p X, V_p)$ and let σ be the mapping $\sigma: J^1Y \to Y': v \mapsto \sigma_p(v)$ where v is an element of $J^1_p Y$. It follows from Proposition 5.1 that (Y', M, η'), σ and $\rho_p = d\vartheta'_p$ are exactly the objects whose existence was postulated in Section 4. For each $p \in M$ the Lagrangian submanifold D_p may be generated by a generating function L_p. We assume that this is the case and that L_p is defined on $J^1_p X$.

DEFINITION 5.1. The function L on J^1X defined by $L(v) = L_p(v)$ if $v \in J^1_p X$ is called a *Lagrangian* of the physical field.

6. AN EXAMPLE

Let (X, M, ξ) be the cotangent fibration (T^*M, M, π_M) of a manifold M. Let $(x^\varkappa)$, $0 \le \varkappa \le n$ be local coordinates of M at a point p, let $(x^\varkappa, A_\lambda)$ be local coordinates of T^*M and $(x^\varkappa, A_\lambda, B_{\mu\nu})$ local coordinates of J^1T^*M. We denote by e the exterior product $dx^1 \wedge \ldots \wedge dx^n$ and by $e_\varkappa$ the exterior product $dx^1 \wedge \ldots \wedge dx^n$ with $dx^\varkappa$ omitted. Elements $e_\varkappa$, $0 \le \varkappa \le n$ form a base of $\overset{n-1}{\wedge} T^*_p M$ and e forms a base of $\overset{n}{\wedge} T^*_p M$. Using these bases we generate local coordinate systems for Y, J^1Y and Y'. The manifold Y has local coordinates $(x^\varkappa, A_\lambda, H^{\mu\nu})$ such that

$$\vartheta_p = \sum_{\varkappa,\lambda} H^{\varkappa\lambda} dA_\varkappa \otimes e_\lambda .$$

There are local coordinates $(x^\varkappa, A_\lambda, H^{\mu\nu}, B_{\omega\pi}, I^{\rho\sigma}{}_\tau)$ of J^1Y such that

$$d_M \vartheta_p = \sum_{\varkappa,\lambda} (I^{\varkappa\lambda}{}_\lambda dA_\varkappa + H^{\varkappa\lambda} dB_{\varkappa\lambda}) \otimes e,$$

and local coordinates $(x^\varkappa, A_\lambda, B_{\mu\nu}, I^\omega, H^{\rho\sigma})$ of Y' such that

$$\vartheta'_p = (\sum_\varkappa I^\varkappa dA_\varkappa + \sum_{\varkappa,\lambda} H^{\varkappa\lambda} dB_{\varkappa\lambda}) \otimes e$$

and

$$\sigma: (x^\varkappa, A_\lambda, H^{\mu\nu}, B_{\omega\pi}, I^{\rho\sigma}{}_\tau) \mapsto (x^\varkappa, A_\lambda, B_{\omega\pi}, I^\rho, H^{\mu\nu})$$

where

$$I^\rho = \sum_\sigma I^{\rho\sigma}{}_\sigma .$$

Let $g^{\varkappa\lambda}$ be components of a symmetric tensor field on M.

Then equations

$$H^{\varkappa\lambda} = \sum_{\mu,\nu} g^{\varkappa\mu} g^{\lambda\nu} (B_{\mu\nu} - B_{\nu\mu}),$$

$$\sum_{\lambda} I^{\varkappa\lambda}{}_{\lambda} = 0$$

define in local terms a submanifold D of J^1Y. The submanifolds D_p are then defined by equations

$$H^{\varkappa\lambda} = \sum_{\mu,\nu} g^{\varkappa\mu} g^{\lambda\nu} (B_{\mu\nu} - B_{\nu\mu}),$$

$$I^{\varkappa} = 0.$$

It is easily seen that D_p are Lagrangian submanifolds generated by the Lagrangian

$$L(x^{\varkappa}, A_{\lambda}, B_{\mu\nu}) = \frac{1}{4} \sum_{\varkappa,\lambda,\mu,\nu} g^{\varkappa\mu} g^{\lambda\nu} (B_{\varkappa\lambda} - B_{\lambda\varkappa})(B_{\mu\nu} - B_{\nu\mu}).$$

The submanifold D determines a system of differential equations for sections of the phase fibration. If $A_{\lambda} = \varphi_{\lambda}(x^{\varkappa})$, $H^{\mu\nu} = \psi^{\mu\nu}(x^{\varkappa})$ is a section of (Y, M, η) then the equations are

$$\psi^{\varkappa\lambda} = \sum_{\mu,\nu} g^{\varkappa\mu} g^{\lambda\nu} (\partial_{\nu}\varphi_{\mu} - \partial_{\mu}\varphi_{\nu}),$$

$$\sum_{\lambda} \partial_{\lambda}\psi^{\varkappa\lambda} = 0.$$

These are the familiar equations of magnetostatics if M is the 3-dimensional physical space or of electrodynamics if M is the 4-dimensional space-time.

CANONICAL TRANSFORMATIONS AND THEIR REPRESENTATIONS IN QUANTUM MECHANICS +

M. Moshinsky ++

Instituto de Física, Universidad de México (UNAM), México, D.F.

1. Introduction

When Prof. Konrad Bleuler invited me to this conference, in which such a prominent part was played by the techniques of geometrical quantization, my first reaction was to refuse as the subject seemed entirely foreign to my current interests. The preceeding expositions, as well as the ones I heard the previous week in Nijmegen, convinced me of the importance of this subject and made me realize that, as in Moliere's play, I had all along been speaking in prose.

When preparing this talk I was reminded of an observation of a brilliant engineer that once came to my house. He saw me go into a basement to disconnect the switch controlling the pump of my swimming pool and smiled sarcastically. "You know", he told me, "I am always amazed how sophisticated you physicists are; we poor engineers would have put the switch of the pump outside, so one could reach it without the contortions you just had to go through".

If you allow me the simile, I would like in my talk to put the switch outside i.e. to discuss the representations of classical canonical transformations in quantum mechanics in a language taken straigth from Dirac's book, rather than in the much more sophisticated one I heard at this conference. That does not mean that I do not appreciate the latter. Turning to the remark of my engineer, I could have answered him that an outside switch can electrocute children having access to it or give rise to short circuits if it rained. No doubt the audience here could predict equally serious consequences from using a non-rigorous formalism in relation with the problem mentioned above. Nevertheless I prefer to stress the physical ideas behind my reasoning rather than its mathematical exactness.

+A more detailed analysis of the problem was given by P.A. Mello and the author in the October 1975 issue of JMP.

++Member of the Instituto Nacional de Energia Nuclear and El Colegio Nacional.

I shall start by discussing some elementary examples of representations of canonical transformations and their applications.

2. Representations in quantum mechanics of simple canonical transformations.

As my intention is to discuss the ideas mentioned above within the simplest possible context, I will analize here mainly canonical transformations in two dimensional phase space. We have then the original coordinates and momenta x, p and the new ones $\bar{x}, \bar{p}$ related functionally by

$$\bar{x} = \bar{x}(x,p), \quad \bar{p} = \bar{p}(x,p) \tag{2.1}$$

where the Poisson bracket

$$\{\bar{x},\bar{p}\} = \frac{\partial \bar{x}}{\partial x}\frac{\partial \bar{p}}{\partial p} - \frac{\partial \bar{x}}{\partial p}\frac{\partial \bar{p}}{\partial x} = 1 \tag{2.2}$$

In quantum mechanics we can take a basis in which the original coordinate x is diagonal i.e. we have, in Dirac's notation[1], the kets

$$x \mid x'\rangle = x' \mid x'\rangle \tag{2.3}$$

We can also have a basis in which the new coordinate $\bar{x}$ is diagonal and to distinguish the states, even when the eigenvalue of $\bar{x}$ happens to be the same as that of x, we shall denote the corresponding kets by round rather than angular brackets i.e.

$$\bar{x} \mid \bar{x}') = \bar{x}' \mid \bar{x}') \tag{2.4}$$

The representation of the canonical transformation (2.1) will then be an operator U that takes us from the angular to the round ket for any given eigenvalue $\bar{x}'$ i.e.

$$\mid \bar{x}') = U \mid \bar{x}'\rangle \tag{2.5}$$

which implies that

$$\langle x' \mid \bar{x}') = \langle x' \mid U \mid \bar{x}'\rangle \tag{2.6}$$

Does U exist and if so how can we determine it explicitly? As Souriau stated it, most physicists consider that the answer to this question is essentially given in Dirac's book, while concerned mathematicians think that with luck and great effort they may give a rigorous formulation to the problem in ten years.

Certainly there are special cases in which the answer is known to any student of elementary quantum mechanics. For example if the classical canonical transformation is

$$\bar{x} = p \quad , \quad \bar{p} = -x \tag{2.7}$$

then its representation in quantum mechanics is well known, as it is associated with the passage to momentum space, and is given by

$$\langle x' \mid \bar{x}') = \langle x' \mid U \mid \bar{x}' \rangle = (2\pi)^{-1/2} \exp (i\, x'\, \bar{x}') \tag{2.8}$$

where we take units in which $\hbar = 1$.

In fact Dirac's discussion[1] allows much more and, as we shall indicate in the next section, we seem in principle able to obtain the representation $\langle x' \mid U \mid \bar{x}' \rangle$ in those cases in which the spectra of $\bar{x}, \bar{p}$ coincide with those of x, p. In particular this last property will hold for any real linear canonical transformation

$$\bar{x} = ax + bp \ , \ \bar{p} = cx + dp \ , \ ad - bc = 1 \tag{2.9}$$

and as is well known[2], (and will also be derived in the next section) for $b \neq 0$, $c = (ad - 1)\, b^{-1}$ we have

$$\langle x' \mid \bar{x}') = (2\pi |b|)^{-1/2} \exp \left[- (i|2b) (ax'^2 - 2x'\, \bar{x}' + d\bar{x}'^2)\right] \tag{2.10}$$

Is the problem of representations of classical canonical transformations in quantum mechanics of interest only for the foundations of the latter as well as for the relations between these two disciplines? The answer is certainly no. For example linear canonical transformations are directly related with the dynamical and symmetry group of the harmonic oscillator. In fact in the three dimensional case, the linear canonical transformations

$$\begin{pmatrix} \vec{r}' \\ \vec{p}' \end{pmatrix} = \begin{pmatrix} \frac{1}{2}(U+U^*) & -\frac{i}{2}(U-U^*) \\ \frac{i}{2}(U-U^*) & \frac{1}{2}(U+U^*) \end{pmatrix} \begin{pmatrix} \vec{r} \\ \vec{p} \end{pmatrix} \tag{2.11}$$

where U is a three dimensional unitary matrix, leave the

Hamiltonian[3]

$$H = \frac{1}{2}(\vec{p}^2 + \vec{r}^2) = \frac{1}{2}(\vec{p}'^2 + \vec{r}'^2) \tag{2.12}$$

invariant. Extensive use of this symmetry has been made since the pioneering work of Elliott[4] in problems of nuclear physics and indirectly it lead to the interest in $U(3)$ for the field of elementary particles.

It is interesting to note also that had we considered the transformation (2.9) in which $b = \varepsilon$ is a small number, the requirement $ad - bc = 1$ up to second order in the ε, would lead to the transformation[5]

$$\begin{pmatrix} \bar{x} \\ \bar{p} \end{pmatrix} = \begin{bmatrix} a + \varepsilon c\,(a + a^{-1}) & \varepsilon \\ c & a^{-1} + \varepsilon c\,(a + a^{-1}) \end{bmatrix} \begin{pmatrix} x \\ p \end{pmatrix} \tag{2.13}$$

Substituting these values in (2.10) and taking the limit $\varepsilon \to 0$ with $a = 1$ and $c = i\gamma$ where γ is real and positive we get

$$\langle x' | U | \bar{x}' \rangle = \exp\left(-\frac{1}{2}\gamma x'^2\right) \delta(x' - \bar{x}') \tag{2.14}$$

which represents a Gaussian potential interaction. Thus an interest develops in complex extensions of linear canonical transformations[5], which has led to many applications ranging from the cluster model of the nucleus to the general understanding of problems of accidental degeneracy[5].

The brief observations of the last paragraphs were intended just to stress the point that the representations of classical canonical transformations in quantum mechanics may have important practical applications. I wish now to return to the question of how to determine these representations in a systematic fashion.

3. Equations that determine the representations in quantum mechanics of classical canonical transformations.

In classical mechanics a canonical transformation may be given explicitly i.e. with $\bar{x}$, $\bar{p}$ being some functions of x, p for which $\{\bar{x}, \bar{p}\} = 1$ or implicitly through the functional relations

$$f_1(x,p) = \bar{f}_1(\bar{x},\bar{p}) \qquad f_2(x,p) = \bar{f}_2(\bar{x},\bar{p}) \tag{3.1}$$

What is the necessary and sufficient condition for (3.1) to define a canonical transformation? To answer this question let us introduce the notation

$$z_1 = x,\ z_2 = p\ ;\quad \bar{z}_1 = \bar{x},\ \bar{z}_2 = \bar{p} \tag{3.2}$$

The Poisson bracket $\{g, h\}$ can then be written as

$$\{g, h\} = \frac{\partial g}{\partial z_\alpha} K_{\alpha\beta} \frac{\partial h}{\partial z_\beta}, \qquad ||K_{\alpha\beta}|| = \begin{pmatrix} 0 & 1 \\ -1 & 0 \end{pmatrix} \tag{3.3}$$

where $\alpha, \beta = 1, 2$ and repeated indices are summed over these two values. Thus a transformation from z_α to $\bar{z}_\alpha$ is canonical if

$$\frac{\partial \bar{z}_\alpha}{\partial z_\gamma} K_{\gamma\delta} \frac{\partial \bar{z}_\beta}{\partial z_\delta} = K_{\alpha\beta} \tag{3.4}$$

Using (3.1) we can express (3.4) as

$$\frac{\partial \bar{z}_\alpha}{\partial \bar{f}_{\lambda'}} \frac{\partial \bar{f}_{\lambda'}}{\partial z_\gamma} K_{\gamma\delta} \frac{\partial \bar{z}_\beta}{\partial \bar{f}_{\mu'}} \frac{\partial \bar{f}_{\mu'}}{\partial z_\delta} = K_{\alpha\beta} \tag{3.5}$$

Multiplying left and right hand sides by $(\partial \bar{f}_\lambda / \partial \bar{z}_\alpha)$, $(\partial \bar{f}_\mu / \partial \bar{z}_\beta)$ and summing over repeated indices we get

$$\frac{\partial f_\lambda}{\partial z_\gamma} K_{\gamma\delta} \frac{\partial f_\mu}{\partial z_\delta} = \frac{\partial \bar{f}_\lambda}{\partial \bar{z}_\alpha} K_{\alpha\beta} \frac{\partial \bar{f}_\mu}{\partial \bar{z}_\beta} \tag{3.6}$$

or in Poisson bracket notation

$$\{f_\lambda, f_\mu\}_z = \{\bar{f}_\lambda, \bar{f}_\mu\}_{\bar{z}} \tag{3.7}$$

where the index outside the curly bracket indicates the variables involved.

It is clear therefore that a necessary condition for the transformation (3.1) to be canonical is that the Poisson bracket between f_λ and f_μ with respect to the z_α be the same as that between $\bar{f}_\lambda$ and $\bar{f}_\mu$ with respect to $\bar{z}_\alpha$ when use is made of the equations (3.1) to relate $\bar{z}_\alpha$ to z_α. That this condition is also sufficient is clear from the fact that the steps leading from (3.4) to (3.7) can be taken in the reverse order.

How can we represent in quantum mechanics the canonical transformation defined implicitly by (3.1) for which the conditions (3.7) hold? We note that from the definition of the operator U of the previous section we expect for the quantum mechanical operators the relations[1]

$$\bar{x} = U\, x\, U^{-1} \quad , \quad \bar{p} = U\, p\, U^{-1} \tag{3.8}$$

or more generally

$$\bar{f}_\alpha\,(\bar{x}, \bar{p}) = U\, \bar{f}_\alpha\,(x, p)\, U^{-1} = f_\alpha\,(x, p) \tag{3.9}$$

where use was made of the equations (3.1). Multiplying by U to the left we have the operator relations

$$\bar{f}_\alpha\,(\bar{x}, \bar{p})\, U = U\, \bar{f}_\alpha\,(x, p) \tag{3.10}$$

We now take the matrix elements of the operators in both sides of (3.10) between states $<x'|$, $|\bar{x}'>$ for which the original position variable is diagonal. We start by discussing the left hand side

$$\begin{aligned} &< x'|\, \bar{f}_\alpha\,(\bar{x}, \bar{p})\, U|\, \bar{x}' > \\ &= \int < x'|\, \bar{f}_\alpha\,(\bar{x}, \bar{p})\, |\, x''>\, dx''\, < x''|\, U|\, \bar{x}'> \end{aligned} \tag{3.11}$$

From (3.1) we note that

$$\begin{aligned} &< x'|\, \bar{f}_\alpha\,(\bar{x}, \bar{p})|\, x''> = < x'|\, f_\alpha\,(x, p)|\, x''> \\ &= f_\alpha\,(x', -i\,\frac{\partial}{\partial x'})\, \delta\,(x' - x'') \end{aligned} \tag{3.12}$$

while in the notation of (2.8)

$$< x'|U|\, \bar{x}'> = < x'|\, \bar{x}') \tag{3.13}$$

Thus we finally obtain

$$\begin{aligned} &< x'|\, \bar{f}_\alpha\,(\bar{x}, \bar{p})\, U|\, \bar{x}' > \\ &= f_\alpha\,(x', -i\,\frac{\partial}{\partial x'})\, < x'|\, \bar{x}') \end{aligned} \tag{3.14}$$

A similar analysis for the right hand side[6] of (3.10) allows us finally to write the following set of differential equations for $<x'|\bar{x}')$:

$$f_\alpha\,(x', -i\,\frac{\partial}{\partial x'})\, < x'|\, \bar{x}') = \left[\bar{f}_\alpha^{+}\,(\bar{x}', -i\,\partial/\partial\bar{x}')\right]^{*} <x'|\, \bar{x}') \tag{3.15}$$

$\alpha = 1,2$

The discussion can be extended immediately to 2n dimensional phase space if $\alpha = 1, 2, \ldots 2n$. We note in particular that for the linear canonical transformation (2.9) we have the equations

$$\left[ax' - i\, b \frac{\partial}{\partial x'}\right] \langle x' | \bar{x}') = \bar{x}' \langle x' | \bar{x}')$$
$$\left[cx' - i\, d \frac{\partial}{\partial x'}\right] \langle x' | \bar{x}') = i \frac{\partial}{\partial \bar{x}'} \langle x' | \bar{x}') \tag{3.16}$$

whose solution for $b \neq 0$ and $c = (ad - 1)\, b^{-1}$ is given precisely by (2.10).

While the meaning of equations (3.15) is clear for linear canonical transformations it is less so in the general case. For example we can ask how to write the corresponding operators for $f_\alpha (x, p)$ $[\alpha = 1, 2]$ if products of x and p are involved. The Wigner-Weyl formalism[7] for translating classical expressions into quantum mechanics could be used in this case. A more difficult situation can occur when $f_\alpha (x, p)$ are algebraic or transcendental functions of <u>both</u> x and p . This situation can be avoided, at least in some cases, by making an appropriate selection of the implicit functions $f_\alpha (x, p)$, $\bar{f}_\alpha (\bar{x}, \bar{p})$ as will be shown in the next section for a specific example.

What seems to be the most serious difficulty concerns the original Dirac operator relations (3.8). If we assume them to be valid in all circumstances, the inverse U^{-1} of the operator must also exist and the spectra of both x and $\bar{x}$ is then the same, a situation that also holds for p and $\bar{p}$. We face then a serious problem if we wish to find the representation in quantum mechanics of transformations such as

$$\bar{x} = -\text{ang tan}\, (p/x), \quad \bar{p} = \frac{1}{2} (p^2 + x^2) \tag{3.17}$$

where for example $\bar{p}$ has the spectrum $-\infty \leq \bar{p}' \leq \infty$, while that of $1/2\, (p^2 + x^2)$ is $(n + 1/2)$ with n being a non-negative integer. An analysis carried out elsewhere[6,8] suggests that rather than use the equations (3.8) or equivalently the more general (3.9), we should consider (3.10) where U^{-1} does not appear explicitly and thus the requirement of equal spectra is no longer necessary. It was in fact from (3.10) that we obtained the differential equations satisfied by $\langle x' | \bar{x}')$.

In the following section we shall solve the equations (3.15) for some specific non linear canonical transformations that take us from one type of Hamiltonian into another and then proceed to outline a general procedure for dealing with this type of problems.

4. Non linear canonical transformations and their representation in quantum mechanics

Among the more interesting canonical transformations are those that transform from one type of Hamiltonian into another. Using the notation of the previous section we can introduce the functions

$$f_1(x, p) = \frac{1}{2} p^2 + V(x), \quad \bar{f}_1(\bar{x}, \bar{p}) = \frac{1}{2} \bar{p}^2 + \bar{V}(\bar{x}) \qquad (4.1ab)$$

corresponding to the two Hamiltonians in question. The classical canonical transformation will be defined in terms of the equations (3.1) in which $f_2(x, p)$, $\bar{f}_2(\bar{x}, \bar{p})$ have to be appropriately defined so that (3.7) holds. To achieve this last point we remember that the canonically conjugate variables to the Hamiltonians (4.1) are the observables associated with the time i.e.[9]

$$T(x, p) = \int_a^x \frac{dy}{\left[2f_1(x, p) - 2V(y)\right]^{1/2}}, \qquad \{T, f_1\}_{x,p} = 1 \qquad (4.2a)$$

$$\bar{T}(\bar{x}, \bar{p}) = \int_{\bar{a}}^{\bar{x}} \frac{dy}{\left[2\bar{f}_1(\bar{x}, \bar{p}) - 2\bar{V}(y)\right]^{1/2}}, \qquad \{\bar{T}, \bar{f}_1\}_{\bar{x}\bar{p}} = 1 \qquad (4.2b)$$

where a, $\bar{a}$ are arbitrary constants.

From (3.7) we see then that the equations

$$f_1(x, p) = \bar{f}_1(\bar{x}, \bar{p}), \quad T(x, p) = \bar{T}(\bar{x}, \bar{p}) \qquad (4.3)$$

define a canonical transformation. We shall not identify though $f_2(x, p)$ with $T(x, p)$ but rather with some convenient function ϕ of $f_1(x, p)$ and $T(x, p)$ i.e.

$$f_2(x, p) = \phi\left[f_1(x, p), \; T(x, p)\right] \qquad (4.4a)$$

For $\bar{f}_2(\bar{x}, \bar{p})$ we shall take the <u>same</u> function ϕ but of

$\bar{f}_1(\bar{x}, \bar{p})$, $\bar{T}(\bar{x}, \bar{p})$ i.e.

$$\bar{f}_2 (\bar{x}, \bar{p}) = \phi \left[\bar{f}_1 (\bar{x}, \bar{p}), \bar{T} (\bar{x}, \bar{p})\right] \tag{4.4b}$$

The equations

$$f_1 (x, p) = \bar{f}_1 (\bar{x}, \bar{p}), \; f_2 (x, p) = \bar{f}_2 (\bar{x}, \bar{p}) \tag{4.5}$$

define then the same canonical transformation as (4.3). We note furthermore from (4.2), that

$$\{f_1(x,p), f_2(x,p)\}_{x,p} = - \frac{\partial\phi(f_1,T)}{\partial f_1} = - \frac{\partial\phi(\bar{f}_1,\bar{T})}{\partial \bar{f}_1} = \{\bar{f}_1(\bar{x},\bar{p}), \bar{f}_2(\bar{x},\bar{p})\}_{\bar{x},\bar{p}} \tag{4.6}$$

thus satisfying the condition (3.7) required for the transformation defined implicitly by (4.5) to be canonical.

We proceed now to discuss the specific example in which

$$V (x) = \frac{1}{2} \frac{\lambda^2}{x^2} , \quad \bar{V} (\bar{x}) = \frac{1}{2} \frac{\bar{\lambda}^2}{\bar{x}^2} \tag{4.7}$$

with λ, $\bar{\lambda}$ being arbitrary real constants. We have then

$$f_1 (x, p) = \frac{1}{2} (p^2 + \frac{\lambda^2}{x^2}) = \frac{1}{2} (\bar{p}^2 + \frac{\bar{\lambda}^2}{\bar{x}^2}) = \bar{f}_1 (\bar{x}, \bar{p}) \tag{4.8a}$$

Furthermore choosing a, $\bar{a}$ appropriately we have that the T, $\bar{T}$ defined by (4.2) take the form

$$T = xp \, (p^2 + \lambda^2 \, x^{-2})^{-1} \quad , \quad \bar{T} = \bar{x}\bar{p} \, (\bar{p}^2 + \bar{\lambda}^2 \, \bar{x}^{-2})^{-1} \tag{4.9}$$

The choice of ϕ we take will be

$$f_2 (x, p) = \phi (f_1 (x, p), T (x,p)) \equiv 2f_1 (x, p) \, T (x,p) = xp \tag{4.10}$$

so that the remaining equation necessary to determine the canonical transformation becomes

$$f_2 (x, p) = xp = \bar{x}\bar{p} = \bar{f}_2 (\bar{x}, \bar{p}) \tag{4.8b}$$

from which we conclude that

$$\bar{x} = x\left[(x^2p^2+\bar{\lambda}^2)\ /\ (x^2p^2+\lambda^2)\right]^{1/2},\ \bar{p} = p\left[(x^2p^2+\lambda^2)\ /\ (x^2p^2+\bar{\lambda}^2)\right]^{1/2} \tag{4.11}$$

While it is very difficult to analyze the representation in quantum mechanics of the highly non linear canonical transformation (4.11), to implement the program we outlined in the previous section for the implicit equations (4.8) is trivial, as we proceed to show. The equations (3.15) take then the form

$$\frac{1}{2}\left(-\frac{\partial^2}{\partial x'^2}+\frac{\lambda^2}{x'^2}\right)\langle x'|\bar{x}') = \frac{1}{2}\left(-\frac{\partial^2}{\partial \bar{x}'^2}+\frac{\bar{\lambda}^2}{\bar{x}'^2}\right)\langle x'|\bar{x}') \tag{4.12a}$$

$$\frac{1}{i}\ x'\frac{\partial}{\partial x'}\langle x'|\bar{x}') = -\frac{1}{i}\frac{\partial}{\partial \bar{x}'}\ \bar{x}'\ \langle x'|\bar{x}') \tag{4.12b}$$

Introducing the change of variable

$$x' = \rho\cosh\theta\quad,\quad \bar{x}' = \rho\sinh\theta \tag{4.13}$$

the second equation becomes

$$(\rho\ \partial/\partial\rho + 1)\ \langle x'|\bar{x}') = 0 \tag{4.14}$$

which implies

$$\langle x'|\bar{x}') = \rho^{-1}\ f\ (\theta) \tag{4.15}$$

From the first equation we obtain then that $f(\theta)$ satisfies the ordinary differential equation[6]

$$\left(\frac{d^2}{d\theta^2}+\frac{\lambda^2}{\cos h^2\theta}-\frac{\bar{\lambda}^2}{\sin h^2\theta}-1\right) f\ (\theta) = 0 \tag{4.16}$$

Taking into account the asymptotic behaviour[6] at $\theta=\infty$ i.e. $x' = \bar{x}'$ we obtain finally that

$$\langle x'|\bar{x}') = \cos\left[(\mu-\bar{\mu})\frac{\pi}{2}\right]\ \delta\ (x'-\bar{x}') + \ldots \tag{4.17}$$

$$+\begin{cases}\dfrac{2\Gamma(\frac{\mu+\bar{\mu}}{2}+1)}{\Gamma(\frac{\bar{\mu}-\mu}{2})\Gamma(\mu+1)}\ \dfrac{1}{\bar{x}'}\left(\dfrac{x'^2}{\bar{x}'^2}\right)^{\frac{\mu}{2}+\frac{1}{4}} F\left(\dfrac{\mu+\bar{\mu}}{2}+1,\ \dfrac{\mu-\bar{\mu}}{2}+1,\ \mu+1;\ \dfrac{x'^2}{\bar{x}'^2}\right), & \text{if } x'<\bar{x}' \\[2ex] \dfrac{2\Gamma(\frac{\mu+\bar{\mu}}{2}+1)}{\Gamma(\frac{\mu-\bar{\mu}}{2})\Gamma(\bar{\mu}+1)}\ \dfrac{1}{x'}\left(\dfrac{\bar{x}'^2}{x'^2}\right)^{\frac{\bar{\mu}}{2}+\frac{1}{4}} F\left(\dfrac{\mu+\bar{\mu}}{2}+1,\ \dfrac{\bar{\mu}-\mu}{2}+1,\ \bar{\mu}+1;\ \dfrac{\bar{x}'^2}{x'^2}\right), & \text{if } x'>\bar{x}'\end{cases}$$

where F is an hypergeometric function[10] and

$$\mu^2 = \lambda^2 + 1/4 \quad , \quad \bar{\mu}^2 = \bar{\lambda}^2 + 1/4 \tag{4.18}$$

We have determined the representation in quantum mechanics of the particular classical non linear canonical transformation (4.11). This representation happens to be unitary[6] and for $\lambda = \bar{\lambda}$ i.e. $\mu = \bar{\mu}$ it reduces to $\delta(x' - \bar{x}')$ as it should, due to the fact that (4.11) gives then $\bar{x} = x$, $\bar{p} = p$.

What can we say in general for the representation of canonical transformations defined by equations of the type (4.8)? To begin with we could look into at the eigenvalues E, $\bar{E}$ and eigenstates $\psi_E(x')$, $\bar{\psi}_{\bar{E}}(\bar{x}')$ of the operators $f_1(x, p)$, $\bar{f}_1(\bar{x}, \bar{p})$, i.e.

$$f_1\left(x', \frac{1}{i}\frac{\partial}{\partial x'}\right)\psi_E(x') = E\psi_E(x'), \bar{f}_1\left(\bar{x}', \frac{1}{i}\frac{\partial}{\partial \bar{x}'}\right)\bar{\psi}_{\bar{E}}(\bar{x}') = \bar{E}\bar{\psi}_{\bar{E}}(\bar{x}') \tag{4.19}$$

We can certainly carry out the expansion

$$\langle x' | \bar{x}' \rangle = \int\int a(E, \bar{E})\, \psi_E(x')\, \bar{\psi}_{\bar{E}}^{*}(\bar{x}')\, dE dE' \tag{4.20}$$

where $a(E, \bar{E})$ are so far arbitrary coefficients and the integration implies also a summation in case the spectrum has a discrete part.

We immediately see then that equation (3.15) for $\alpha = 1$ implies that

$$(E - \bar{E})\, a(E, \bar{E}) = 0 \tag{4.21}$$

If the spectra E, $\bar{E}$ happen to be both continuous and in the same range $0 \leq E, \bar{E} \leq \infty$ then equation (4.21) implies that

$$a(E, \bar{E}) = b(E)\, \delta(\bar{E} - E) \tag{4.22}$$

but even in other cases e.g. if $0 \leq E \leq \infty$ and $\bar{E} = n + 1/2$ with n integer, there is a subset of values of E for which a solution exists and this may be relevant for problems such as those indicated in (3.17).

Restricting ourselves to problems in which $f_1(x, p)$, $\bar{f}_1(\bar{x}, \bar{p})$ have the same continous spectra E , we see that the equation (3.15) for $\alpha = 2$ implies that

$$\int\left\{f_2\left(x', \frac{1}{i}\frac{\partial}{\partial x'}\right) - \left[\bar{f}_2^{+}\left(\bar{x}', \frac{1}{i}\frac{\partial}{\partial \bar{x}'}\right)\right]^{*}\right\} b(E)\, \psi_E(x')\bar{\psi}_E^{*}(\bar{x}') dE = 0 \tag{4.23}$$

which is a complicated transform[6] whose vanishing will in principle determine the $b(E)$. Knowing the latter we can determine from (4.20), (4.22) the representation $< x'|\bar{x}') = < x'|U|\bar{x}' >$.

While for problems in which $f_\alpha(x, p)$ and $\bar{f}_\alpha(\bar{x}, \bar{p})$ have the same spectra, we could expect the representations to be unitary and thus have also the reciprocal of the operator U, for the cases when the spectra are different this would not be the case as the transformation $\bar{x} = UxU^{-1}$, $\bar{p} = UpU^{-1}$ conserves the spectra. Yet the determination of a $< x'|U|\bar{x}' >$ through the equations (3.15) could still lead to definite results. Furthermore if one is dealing with canonical transformations depending on a set of parameters τ, then the operators $U(\tau)$ will follow the same multiplication law as the corresponding classical transformations, so they constitute a representation of the latter in the usual group theoretical sense. However, the absence of an inverse indicates that we are now dealing with a semigroup rather than with a group of transformations.

Many points are still open to question in this discussion, but I hope that the actual determination of unitary representations of some complex non-linear canonical transformations, as well as the possibilities of analyzing cases in which the spectra change, will stimulate also those that have a more rigorous approach to the subject.

REFERENCES

[1] P.A.M. Dirac, Quantum Mechanics (Clarendon Press, Oxford, Fourth Edition 1958) pp. 103-107.

[2] M. Moshinsky and C. Quesne, J. Math. Phys. 12, 1772 (1971)
C. Itzykson, Commun. Math. Phys. 4, 92 (1967)
V. Bargmann "Group Representations on Hilbert Spaces of Analytic Functions" in Analytic Methods of Mathematical Physics edited by Gilbert and Newton (Gordon & Breach, N.Y. 1968).

[3] M. Moshinsky, SIAM, J. Appl. Math. 25, 193 (1973).

[4] J.P. Elliott, Proc. Roy. Soc. London A 245, 128 (1958).

[5] P. Kramer, M. Moshinsky and T.H. Seligman, "Complex Extensions of Canonical Transformations and Quantum Mechanics" in Group Theory and its Applications Vol. III, edited by E.M. Loebl (Academic Press, New York 1975).

[6] P.A. Mello and M. Moshinsky, J. Math. Phys. October 1975.

[7] S.R. de Groot and L.G. Suttorp, Foundations of Electrodynamics (North Holland Publishing Co., Amsterdam 1972) Chapter VI.

[8] P. Kramer and M. Moshinsky (Paper in preparation).

[9] H. Goldstein, Classical Mechanics (Addison-Wesley Publishing Co., Reading, Mass., 1959), p. 243.

[10] I.S. Gradshteyn and I.W. Ryzhik, Tables of Integrals Series and Products, (Academic Press, New York and London 1965) p. 1039.

ON A SYMPLECTIC STRUCTURE OF GENERAL RELATIVITY

Wiktor Szczyrba

Institute of Mathematics, Polish Academy of Sciences
00-950 Warsaw, Poland

The concept of the canonical quantization plays an important part in passing from classical to quantum systems. This procedure, well known in mechanics is based on the Hamilton (canonical) formulation of physical systems with finite degrees of freedom [27], [25]. In recent fifteen years an elegant geometric approach to the Hamilton formalism has been found [1], [31]. In this formulation we consider a 2n-dimensional manifold $\mathcal{P}_{2n}$ – the phase space of a system and a closed non-degenerate differential 2-form Ω on $\mathcal{P}_{2n}$. The form Ω defines a Lie algebra structure in the set $\mathcal{F}$ of all smooth functions on $\mathcal{P}_{2n}$. In many examples of this theory the manifold $\mathcal{P}_{2n}$ is simply the cotangent bundle of an n-dimensional manifold V (the configuration space). In this case Ω is the canonical 2-form on T^*V (cf. [1], [31]) and if (q^i) are local coordinates in V, (p_i, q^i) are local coordinates in $\mathcal{P}_{2n} = T^*V$ then

$$\Omega = \sum_{i=1}^{n} dp_i \wedge dq^i \tag{1.1}$$

The form Ω defines an isomorphism between the bundles $T(\mathcal{P}_{2n})$ and $T^*(\mathcal{P}_{2n})$. If $X \in T_p(\mathcal{P}_{2n})$ then this isomorphism is given by:

$$T_p(\mathcal{P}_{2n}) \ni X \longrightarrow X^\flat \in T_p^*(\mathcal{P}_{2n}), \text{ where } \langle Y, X^\flat\rangle = -\Omega(X,Y), \quad Y \in T_p(\mathcal{P}_{2n}) \tag{1.2}$$

We denote the inverse mapping by #.

For $f_1, f_2 \in \mathcal{F} = C^\infty(\mathcal{P}_{2n})$

$$\{f_1, f_2\} = \Omega\left((df_1)^\#, (df_2)^\#\right) \tag{1.3}$$

In local coordinates we have the known classical formula:

$$\{f_1, f_2\} = \sum_{i=1}^{n} \left(\frac{\partial f_1}{\partial p_i}\frac{\partial f_2}{\partial q^i} - \frac{\partial f_2}{\partial p_i}\frac{\partial f_1}{\partial q^i} \right) \tag{1.4}$$

The above formulation of mechanics is not convenient for a generalization for the case of field theory. A more suitable is the homogeneous approach. We consider an 2n+1 dimensional submanifold $\mathcal{P}_{hom}$ of the bundle $T^*(V \times \mathbb{R})$ given by the constraint equation

$$H = H(p_i, q^j, t) \tag{1.5}$$

where t is the global coordinate in $\mathbb{R}$ and $-H$ is the conjugate coordinate in $T^*(V \times \mathbb{R})$. The canonical 2-form on $T^*(V\times\mathbb{R})$ generates a closed differential 2-form ω_{hom} on $\mathcal{P}_{hom}$. In local coordinates

$$\omega_{hom} = \sum_{i=1}^{n} dp_i \wedge dq^i - dH \wedge dt \tag{1.6}$$

The form ω_{hom} is degenerate and therefore we can distinguish a family $\mathcal{H}$ of 1-dimensional trajectories (submanifolds) of $\mathcal{P}_{hom}$ such that for $C \in \mathcal{H}$ and any vector field X tangent to $\mathcal{P}_{hom}$ and defined on C we have

$$(X \lrcorner\, \omega_{hom})|C = 0 \tag{1.7}$$

(| denotes the pull-back of the 1-form $X \rfloor \omega_{hom}$ onto manifold C).
It is easy to show that elements of $\mathcal{H}$ are in one to one correspondance with trajectories in V. If C is parametrized by the coordinate t
$C = \{ (p_i(t), q^i(t), t) \in \mathcal{P}_{hom} \; : \; t \in \mathbb{R} \}$ then

$$\frac{dq^i}{dt} = \frac{\partial H}{\partial p_i} \;, \qquad \frac{dp_i}{dt} = - \frac{\partial H}{\partial q^i} \qquad i = 1, \dots, n \tag{1.8}$$

The space $\mathcal{H}$ is 2n-dimensional because it can be parametrized by initial values of coordinates and momenta and it is diffeomorphic to $\mathcal{P}_{2n}$ (different instants of time give different diffeomorphisms). Therefore the space $\mathcal{H}$ carries a natural symplectic structure induced by the 2-form Ω in $\mathcal{P}_{2n}$. This structure is independent of a choice of an initial instant of time (motion is a symplectic diffeomorphism).

The homogeneous formulation of mechanics can be generalized for multidimensional cases i.e. for field theories. This generalization is based on the geometric theory of the calculus of variations (cf. [8], [20], [22], [32]). It turns out that for any variational problem with a fixed boundary in the space - time M there exists a bundle $\mathcal{P}$ over M and a closed 5-form $\overset{5}{\omega}$ on $\mathcal{P}$.
The equation

$$(X \rfloor \overset{5}{\omega}) | C = 0 \tag{1.9}$$

where C is a 4-dimensional submanifold of $\mathcal{P}$ (the image of a section of $\mathcal{P}$) and X is an arbitrary vector field on C (tangent to $\mathcal{P}$), gives a family $\mathcal{H}$ of submanifolds in $\mathcal{P}$. The elements of $\mathcal{H}$ are in one to one correspondance with the set of solutions of field equations of our variational problem. The multisymplectic approach to the field theory has been investigated in papers [13], [20], [22], where it was pointed out that the multisymplectic formulation is more useful than the Lagrangian one. In the present paper we construct the multisymplectic structure for General Relativity despite of non-existence of a covariant lagrangian density depending on first derivatives of the metric tensor.

Having the multisymplectic structure of the given field theory i.e. the manifold $\mathcal{P}$ and the closed 5-form $\overset{5}{\omega}$ one can define a presymplectic structure in the space of states $\mathcal{H}$. Such a construction has been recently done by J.Kijowski and the author [23]. It means that we endow the space $\mathcal{H}$ with a pseudo-differential structure and define on $\mathcal{H}$ a closed differential 2-form Ω. In theories with a gauge the form Ω is degenerate. Its degeneracy distribution W is involutive and therefore we can try to construct the space $\tilde{\mathcal{H}}$ such that $T(\tilde{\mathcal{H}}) = T(\mathcal{H})/W$. Projecting Ω onto $\tilde{\mathcal{H}}$ we obtain a non-degenerate closed 2-form $\tilde{\Omega}$. Such a structure enebles to define a Lie algebra structure in the set of smooth functionals (physical quantities) on $\tilde{\mathcal{H}}$. But the degeneracy of Ω implies restrictions on the set of functionals to be considered. We can consider only gauge independent functionals i.e.

functionals which can be projected on $\tilde{\mathcal{H}}$. For instance in electrodynamics the degeneration of Ω corresponds to an invariance of the Maxwell equations with respect to the gradient gauge $A_\mu \to A_\mu + \partial_\mu \varphi$ and the potentials A_μ do not define physical quantities but field strenghts $\underline{B}$ and $\underline{E}$ do (cf. [23]).

In the present paper we give the construction of the form Ω in gravidynamics. First at all, we construct a multisymplectic manifold ($\mathcal{P}, \overset{5}{\omega}$) such that sections of the bundle $\tau : \mathcal{P} \to M$ satisfying (1.9) are in one to one correspondance with the set of solutions of the Einstein equations:

$$\Gamma^\lambda_{\mu\nu} = \tfrac{1}{2} g^{\lambda\sigma}(\partial_\mu g_{\nu\sigma} + \partial_\nu g_{\mu\sigma} - \partial_\sigma g_{\mu\nu}) \tag{1.10}$$

$$R_{\mu\nu} - \tfrac{1}{2} g_{\mu\nu} R + \lambda g_{\mu\nu} = 0 \tag{1.11}$$

Having this structure we define the presymplectic form Ω according to the general formula given in [23] . It turns out that the diagonalization of the form Ω distinguishes the ADMW coordinate system (cf. [3],[33]).
If $\sigma \subset M$ is a space-like surface and g_{ij}, K_{ij} are its first and second fundamental forms then the form Ω has a diagonal expression in terms of the infinitesimal translations δg_{ij} , $\delta \pi^{ij}$. The quantities π^{ij} are components of a tensor density on σ and are defined by :

$$\pi^{ij} = - \sqrt{g}\,(K_{pq} - g_{pq}\,\mathrm{tr}K\,)\, g^{ip} g^{jq} \tag{1.12}$$

Using the diagonal expression for Ω we prove that its degeneration distribution W is determined by an action of the diffeomorphism group of M in the space $\mathcal{H}$. If we divide $\mathcal{H}$ by this action we obtain the superphase space for General Relativity.

The presented approach elucidates some classical problems of General Relativity. In particular, the geometric character of the ADMW coordinates is proved. We discuss also the problem of degrees of freedom for gravitational field and show that their numbers is equal to four (in the phase space).

The considered here problems are classical and were investigated earlier by Bergmann, Dirac, Wheeler, Arnowiitt-Deser-Misner, De Witt ,Fadeev, Fischer and Marsden. In the next paper we shall discuss differences between their and our approaches.

This paper summarizes only our main results. The complete exposition with proofs will appear in Communications in Math. Physics.

2. The multisymplectic structure of General Relativity

Let M be a 4-dimensional smooth manifold - the space-time and λ be a real number - the cosmological constant. Let $S^2_*T^*M$ be the bundle of symmetric 2-covariant non-degenerate tensors (metrics) on M with the negative determinant $g = \det g_{\mu\nu}$. We construct over $S^2_*T^*M$ the bundle of coefficients of the affine connection Aff. If $(x^\lambda, g_{\mu\nu})$ are local coordinates in $S^2_*T^*M$ then we have local coordinates $(x^\lambda, g_{\mu\nu}, \Gamma^\tau_{\mu\nu})$ $(\Gamma^\tau_{\mu\nu} = \Gamma^\tau_{\nu\mu})$ in Aff with known transformation properties:

$$g_{\mu'\nu'} = \frac{\partial x^\mu}{\partial x^{\mu'}} \frac{\partial x^\nu}{\partial x^{\nu'}} g_{\mu\nu}$$
$$\Gamma^{\tau'}_{\mu'\nu'} = \frac{\partial x^{\tau'}}{\partial x^\tau} \frac{\partial x^\mu}{\partial x^{\mu'}} \frac{\partial x^\nu}{\partial x^{\nu'}} \Gamma^\tau_{\mu\nu} + \frac{\partial^2 x^\lambda}{\partial x^{\mu'} \partial x^{\nu'}} \frac{\partial x^{\tau'}}{\partial x^\lambda} \tag{2.1}$$

We put $\mathcal{P}$ = Aff and

$$\overset{4}{\theta} = g^{\alpha\rho}\sqrt{-g}\, dx^0 \wedge \ldots \wedge \underset{\tau}{\underline{d\Gamma^\tau_{\alpha\rho}\wedge}} \ldots \wedge dx^3 - g^{\alpha\tau}\sqrt{-g}\, dx^0 \wedge \ldots \wedge \underset{\tau}{\underline{d\Gamma^\rho_{\rho\alpha}\wedge}} \ldots \wedge dx^3 +$$
$$- \left(g^{\nu\varsigma}(\Gamma^\mu_{\nu\tau}\Gamma^\tau_{\mu\varsigma} - \Gamma^\mu_{\mu\tau}\Gamma^\tau_{\nu\varsigma}) + 2\lambda \right)\sqrt{-g}\, dx^0 \wedge \ldots \wedge dx^3 \tag{2.2}$$

Proposition 1

The formula (2.2) is covariant with respect to the coordinate transformations (2.1) and therefore it defines a differential 4-form on $\mathcal{P}$.

We define $\overset{5}{\omega} = d\overset{4}{\theta}$ (2.3)

The couple $(\mathcal{P}, \overset{5}{\omega})$ determines a multisymplectic structure of General Relativity. Let $f: M \to \mathcal{P}$ be a section of the bundle $\tau : \mathcal{P} \to M$ and X be a τ-vertical vector field on $C = f(M)$. We look for such sections of $\mathcal{P}$ that for every X

$$(X \rfloor \overset{5}{\omega})|C = 0 \tag{2.4}$$

($|$ denotes the pull-back of the differential 4-form $X \rfloor \overset{5}{\omega}$ onto C).

In local coordinates we have $f(x^\lambda) = (x^\lambda, g_{\mu\nu}(x^\lambda), \Gamma^\tau_{\mu\nu}(x^\lambda))$ (2.5)

$X = \sum_{\alpha\le\rho} Q_{\alpha\rho}\frac{\partial}{\partial g_{\alpha\rho}} + \sum_{\mu\le\nu} P^\tau_{\mu\nu}\frac{\partial}{\partial\Gamma^\tau_{\mu\nu}}$. It is easy to see that (2.4) implies

$$\Gamma^\lambda_{\mu\nu} = \tfrac{1}{2} g^{\lambda\sigma}(\partial_\mu g_{\nu\sigma} + \partial_\nu g_{\mu\sigma} - \partial_\sigma g_{\mu\nu}) \tag{2.6}$$

$$G_{\mu\nu} = R_{\mu\nu} - \tfrac{1}{2} g_{\mu\nu} R + \lambda g_{\mu\nu} = 0 \tag{2.7}$$

Where
$$R^\rho_{\mu\alpha\nu} = \partial_\alpha\Gamma^\rho_{\mu\nu} - \partial_\nu\Gamma^\rho_{\mu\alpha} + \Gamma^\rho_{\alpha\tau}\Gamma^\tau_{\mu\nu} - \Gamma^\rho_{\tau\nu}\Gamma^\tau_{\mu\alpha}$$
$$R_{\mu\nu} = R^\alpha_{\mu\alpha\nu} \; ; \; R = g^{\mu\nu}R_{\mu\nu} \tag{2.8}$$

We have obtained

Proposition 2

Sections of $\tau : \mathcal{P} \to M$ satisfying (2.4) are in one to one to one correspondance with the set of solutions of the Einstein equations (2.6) and (2.7).

We can consider the equation (2.4) as a geometric formulation of the Hamilton equations. It is known (cf. [20], [22], [32],) that for a field theory with a lagrangian function it is always possible to construct a multisymplectic manifold by means of the Legendre transformation. In General Relativity we have no Lagrangian depending on first derivatives of a metric tensor $g_{\mu\nu}$ and therefore the multisymplectic approach have to be done axiomatically. We see that the Hamilton formulation is more appropriate for General Relativity than the Lagrangian formulation.

However, it is possible to give a non-covariant construction of the multisymplectic 5-form $\overset{5}{\omega}$. Let $\mathcal{L}(g_{\mu\nu}, \Gamma^{\tau}_{\mu\nu})$ =

$$= \sqrt{-g}\ \left(g^{\nu\varsigma}(\Gamma^{\mu}_{\tau\nu}\Gamma^{\tau}_{\mu\varsigma} - \Gamma^{\mu}_{\mu\tau}\Gamma^{\tau}_{\nu\varsigma}) + 2\lambda \right) \tag{2.9}$$

be a non-covariant lagrangian density for gravidynamics (cf. [2], [29]).

Using the Dedecker formula (cf. [32])

$$\Psi = \sum_{\mu\leq\nu} \frac{\partial\mathcal{L}}{\partial g_{\mu\nu,\lambda}}\, dx^0\wedge \dots \wedge \underset{\lambda}{\underbrace{dg_{\mu\nu}}} \wedge \dots \wedge dx^3 - \Big(\sum_{\mu\leq\nu} \frac{\partial\mathcal{L}}{\partial g_{\mu\nu,\lambda}} g_{\mu\nu,\lambda} - \mathcal{L}\Big) dx^0\wedge \dots \wedge dx^3 \tag{2.10}$$

we obtain :

$$\Psi = \Big(g^{\omega\mu} g^{\rho\nu}\Gamma^{\lambda}_{\mu\nu} - \tfrac{1}{2}\big(g^{\omega\mu} g^{\rho\lambda}\Gamma^{\tau}_{\tau\mu} + g^{\omega\lambda} g^{\rho\mu}\Gamma^{\tau}_{\tau\mu}\big) - \tfrac{1}{2} g^{\omega\rho} g^{\mu\nu}\Gamma^{\lambda}_{\mu\nu} + \tfrac{1}{2} g^{\omega\rho} g^{\lambda\mu}\Gamma^{\tau}_{\tau\mu}\Big) \sqrt{-g}\, dx^0\wedge \dots \wedge \underset{\lambda}{\underbrace{dg_{\omega\rho}}}\wedge \dots \wedge dx^3 +$$

$$- \left(g^{\nu\varsigma}(\Gamma^{\mu}_{\tau\nu}\Gamma^{\tau}_{\mu\varsigma} - \Gamma^{\mu}_{\mu\tau}\Gamma^{\tau}_{\nu\varsigma}) + 2\lambda \right)\sqrt{-g}\, dx^0\wedge \dots \wedge dx^3 \tag{2.11}$$

It turns out, that the formal exterior differentiation of the non-covariant expression (2.11) gives the 5-form $\overset{5}{\omega}$ i.e. $d\Psi = \overset{5}{\omega}$ (2.12)

The formula (2.12) together with proposition 2 give a connection between the classical formulation of General Relativity and the multisymplectic approach presented above. It is interesting that the equation (2.4) can be obtained from the Palatini - variational principle (cf. [29]) for the action integral

$$S(f) = \int_{f(M)} \overset{4}{\theta} = \int_M f^{*}\overset{4}{\theta} \tag{2.13}$$

where $f: M \to \mathcal{P}$ is a section of τ.
(Note that $f^{*}\overset{4}{\theta} = (R - 2\lambda)\sqrt{-g}\; dx^{0} \wedge \ldots \wedge dx^{3}$). The detailed discussion of this fact will be given elsewhere.

Having the multisymplectic manifold ($\mathcal{P}$, $\overset{5}{\omega}$) we need an additional structure - a set $\mathcal{C}$ of 3-dimensional submanifolds in $\mathcal{P}$ which determine initial data for the field equations (2.6) and (2.7). We call elements of admissible initial surfaces in $\mathcal{P}$ (cf. [22]). They are lifts of space-like surfaces in M (for a given metric $g_{\mu\nu}$ satisfying the Einstein equations) to the bundle $\mathcal{P}$. The description of the set $\mathcal{C}$ in a special coordinate system is given in the section 3.

The set $\mathcal{N}$ of 4-dimensional submanifolds of $\mathcal{P}$ which satisfy (2.4) is called the space of states in gravidynamics or the pre-phase space.

3. The Cauchy problem and ADMW coordinates in General Relativity

The Einstein equations form a system of second order differential equations for components of the metric tensor $g_{\mu\nu}$ on the manifold M . These equations have the hyperbolic character and we can consider the Cauchy problem for them. The discussion of the problem will be given in the ADMW coordinate system (cf. [3], [29], [33]). Let $g_{\mu\nu}$ be a metric tensor on M with a signature (-1, +3) and σ be a 3-dimensional space-like submanifold of M. We choose such a coordinate system in M that $\sigma = \{ x \in M : x^{0} = \text{const} \}$. The metric $g_{\mu\nu}$ on M generates on σ a positively defined metric tensor g_{ij} . We denote by $\bar{g}^{ij}$ the inverse matrix to g_{ij}. The inequality $g = \det g_{\mu\nu} < 0$ implies $g^{00} < 0$. Therefore we put :

$$N = (-g^{00})^{-\frac{1}{2}} \; ; \quad N_{k} = g_{0k} \; ; \quad N^{k} = \bar{g}^{kj} N_{j} \; ;$$
$$\bar{g} = \det g_{ij} > 0 \tag{3.1}$$

It is known that N is a scalar function on σ (a lapse function) and N^{k} are components of a vector field on σ (a shift vector). We have also formulas:
$$g^{0k} = N^{k}/N^{2} \; ; \quad g_{00} = -N^{2} + N^{k}N_{k} \; ; \quad g^{sp} = \bar{g}^{sp} - N^{s}N^{p}/N^{2} ;$$

$$\sqrt{-g} = N\sqrt{\bar{g}} \tag{3.2}$$

The normal unit vector to σ is given by $n^{\mu} = \left(\frac{1}{N}, -\frac{N^k}{N}\right)$ (3.3)

and the second fundamental form is defined by (cf [24]) : $K_{ij} = -g_{j\mu}\nabla_i n^{\mu}$ (3.4)

In our special coordinate system $K_{ij} = -N\Gamma^{o}_{ij}$ (3.5)

The metric tensor g_{ij} on σ defines the Christoffel symbols $\bar{\Gamma}^{k}_{ij}$ and the covariant derivative $\bar{\nabla}_k$ on σ . We have

$$\partial_o g_{ij} = \bar{\nabla}_i N_j + \bar{\nabla}_j N_i + 2N^2\Gamma^{o}_{ij} \tag{3.6}$$

In our further considerations an important role plays the ADMW tensor density on σ defined by :

$$\pi^{ij} = -\sqrt{\bar{g}}\,(K_{pq} - g_{pq}K_{rs}\bar{g}^{rs})\bar{g}^{ip}\bar{g}^{jq} \tag{3.7}$$

It is known (cf. [2],[3]) that the system of the Einstein equations (2.7) is equivalent to the system:

$$R_{ks} = \lambda g_{ks} \quad \text{on } M \tag{3.8a}$$

$$G^{o}_{\mu} = 0 \quad \text{on } \sigma \tag{3.8b}$$

We consider a neighbourhood of σ in M which is diffeomorphic to $\sigma \times \mathbb{R}$ and take the system of coordinates $(N, N_k, M_\mu, M_{\mu k}, g_{ij}, \pi^{ij}, \bar{\Gamma}^{k}_{ij})$ (where $M_\mu = \partial_\mu N$, $M_{\mu k} = \partial_\mu N_k$) in the corresponding subset of $\mathcal{P}$. The equations (3.8) read:

$$\begin{aligned}\partial_o\pi^{ij} = & -N\sqrt{\bar{g}}\,(\bar{R}^{ij} - \bar{g}^{ij}\bar{R}) - \frac{2N}{\sqrt{\bar{g}}}(\pi^{i}_{q}\pi^{qj} - \tfrac{1}{2}\,\mathrm{tr}\pi\,\pi^{ij}) + \\ & + \sqrt{\bar{g}}\,(\bar{\nabla}^i\bar{\nabla}^j N - \bar{g}^{ij}\bar{\nabla}^s\bar{\nabla}_s N) + \bar{\nabla}_u(N^u\pi^{ij}) + \\ & - \bar{\nabla}_s N^i\pi^{sj} - \bar{\nabla}_s N^j\pi^{si} - 2\lambda N\sqrt{\bar{g}}\,\bar{g}^{ij}\end{aligned} \tag{3.9a}$$

$$\bar{\nabla}_i\pi^{ij} = 0 \quad \text{on } \sigma \tag{3.9b'}$$

$$\bar{R} - 2\lambda - \frac{1}{\bar{g}}\left(\pi_{pq}\pi^{pq} - \tfrac{1}{2}(\mathrm{tr}\pi)^2\right) = 0 \quad \text{on } \sigma \tag{3.9b''}$$

The equations (3.9b) are simply the constraints imposed on initial data g_{ij}, π^{ij} on σ . These equations do not involve N and N_k therefore we can assign N and N_k in an arbitrary way on σ . The dynamical equations (3.9a) do not contain the time-derivatives of N and N_k. Therefore having a positive defined metric g_{ij} on σ and a tensor density π^{ij} on σ which satisfy the constraints equations (3.9b) we must choose

N and N_k in ax neighbourhood of σ in M. In this case we have a unique solution for g_{ij} and π^{ij} solving the equations (3.6) and (3.9a). Technical details concerning this problem can be found in [6,7,15,16,21,28].

We see that the surface σ in M together with quantities g_{ij}, π^{ij} N, N_k on it determine an admissible initial surface in $\mathcal{P}$. Such a surface generates a family of states i.e. a family of solutions of the Einstein equations. In the following we shall consider two cases:

a) σ - is a compact manifold without boundary

b) M is asymptotically flat and in spatially distant points σ is a surface x^0 = const in the cartesian coordinate system $(x^0; x^1, x^2, x^3)$. (It implies that $N_k = 0$, $N = 1$ in spatially remote points).

4. A symplectic structure of the set of Einstein metrics

We consider the space $\mathcal{H}$ of all 4-dimensional submanifolds of $\mathcal{P}$ satisfying (2.4). This set is simply the set of all Einstein metrics in M (for a given λ). According to the general approach to infinite dimensional manifolds (cf. [11],[23]) the tangent space $T_C(\mathcal{H})$ to $\mathcal{H}$ at $C \in \mathcal{H}$ can be defined by means of curves in $\mathcal{H}$. Every such a curve generates a τ-vertical vector field X on C which satisfies the linearized system of Einstein equations. If X is given by

$$X = \sum_{\mu \le \nu} \delta g_{\mu\nu} \frac{\partial}{\partial g_{\mu\nu}} + \sum_{\mu \le \nu} \delta \Gamma^{\lambda}_{\mu\nu} \frac{\partial}{\partial \Gamma^{\lambda}_{\mu\nu}} \tag{4.1}$$

then the components $\delta g_{\mu\nu}$ and $\delta \Gamma^{\lambda}_{\mu\nu}$ satisfy:

$$\delta \Gamma^{\lambda}_{\mu\nu} = \tfrac{1}{2} g^{\lambda\sigma} (\nabla_\mu \delta g_{\nu\sigma} + \nabla_\nu \delta g_{\mu\sigma} - \nabla_\sigma \delta g_{\mu\nu}) \tag{4.2a}$$

$$\delta (R_{\mu\nu} - \lambda g_{\mu\nu}) = \sum_{\alpha \le \rho} \frac{\partial R_{\mu\nu}}{\partial g_{\alpha\rho}} \delta g_{\alpha\rho} + \sum_{\alpha \le \rho} \frac{\partial R_{\mu\nu}}{\partial \Gamma^{\tau}_{\alpha\rho}} \delta \Gamma^{\tau}_{\alpha\rho} - \lambda \delta g_{\mu\nu} = 0 \tag{4.2b}$$

The set of vector fields on C of the type (4.1) satisfying (4.2) is denoted by $\tilde{T}_C(\mathcal{H})$. Of course, $T_C(\mathcal{H}) \subset \tilde{T}_C(\mathcal{H})$ and the problem of equality of these spaces is the problem of the linearization stability of the Einstein equations (cf. [17]). We do not consider it here.

Let $C \in \mathcal{H}$, $\sigma \subset M$ be an arbitrary space-like surface (with respect to the metric $g_{\mu\nu}$ determined by C) and $c \subset C$ be the admissible initial surface determined by σ and C . Let $\hat{X}_1, \hat{X}_2 \in \tilde{T}(\mathcal{H})$ and X_1, X_2 are vector fields of the type (4.1) representing $\hat{X}_1, \hat{X}_2$. We define a skew-symmetric mapping :

$$\tilde{T}_C(\mathcal{H}) \times \tilde{T}_C(\mathcal{H}) \ni (\hat{X}_1,\hat{X}_2) \to \Omega(\hat{X}_1,\hat{X}_2) = \int_c (X_1 \wedge X_2) \lrcorner \tilde{\omega} \tag{4.3}$$

Of course, to provide the convergence of the integral in (4.3) we have to assume that appropriate boundary conditions in the spatial infinity are fulfilled (e.g. one of X_1, X_2 has a compact support on c).

The above construction for an arbitrary hyperbolic field theory has been discussed in [23] to which we refer for details.

Proposition 3

The integral (4.3) does not depend on a choice of a space-like surface $\sigma \subset M$ (of a choice of an admissible initial surface $c \subset C$).

It has been shown in the paper [23] that the space $\mathcal{H}$ can be endowed with a pseudo-differential structure and the formula (4.3) defines a differential 2-form on $\mathcal{H}$.

Proposition 4

The form Ω is closed i.e. $d\Omega = 0$.

It can be also proved that Ω is exact. (This result will be published elsewhere)

In the ADMW coordinate system a vector $\hat{X}$ at C is represented by a vector field X of the type :

$$X = \delta N \frac{\partial}{\partial N} + \delta N_k \frac{\partial}{\partial N_k} + \sum_{i\le j} \delta g_{ij} \frac{\partial}{\partial g_{ij}} + \sum_{i\le j} \delta \pi^{ij} \frac{\partial}{\partial \pi^{ij}} + \delta M_\mu \frac{\partial}{\partial M_\mu} +$$
$$+ \delta M_{\mu k} \frac{\partial}{\partial M_{\mu k}} + \sum_{k\le s} \delta \bar{\Gamma}^i_{ks} \frac{\partial}{\partial \bar{\Gamma}^i_{ks}} \tag{4.4}$$

where the following equations hold:

$$\delta M_\mu = \partial_\mu \delta N \quad ; \quad \delta M_{\mu k} = \partial_\mu \delta N_k \tag{4.5}$$

$$\delta \bar{\Gamma}^j_{ks} = \tfrac{1}{2} \bar{g}^{ja}(\bar{\nabla}_k \delta g_{sa} + \bar{\nabla}_s \delta g_{ka} - \bar{\nabla}_a \delta g_{ks}) \tag{4.6}$$

$$\partial_0 \delta g_{ij} = \delta\Big(\bar{\nabla}_i N_j + \bar{\nabla}_j N_i + \frac{2N}{\sqrt{g}}(g_{ip}g_{jq}\pi^{pq} - \tfrac{1}{2}g_{ij}\mathrm{tr}\,\pi) \Big) \tag{4.7}$$

$$\partial_0 \delta \pi^{ij} = \delta\Big(- N\sqrt{g}\,(\bar{R}^{ij} - \bar{g}^{ij}\bar{R}) - \frac{2N}{\sqrt{g}}(\pi^i_{\ q}\pi^{qj} - \tfrac{1}{2}\mathrm{tr}\,\pi\, \pi^{ij}) \Big) +$$

$$+ \quad \delta\big(\sqrt{\bar g}\,(\bar\nabla^i \bar\nabla^j N - \bar g^{ij}\bar\nabla^s \bar\nabla_s N) + \bar\nabla_u (N^u \pi^{ij}) \big) \quad +$$

$$+ \quad \delta\big(-\bar\nabla_s N^i \pi^{sj} - \bar\nabla_s N^j \pi^{si} - 2\lambda N \sqrt{\bar g}\,\bar g^{ij} \big) \qquad (4.8a)$$

$$\delta(\bar\nabla_j \pi^{ij}) = \bar\nabla_j \delta\pi^{ij} + \delta \bar\Gamma^i_{kj} \pi^{kj} = 0 \ \text{ on } \sigma \qquad (4.8b')$$

$$\delta(\bar R - 2\lambda - \frac{1}{\bar g}(\pi_{\rho\nu}\pi^{\rho\nu} - \tfrac{1}{2}(\mathrm{tr}\,\pi)^2)) = 0 \ \text{ on } \sigma \qquad (4.8b'')$$

Therefore a vector $\hat X \in \tilde T_C(\mathcal{M})$ determines 12 quantities $(\delta\pi^{ij}, \delta g_{ij})$ on $c \subset C$ (on $\sigma \subset M$) which satisfy the constraints equations (4.8b) and four arbitrary quantities δN, δN_k given in a neighbourhood of c in C (in a neighbourhood of σ in M). On the other hand, if we have 12 quantities $(\delta\pi^{ij}, \delta g_{ij})$ on c satisfying the constraints equations (4.8b) and four quantities δN, δN_k in a neighbourhood of c in C we can determine the vector field X on C solving the equations (4.7) and (4.8a).

It turns out that in the ADMW coordinates the form Ω is diagonal:

Theorem 1

Let $\hat X_1, \hat X_2 \in \tilde T_C(\mathcal{M})$ be represented by vector fields X_1, X_2 of the type (4.4) then

$$\Omega(\hat X_1, \hat X_2) = \int_C (\underset{1}{\delta}\pi^{ij} \underset{2}{\delta} g_{ij} - \underset{2}{\delta}\pi^{ij} \underset{1}{\delta} g_{ij}) \, dx^1 \wedge dx^2 \wedge dx^3 \qquad (4.9)$$

This theorem shows that the components of the ADMW density π^{ij} can be treated as the conjugate varibles to the spatial components of the metric tensor $g_{\mu\nu}$. But these quantities are not independent, they satisfy the constraints equations (3.9b). These four equations cause the degeneration of the form Ω. We investigate this problem in the next section.

5. The gauge distribution, an action of the diffeomorphism group and the superphase space for General Relativity

We know that in mechanics the symplectic form (1.1) provides an isomorphism between the tangent and the cotangent bundle of the phase space $\mathcal{P}_{2n}$. The form Ω in General Relativity is degenerate and therefore the correspondance between tangent and cotangent vectors in $\mathcal{H}$ is not one to one.

We define the gauge distribution of the form

$$W_C = \{ \hat{Y} \in \tilde{T}_C(\mathcal{H}) : \Omega(\hat{X},\hat{Y}) = 0 , \ \hat{X} \in \tilde{T}_C(\mathcal{H}) \} \ ; \quad C \in \mathcal{H} \qquad (5.1)$$

and two subspaces of $\tilde{T}_C(\mathcal{H})$:

Definition: $\overset{\circ}{\tilde{T}}_C(\mathcal{H},c)$ is a linear subspace of $\tilde{T}_C(\mathcal{H})$ consisting of these $\hat{Y} \in \tilde{T}_C(\mathcal{H})$ which are represented by vector fields Y on C of the type (4.4) such that:

1^0 δN , δN_k , δM_o , δM_{ok} are arbitrary on c

2^0 $\delta M_k = \partial_k \delta N$, $\delta M_{sk} = \partial_s \delta N_k$ on c

3^0 $\delta \pi^{ij} = 0$, $\delta g_{ij} = 0$ on c (5.2)

Definition: $\hat{\tilde{T}}_C(\mathcal{H},c)$ is the linear subspace of $\tilde{T}_C(\mathcal{H})$ consisting of vectors which are represented by vector fields on C of the type (4.4) such that : $\delta N = 0$, $\delta N_k = 0$ on C (5.3)

Proposition 5

$$\tilde{T}_C(\mathcal{H}) = \overset{\circ}{\tilde{T}}_C(\mathcal{H},c) \oplus \hat{\tilde{T}}_C(\mathcal{H},c) \qquad \text{(a direct sum)}$$

Proposition 6

$$\overset{\circ}{\tilde{T}}(\mathcal{H},c) \subset W_C \ .$$

By vitrue of propositions 5 and 6 we can consider only the subspace $\hat{\tilde{T}}(\mathcal{H},c)$. Let $C_\sigma = C^\infty(\mathrm{den}S^2T(\sigma) \oplus S^2T^*(\sigma))$ be a subspace consisting of couples $\mathfrak{X} = (\delta\pi^{ij}, \delta g_{ij})$, where δg_{ij} is a 2-covariant symmetric tensor field on σ and $\delta\pi^{ij}$ is a 2-contravariant symmetric tensor density on σ . This space has a natural scalar product:

$$g(\sigma,C)(\mathfrak{X}_1,\mathfrak{X}_2) = \int_\sigma \Big(\frac{1}{\sqrt{g}} \underset{1}{\delta}\pi^{ij} g_{ip} g_{jq} \underset{2}{\delta}\pi^{pq} + \sqrt{g}\ \underset{1}{\delta} g_{ij} g^{ip} g^{jq} \underset{2}{\delta} g_{pq} \Big) dx^1 \wedge dx^2 \wedge dx^3 \qquad (5.4)$$

Let J be the operator $C_\sigma \to C_\sigma$; $J(\delta\pi^{ij}, \delta g_{ij}) =$

$$= \Big(\sqrt{g}\ \delta g_{pq}\, g^{pi} g^{qj} , -\frac{1}{\sqrt{g}} g_{ip} g_{jq} \delta\pi^{pq} \Big) \qquad (5.5)$$

The operator J has a property $J^2 = -\mathrm{id}$.

Let $g_1(\sigma,C)$ be the scalar product in the space $C^\infty(T(\sigma) \oplus \mathbb{R})$ consisting of couples $U = (u^j, \chi)$, where u^j are components of a vector field on σ and χ is a scalar function on σ.

$$g_1(\sigma, C)(U_1, U_2) = \int_\sigma (\underset{1}{u}_j \underset{2}{u}^j \sqrt{\bar{g}} + \underset{1}{\chi}\underset{2}{\chi} \sqrt{\bar{g}}\,) dx^1 \wedge dx^2 \wedge dx^3 \qquad (5.6)$$

The constraints equations (4.8b) generate a differential operator A : $C^\infty(\mathrm{dens} S^2T(\sigma) \oplus S^2T^*(\sigma)) \to C^\infty(T(\sigma) \oplus \mathbb{R})$. For $\mathfrak{X} = (\delta\pi^{ij}, \delta g_{ij})$ we have:

$$A\mathfrak{X} = \left(\frac{1}{\sqrt{\bar{g}}} (\bar{\nabla}_j \delta\pi^{ij} + \delta\bar{\Gamma}^i_{ks} \pi^{ks}), \delta\bar{R} + \frac{1}{\bar{g}} (\bar{R} - 2\lambda) \delta\bar{g} - \frac{1}{\bar{g}} \delta(\pi^{ks}\pi_{ks} - \tfrac{1}{2}(\mathrm{tr}\,\pi)^2) \right) \qquad (5.7)$$

By means of the scalar products (5.4) and (5.6) we define the adjoint operator

$A^* : C^\infty(T(\sigma) \oplus \mathbb{R}) \longrightarrow C^\infty(\mathrm{dens} S^2T(\sigma) \oplus S^2T^*(\sigma))$

$A^*(u^j, \chi) = (\delta\pi^{ij}, \delta g_{ij})$, where

$$\delta\pi^{ij} = -\tfrac{1}{2}(\bar{\nabla}^i u^j + \bar{\nabla}^j u^i) \sqrt{\bar{g}} - 2(\pi^{ij} - \tfrac{1}{2}\mathrm{tr}\,\pi\, \bar{g}^{ij}) \chi$$

$$\begin{aligned} \delta g_{ij} = & - \frac{1}{2\sqrt{\bar{g}}} (\pi_{ia} \bar{\nabla}^a u_j + \pi_{ja} \bar{\nabla}^a u_i - \bar{\nabla}_a (\pi_{ij} u^a)) + \\ & - \frac{2}{\bar{g}} (\pi_{is} \bar{g}^{sk} \pi_{kj} - \tfrac{1}{2}\mathrm{tr}\,\pi\, \pi_{ij}) \chi + \bar{\nabla}_i \bar{\nabla}_j \chi - g_{ij} \bar{\nabla}^k \bar{\nabla}_k \chi - \bar{R}_{ij} \chi + \\ & + g_{ij} (\bar{R} - 2\lambda) \chi \end{aligned} \qquad (5.8)$$

<u>Proposition 7</u>

im $JA^* \subset$ ker A .

The above proposition enables to construct vectors belonging to $\tilde{\overset{1}{T}}_C(\mathcal{H}, c)$. Every element $U = (u^j, \chi)$ generates such a vector.

<u>Definition:</u> $\overset{\circ}{W}_C(c)$ is a linear subspace of $\tilde{\overset{1}{T}}_C(\mathcal{H}, c)$ consisting of vectors which are represented by vector fields on C generated by im JA^* .

<u>Proposition 8</u>

$\overset{\circ}{W}_C(c) \subset W_C$

<u>Definition:</u> $\overset{1}{W}_C(c) = \overset{\circ}{W}_C(c) \oplus \tilde{\overset{\circ}{T}}_C(\mathcal{H}, c)$ (a direct sum) .

<u>Theorem 2</u>

If σ is a compact manifold without boundary then

1° ker A = (ker A $\cap$ ker AJ) $\oplus$ imJA^* (an othogonal sum)

2° $W_C = \overset{1}{W}_C(c)$

3° $\tilde{T}_C(\mathcal{H}) = F_C(c) \oplus W_C$ (a direct sum) , where $F_C(c)$ is a subspace of $\tilde{\hat{T}}_C(\mathcal{H}, c)$ generated by ker A $\cap$ ker AJ .

The proof of this theorem is based on the theory of differential operators with injective symbols (the operator A^* has an injective symbol) (cf. [4], [30]). Probably this theorem can be also proved in a non-compact case but it is necessary to impose appropriate boundary conditions at the spatial infinity.

We shall explain now the relation between the gauge distribution W and an action of the diffeomorphism group of M in the space $\mathcal{H}$. The group Diff M acts on the right in the set of all metrics on M having the signature (-1,+3)

$$\mathrm{Diff}\, M \times C^{\infty}(S^2T^*M) \ni (\varphi, \underline{g}) \longrightarrow R_{\varphi}\underline{g} = \varphi^* \underline{g} \in C^{\infty}(S^2T^*M) \tag{5.9}$$

The action (5.9) generates a left action in the bundle $\mathcal{P}$ and the forms $\overset{\mu}{\theta}$ and $\overset{5}{\omega}$ are invariant with respect to it. Therefore we have a right action in the space $\mathcal{H}$:

$$\mathrm{Diff}\, M \times \mathcal{H} \ni (\varphi, C) \rightarrow \hat{R}_{\varphi}(C) = \varphi^* C \in \mathcal{H} \tag{5.10}$$

The Lie algebra of Diff M can be identified with the Lie algebra of smooth vector fields on M (cf. [11]) with the commutator as the Lie bracket.
The action (5.10) generates an action of the Lie algebra:

$$C^{\infty}(TM) \ni v \longrightarrow d\hat{R}_{id}(C)v \in \tilde{T}_C(\mathcal{H}) \tag{5.11}$$

where $d\hat{R}_{id}(C)$ is the derivative of (5.10) with respect to the first variable at the point $(id, C) \in \mathrm{Diff}\, M \times \mathcal{H}$.

<u>Proposition 9</u>

$$\mathrm{im}\, d\hat{R}_{id}(C) = \hat{W}_C(c)$$

<u>Theorem 3</u>

If the manifold M has compact spatial sections i.e. if σ is compact then $\mathrm{im}\, d\hat{R}_{id}(C) = W_C$.

We see that the action (5.10) determines the gauge distribution of Ω.

It has been proved in [23] that the gauge distribution W is involutive, that means the commutator of two vector fields with values in W is also in W. This fact is simply the Frobenius integrability condition and therefore we can look for a space $\tilde{\mathcal{H}}$ for which $T(\tilde{\mathcal{H}}) = T(\mathcal{H})/W$. The theorem 3 suggests to divide the space $\mathcal{H}$ by the action of Diff M and take $\tilde{\mathcal{H}} = \mathcal{H}/\mathrm{Diff}\, M$ (5.12

The other possibility is to define $\tilde{\mathcal{H}}$ axiomat-ically in the following way: let $(\delta\pi^{ij}, \delta g_{ij}) \in C_{\sigma}$ (for a given $\sigma \subset M$) satisfy the constraint

equations (4.8b) and four equations obtained from (4.8b) by the transformation $\delta\pi^{ij} \to \sqrt{g}\; g^{ip}g^{jq}\;\delta g_{pq}$, $\delta g_{ij} \to -\frac{1}{\sqrt{g}}\, g_{ip}g_{jq}\,\delta\pi^{pq}$ (5.13)
In this case the system ($\delta\pi^{ij}$, δg_{ij}) determines a vector belonging to $F_C(c)$. Eight equations for ($\delta\pi^{ij}$, δg_{ij}) reduce the number of independent variables to four but these four variables are given in an implicit way. To carry these consideration on the space $\mathcal{H}$ we take on account 12 quantities (π^{ij}, g_{ij}) satisfying (3.9b) on σ . Solving the dynamical equations (3.6) and (3.9a) (for $N = 1$, $N_k = 0$) we obtain an Einstein metric on M. By means this metric we define an action of Diff M in the systems (π^{ij} ,g_{ij}). If we denote by Cd (Cauchy data) the set of couples (π^{ij},g_{ij}) which satisfy (3.9b) we have

$$\tilde{\mathcal{H}} = \mathrm{Cd/Diff\ M} \tag{5.14}$$

Remarks:

1° In our considerations we have put $N = 1$, $N_k = 0$. If we take another lapse function and another shift vector on σ we obtain another action of Diff M in Cd. It can be proved that the corresponding actions of the Lie algebra in the tangent space are isomorphic but it is difficult to prove it for the space Cd.

2° It can be proved that the tangent spaces of two possible choices of $\tilde{\mathcal{H}}$ ((5.12) and (5.14)) are isomorphic. The qu stion arises whether these spaces are isomorphic?

An object of the type (5.14) has been recently proposed by A.Fischer and J.Marsden [15], [18] as a possible choice of a superphase space for the Einstein dynamics. These authors have pointed out that the superphase space is a very complicated object (cf. also [14]) and maybe therefore the problem of the quantization of gravitation is so difficult.

R E F E R E N C E S

1. Abraham,R. Foundations of mechanics,New York: Benjamin 1967.

2. Adler,R, Bazin,M, Schiffer,M. Introduction to General Relativity. New York: Mc Graw Hill 1965.

3. Arnowitt,R, Deser,S, Misner,C.W . The dynamics of General Relativity. In: Witten,L (ed), Gravitation-an introduction to current research, New York: John Wiley 1962.

4. Berger,M, Ebin,D, Some decompositions of the space of symmetric tensors on a Riemannian manifold , Journ. of Differential Geometry 3 (1969) 379-392.

5. Bergmann,P.G, Komar,A.B , Status report on the quantization of the gravitational field. In: Recent Developments in General Relativity. London-Warsaw. Pergamon Press- PWN: London - Warsaw 1962.

6. Choquet-Bruhat,Y, The Cauchy problem. In: Witten,L(ed) Gravitation - an introduction to current research. New York: J.Wiley 1962.

7. Choquet-Bruhat,Y, Geroch,R, Global aspects of the Cauchy problem in G.R. Comm. Math. Phys. 14 (1969) 329-335.

8. Dedecker,P, Calcul des variations, formes differentieles et champs geodesiques. In: Coll. Intern. Geometrie Differ. Strasbourg 1953.

9. De Witt,B, Quantum theory of gravity I. The canonical theory. Phys. Rev.160 (1967) 1113-1148.

10. Dirac,P.A.M. Generalized Hamiltonian dynamics, Proc.Roy.Soc.(London) A246 (1958) 326-332; The theory of gravitation in Hamiltonian form, Proc. Roy.Soc. A246 333-346.

11. Ebin,D, Marsden,J, Group of diffeomorphisms and the motion of an incompressible fluid. Ann of Mathematics 92 (1970) 102-163.

12. Fadeev,L, Symplectic structure and quantization of the Einstein gravitation theory. In: Actes du Congres Intern. d. Math. Nice 1970 35-39.(vol.3)

13. Fadeev,L, Popov,V, A covariant quantization of the gravitational field, Uspechi fiz. Nauk 111 (1973) 427-450 (in Russian).

14. Fischer,A, The theory of superspaces. In: Carmell,M, Fickler,S, Witten,L (ed) Relativity . New York: Plenum Press 1970.

15. Fischer,A, Marsden,J, The Einstein equations of evolution. A geometric approach, Journ. Mat.Phys. 13 (1972) 546-568.

16. Fischer,A, Marsden,J, The Einstein evolution equations as a quasi-linear first order symmetric hyperbolic system I. Comm. Math.Phys. 28 (1972) 1-38.

17. Fischer,A,Marsden,J, Linearization stability of the Einstein equations, Bull. Am.Math.Soc.79 (1973) 997-1003.

18. Fischer,A, Marsden,J, General Relativity as a Hamiltonian system. Symposia Mathematica XIV (published by Istituto di Alta Matematica Roma) London-New York Academic press 1974, 193-205.

19. Gawędzki,K, On the geometrization of the canonical formalism in the classical field theory , Reports on Math. Phys. 3 (1972) 307-326.

20. Goldschmidt,H, Sternberg,S The Hamilton-Cartan formalism in the calculus of variations, Ann.Inst.Fourier 23 (1973) 203-267.

21. Hawking,S.W. Ellis, G,F.R, The large scale structure of space-time, Cambridge University Press 1973.

22. Kijowski,J, A finite dimensional canonical formalism in the classical field theory, Comm. Math.Phys. 30 (1973) 99-128.

23. Kijowski,J, Szczyrba,W,A canonical structure of classical field theories (to appear in Comm. Math.Phys.).

24. Kobayashi,S, Nomozu,K, Foundations of differential geometry, New York:Interscience Publ. vol.1 1963,vol.2 1969.

25. Kostant;B Quantization and unitary representations, in: Lecture Notes in Mathematics 170, Berlin:Springer-Verlag 1970.

———, Symplectic spinors, in Symposia Mathematica vol.XIV London -New York: Academmic Press 1974.

26. Kuchar,K A buble-time canonical formalism for geometrodynamics, Journ. Math. Phys. 13 (1972) 768-781.

27. Kundt,W, Canonical quantization of gauge invariant field theories, Springer tracts in modern physics 40. Berlin - Heidelberg -New York: Springer-Verlag 1966.

28. Lichnerowicz,A Relativistic hydrodynamics and magmetohydrodynamics, New York:Benjamin 1967.

29. Misner,Ch.W, Thorne,K.S, Wheeler,J.A, Gravitation, San fransicso: W.H.Freeman and Co 1973.

30. Narasimhan,R Analysis on real and complex manifolds, Paris:Masson ans Cie 1968.

31. Souriau,J.M, Structure des systemes dynamiques, Paris:Dunod 1969.

32. Szczyrba, Lagrangian formalism in the classical field theory, Ann, Pol.Math. 32 (1975) (to appear).

33. Wheeler,J.A , Geometrodynamics and the issue of the final state. In: DeWitt B, DeWitt C (ed), Relativity, Groups and Topology, New York:Gordon and Breach 1964.

ON THE SYMPLECTIC FORMULATION OF THE EINSTEIN SYSTEM OF EVOLUTION IN PRESENCE OF A SELF-GRAVITATING SCALAR FIELD

Mauro Francaviglia

Istituto di Fisica Matematica, Università di Torino

SUMMARY

In uno spazio-tempo $M \times \mathbb{R}$, con M varietà chiusa tridimensionale, descritto da coordinate Gaussiane si studia la formulazione simplettica della relatività generale in presenza di un campo scalare autogravitante. Si deducono analiticamente le equazioni di Lagrange e di Hamilton, mostrandone la piena equivalenza, e sottolineandone il significato di equazioni delle geodetiche rispetto a una opportuna metrica indefinita in una varietà a infinite dimensioni. Si deducono le leggi di conservazione del sistema Lagrangiano adottato, mostrando come esse assegnino un significato geometrico ai classici dati iniziali di Cauchy e ritrovando incidentalmente risultati di Lichnerowicz e Bruhat; tali leggi sono interpretate fisicamente come conservazione della densità di energia e del flusso di densità di energia, ovvero delle relative componenti del tensore energetico $T_{\mu\nu}$. Si mostra infine come non sia restrittivo supporre nulla la dansità di energia $\mathcal{H}$ provando esplicitamente l'equivalenza (sotto i dati iniziali su di una ipersuperficie spaziale) tra i due sistemi di evoluzione descriventi il problema di Einstein nel caso completo e nel caso ridotto (dedotto dal precedente annullando $\mathcal{H}$).

1. Introduction :

This paper was suggested by the article [14] of Fischer-Marsden (here FM), where they deduced the symplectic formulation of the Einstein system in an empty space by means of geometrical methods; and by the recent paper of Christodoulou [8] , in which he studied the canonical quantization of general relativity, under external fields of spin 0 and 1. Here, making use of the same techniques of FM and of many results of theirs, we study the more general problem in which we consider an additional self-gravitating scalar field; it produces additional terms for FM's equations referring to the empty-space case.

The so-called "Einstein problem" is the study of the partial differential system $G_{\mu\nu}=\chi T_{\mu\nu}$, where $G_{\mu\nu}=R_{\mu\nu}-\frac{1}{2}Rg_{\mu\nu}$ is the classical Einstein tensor and T is the stress-energy tensor. From extensive works of Lichnerowicz and Bruhat (see e.g. [5]) it is well known that, given an initial space-like hypersurface Σ with well-defined initial Cauchy data on it, the equations are solved (at least for a finite time) and the "constraints" imposed on Σ are conserved in time. Making use of a canonical theory developed by Arnowitt-Deser-Misner (ADM, [2]); Wheeler, De Witt and others ([21], [9]) traduced this general situation in Hamiltonian formalism with methods different from ours; in particular, they showed that in the empty-space case a tensor π , directly related to the 2nd fundamen= tal form of the embedded Σ , plays a central role.

In the present work, the scalar field is described by the tensor:

$$T_{\mu\nu} = -\beta\left[2\varphi_{,\mu}\varphi_{,\nu} - g_{\mu\nu}(\varphi_{,\rho}\varphi^{,\rho} + m^2\varphi^2)\right]$$

β , m^2 being two positive constants respectively related to the mass of the field and to the choice of units; if we let $\beta = 0$ we of course re-obtain all the results given in [14] . We restrict our study to space-times of the form $M\times\mathbb{R}$, where M is a 3-dimensional closed orientable Riemannian manifold, and to Lorentz metrics expressed in Gaussian form. These simplifications allow us to transform the problem into the problem of finding the possible evolutions of a pair (g_t,φ_t), where g_t is a time-dependent C^∞-metric and φ_t a time-dependent C^∞-map on M, t being the fourth coordinate x^0 of space-time itself. We give a de= tailed discussion of the configuration space, making both the Lagrangian and the Hamiltonian formulation, emphasizing their full equivalence.

We define a so-called "Christodoulou-De Witt's kinetic Lagrangian" (since it reduces itself to that of De Witt in the empty space) on the configuration space $\mathcal{B} = \mathcal{M}\times\mathcal{F}$ (the space of pairs of metrics and maps on M) and we show that it is weakly non-degenerate in the sense of Marsden. In such a situation the Lagrangian canonically induces a symplectic structure on the tangent bundle. Thus we can deduce the explicit form of the Lagrange's equations by rather long but interesting calculations, obtaining solutions for a generally finite time. The reason for this restriction to local-time solutions is explained and related to the fact that they are in general extendible only in a larger manifold, as De Witt observed in a non rigorous manner ([9], 1967). We notice also that the La= grangian field exists, and it is in fact the geodesic spray inducing the geode= sic equations found by Christodoulou in [8] by means of completely different methods.

Then we give the moments canonically conjugated to the pair (g,φ) , prefer= ring for them an equivalent form in terms of tensor densities of weight $\frac{1}{2}$, rela= ted to the π of ADM and to the gradient of the field. The final form of the evo= lution system is obtained introducing a general potential V_N (where N is the classical "lapse function") and evaluating its gradient with respect to the Christodoulou-De Witt's metric. We can thus find the total Hamiltonian for our problem and translate the system into its dual Hamiltonian formulation, by also incidentally re-obtaining some of the results given in [7] by means of different formalism and notations.

This part is preliminary to our central results, which extend to the present case the results of FM. Starting from the Einstein system in its two equivalent formulations, we show the validity of two interesting conservation laws intimately related to the initial Cauchy data. Making reference to classical results of Bruhat ([5]) for the initial-value problem and to the "constraints" for this theory (as, for instance, in [8]), we show that the moments-constraint corresponds to a quantity conserved along the integral curves, wich describes the total energy-density flow. We then evaluate the time derivative of the total energy density $\mathcal{H}\mu_g$, observing that it equals the divergence of the first constraint in a kind of a continuity equation. We can so state that the conservation of energy is a consequence of the first constraint and, according to classical results, the conditions "energy-density flow" = 0 and $\mathcal{H} = 0$ are conserved in time along the integral curves. However, in this context, they assume the role of conservation theorems, reaching a deeper geometrical meaning.

We finally drop the role of $\mathcal{H}$ and study the "reduced Einstein system", i.e. the system obtained by posing $\mathcal{H} = 0$ in the full one. We first note that under the usual Cauchy data every solution to the full system satisfies also the reduced one, because of the given conservation laws. But we show also that the converse is true, i.e. the reduced system joined to the Cauchy data is equivalent to the full one. This last beautiful result, which extends the analogous one of [14], is not trivial because the previous conservation laws are a priori valid only for the full system; so we are compelled to prove the stated equivalence making use of a system of partial differential equations.

The extension of this method that goes back to Fischer-Marsden to fields of spin 1 (e.g. the electromagnetic field) will be the subject of a future paper. The present case applies, for instance, to π^0-mesons ($m \neq 0$) and to the Brans-Dicke field ($m=0$; see [10]). If we let also β take negative values, we soon extend the validity of the given equations to the C-field proposed in 1963 by Hoyle-Narlikar [18], which is related to the so-called "steady-state universe".

2. <u>Preliminaries</u> :

Let M be a Riemannian orientable closed (i.e. compact without boundary)

3-dimensional manifold; any expression in coordinates is written using the usual "Einstein convenction", and we always intend latin indexes to vary between 1 and 3 and greek ones between 0 and 3. We assume for Minkowski space and Lorentz metrics the signature (−,+,+,+); all the geometrical quantities used in the sequel are quantities on M, evaluated with respect to the metric defining the structure of M itself. We make a sistematic use of the notations appearing in [14].

For any linear (or multilinear) operator H, defined in a point e, we generally use the notation H·e instead of H(e); if E is a vector space with a topology, we denote by E* its topological dual space. The general properties of tensor calculus may be found e.g. in [16]; we make also reference, for the basic techniques in Hamiltonian mechanics and symplectic geometry, to [1], [4], [14], [20].

In the present paper we make use of the symplectic formalism in an infinite dimensional configuration space (see [4]); we assume the reader is familiar with differential calculus on Banach and Frechet manifolds, following the treatment of [11] and [19]. In particular, we use letters D, D_1, ... to denote total and partial derivatives in Banach spaces.

Let B be a C^∞-manifold modelled on a Banach space $\mathcal{E}$; by a Lagrangian (or action integral) on B we mean any C^∞-map from the tangent bundle TB into $\mathbb{R}$ with its ordinary manifold structure. Let now ω be a 2-form on TB; recall that locally (in an open set $U \subseteq B$) we can identify TU with $\mathcal{E} \times \mathcal{E}$ and the restriction $\omega|_{TU}$ with a bilinear skew-symmetric form from $(\mathcal{E}\times\mathcal{E})\times(\mathcal{E}\times\mathcal{E})$ into $\mathbb{R}$. The form ω is said weakly non-degenerate if (given $U \subseteq B$) we have:

(1) $$\omega(u,e)\cdot((e_1,e_2),(e_3,e_4)) = 0 \Rightarrow e_1 = e_2 = 0$$
$$\forall u \in U, \quad \forall e \in \mathcal{E}, \quad \forall (e_1,e_2),(e_3,e_4) \in \mathcal{E}\times\mathcal{E}.$$

By a symplectic form on B we mean a closed non-degenerate 2-form ω. Recall also that, given a Lagrangian L on B, the collection of the restrictions $DL|T_xB$ defines a map FL : TB ⟶ TB* (fiber derivative or Legendre transformation) which, as exposed in [1] p. 116, induces a sympleetic form ω_L on TB by pulling back the canonical symplectic form on TB* (see [1] p. 96); if ω_L is non-degenerate, we usually say that L is non-degenerate (or regular).

Let Z be a vectorfield on TB; we say it is a Lagrangian vectorfield for L if:

(2) $$2\,\omega_L(v)\cdot(Z_v,\xi) = dE(v)\cdot\xi, \quad \forall v \in TB, \quad \forall \xi \in T_v(TB),$$

E being the energy of L ([4], §5) and d denoting the differential. The possibility of finding such Z is deeply important: if it exists, in fact, we are able to find the "equations of motion" for L (i.e. the integral curves of Z); we must besides emphasize that when ω_L is non-degenerate the existence of Z implies its uniqueness. Recall that the field Z is said of second order if, in a local chart $U \times \mathcal{E}$ of TB, we have $Z(u,e) = (u, Z_2(u,e)) \in U \times \mathcal{E}$, Z_2 being a C^∞-map from $U \times \mathcal{E}$ into $\mathcal{E}$; the connection with second order differential eqs. becomes clear if one express Z in local coordinates (see[4]p. 102-103). If Z is a second order vectorfield we can easily show ([14],prop. 2.2) that the integral curves $(u(t),v(t)) \in U \times \mathcal{E}$ of Z satisfy the equations:

$$(3) \quad \frac{du}{dt} = v \quad ; \quad \frac{d}{dt} D_2 L(u(t), v(t)) \cdot e = D_1 L(u(t), v(t)) \cdot e \quad , \quad \forall e \in \mathcal{E} \quad ,$$

which are Lagrange's equations.

By an indefinite metric Φ on B we mean a (smooth) family $\{\Phi_x\}$ of symmetric bilinear forms (one for each $x \in B$) on the tangent space $T_x B$; the Riemannian metrics are of course particular cases. To the metric $\Phi : x \longrightarrow \Phi_x$ we canonically associate a "kinetic Lagrangian" $T(v) = \frac{1}{2}\Phi_x(v,v)$, $\forall v \in T_x B$. If the forms Φ_x are non-degenerate, we can easily recognize that the symplectic form ω_T associated to T is non-degenerate. If a Lagrangian vectorfield Z for T exists, it is thus a spray (i.e. a second order vectorfield quadratically homogeneous in the "velocity" variable): like the Riemannian case, we use to say it is the geodesic spray of the metric Φ , and its base-integral curves the geodesics.

Let finally V be a C^∞-map from B into $\mathbb{R}$; we can consider it as a "potential term" and consider the Lagrangian:

$$L : (x,v) \longmapsto \tfrac{1}{2}\Phi_x(v,v) - V(x) \quad , \quad \forall x \in B \; , \; \forall v \in T_x B \; .$$

The duality defined in the tangent bundle TB by the metric Φ , allow us to consider the vectorfield gradV dual of the 1-form dV ,i.e. the field defined by:

$$(4) \quad \Phi_x(\mathrm{grad}V, v) = dV(x)\cdot v \quad , \quad \forall x \in B \quad , \quad \forall v \in T_x B \quad .$$

It is well known ([4],p. 107) that if the Lagrangian vectorfield Z for L exists it must be the vectorfield locally expressed as $Z(u,e) = (e, S_2(u,e) - \mathrm{grad}V(u))$, where $S_2(u,e)$ is the geodesic spray of T; Z is clearly of second order.

We conclude with a remark on notations: following the use of Abraham ([1], p. 86,92) we shall denote by the symbols $\flat$, $\#$ the lowering and raising actions associated to a non-degenerate bilinear form (like, for instance, the actions induced by an indefinite metric: the usual techniques of moving the indexes in Riemannian geometry is an example of this procedure).

3. The Symplectic Form Of The Einstein System :

In this section, starting from a non-degenerate Lagrangian, we deduce the Einstein equations in presence of a self-gravitating scalar field, writing them in Gaussian coordinates. Since the metric takes the form $ds^2 = -dt^2 + g_{ij}dx^i dx^j$ we assume that the physical system in which we are interested is described by the evolution in time (the 4th coordinate x^0 itself) of a pair (g_t, φ_t), g_t being a C^∞-metric on M and φ_t a C^∞-map from M into $\mathbb{R}$. We thus consider the configuration space $\mathcal{B} = \mathcal{M} \times \mathcal{F}$, in which $\mathcal{M}$ is the space of metrics and $\mathcal{F}$ the space of maps on M, each one taken with the topology of uniform convergence in the derivatives of any order. It is known (see [12], [13]) that $\mathcal{M}$ is a manifold modelled on the space $\mathcal{S}_2(M) = \{$twice-covariant symmetric tensorfields on M$\}$, being an open convex cone in $\mathcal{S}_2(M)$ itself; in particular, we have $T\mathcal{M} = \mathcal{M} \times \mathcal{S}_2(M)$. Since $\mathcal{F}$ is a linear space, it has a canonical structure of $\mathcal{F}$-manifold, in which $T\mathcal{F} = \mathcal{F} \times \mathcal{F}$. The product $\mathcal{B}$ is a manifold under the ordinary product structure.

Let (g,φ) be any point in $\mathcal{B}$. We introduce the following notations: $\mathrm{tr}k$ = trace of k $= g_{ij}k^{ij}$; $h \cdot k = h_{ij}k^{ij}$; $h \times k = h_{i\ell}k_j{}^{\ell}$; det(g) = determinant of g ; $\mu_g = |\det(g)|^{\frac{1}{2}} dx^1 dx^2 dx^3$. Define the bilinear form on the tangent space $T_{(g,\varphi)}\mathcal{B}$:

$$(5) \quad \Phi_{(g,\varphi)}((k,\nu),(h,\rho)) = \int_M [(\mathrm{tr}h)(\mathrm{tr}k) - h \cdot k - 4\beta\nu\rho]\mu_g \ , \quad \forall h,k \in \mathcal{S}_2(M), \ \forall \nu,\rho \in \mathcal{F} \ ,$$

in which β is a real positive number. The assignment $(g,\varphi) \longmapsto \Phi_{(g,\varphi)}$ defines an indefinite metric on $\mathcal{B}$, whose kinetic Lagrangian is given by:

$$(6) \quad T(k,\nu) = \frac{1}{2}\int_M [(\mathrm{tr}k)^2 - k \cdot k - 4\beta\nu^2]\mu_g \ , \quad \forall (k,\nu) \in \mathcal{S}_2(M) \times \mathcal{F} \ .$$

We shall prove the following:

PROPOSITION 1 / The Lagrangian T is weakly non-degenerate; its Lagrangian vectorfield Z : $T\mathcal{B} \longrightarrow \mathcal{S}_2(M) \times \mathcal{F} \times \mathcal{S}_2(M) \times \mathcal{F}$ exists and it is given by

$$(7) \quad Z((g,\varphi),(k,\nu)) = \left(k, \nu, k \times k - \tfrac{1}{2}k(\mathrm{tr}k) + \tfrac{1}{8}[(\mathrm{tr}k)^2 - k \cdot k]g - \tfrac{1}{2}\beta\nu^2 g, -\tfrac{1}{2}(\mathrm{tr}k)\nu\right)$$

Given any initial condition $(\bar{g},\bar{\varphi},\bar{k},\bar{\nu}) \in \mathcal{M} \times \mathcal{F} \times \mathcal{S}_2(M) \times \mathcal{F}$ there exist a $\tau > 0$ and a C^∞-curve $(g_t,\varphi_t,k_t,\nu_t)$ defined on $]-\tau,\tau[$ such that the curve satisfies Z for $|t| < \tau$ and $(g_0,\varphi_0,k_0,\nu_0) = (\bar{g},\bar{\varphi},\bar{k},\bar{\nu})$.

Proof. a) T is non-degenerate. Suppose in fact $\Phi((k,\nu),(h,\rho)) = 0$, $\forall (h,\rho) \in \mathcal{S}_2(M) \times \mathcal{F}$.

Hence, choosing $h' = \frac{1}{2}[(\mathrm{tr}k)g - k] \in \mathcal{S}_2(M)$, $\rho' = -\nu \in \mathcal{F}$, we find:

$\Phi((k,\nu),(h',\rho')) = \int_M [k \cdot k + 4\beta\nu^2]\mu_g = 0$. Since $\beta > 0$, it follows that $k = 0$, $\nu = 0$.

b) We fix $\omega \in \mathcal{S}_2(M)$, $\varepsilon \in \mathcal{F}$, and denote by D_1, D_2 the partial derivatives with respect to the 1st pair (g,φ) and to the 2nd pair (k,ν).

Since T operates from $\mathcal{K}\times\mathcal{F}\times S_2(M)\times\mathcal{F}$ into $\mathbb{R}$, we have of course $D_i T : S_2(M)\times\mathcal{F}\longrightarrow\mathbb{R}$ (i=1,2); so, (3) assumes the following form:

$$(8)\quad \frac{d}{dt}(g(t),\varphi(t)) = (k(t),\nu(t)) \quad ; \quad \frac{d}{dt}D_2T[(g(t),\varphi(t)),(k(t),\nu(t))]\cdot(\omega,\epsilon) = D_1T((g,\varphi),(k,\nu)).(\omega,\epsilon)$$

Define $\mathcal{H}$ (as in [14]) by: $\mathcal{H}(g,k) = \frac{1}{2}[(\operatorname{tr}k)^2 - k\cdot k]$. We fix then an instant t_0, and denote by (g,φ) the pair $(g(t_0),\varphi(t_0))$, and by (k,ν) the pair $(k(t_0),\nu(t_0))$. Let (g_s,φ_s) be a curve in $\mathcal{B}$ such that $(g_0,\varphi_0) = (g,\varphi)$ and $(\dot g_0,\dot\varphi_0) = (\omega,\epsilon)$. By the chain rule we find: (1)

$$(9)\quad D_1T((g,\varphi),(k,\nu))\cdot(\omega,\epsilon) = \frac{d}{ds}T((g_s,\varphi_s),(k,\nu))\Big|_{s=0} = \int_M \frac{d}{ds}\left[(\mathcal{H}-2\beta\nu^2)\mu_{g_s}\right]\Big|_{s=0} .$$

In order to transform (9), notice that $\mathcal{H}$ depends on s only through g_s. Making use of the following relations:

$$(10)\quad \frac{dg^{\#}}{ds} = -\omega^{\#} \quad ; \quad \frac{1}{2}\frac{d}{ds}(k\cdot k) = -(k\times k)\cdot\omega^{\#} \quad ; \qquad (2)$$

$$(10)'\quad \frac{d\mu_g}{ds} = \frac{1}{2}\left(\operatorname{tr}\frac{dg}{ds}\right)\mu_g \quad ;$$

we may easily obtain:

$$(11)\quad D_1T((g,\varphi),(k,\nu))\cdot(\omega,\epsilon) = \int_M \left[k\times k - k(\operatorname{tr}k) + \frac{1}{2}\mathcal{H}g - \beta\nu^2 g\right]\cdot\omega^{\#}\mu_g .$$

Let then (k_s,ν_s) be a curve in $S_2(M)\times\mathcal{F}$ such that $(k_0,\nu_0) = (k,\nu)$ and $(\dot k_0,\dot\nu_0) = (\omega,\epsilon)$ (1). Now $\mathcal{H}$ depends on s only through k_s, and we have:

$$\frac{d\mathcal{H}}{ds} = [g(\operatorname{tr}k) - k]\cdot\omega^{\#} \quad ; \quad \frac{d\nu^2}{ds} = 2\nu\epsilon .$$

Performing calculations analogous to that of (9) we find:

$$(12)\quad D_2T((g,\varphi),(k,\nu))\cdot(\omega,\epsilon) = \int_M \{[g(\operatorname{tr}k) - k]\cdot\omega^{\#} - 4\beta\nu\epsilon\}\mu_g .$$

From (11) and (12), rewritten for a generic t, we get Lagrange's eqs. :

$$(13)\quad \frac{d}{dt}\int_M \{[(\operatorname{tr}k)g - k]\cdot\omega^{\#} - 4\beta\nu\epsilon\}\mu_g = \int_M \left[k\times k - k(\operatorname{tr}k) + \frac{1}{2}\mathcal{H}g - \beta\nu^2 g\right]\omega^{\#}\mu_g$$

from which, differentiating under the sign of integral, it follows:

$$\int_M \left\{\frac{d}{dt}[g(\operatorname{tr}k)-k]\mu_g - \left[k\times k - k(\operatorname{tr}k) + \frac{1}{2}\mathcal{H}g - \beta\nu^2 g\right]\mu_g\right\}\cdot\omega^{\#} - 4\beta\int_M \left[\frac{1}{2}(\operatorname{tr}k)\nu + \frac{d\nu}{dt}\right]\epsilon\,\mu_g = 0 .$$

Since (ω,ϵ) is arbitrary, we finally have:

$$(14)\quad \begin{cases} \dfrac{d\nu}{dt} = -\dfrac{1}{2}(\operatorname{tr}k)\nu \\[2ex] \dfrac{d}{dt}\left\{[g(\operatorname{tr}k)-k]^{\#}\mu_g\right\} = \left[k\times k - k(\operatorname{tr}k) + \dfrac{1}{2}\mathcal{H}g - \beta\nu^2 g\right]^{\#}\mu_g \end{cases}$$

Equation $(14)_{II}$ already represents the part of (8) which refers to the cotangent bundle $T\mathcal{M}^*$; its explicit deduction, its following development and some other relations in the sequel complete the corresponding ones in [14]. They are reported here to make easier the reading and for reasons of clearness in the exposition.

In order to obtain the explicit form of Z, we must carry $(14)_{II}$ in its equivalent form in $T\mathfrak{M}$, solving it with respect to $\frac{dk}{dt}$. Making use of the relations:

$$(15) \quad \frac{dg^{\#}}{dt} = -k^{\#} \quad ; \quad \frac{d}{dt}(\mathrm{tr}\, k) = k\cdot k + \mathrm{tr}\left(\frac{dk^{\#}}{dt}\right) , \qquad (^3)$$

we easily find (eliminating a common factor μ_g):

$$(16) \quad \frac{d}{dt}\left[(\mathrm{tr}\, k)g - k\right]^{\#} = \left[k\times k - \tfrac{1}{2}k(\mathrm{tr}\, k) - \tfrac{1}{2}g(\mathrm{tr}\, k)^2 + \tfrac{1}{2}\mathcal{X}g - \beta\nu^2 g\right]^{\#} .$$

Taking the traces in (16) and solving with respect to $\frac{d}{dt}(\mathrm{tr}\, k)$, we have:

$$2\frac{d}{dt}(\mathrm{tr}\, k) = -(\mathrm{tr}\, k)^2 + \tfrac{3}{2}\mathcal{X} - 3\beta\nu^2 .$$

By substitution in (16), we can evaluate $\frac{dk^{\#}}{dt}$ and find:

$$(17) \quad -\frac{dk^{\#}}{dt} = \left[k\times k + \tfrac{1}{2}k(\mathrm{tr}\, k) - \tfrac{1}{4}\mathcal{X}g + \tfrac{1}{2}\beta\nu^2 g\right]^{\#} .$$

We ought now to lower the indexes in (17), using the following relation:

$$(18) \quad \frac{dk^{\#}}{dt} = \left(\frac{dk}{dt}\right)^{\#} - 2(k\times k)^{\#} \qquad (^4)$$

and we immediately find:

$$(19) \quad \frac{dk}{dt} = k\times k - \tfrac{1}{2}k(\mathrm{tr}\, k) + \tfrac{1}{4}\mathcal{X}g - \tfrac{1}{2}\beta\nu^2 g .$$

The group of equations $(8)_I$, $(14)_I$, (19) will be called in the sequel "Lagrange's equations" associated to T; as they traduce (3) in a different manner, we have implicitly got the expression for the field Z: it is in fact the vectorfield on $T\mathcal{B}$ that sends $((g,\varphi),(k,\nu))$ into $((k,\nu), Z_2(k,\nu))$, where $Z_2(k,\nu) = (\frac{dk}{dt}, \frac{d\nu}{dt})$; so it must be given by equation (7).

c) <u>existence of solutions:</u> We fix now $(\bar{g},\bar{\varphi},\bar{k},\bar{\nu})$ and we first observe that if $\beta = 0$ our Lagrangian (6) reduces itself to that appearing in [14]; consequently (19) reduces to the corresponding one in the empty-space. Since the new terms $(-\frac{1}{2}\beta\nu^2 g$ and $(14)_I$ itself) are algebraic in (g,k,ν) and the terms till appearing in [14] are also algebraic in (g,k), the equations obtained are globally algebraic. The same considerations of [14] allow us to state the existence and the uniqueness of a solution in an interval $]-\tau,\tau[$, where $\tau > 0$ is generally finite. $(^5)$ ■

Notice that Z is of second order, and precisely it is a spray (the proof is straightforward). So Z is the geodesic spray of Φ , and its integral curves are the geodesics of Φ . We can easily check the equivalence between our formulation and that appearing in [7],[8], as we have till studied (see [15]) in the empty-space; introducing indeed the geometric object:

$$(20) \quad G^{ijmn} = \tfrac{1}{4}|\det(g)|^{\frac{1}{2}}\{-g^{ij}g^{mn} + \tfrac{1}{2}(g^{in}g^{jm} + g^{im}g^{jn})\} ,$$

(which appears, with a different factor, in [9]), we find for any $k \in S_2(M)$:

$$(21) \quad G^{ijmn}k_{ij}k_{mn} = -\tfrac{1}{4}|\det(g)|^{\frac{1}{2}}\left[(\mathrm{tr}\, k)^2 - k\cdot k\right]$$

Thus, keeping in mind $(8)_I$, the Lagrangian (6) assumes the following form:

$$(22)\qquad T = -2\left\{ \int_M G^{ijmn} \frac{dg_{ij}}{dt}\frac{dg_{mn}}{dt}\, dx^1dx^2dx^3 + \beta \int_M \left(\frac{d\varphi}{dt}\right)^2 |\det(g)|^{\frac{1}{2}}\, dx^1dx^2dx^3 \right\} .$$

If we compare (22) with the Lagrangian of [8], we notice that our L is minus twice Christodoulou's one, with of course $\beta = 16\pi$; the constant β is related to the choice of units, because if we let $\beta = 1$ we get the usual units in which G = $= 16\pi c$ = 1 (c=velocity of light in the vacuum, G=gravity constant). Equation (21) can be interpreted saying that G^{ijmn} is a kind of "metric tensor" on $\mathcal{M}$, which induces the part of Φ corresponding to the space of metrics: it is of course the same of the metric used by FM; by these reasons we shall call it "geometric" part of Φ, and Φ with the name of "Christodoulou-De Witt's metric". Since the metric Φ is weakly non-degenerate, the existence of geodesics is not assured: but we have checked it in prop. 1 ; these geodesics could be explicitly deduced, as partially made in [8]: from the formulas appearing in that paper, we can notice that they generally exist in $\mathcal{M}\times\mathcal{F}$ only for a finite time, in the sense that they evolve for larger values of time into curves of $S_2(M)\times\mathcal{F}$.

Let's now define the raising and lowering actions $\Phi^{\#}$ and Φ^{b}, associated to the metric Φ (see [1], p. 82). $\Phi^b : T\mathcal{B} \longrightarrow T\mathcal{B}^*$ is defined by:

$$(23)\qquad \Phi^b_{(g,\varphi)}(h,\nu) : (h',\nu') \mapsto \Phi_{(g,\varphi)}((h,\nu),(h',\nu')) \quad , \quad \forall (g,\varphi)\in\mathcal{B},\ \forall h,h'\in\mathcal{M},\ \forall \nu,\nu'\in\mathcal{F} .$$

Since Φ is weakly non-degenerate, Φ^b must be injective (it is not an isomorphism between the bundles $T\mathcal{B}$ and $T\mathcal{B}^*$, because it is not strongly non-degenerate; see [4], § 1). But $T\mathcal{B}^* = (T\mathcal{M}\times\mathcal{F})^*$; fix then a point $(g,\varphi)\in\mathcal{M}\times\mathcal{F}$, and observe that $T_{(g,\varphi)}\mathcal{B}^* = T_g\mathcal{M}^* \times \mathcal{F}^*$. Here $T_g\mathcal{M}^* = \{$space of continuous functionals on $T_g\mathcal{M}\}$ and $\mathcal{F}^* = \{$space of cont. functionals on $\mathcal{F}\}$: thus $T_{(g,\varphi)}\mathcal{B}^*$ contains pairs of (Schwartz's) distributions on M, the first one tensor-valued and the second one scalar-valued. Recall then the canonical injection of $T_{(g,\varphi)}\mathcal{B}$ into $T_{(g,\varphi)}\mathcal{B}^*$, defined by identifying $h\in S_2(M)$ and $\nu\in\mathcal{F}$ to the distributions:

$$(24)\qquad h' \mapsto \int_M (h\cdot h')\mu_g \quad , \quad \nu' \mapsto \int_M \nu\nu'\mu_g \quad , \quad \forall h'\in S_2(M) \ , \ \forall \nu'\in\mathcal{F} .$$

Hence, we may identify $\Phi^b_{(g,\varphi)}(k,\nu)$ to the pair (π,σ), where:

$$(25)\qquad \begin{cases} \pi = [g(\operatorname{tr} k) - k]^{\#}\mu_g \\ \sigma = -4\beta\nu\mu_g \end{cases} \qquad\qquad (^6)$$

The pair (π,σ) is the pair of moments conjugated to (k,ν). Define $(S^2(M)\times\mathcal{F})\otimes\mu_g$ be the range of $\Phi^b_{(g,\varphi)}$, i.e. the set $\{(h\mu_g, \rho\mu_g) \mid h\in S^2(M)$ = symmetric twice contravariant tensors , $\rho\in\mathcal{F}\}$; so $\Phi^b_{(g,\varphi)}$ can be considered as a one-one mapping between $T_g\mathcal{M} \times \mathcal{F}$ and $(S^2(M) \times \mathcal{F}) \otimes \mu_g$. If we set then $\Phi^{\#}_{(g,\varphi)} = [\Phi^{b}_{(g,\varphi)}]^{-1}$, we easily find:

(26) $$\Phi^{\#}_{(g,\varphi)}(\pi, \rho\mu_g) = \left[\tfrac{1}{2}(tr\,\pi')g - (\pi')^{b} , -\tfrac{1}{4\beta}\rho \right] ,$$

in which π' is the tensor part of π (i.e. $\pi = \pi' \otimes \mu_g$).

4. The Potential Of The Self-Gravitating Scalar Field :

Let's agree to denote by commas ordinary derivatives; we define the covariant tensor $\tilde{\varphi} \in S_2(M)$ by the following:

(27) $$\tilde{\varphi}_{ij} = \varphi_{,i}\,\varphi_{,j} \qquad , \quad i,j = 1,2,3 \quad .$$

Let m be a real constant (possibly also zero, as explained in the introduction) representing the mass related to the field and $A(\varphi) = \varphi_{,i}\,\varphi^{,i} + m^2\varphi^2$ be the density of external potential energy. We define the potential as follows:

(28) $$V(g,\varphi) = 2\int_M [R(g) - A(\varphi)]\mu_g \qquad ,$$

where R(g) denotes the scalar of curvature of the metric g. Observe that the part $\int R(g)\mu_g$ is the geometric part introduced by Wheeler and De Witt ([21],[9]). We make use of the following notations: $|i$ = covariant derivative in M ; ∇ = scalar gradient in M ; $\Delta\varphi = -g^{ij}\varphi_{|ij}$ = Laplace-Beltrami operator on scalars ; $\delta k = -(k_i{}^j{}_{|j})$ = covariant divergence of the tensor k ; HessN = $N_{|ij}$ = Hessian of the function N ; Ric(g) = Ricci tensor of the metric g .

We begin to study V from the following:

Lemma 1 / The rate of change of $A(\varphi)$ is given by:

(29) $$\frac{dA}{dt} = 2m^2\varphi\frac{d\varphi}{dt} + 2(\nabla\varphi)^{\#}\cdot\nabla\left(\frac{d\varphi}{dt}\right) + \tilde{\varphi}\cdot\frac{dg^{\#}}{dt} \quad .$$

The proof is straightforward and we omit it.

In this paper we work only in Gaussian coordinates; but thinking of a possible extension to general coordinates, let we study the slightly more general:

(28)' $$V_N(g,\varphi) = 2\int_M N(x)[R(g) - A(\varphi)]\mu_g \qquad , \quad x \in M,$$

in which N is a generic C^{∞}-map on M, taking nonnegative values. The potential (28)' is needed if we express the Lorentz metric of space-time in the so-called "Lapse and Shift form" (see [2],[21]):

$$ds^2 = (X_i X^i - N^2)dt^2 + 2X_i\,dx^i dt + g_{ij}(x^k;t)\,dx^i dx^j \qquad ,$$

where $\{X_i\}$ is a covariant vector ("shift vector") on M.

So, we shall show that:

PROPOSITION 2 / The Φ-gradient of V_N is given by :

(30) $$\left(2\left\{-\mathrm{Hess}N + N\left[\mathrm{Ric}(g) - \frac{1}{4}R(g)g\right] - \frac{\beta N}{2}\left[2\tilde{\varphi} + \frac{1}{2}gB(\varphi)\right]\right.\right., + \left[N(\Delta\varphi + m^2\varphi) - \nabla N\cdot\nabla\varphi\right]\Big),$$

in which we have posed $B(\varphi) = A(\varphi) - 2\varphi_{,i}\varphi^{,i}$.

Proof : By duality, eq. (4) assures that $\mathrm{grad}V_N = \Phi^{\#}(dV_N)$. Since $\mathcal{F}$ and $S_2(M)$ are both linear spaces, we can identify dV_N with the total derivative DV_N. Fix $(g,\varphi) \in \mathcal{M}\times\mathcal{F}$ and $(\omega, \varepsilon) \in S_2(M)\times\mathcal{F}$; let (g_s, φ_s) be a C^∞-curve in $\mathcal{M}\times\mathcal{F}$ such that $(g_0, \varphi_0) = (g,\varphi)$ and $(\dot{g}_0, \dot{\varphi}_0) = (\omega, \varepsilon)$ (see footnote 1). So we find:

(31) $$DV_N(g,\varphi)\cdot(\omega,\varepsilon) = 2\int_M N\frac{dR(g)}{ds}\mu_g + 2\int_M NR(g)\frac{d\mu_g}{ds} - 2\beta\int_M N\frac{d}{ds}\left[A(\varphi)\mu_g\right] .$$

It is also known (see [3]) that:

(32) $$\frac{dR(g)}{ds} = \Delta(\mathrm{tr}\,\omega) + \delta\delta\omega - \mathrm{Ric}(g)\cdot\omega .$$

Inserting (32) into (31) we easily get (taking into account lemma 1):

(33) $$DV_N(g,\varphi)\cdot(\omega,\varepsilon) = 2\int_M N\left[\Delta(\mathrm{tr}\,\omega) + \delta\delta\omega - \mathrm{Ric}(g)\cdot\omega + \frac{1}{2}(\mathrm{tr}\,\omega)R(g)\right]\mu_g - $$
$$- 2\beta\int_M N\left[\frac{1}{2}A(\varphi)\,\mathrm{tr}\,\omega + 2m^2\varphi\varepsilon + 2(\nabla\varphi)^{\#}\cdot\nabla\varepsilon - \tilde{\varphi}\cdot\omega^{\#}\right]\mu_g .$$

Use now the fact that M is compact without boundary in order to transform the integrals of $N\Delta(\mathrm{tr}\,\omega)$, $N\,\delta\delta\omega$ and $N\nabla\varphi\nabla\varepsilon$; because the boundary ∂M is empty, we get:

$$\int_M N\left[(\Delta\,\mathrm{tr}\,\omega) + \delta\delta\omega\right]\mu_g = \int_M (g\Delta N + \mathrm{Hess}N)\mu_g ,$$

$$\int_M N\nabla\varphi\nabla\varepsilon\,\mu_g = -\int_M\left[N\Delta\varphi - \nabla N\nabla\varphi\right]\varepsilon\,\mu_g , \qquad (7)$$

finally obtaining the total derivative as:

(34) $$DV_N(g,\varphi)\cdot(\omega,\varepsilon) = 2\int_M\left\{g\Delta N + \mathrm{Hess}N - N\left[\mathrm{Ric}(g) - \frac{1}{2}R(g)g\right] + \frac{\beta N}{2}\left[2\tilde{\varphi} - gA(\varphi)\right]\right\}\cdot\omega\,\mu_g +$$
$$+ \int_M\left\{4\beta\left[N(-\Delta\varphi - m^2\varphi) + \nabla N\cdot(\nabla\varphi)^{\#}\right]\varepsilon\right\}\mu_g .$$

Following Dieudonné [11], let's now identify the total derivative $DV_N(g,\varphi)$: $S_2(M)\times\mathcal{F} \longrightarrow \mathbb{R}$ with the pair of partial derivatives $D_1V_N(g,\varphi) : S_2(M) \longrightarrow \mathbb{R}$ and $D_2V_N(g,\varphi) : \mathcal{F} \longrightarrow \mathbb{R}$.(8) Since we have supposed N to be a C^∞-map, these partial derivatives are distributions on M; with the identifications (24) we can so write:

(35) $$\begin{cases} D_1V_N(g,\varphi) = 2\left\{g\Delta N + \mathrm{Hess}N - N\left[\mathrm{Ric}(g) - \frac{1}{2}R(g)g\right] + \frac{\beta N}{2}\left[2\tilde{\varphi} - gA(\varphi)\right]\right\}^{\#}\mu_g \\ D_2V_N(g,\varphi) = -4\beta\left\{N(\Delta\varphi + m^2\varphi) - \nabla N\,\nabla\varphi\right\}\mu_g \end{cases}$$

Now, in order to find $\mathrm{grad}V_N$, we need only to apply (25) to (35). But :

(36) $\mathrm{tr}\,[g\,\Delta N + \mathrm{Hess}\,N - N(\mathrm{Ric}(g) - \frac{1}{2}R(g)g)] = 2\,\Delta N + \frac{1}{2}N R(g)$,

(37) $\mathrm{tr}\,(2\tilde{\varphi} - g A(\varphi)) = 2\,\varphi_{,i}\,\varphi^{,i} - 3A(\varphi)$.

So, performing some calculations, we finally get (30). ∎

We obtain the Gaussian coordinates case simply by posing N = 1 ; subtracting then the result obtained from the field Z given by (7) we find (as previously explained in § 2) the following proposition:

PROPOSITION 3 / Let L be the Lagrangian T – V . The dynamical equations associated to it on the tangent bundle $T(\mathcal{M}\times\mathcal{F})$ are :

$$(E)\quad\begin{cases} \dfrac{\partial g}{\partial t} = k \quad , \quad \dfrac{\partial \varphi}{\partial t} = \nu \quad , \\ \dfrac{\partial k}{\partial t} = k \times k - \frac{1}{2}k(\mathrm{tr}\,k) + \frac{1}{4}\varkappa g - \frac{1}{2}\beta\nu^2 g - 2\left[\mathrm{Ric}(g) - \frac{1}{4}R(g)g\right] + \beta\left[2\tilde{\varphi} + \frac{1}{2}g\,B(\varphi)\right] \\ \dfrac{\partial \nu}{\partial t} = -\frac{1}{2}(\mathrm{tr}\,k)\nu - (\Delta\varphi + m^2\varphi) \end{cases}$$

We ought to notice that the system (E) ("Einstein system" in the sequel) gives of course solutions only for the generally finite time τ , and that these solutions describe the space–time $M \times]-\tau,\tau[$ by means of Gaussian coordinates.

5. Energy And Conservation Laws :

Let E be the "energy" of L ; we intend here to evaluate the Hamiltonian $H = E\,(FL)^{-1}$ (see [1] ; FL is the fiber derivative). Since our formalism is a canonical formalism, we can find H subtracting the Lagrangian L from the integral

$$\int_M (k\cdot\pi' + \nu\sigma')\,\mu_g \quad , \text{ where } \quad \pi = \pi'\otimes\mu_g \quad \text{and} \quad \sigma = \sigma'\mu_g \; .$$

This procedure corresponds in fact to the classical formula of analytic mechanics $H(q,\dot{q}) = \sum_i p_i\,\dot{q}^i - L(q,\dot{q})$; here the sum is replaced by the integral (we should notice that under the identifications (24) the integral coincides with the sum of the formal products : $\int_M (k\cdot\pi')\mu_g$, $\int_M (\nu\sigma')\mu_g$). Hence we get:

(38) $H((g,\varphi),(k,\nu)) = \int_M (\mathcal{H}_G + \mathcal{H}_F)\,\mu_g$, where:

(38)' $\mathcal{H}_G = \frac{1}{2}\left[(\mathrm{tr}\,k)^2 - k\cdot k\right] + 2R(g)$;

(38)" $\mathcal{H}_F = -2\beta\left[\nu^2 + A(\varphi)\right]$

Here $\mathcal{H}_G$ is the gravitational energy density and $\mathcal{H}_F$ the field energy density; their sum $\mathcal{H}$ is indeed the total energy density. We should notice that it is possible to get (38) by other ways; we can in fact apply prop. 18.11 of [1] ([9]), deducing H as follows:

(39) $$H((g,\varphi),(\pi,\sigma)) = \frac{1}{2}\Phi(\Phi^{\#}(\pi,\sigma),\Phi^{\#}(\pi,\sigma)) - V(g,\varphi) =$$
$$= \frac{1}{2}\int_H \left[\frac{1}{2}(\mathrm{tr}\,\pi')^2 - \pi'\cdot\pi' - \frac{1}{4\beta}(\sigma')^2\right]\mu_g - V(g,\varphi) \quad .$$

This gives H expressed on the cotangent bundle; going down to $T\mathcal{B}$ by means of (25) we find:

(40)' $$\frac{1}{2}(\mathrm{tr}\,\pi')^2 - \pi'\cdot\pi' = (\mathrm{tr}\,k)^2 - k\cdot k \quad ,$$

(40)" $$-\frac{1}{4\beta}(\sigma')^2 = -4\beta\nu^2 \quad ,$$

and the substitution of (40)' and (40)" into (39) gives the desired H, expressed in form of a functional depending upon g,φ,k, and ν .

In order to complete the full equivalence between the formulation in the "phase-moments space" $T\mathcal{B}^*$ and in the "phase-velocity space" $T\mathcal{B}$, we need to translate the Einstein system (E) into the cotangent bundle, evaluating the time-derivatives of π and σ. By (15),(18),(24) and footnote (5) we obtain:

(41) $$\frac{\partial\pi}{\partial t} = \left\{-\frac{3}{2}k(\mathrm{tr}\,k) - g(k\cdot k) + g\,\mathrm{tr}\left(\frac{\partial k}{\partial t}\right) - \frac{\partial k}{\partial t} + 2(k\times k) + \frac{1}{2}g(\mathrm{tr}\,k)^2\right\}^{\#}\mu_g \quad .$$

Consider now the expression for $\frac{\partial k}{\partial t}$ given by (E) and put it into the previous result, obtaining after some calculations:

(42) $$\frac{\partial\pi}{\partial t} = \left\{k\times k - k(\mathrm{tr}\,k) + \frac{1}{2}\mathcal{H}g - \beta\nu^2 g + 2\left[Ric(g) - \frac{1}{2}R(g)g\right] - \beta(2\tilde{\varphi} - gA(\varphi))\right\}^{\#}\mu_g \quad .$$

Analogously, by (24) and (E) we get:

(43) $$\frac{\partial\sigma}{\partial t} = \frac{\partial}{\partial t}(-4\beta\nu\mu_g) = 4\beta(\Delta\varphi + m^2\varphi)\mu_g \quad .$$

The set of equations $(E)_{I,II}$,(42),(43) represents the canonical form of the Einstein system on the cotangent bundle $T\mathcal{B}^*$: they coincide, though written in a very different manner (and apart from the numerical factor −2 introduced), with the corresponding equations variationally deduced by Christodoulou ([8]). We could notice that these equations correspond to Lagrange's eqs. written as follows:

(44) $$D_2L((g,\varphi),(k,\nu)) = -D_1H((g,\varphi),(k,\nu)) \quad , \quad \pi = \pi(k) \quad , \quad \nu = \nu(k) \quad .$$

Before deducing the conservation laws of our Lagrangian system, we need a lemma. Consider the quantity $(\nu^2 g - gA(\varphi) + 2\tilde{\varphi})$: as we shall see in the sequel, it is the purely spatial part of the stress-tensor. Now, we have:

Lemma 2 / <u>In the previous hypothesis it is:</u>

(45) $$\delta\left(\nu^2 g - gA(\varphi) + 2\tilde{\varphi}\right) = 2\left[(\Delta\varphi + m^2\varphi)\nabla\varphi - \nu\nabla\nu\right]$$

The proof is straightforward.

We are now able to state the main theorem:

THEOREM 1 / Let $(g_t, \varphi_t, k_t, \nu_t)$ be a curve in $T\mathcal{B}$ satisfying (E) and denote by (π, σ) the moments conjugated to (k, ν). The Lagrangian L considered admits the following conservation laws:

(46) $\frac{\partial}{\partial t}(2\delta\pi^b + \nabla\varphi\,\sigma) = 0$;

(47) $\frac{\partial}{\partial t}(\mathcal{H}\mu_g) + \delta(2\delta\pi + \sigma(\nabla\varphi)^{\#}) = 0$.

If the solutions of (E) exist in $]-\tau,\tau[$ and if t = 0 implies $-\delta\pi^b = \frac{1}{2}\sigma\nabla\varphi$ this condition is mantained for any $t \in]-\tau,\tau[$. If also $\mathcal{H}$ is zero when t = 0, $\mathcal{H}$ is zero for any $t \in]-\tau,\tau[$.

Proof : Define F by F = $2\delta\pi^b + \sigma\nabla\varphi$. We find by (7) and (43) :

$$\frac{\partial F}{\partial t} = 2\frac{\partial}{\partial t}(\delta\pi^b) - 4\beta\left[\nu\nabla\nu - \nabla\varphi(\Delta\varphi + m^2\varphi)\right]\mu_g .$$

By the previous lemma we have then:

(48) $$\frac{\partial F}{\partial t} = 2\frac{\partial}{\partial t}(\delta\pi^b) + 2\beta\,\delta(\nu^2 g - gA(\varphi) + 2\tilde{\varphi})\mu_g$$

Recall then that ([14] ,p. 553):

(49) $$\frac{\partial}{\partial t}(\delta\pi^b) = \delta\left(\frac{\partial\pi}{\partial t}\right)^b + \delta\left[k(\mathrm{tr}\,k) - k\times k - \frac{1}{2}\mathcal{K}g\right] ,$$

and also that the divergence of the tensor $[\mathrm{Ric}(g) - \frac{1}{2}R(g)g]$ is zero. We can then lower the indexes in (42) and, by substitution in (49), get:

$$\frac{\partial F}{\partial t} = 2\delta\Big[k\times k - k(\mathrm{tr}\,k) + \frac{1}{2}\mathcal{K}g - \beta\nu^2 g + 2\left(\mathrm{Ric}(g) - \frac{1}{2}R(g)g\right) - 2\beta\tilde{\varphi} + \beta gA(\varphi) +$$
$$+ k(\mathrm{tr}\,k) - k\times k - \frac{1}{2}\mathcal{K}g + \beta\nu^2 g - \beta gA(\varphi) + 2\beta\tilde{\varphi}\Big] = 0 .$$

In order to evaluate $\frac{\partial}{\partial t}(\mathcal{H}\mu_g)$ we initially observe that $\mathcal{K}\mu_g = \frac{1}{2}k\cdot\pi$, and thus:

$$\frac{\partial}{\partial t}(\mathcal{K}\mu_g) = \frac{1}{2}\left\{\frac{\partial k}{\partial t}\cdot\pi + k\cdot\frac{\partial\pi}{\partial t}\right\} .$$

By (42) and $(E)_{II}$, performing some calculations that we omit, we get:

(50) $$\frac{\partial}{\partial t}(\mathcal{K}\mu_g) = 2\left[\mathrm{Ric}(g) - \frac{1}{2}R(g)g\right]\cdot k\,\mu_g - \beta\left[\nu^2 g + (2\tilde{\varphi} - gA(\varphi))\right]\cdot k\,\mu_g .$$

It is also known (see [14],p. 553) that:

(51) $$\frac{\partial}{\partial t}(2R(g)\mu_g) = -2\delta\delta\pi - 2\left[\mathrm{Ric}(g) - \frac{1}{2}R(g)g\right]\cdot k\,\mu_g .$$

By (29) and $(E)_{IV}$ we get finally:

(52) $$\frac{\partial}{\partial t}(\mathcal{H}_F\mu_g) = -2\beta\left\{-\frac{1}{2}\nu^2(\mathrm{tr}\,k) - 2\nu\Delta\varphi + 2\nabla\varphi\nabla\nu - \frac{1}{2}\left[2\tilde{\varphi} - gA(\varphi)\right]\cdot k\right\}\mu_g ;$$

and it is easy to deduce the following:

(53) $$\delta((\nabla\varphi)^{\#}\sigma) = \sigma\Delta\varphi - \nabla\sigma\nabla\varphi .$$ [10]

Combining the relations from (50) to (53) we find so:

$$\frac{\partial}{\partial t}(\mathcal{H}\mu_g) = -2\delta\delta\pi - \sigma\Delta\varphi + \nabla\sigma\nabla\varphi = -\delta(2\delta\pi + (\nabla\varphi)^{\#}\sigma) .$$ ∎

Looking at (46) and (47) we soon notice that the vector $(2\delta\pi^b + \sigma\nabla\varphi)$ is always conserved along the integral curves of Z (i.e. the solutions to (E)); the conservation of energy, on the contrary, follows only if at time zero (on the initial hypersurface $\Sigma'_0 = M\times\{0\} \subseteq M\times]-\tau,\tau[$) the previous vector is zero. This last condition, that in components is written as follows:

(54) $\quad -2\pi_i{}^j{}_{|j} + \sigma\varphi_{,i} = 0 \qquad (i=1,2,3)$,

is the classical "moments-constraint" on Σ'_0 (look e.g. at [8], where it appears with the indexes raised) while $\mathcal{H} = 0$ is the so-called "Hamiltonian constraint". We can so translate the theorem saying that the so-called "constraints" are much more, because they traduce conservation laws. This is in perfect accordance with the statements about the initial-value problem given (by means of different tec= hniques) by Y.C. Bruhat ([5],p. 154) and Wheeler ([21], p. 359); we may notice in particular that $\frac{1}{2}(\nabla\varphi)\sigma = -2\beta\nu\nabla\varphi\mu_g$ can be physically interpreted as the vector $\{S_i\}$ describing the <u>external energy density flow</u> : we shall see later that it is in fact componed by the three components T_{0i} of the stress-energy tensor inducing the Lagrangian L. To recognize this equivalence, we begin to write $\varphi_{,0}$ instead of $\nu = \frac{\partial\varphi}{\partial t}$, in order to translate the problem into its full space-time formulation, rather than in the canonical (3+1) splitting. Hence, we deduce by (6) and (28) the following form for the field part of L:

(55) $$L_F = 2\beta(-\varphi_{,0}{}^2 + \varphi_{,k}\varphi^{,k} + m^2\varphi^2) = 2\beta(\varphi_{,\rho}\varphi^{,\rho} + m^2\varphi^2) \qquad ('')$$

It is well known that it is possible to start from the Lagrangian to obtain the stress tensor that induces it (see e.g. [17],p. 68); in the present case we have:

(56) $$T_{\mu\nu} = -2\beta\left[2\varphi_{,\mu}\varphi_{,\nu} - {}^4g_{\mu\nu}(\varphi_{,\rho}\varphi^{,\rho} + m^2\varphi^2)\right] \quad ,$$

where ${}^4g_{\mu\nu}$ is the 4-dimensional metric $\quad ds^2 = -dt^2 + g_{ij}\,dx^i dx^j$.

From (56) immediately follows:

(57) $$\begin{cases} T_{0i} = -4\beta\nu\varphi_{,i} \\ T_{00} = -2\beta\left[\nu^2 + A(\varphi)\right] = \mathcal{H} \end{cases}$$

Equations (57) show that $\mathcal{H}$ is just the (0,0) component of the stress tensor, and that $\frac{1}{2}(\nabla\varphi)\sigma$ differs from $\{T_{0i}\}_{i=1,2,3}$ only by the factor $(2\mu_g)$.

6. <u>The Reduced Evolution System</u> :

Let τ , Σ'_0 are as before; let γ_t be a C^∞-curve $(g_t, \varphi_t, k_t, \nu_t)$ in $\mathcal{M}\times\mathcal{F}\times S_2(M)\times\mathcal{F}$ defined in $]-\tau,\tau[$. Following the usual notations, the <u>Cauchy data for γ_t</u> are the system on Σ'_0 :

$$(C)\begin{cases} 2\delta(g_0\, \mathrm{tr} k_0 - k_0) - 4\beta\dot{\varphi}_0\varphi_0 = 0 & , \\ \mathcal{H}((g_0,\varphi_0),(k_0,\dot{\varphi}_0)) = 0 & . \end{cases}$$

Define the "De Witt spray" $S_g(k)$ (see [14]) and the tensor $P(\varphi)$ by:

$$S_g(k) = k \times k - \tfrac{1}{2}k(\mathrm{tr}k) \quad ; \quad P(\varphi) = \beta\left[2\tilde{\varphi} + g(m^2\varphi^2)\right] \quad .$$

Performing some easy calculations on eq. $(E)_{III}$ we transform it into the following:

$$\frac{\partial k}{\partial t} = S_g(k) - 2\,Ric(g) + \tfrac{1}{4}\mathcal{H}g + P(\varphi) \quad .$$

Because we have seen that for the Einstein system, $\mathcal{H} = 0$ is conserved if $(C)_I$ is satisfied, it can be interesting to study the system obtained simply making $\mathcal{H} = 0$. It is the system:

$$(\tilde{E})\begin{cases} \dfrac{\partial g}{\partial t} = k \quad , \quad \dfrac{\partial\varphi}{\partial t} = \dot{\varphi} \\ \dfrac{\partial k}{\partial t} = S_g(k) - 2\,Ric(g) + P(\varphi) & , \\ \dfrac{\partial\dot{\varphi}}{\partial t} = -\tfrac{1}{2}(\mathrm{tr}k)\dot{\varphi} - (\Delta\varphi + m^2\varphi) & ; \end{cases}$$

and we may call it with the name of "reduced Einstein system".

So, we shall prove the following valuable theorem:

THEOREM 2 / <u>If (C) is satisfied on the initial space-like hypersurface Σ_0, the systems (E) and $(\tilde{E})$ are equivalent, i.e. any γ_t satisfying one of them satisfies both the systems.</u>

Proof : Let γ_t be a solution of (E); if (C) is valid, the conservation laws (46) and (47) assure us that $\mathcal{H} = 0$ is conserved along γ_t ; so (E) reduces to $(\tilde{E})$ along the solution γ_t , which then solves also $(\tilde{E})$.

Conversely, let γ_t be a solution of $(\tilde{E})$: at a first sight, we could think to reverse the previous argument. But this would be not correct at all, because (at least a priori) the laws (46) and (47) are not conservation laws also for the system $(\tilde{E})$. So we need to deal with the problem as follows, adapting to the present case the way followed in [14], p. 554. Taking into account $(\tilde{E})_{III}$ we get:

$$(58) \quad \frac{\partial\pi^b}{\partial t} = \left\{k(\mathrm{tr}k) - k\times k + 2\left[Ric(g) - R(g)\,g\right] - 2\beta\tilde{\varphi} + 2\beta g\,A(\varphi)\right\}\mu_g \quad .$$

Making then use of (48), (49) and lemma 2 we finally get the relations:

$$(59) \quad \frac{\partial}{\partial t}(2\delta\pi^b + \sigma\nabla\varphi) + \delta(\mathcal{H}\mu_g)g = 0 \quad .$$

$$(60) \quad \frac{\partial}{\partial t}(\mathcal{H}\mu_g) + \tfrac{1}{2}\mathcal{H}(\mathrm{tr}k)\mu_g + \delta(2\delta\pi^b + \sigma\nabla\varphi) = 0 \quad .$$

Now, we can consider the pair (59), (60) as a system of partial differential equations in the unknowns $(\mathcal{H}\mu_g)$ and $(2\delta\pi^b + \nabla\varphi\sigma)$. This system is of first order, linear and homogeneous; since the Cauchy data assure that $\partial\ell\mu_g$ and $2\delta\pi^b + \sigma\nabla\varphi$ are zero for t = 0, the system implies that these conditions are mantained for any t along γ_t. So (E) reduces again to $(\tilde{E})$ along γ_t, which then also satis= fies (E). ■

N.B.: We needed to use such a method to get the stated equivalence, because (like in the empty-space case) we have no direct theorems of uniqueness for $(\tilde{E})$. We refer to [14] for this topic.

ACKNOWLEDGMENTS :

The author gratefully acknowledges the ROTARY CLUB of Genova and Mondovì for their financial assistance with a partecipation grant for the Symposium of Bonn; he thanks Prof. K. Bleuler and his colleagues for the kind hospitality at the "Institut für theoretische kernphysik der Universität Bonn".

---

footnotes :

(1) – We have used the letter s to denote the parameter, because (as explained above) (g,φ) represents for us the istantaneous pair $(g(t_0),\varphi(t_0))$. The curve (g_s,φ_s) is thus a curve through the point (g,φ), which generally will have nothing to do with the curve $(g(t),\varphi(t))$ satisfying eqs. (8). Analogously for (k,γ).

(2) – $(10)_I$ follows from $g^{ij}g_{jk} = \delta_k^{\ i}$. We find indeed by derivation:

$$\left(\frac{dg}{ds}\right)^{ij} = \frac{dg^{ij}}{ds} = -g^{im}g^{jn}\frac{dg_{mn}}{ds} = -\omega^{ij} \quad .$$

The symmetry of g and k implies $(10)_{II}$; in fact we have:

$$\frac{d}{ds}\left(k_{ij}g^{ai}g^{bj}k_{ab}\right) = 2k_{ij}k_{ab}g^{ai}\frac{dg^{bj}}{ds} = -2k_{ij}k_{ab}g^{ai}\omega^{bj} = -2(k\times k)\cdot\omega^{\#} \,. \quad ■$$

(3) – $(15)_{II}$ follows from $\frac{d}{dt}\left(g_{ij}k^{ij}\right) = k_{ij}k^{ij} + tr\left(\frac{dk^{ij}}{dt}\right)$.

(4) – We have indeed $\frac{dk^{ij}}{dt} = \frac{d}{dt}\left(g^{ai}g^{bj}k_{ab}\right) = \cdots = \left(\frac{dk_{ij}}{dt}\right)^{\#} - 2(k\times k)^{\#}$.

Notice that taking the traces in (18) we have $tr\left(\frac{dk^{\#}}{dt}\right) = tr\left(\frac{dk}{dt}\right) - 2k\times k$. Combining this with (15) we get the relation useful later: $\frac{d}{dt}(tr\,k) = -k\cdot k + tr\left(\frac{dk}{dt}\right)$.

(5) – Sketch of the proof (for details see [14]):

Define $\mathcal{M}^n$ = C^n-metrics on M and $\mathcal{F}^n$ = C^n-maps on M, which are Banach spaces in the C^n-topology; $\mathcal{M} = \bigcap_n \mathcal{M}^n$ and $\mathcal{F} = \bigcap_n \mathcal{F}^n$ are Frechet spaces. Since the product $T\mathcal{M}^n \times \mathcal{F}^n$ is a Banach space (see [13]) and in our hypothesis the field Z is de= fined for any integer n, the Picard's method assures us the existence of a C^n-flow (for any $n < \infty$) defined in a certain interval $]-\tau_n, \tau_n[$ ($\tau_n > 0$).

Like in [14] p. 551, making use of the methods exposed in [6], we can immediately recognize that $\lim_{n\to\infty} \tau_n = \tau > 0$ and thus check the existence of a C^∞-flow for Z in $]-\tau,\tau[$ ■ .

(6) - Eqs. (25) are not completely correct, in the following sense: $\mu_g = |\det(g)|^{\frac{1}{2}} dx^1 \cdots dx^3$ while we ought to assume π and σ defined without differentials, i.e. like usual tensor densities of weight $\frac{1}{2}$. We prefer this slightly incorrect form (as appears also in [14]) for reasons of graphical necessity.

(7) - The 1st formula is deduced in FM p. 552.

The 2nd one is a consequence of the Green-Gauss theorem in its form:

$$\int_H \nabla\psi\nabla\varphi\,\mu_g - \int_H \psi\Delta\varphi\,\mu_g = 0$$, in which we pose $\psi = N\varepsilon$.

(8) - Let's in substance apply the relation:

$$DV_N(g,\varphi)\cdot(\omega,\varepsilon) = D_1V_N(g,\varphi)\cdot\omega + D_2V_N(g,\varphi)\cdot\varepsilon \quad .$$

(9) - The proposition is the following:

"Let B be a manifold, m be an indefinite metric ($m : TB \longrightarrow \mathbb{R}$), T its kinetic energy defined by $T : v \mapsto \frac{1}{2}m(v,v)$, $\forall v \in TB$. The Hamiltonian of T is the map $H_T : TB^* \longrightarrow \mathbb{R}$ given by:

$H_T : \omega \longmapsto \frac{1}{2}m(m^\#(\omega), m^\#(\omega))$, $\forall \omega \in TB^*$."

(10) - In fact: $-\delta(\sigma\nabla\varphi^\#) = (\sigma\varphi^{,i})_{|i} = \sigma\varphi^{,i}{}_{|i} + \sigma_{,i}\varphi_{,i} = -\sigma\Delta\varphi + \nabla\varphi\nabla\sigma$. ■

(11) - We have used the fact that, being the metric a Gaussian one, $g^{0\mu} = -\delta^{0\mu}$ and thus: $\varphi^{,0} = g^{0\mu}\varphi_{,\mu} = g^{00}\varphi_{,0} = -\varphi_{,0} \Rightarrow -\varphi_{,0}{}^2 = \varphi_{,0}\varphi^{,0}$.

(*) - This work, made inside the GNFM (Gruppo Nazionale per la Fisica Matematica) of CNR, have been the subject of a communication during the "Symposium on Differential Geometrical Methods in Mathematical Physics", which was holded at the University of Bonn in the days 1 - 4 July 1975.

- REFERENCES -

[1] - ABRAHAM,R. - "Foundations of mechanics", Benjamin (1967)

[2] - ARNOWITT R. ,DESER S. ,MISNER C.W. - "The dynamics of general relati= vity", chap. 7 in "Gravitation: an introduction to current research" Wiley & Sons (1962)

[3] - BERGER M. ,EBIN D. - "Some decomposition of the space of symmetric tensors on a riemannian manifold", Journ. of Diff. Geom. III (1969) p. 379-392

[4] - CHERNOFF P.,MARSDEN J. - "Properties of infinite dimensional Hamil= tonian systems", Lect. Notes in Math. ,Springer-Verlag (1974)

[5] - CHOQUET BRUHAT Y. - "The Cauchy problem", chap. in ref. 2 p. 130-168

[6] - CHOQUET BRUHAT Y. - "Equations aux derivees partielles: solutions C d'equations hyperboliques non lineares", C. Rendus Acad. Sc. Paris, 272 (1971) p. 386-388

[7] - CHRISTODOULOU D. - "On quantum geometrodynamics", Proc. Ac. Athens 47 (1972) p. 1-27

[8] - CHRISTODOULOU D. - "On the quantization of self-gravitating fields of integral spin: scalar and vector fields", Nuovo Cimento B 26 (1975) p. 335-369

[9] - DE WITT B.S. - "Quantum theory of gravity; I -the canonical theory", Phys. Rev. 160 (1967) p. 1113-1148

[10] - DICKE R.H. - "The theoretical significance of experimental Relati= vity", Blackie (1964)

[11] - DIEUDONNE' J. - "Elements d'analyse" I and III tome, Gauthier-Villars (1969,1970)

[12] - EBIN D. - "The manifold of riemannian metrics", Proc. Symp. Pure Math. AMS (1970), vol XV p. 11-40

[13] - FISCHER A.E. - "The theory of Superspace", in "Relativity" edited by Carmely-Fickler-Witten , Plenum Press (1970) p. 303-357

[14] - FISCHER A.E. ,MARSDEN J. - "The Einstein equations of evolution - A Geometric approach", Journ. Math. Phys. 13 (1972) p. 546-568

[15] - FRANCAVIGLIA M. - "Metodi globali della geometria differenziale in Relatività Generale", thesis (unpublished), Univ. of Turin (1975)

[16] - GERRETSEN J.H.C. - "Tensor calculus and differential geometry", P. Noordhoff N.V. (1962)

[17] - HAWKING S.W. ,ELLIS G.F.R. - "The large scale structure of space time", Cambridge Univ. Press (1973)

[18] - HOYLE F. ,NARLIKAR J. - "Time symmetric electrodynamics and the arrow of time in cosmology", Proc. Roy. Soc. London A 282 (1963) p. 191-207

[19] - LANG S. - "Differential manifolds", Addison Wesley (1972)

[20] - MARSDEN J. - "Hamiltonian one parameter groups", Arch. Rat. Mech. Analysis 28 (1968) p. 362-396

[21] - WHEELER J.A. - "Geometrodynamics and the issue of the final state", Proc. Les Houches 1963 , Gordon & Breach (1964) p. 317-520

INVERTIBLE FOLIATIONS AND TYPE D-SPACES

R. Debever

Université Libre, Bruxelles

§ 1. Some remarks on type D electrovac spaces.

Electrovac spaces are solutions of Maxwell-Einstein equations

$$R_{ij} + \Lambda\, g_{ij} = \kappa\, T_{ij} \qquad (1.1)$$

of general relativity where

$$T_{ij} = \frac{1}{4}\, g_{ij}\, F_{rs}\, F^{rs} - F_{ir}\, F_j{}^r \qquad (1.2)$$

and

$$F_{[ij;k]} = 0 \quad , \qquad F^i{}_{j;i} = 0 \qquad (1.3)$$

and Λ is the cosmological constant.

We further suppose the space to be of Petrov type D and the Maxwell-field F_{ij} non singular and coïncident.

Non singular means

$$F_{rs}\, F^{rs} + F_{rs}\, \overset{*}{F}{}^{rs} \neq 0 \qquad (1.4)$$

$$\overset{*}{F}_{rs} = \frac{1}{2}\, \eta_{rstu}\, F^{tu} \qquad (1.5)$$

η_{rstu} being the volume element tensor.

Type D coïncident reads, there exist two null vectors $\ell_{(A)}$ (A = 1, 2),

$$\ell^i_{(A)}\, \ell_{(A)i} = 0$$

such that

$$F_{ij}\, \ell^{j}_{(A)} = E_A\, \ell_{(A)i}$$

$$C_{ijkl}\, \ell^{k}_{(A)}\, \ell^{j}_{(A)} = \psi_2\, \ell_{(A)i}\, \ell_{(A)l}$$

where C_{ijkl} is the Weyl tensor.

One knows from W.Kinnersley, J.Plebanski and M.Demianski [1] that the metric tensor g_{ij} solution of equations (1.1) and auxiliary conditions is such that there exist coordinate charts t, z, x, y such that

$$g_{ij} = \begin{array}{c} \begin{array}{cccc} t & z & x & y \end{array} \\ \left(\begin{array}{c|c} g_{i'j'} & 0 \\ \hline 0 & g_{i''j''} \end{array} \right) \end{array}$$

the g_{ij} being functions of two variables only : x, y.

This can be expressed by the fact that there exists an abelian, two-dimensional, invertible group of isometries. [2]

In this context the group of isometries is given by

$$t \to t + t_o\ ,\ z \to z + z_o\ ,\ x \to x\ ,\ y \to y$$

and the invertibility expresses the fact that

$$t \to -t\ ,\quad z \to -z\ ,\quad x \to x\ ,\ y \to y$$

is an isometry.

In general (M, g) being a (pseudo) riemannian manifold an invertible G_p group of isometries in x such that

a) the orbits of the group are non singular

b) there exists in x an involution L_x which keeps fixed the vectors normal to the orbit and inverts the direction of vectors tangent to the orbit. L_x operating in the tangent space $T_x(M)$ one can speak of invertible tensors, it gives sense to say that the riemannian structure is preserved by L.

§ 2. A research program.

It turns out that the electrovac type D spaces admit an invertible, two-dimensional group of isometries. One can ask for some general conditions which lead to the existence of such a group, the conditions being such that they are necessarily fullfilled in the case of electrovac type D spaces. We give in what follows a sufficient condition for a positive answer which is very close to the type D situation, a more general situation is also described.

§ 3. Invertible foliations.

If (M_4, g) is a lorentzian manifold and F a Maxwell-field then there exist null-frames such that

$$g_{ij} = \ell_i n_j + \ell_j n_i - m_i \bar{m}_j - m_j \bar{m}_i \tag{3.1}$$

$$F_{ij} = A Z_{ij} + \text{conj.} \tag{3.2}$$

$$Z_{ij} = \ell_j n_i - \ell_i n_j + m_i \bar{m}_j - m_j \bar{m}_i \tag{3.3}$$

ℓ, n are real null vectors, m, $\bar{m}$ complex conjugate null vectors; Z_{ij} is self dual and

$$Z_i{}^j Z_j{}^k = \delta_i^k \tag{3.4}$$

Let

$$Z = \frac{1}{2} Z_{ij} \, dx^i \wedge dx^j \tag{3.5}$$

be the 2-form associated to Z_{ij}; then one knows from Th.Lepage [3] that

$$dZ = \theta \wedge Z \tag{3.6}$$

where $\underline{\theta}$ is a well defined, complex valued 1-form.
The Maxwell equations (1.3) amount to

$$d\theta = 0 \tag{3.7}$$

The existence of the closed 1-form θ gives rise to a <u>foliation</u>.
This situation can be generalized as follows : Let (M_{2n}, g) be an even dimensional pseudo-riemannian manifold, and Z a non-singular 2-form.

We postulate that

1) there exists a null frame such that

$$\left.\begin{aligned} ds^2 &= 2(\theta^1\ \theta^{n+1} + \theta^2\ \theta^{n+2} + \dots + \theta^n\ \theta^{2n}) \\ Z &= \theta^1 \wedge \theta^{n+1} + \theta^2 \wedge \theta^{n+2} + \dots + \theta^n \wedge \theta^{2n} \end{aligned}\right\} \tag{3.8}$$

2) $$\left.\begin{aligned} dZ &= \theta \wedge Z \\ d\theta &= 0 \end{aligned}\right\} \tag{3.9}$$

Let us remind, from Th.Lepage [4] that Z being non singular, there exist a 3-form ρ_3 and a 1-form θ such that

$$dZ = \rho_3 + \theta \wedge Z \tag{3.10}$$

iff
$$\rho_3 \wedge Z^{n-2} = 0 \ . \tag{3.11}$$

θ again gives rise to a <u>foliation</u>.

We now specialize the structure by the following axioms

3) there exists in the tangent space an involution L,

$$L^2 = I \tag{3.12}$$

4) $$g(LX, LY) = g(X, Y) \tag{3.13}$$

5) L preserves the riemannian connexion, v being any vector

$$DLv = LDv \tag{3.14}$$

if D is the covariant differentiation.

6) Let

$$Z(X, Y) = g(\Omega X, Y) = g(X, \Omega Y) \tag{3.15}$$

$$L\Omega + \Omega L = 0 \tag{3.16}$$

Conditions 1 to 6 define an invertible foliation, associated to what we may call a Riemannian-Maxwellian invertible structure.

§ 4. Riemannian-Maxwellian invertible structures on lorentzian manifolds.

Theorem : A Riemannian-Maxwellian invertible structure on a lorentzian manifold admits a two-dimensional abelian invertible isometry group. The conditions 1-6 and especially 6, imply that with respect to a null frame ℓ, n, m, $\bar{m}$ where (3.1) and (3.3) hold the spin coefficients and the components of the Riemann tensor verify the following conditions [5] :

$$\kappa = \nu \ , \ \tau = \pi \ , \ \sigma = \lambda \ , \ \rho = \mu \ , \ \varepsilon = \gamma \ , \ \beta = \alpha \ . \tag{4.1}$$

$$\psi_o = \psi_4 \quad , \quad \psi_1 = \psi_3 \tag{4.2}$$

$$\phi_{oo} = \phi_{22} \quad , \quad \phi_{o2} = \phi_{2o} \quad , \quad \phi_{o1} = \phi_{21} \tag{4.3}$$

From (3.7) where $\theta = -\ 2\rho\ \theta^1 + 2\mu\ \theta^2 - 2\tau\ \theta^3 + 2\pi\ \theta^4$ (4.4)

one deduces

$$2(\beta + \bar{\beta}) = \tau + \bar{\tau} + \kappa + \bar{\kappa} \quad , \quad 2(\varepsilon - \bar{\varepsilon}) = \rho - \bar{\rho} + \sigma - \bar{\sigma} \tag{4.5}$$

$$\left.\begin{aligned} &(D + \Delta)\rho = (\delta + \bar{\delta})\rho = (D + \Delta)\tau = (\delta + \bar{\delta})\tau = 0 \\ &(D + \Delta)\sigma = (\delta + \bar{\delta})\sigma = (D + \Delta)\kappa = (\delta + \bar{\delta})\kappa = 0 \\ &\delta(\varepsilon + \bar{\varepsilon}) = (D + \Delta)\beta \end{aligned}\right\} \tag{4.6}$$

D, Δ, δ are the derivations in the direction of ℓ, n, m respectively. From (4.4) follows that

$$[D + \Delta, \delta + \bar{\delta}] = 0 \tag{4.7}$$

which, with conditions (4.5) suggest that the Killing vectors

$$k = k_1\, n + k_2\, \ell - k_3\, \bar{m} - k_4\, m \tag{4.8}$$

are such that

$$k_1 = k_2 = k \qquad\qquad k_3 = k_4 = j \tag{4.9}$$

Killing equations then read

$$\left.\begin{aligned} Dk &= (\varepsilon + \bar{\varepsilon})k - (\kappa + \bar{\kappa})j \\ \delta k &= \tfrac{1}{2}(\bar{\tau} - \tau + \bar{\kappa} - \kappa)k + (\rho - \bar{\rho})j \\ Dj &= (\tau + \bar{\tau})k - \tfrac{1}{2}(\rho + \bar{\rho} + \sigma + \bar{\sigma})j \\ \delta j &= (\bar{\sigma} - \sigma)k + (\beta - \bar{\beta})j \end{aligned}\right\} \tag{4.10}$$

Using conditions (4.2), (4.3), (4.5) one can verify that the commutation condition

$$(\delta D - D\delta)(.) = (\beta + \bar{\beta} - \bar{\tau} - \kappa)D(.) - (\varepsilon - \bar{\varepsilon} - \bar{\rho} - \sigma)\delta(.) \tag{4.11}$$

applied to the system (4.10) is satisfied.

As a consequence the space of solutions of (4.10) is two-dimensional, giving rise to a 2-dimensional group of isometries, abelian because of (4.7).

Finally if the Killing vectors are written as

$$\frac{\partial}{\partial z} \quad , \quad \frac{\partial}{\partial t} \tag{4.12}$$

the ds^2 reads

$$ds^2 = (Ldt + Mdz)^2 - (Ndt + Pdz)^2 - S^2(dx^2 + dy^2) \qquad (4.13)$$

where L, M, N, P, S are functions of x, y only.
An example of such a situation, besides the type D, to be considered later has recently been given by R.Mc Lenaghan and N.Tarig [6].

§ 5. Riemannian-premaxwellian invertible structures.

Riemannian-premaxwellian structures are a generalization based on the substitution to $d\theta = 0$ of the condition :

$$d(\theta + \bar{\theta}) = 0 \qquad (5.1)$$

We have seen that (3.9) is equivalent to Maxwell's equations.
(5.1), or premaxwellian is equivalent to state that the tensor $T^i{}_j$ defined by (1.2), where F_{ij} is given by (3.2) and θ by (3.6) is such that $T^i{}_j$ is conservative : [7]

$$T^i{}_{j;i} = 0$$

The theorem of § 4 can be extended to the premaxwellian case following the same line of argument.

§ 6. The type D electrovac solutions. [8]

The theorem of § 4 has a direct bearing on the type D electrovac solutions defined at the beginning.
The principal vectors $\ell_{(A)}$ (1.6) play a symmetric rôle, if there exists an electromagnetic non singular, coïncident field we have the situation described as invertible Riemannian-Maxwellian and the abelian G_2 exists.
The integration of the general solution can be performed as was done in R.Debever. [9]

If the space is empty :

$$R_{ij} = 0 \tag{6.1}$$

because of Bianchi identities the type D is such that, with respect to a null tetrad associated with $\ell_{(A)}$:

$$\theta = -\frac{2}{3}\frac{d\psi_2}{\psi_2} \tag{6.2}$$

so that the Riemannian invertible structure is again Maxwellian.

REFERENCES

[1] KINNERSLEY, W. J.M.P. 10, 1195, 1969
PLEBANSKI, J. - DEMIANSKI, M. J.M.P. to appear.

[2] CARTER, B. J.M.P., 10, 70, 1969

[3] LEPAGE, Th. Bull. Cl. Sc. Acad. r. Belgique, XXXIII, 288, 1947

[4] Loc. cit.

[5] NEWMAN E. & PENROSE,R. J.M.P., 3, 566, 1962

[6] Mc LENAGHAN, R.G. & TARIG, N. A new solution of the Einstein-Maxwell equations. Preprint. (University of Waterloo, Ontario, Canada).

[7] MISNER, J. & WHEELER, A. Ann. Phys. 2, 525, 1957

[8] The relationships between type D solutions and the abelian G_2 group have been investigated by
HUGSTON, L. and SOMMERS, P. Commun. Math..Phys., 32, 147, 1973 and 33, 129, 1973.

[9] DEBEVER, R. Bull. Cl. Sc. Acad. r. Belgique, LV, 8, 1969.

DEFORMATIONS OF THE EMBEDDED EINSTEIN SPACES

Richard Kerner

Département de Mécanique Relativiste, Université Pierre et Marie Curie
4, Place Jussieu, Paris, France

Abs act. We discuss a method of studying the stability of solutions of Einstein's equations, which can be outlined as follows: Consider an embedding of a given Einstein space V_4 into a pseudo-Euclidean space $E^N_{p,q}$ $(N>4,\ p+q = N)$, (p,q) describing the signature of the space $E^N_{p,q}$. Then all the geometrical objects of V_4 can be expressed in terms of the embedding functions, $Z^A(x^i)$, $A = 1,2,\dots,N$, $i = 0, 1,2,3$.

Then let us deform the embedding: $Z^A \to Z^A + \varepsilon v^A$, ε being an infinitesimal parameter. The Einstein equations can be developed then in the powers of ε; we study the equations arising by requirement of the vanishing of the first or second order terms. Some partial results concerning the de Sitter, Einstein and Minkowskian spaces are given.

I Introduction.

In the recent few years the problem of the topology of the set of solutions of Einstein's equations has been largely discussed. The essential question could be stated as follows : given an exact solution $\overset{\circ}{g}_{jk}$ of the system:

$$R_{jk} - \frac{1}{2} g_{jk} R = \lambda g_{jk} \tag{0}$$

(here R_{jk} is the Ricci tensor, g_{jk} the metric tensor, T_{jk} the energy-momentum tensor, R the scalar curvature and λ the cosmological constant), do there exist other exact solutions which in some suitable topology introduced in the space of the symmetric tensor fields could be regarded as being near the given solution $\overset{\circ}{g}_{jk}$? Then, do there exist any isolated points in the set of all solutions of the system (0) ?

It seems unnecessary to underline here the complexity of such a problem. A lot of interesting partial results have been obtained recently by Choquet-Bruhat, Marsden and Fischer, Deser, Montcrief, Geroch and others.[1]-[9] One of the weak points of these results is that they give information in terms of the components of the metric tensor, but in general one can not be sure if the two metrics arbitrarily near such other in some suitable topology can be realized on the four-dimensional manifold of the same kind. For example, the Minkowskian metric g_{ij} = diag (+ - - -) can be realized on R^4, or on a cylinder C_4 embedded in R^5 as

$$Z_1 = t$$

$$Z_2 = x$$

$$Z_3 = y$$

$$Z_4 = R \cos \frac{z}{R}$$

$$Z_5 = R \sin \frac{z}{R}$$

Both manifolds represent exact solutions of the system (0), but it seems unnatural to consider them as being "near" each other because of the different topological properties of $\mathbb{R}^4$ and C_4 .

Here we propose to investigate the problems stated at the beginning by a slightly modified method, which shall in a way take naturally into account the differences coming from the global properties of the Einstein spaces. The method proposed has also another advantage, which is to eliminate from the beginning a number of supercifial degrees of freedom contained in g_{jk}. The method is based on the deformations of the embeddings of the Einstein spaces.

§ 2. The equations describing the deformations of the embeddings.

2.1) Notations :

$E^N_{p,q}$: the pseudo-Euclidean space of dimension N, with the metric tensor $\eta_{AB} = \text{diag}\,(p^+, q^-)$; A, B = 1, 2, ... N; a point in $E^N_{p,q}$ is given by N numbers Z^A (an N-dimensional vector). $p + q = N$, $p \geq 1$, $q \geq 3$.

$\overset{\circ}{V}_4$: a given Einstein manifold, i.e. a pseudo-Riemannian 4-dimensional manifold with a metric $\overset{\circ}{g}_{ij}$ (i,j = 0, 1, 2, 3) verifying :

$$\overset{\circ}{R}_{ij} - \tfrac{1}{2}\overset{\circ}{g}_{ij}\overset{\circ}{R} = \overset{\circ}{\lambda}\,\overset{\circ}{g}_{ij} \qquad \overset{\circ}{\lambda} = \text{Const} \tag{1}$$

An embedding is a regular (C^∞) injection mapping

$$\overset{\circ}{V}_4 \xrightarrow{\;Z\;} E^N_{p,q} \tag{2}$$

given by N functions $Z^A(x^k)$ such that :

$$\overset{\circ}{g}_{ij} = \eta_{AB}\,\partial_i Z^A\,\partial_j Z^B \tag{3}$$

An infinitesimal deformation of the embedding produces a new manifold V_4 defined by the new embedding functions :

$$w^A(x) = Z^A(x) + \epsilon\, v^A(x) \tag{4}$$

ϵ being an infinitesimal parameter.

Therefore

$$V_4 \xrightarrow{\;w\;} E^N_{p,q} \tag{5}$$

and the new metric is given by :

$$g_{ij} = \overset{\circ}{g}_{ij} + 2\epsilon\,\eta_{AB}\,\partial_i Z^A \partial_j v^B + \epsilon^2\,\eta_{AB}\,\partial_i v^A \partial_j v^B \tag{6}$$

If the vector field v^A is tangent to the $\overset{\circ}{V}_4$ in $E^N_{p,q}$, then the deformation :

$$Z^A \longrightarrow Z^A + \varepsilon\, v^A(x) \tag{7}$$

can be implemented by a coordinate transformation in $\overset{\circ}{V}_4$:

$$Z^A(x) + \varepsilon\, v^A(x) = Z^A\left(x + \varepsilon\varphi(x)\right) \tag{8}$$

and as such is of no interest. Therefore the deformation we shall investigate will be subject to the condition of orthogonality :

$$\forall_i \quad \eta_{AB}\, v^A\, \partial_i Z^B = 0 \tag{9}$$

For a given $\overset{\circ}{V}_4$ embedded in $E^N_{p,q}$ there exist $N - 4$ independant fields X^B_α ($\alpha = 1, 2, \dots N-4$) verifying (9); so the deformations which are of interest have the form $v^B = v^\alpha X^B_\alpha$ (10)

Suppose the embedding (2) is <u>minimal</u> , i.e. there do not exist any $E^{N'}_{p'q'}$, $N' < N$ for which the embedding is possible. But there exist many $E^{N'}_{p'q'}$, $N' > N$ admitting non-minimal embeddings. $E^N_{p,q} \subset E^{N'}_{p'q'}$; then, let the metric tensor of $E^{N'}_{p'q'}$ be

$$\begin{pmatrix} \eta_{AB} & \\ & \eta_{\alpha\beta} \end{pmatrix} \qquad \begin{array}{l} A, B = 1, 2, \dots N \\ \alpha, \beta = N+1, N+2, \dots N' \end{array} \tag{11}$$

The deformations in $E^{N'}_{p'q'}$ can be divided in two categories :

1) deformations leaving V_4 embedded in $E^N_{p,q}$

2) deformations in the directions orthogonal to the subspace $E^N_{p,q}$ in the $E^{N'}_{p'q'}$

Any deformation of $\overset{\circ}{V}_4$ embedded in $E^{N'}_{p'q'}$ can be regarded as a sum of the deformation of the 1st type and of the deformation of the 2nd type.

We shall remark that for the deformation of the second type the induced metric g_{ij} <u>does not</u> change in the first order :

$$g_{ij} = \overset{\circ}{g}_{ij} + \varepsilon^2\, \eta_{\alpha\beta}\, \partial_i v^\alpha\, \partial_j v^\beta \tag{12}$$

because $\partial_i v^\alpha$ is always orthogonal to $\partial_k Z^A$ (11).

∇_i means the covariant derivation with respect to the Christoffel symbols of the non-perturbed metric $\overset{\circ}{g}_{ij}$; of course

$$\partial_i z^A = \nabla_i z^A \ , \quad \nabla_i v^B = \partial_i v^B \tag{13}$$

Finally , if the deformed metric is :

$$g_{ij} = \overset{\circ}{g}_{ij} + \varepsilon \overset{1}{g}_{ij} + \varepsilon^2 \overset{2}{g}_{ij} \tag{14}$$

with

$$\begin{aligned} \overset{1}{g}_{ij} &= 2\eta_{AB} \nabla_i z^A \nabla_j v^B \\ \overset{2}{g}_{ij} &= \eta_{AB} \nabla_i v^A \nabla_j v^B \end{aligned} \tag{15}$$

the equation (1) can be also developed in orders of magnitude $\varepsilon^0, \varepsilon^1, \varepsilon^2, \varepsilon^3$, etc, as follows : (we shall also admit that the cosmological constant can vary with the deformation : $\overset{\circ}{\lambda} \rightarrow \overset{\circ}{\lambda} + \varepsilon \overset{1}{\lambda} + \varepsilon^2 \overset{2}{\lambda} + \dots$)

Because (1) can be written as

$$\overset{\circ}{R}_{ij} = \frac{\overset{\circ}{R}}{4} \overset{\circ}{g}_{ij} \qquad \left(\overset{\circ}{\lambda} = - \frac{\overset{\circ}{R}}{4} \right) \tag{16}$$

we can develop :

$$R_{ij} = \overset{\circ}{R}_{ij} + \varepsilon \overset{1}{R}_{ij} + \varepsilon^2 \overset{2}{R}_{ij} + \dots$$

$$R = \overset{\circ}{R} + \varepsilon \overset{1}{R} + \varepsilon^2 \overset{2}{R} + \dots$$

where

$$\overset{1}{R} = \overset{\circ}{g}{}^{kl} \overset{1}{R}_{kl} + \overset{1}{g}{}^{kl} \overset{\circ}{R}_{kl} \tag{17}$$

$$\overset{2}{R} = \overset{\circ}{g}{}^{kl} \overset{2}{R}_{kl} + \overset{1}{g}{}^{kl} \overset{1}{R}_{kl} + \overset{2}{g}{}^{kl} \overset{\circ}{R}_{kl} \ , \ etc.$$

and

$$\overset{1}{g}{}^{kl} = - \overset{\circ}{g}{}^{ki} \overset{\circ}{g}{}^{lj} \overset{1}{g}_{ij} \tag{18}$$

$$\overset{2}{g}{}^{kl} = - \overset{\circ}{g}{}^{ki} \overset{\circ}{g}{}^{lj} \overset{2}{g}_{ij} + \overset{\circ}{g}{}^{ki} \overset{\circ}{g}{}^{lj} \overset{\circ}{g}{}^{mn} \overset{1}{g}_{im} \overset{1}{g}_{jn}$$

Finally we can develop (16) as follows :

$$R_{ij} - \frac{R}{4} g_{ij} = \overset{o}{R}_{ij} - \frac{\overset{o}{R}}{4}\overset{o}{g}_{ij} + \varepsilon\left[\overset{1}{R}_{ij} - \frac{\overset{1}{R}}{4}\overset{o}{g}_{ij} - \frac{\overset{o}{R}}{4}\overset{1}{g}_{ij}\right] + \tag{19}$$
$$+ \varepsilon^2\left[\overset{2}{R}_{ij} - \frac{\overset{2}{R}}{4}\overset{o}{g}_{ij} - \frac{\overset{1}{R}}{4}\overset{1}{g}_{ij} - \frac{\overset{o}{R}}{4}\overset{2}{g}_{ij}\right] + \ldots$$

We define the deformation preserving in the <u>first order</u> the property of $\overset{o}{V}_4$ being Einsteinian as a set of functions $\nu^B = \nu^\alpha X^B_\alpha$ (10) verifying the equation :

$$\overset{1}{R}_{ij} - \frac{\overset{1}{R}}{4}\overset{o}{g}_{ij} - \frac{\overset{o}{R}}{4}\overset{1}{g}_{ij} = 0 \tag{20}$$

Preserving Einstein equations up to the second order means that ν^B verifies (20) <u>and</u> :

$$\overset{2}{R}_{ij} - \frac{\overset{2}{R}}{4}\overset{o}{g}_{ij} - \frac{\overset{1}{R}}{4}\overset{1}{g}_{ij} - \frac{\overset{o}{R}}{4}\overset{2}{g}_{ij} = 0 \tag{21}$$

etc .

Now we shall proceed to the explicit form of the equations (20) and (21).

2.2) <u>Derivation of the equations.</u>

We use the following convention :

$$\overset{o}{R}_{ijk}{}^{m} = \nabla_i \Gamma^m_{jk} - \nabla_j \Gamma^m_{ik} \tag{22}$$

$$\nabla_i \nabla_j u_k - \nabla_j \nabla_i u_k = \overset{o}{R}_{ijk}{}^{m} u_m \tag{23}$$

$$\overset{o}{R}_{ijkm} = \overset{o}{g}_{ml} \overset{o}{R}_{ijk}{}^{l} \tag{24}$$

$$\Gamma^l_{jk} = \frac{1}{2}\overset{o}{g}{}^{lm}\left(\partial_j \overset{o}{g}_{km} + \partial_k \overset{o}{g}_{jm} - \partial_m \overset{o}{g}_{jk}\right) \tag{25}$$

$$\nabla_i u_l = \partial_i u_l - \Gamma^m_{il} u_m \tag{26}$$

$$\nabla_i \overset{o}{g}_{kl} = \nabla_i \overset{o}{g}{}^{kl} = 0 \tag{27}$$

Let us express Γ^l_{jk} in terms of Z^A. We have :

$$\overset{\circ}{g}_{jk} = \eta_{AB} \nabla_j Z^A \nabla_k Z^B$$

$$\partial_i \overset{\circ}{g}_{jk} = \eta_{AB} \left(\partial^2_{ij} Z^A \partial_k Z^B + \partial_j Z^A \partial^2_{ik} Z^B \right) \tag{28}$$

and
$$\Gamma^l_{jk} = \eta_{AB} \overset{\circ}{g}{}^{lm} \partial_m Z^A \partial^2_{jk} Z^B \tag{29}$$

But
$$\partial^2_{jk} Z^B = \nabla_j \nabla_k Z^B - \Gamma^l_{jk} \nabla_l Z^B \tag{30}$$

so that
$$\Gamma^l_{jk} = \eta_{AB} \overset{\circ}{g}{}^{lm} \nabla_m Z^A \nabla_j \nabla_k Z^B + \Gamma^l_{jk} \tag{31}$$

and we have the useful identity :

$$\eta_{AB} \overset{\circ}{g}{}^{lm} \nabla_m Z^A \nabla_j \nabla_k Z^B = 0$$

or

$$\eta_{AB} \nabla_m Z^A \nabla_j \nabla_k Z^B = 0 \tag{32}$$

also
$$\nabla_i \left(\eta_{AB} \nabla_m Z^A \nabla_j \nabla_k Z^B \right) = 0$$

and we have :

$$\eta_{AB} \left[\nabla_i \left(\nabla_m Z^A \nabla_j \nabla_k Z^B \right) - \nabla_j \left(\nabla_m Z^A \nabla_i \nabla_k Z^B \right) \right] = 0 \tag{33}$$

But :
$$\eta_{AB} \left[\nabla_i \left(\nabla_m Z^A \nabla_j \nabla_k Z^B \right) - \nabla_j \left(\nabla_m Z^A \nabla_i \nabla_k Z^B \right) \right] =$$

$$= \eta_{AB} \left[\nabla_i \nabla_m Z^A \nabla_j \nabla_k Z^B - \nabla_j \nabla_m Z^A \nabla_i \nabla_k Z^B \right] +$$

$$+ \eta_{AB} \left[\nabla_i \nabla_j \nabla_k Z^B - \nabla_j \nabla_i \nabla_k Z^B \right] \nabla_m Z^A =$$

$$= \eta_{AB} \left[\nabla_i \nabla_m Z^A \nabla_j \nabla_k Z^B - \nabla_j \nabla_m Z^A \nabla_i \nabla_k Z^B \right] +$$

$$+ \eta_{AB} \overset{\circ}{R}_{ijk}{}^{l} \overset{\circ}{g}_{ml} \tag{34}$$

So that we can write :

$$\overset{\circ}{R}_{ijkm} = - \eta_{AB} \left[\nabla_i \nabla_m Z^A \nabla_j \nabla_k Z^B - \nabla_j \nabla_m Z^A \nabla_i \nabla_k Z^B \right] \tag{35}$$

Let us introduce the notation :

$$\Theta_{ijkm}(u^A, v^B) = -\eta_{AB}\left[\nabla_i \nabla_m u^A \nabla_j \nabla_k v^B - \nabla_j \nabla_m u^A \nabla_i \nabla_k v^B\right] \tag{36}$$

and

$$\Theta_{jk}(u^A, v^B) = \overset{\circ}{g}{}^{im}\, \Theta_{ijkm}(u^A, v^B)$$

It is trivial now to see that :

$$\overset{1}{R}_{jk} = \Theta_{jk}(z^A, v^B) + \Theta_{jk}(v^A, z^B) - 2\eta_{AB}\, \overset{\circ}{g}{}^{im}\, \overset{\circ}{R}_{ijk}{}^{l}\, \nabla_m z^A \nabla_l v^B \tag{37}$$

and the equation (20) becomes :

$$\Theta_{jk}(z^A, v^B) + \Theta_{jk}(v^A, z^B) - 2\eta_{AB}\, \overset{\circ}{g}{}^{im}\, \overset{\circ}{R}_{ijk}{}^{l}\, \nabla_m z^A \nabla_l v^B = \\ = -\overset{1}{\lambda}\, \overset{\circ}{g}_{jk} - 2\overset{\circ}{\lambda}\, \eta_{AB} \nabla_j z^A \nabla_k v^B \tag{38}$$

For the general deformation :

$$z^A \longrightarrow z^A + \varepsilon \overset{1}{v}{}^A + \varepsilon^2 \overset{2}{v}{}^A + \cdots \tag{39}$$

The eqyation (21) becomes :

$$\Theta_{jk}(z^A, \overset{2}{v}{}^B) + \Theta_{jk}(\overset{2}{v}{}^A, z^B) + 2\Theta_{jk}(\overset{1}{v}{}^A, \overset{1}{v}{}^B) - \\ - 2\eta_{AB}\, \overset{\circ}{g}{}^{im}\, \overset{\circ}{R}_{ijk}{}^{l}\left(\nabla_m z^A \nabla_l \overset{2}{v}{}^B + \nabla_m \overset{1}{v}{}^A \nabla_l \overset{1}{v}{}^B\right) + \\ + 2\eta_{AB}\eta_{CD}\, \overset{\circ}{g}{}^{is}\, \overset{\circ}{g}{}^{mp}\, \overset{\circ}{R}_{ijk}{}^{l}\, \nabla_s z^C \nabla_m z^A \nabla_p \overset{1}{v}{}^D \nabla_l \overset{1}{v}{}^B - \\ - 2\eta_{AB}\, \overset{\circ}{g}{}^{im}\, \overset{1}{R}_{ijk}{}^{l}\, \nabla_m z^A \nabla_l \overset{1}{v}{}^B = \\ = -\overset{\circ}{\lambda}\left(2\eta_{AB} \nabla_j z^A \nabla_k \overset{2}{v}{}^B + \eta_{AB} \nabla_j \overset{1}{v}{}^A \nabla_k \overset{1}{v}{}^B\right) - \overset{2}{\lambda}\, \overset{\circ}{g}_{jk} - \\ - 2\overset{1}{\lambda}\, \eta_{AB} \nabla_j z^A \nabla_k \overset{1}{v}{}^B \tag{40}$$

§ 3. Applications : Einsteinian deformations of some simple Einstein spaces.

3.1. We begin with two very simple examples : the de Sitter space and the Einstein space. Both spaces can be embedded globally (the minimal embedding) in $E^5_{1,4}$:

de Sitter space :

$$\eta_{AB} z^A z^B = R^2$$

Einstein space :

$$z^1 = t, \quad (z^2)^2 + (z^3)^2 + (z^4)^2 + (z^5)^2 = R^2$$

The non-trivial (i.e. orthogonal to $\overset{\circ}{V}_4$) deformations are given by :

$\nu^A = \nu z^A$, $A = 1, 2, \dots 5$ for the de Sitter space.

$A = 2, 3, \dots 5$ for the Einstein space.

indeed, we have :

$$\eta_{AB} z^A \nabla_j z^B = 0 \tag{41}$$

Let us consider the deformations which do not go beyond the $E^5_{1,4}$; then, in the first-order approximation $\overset{1}{g}_{jk}$ is conformal to $\overset{\circ}{g}_{jk}$:

$$\overset{1}{g}_{jk} = 2\eta_{AB}\,\partial_j z^A \partial_k \nu^B = 2\nu\,\eta_{AB}\,\partial_j z^A \partial_k z^B + \tag{42}$$

$$+ 2\eta_{AB} z^A \partial_j \nu\, \partial_k z^B = 2\nu\, \overset{\circ}{g}_{jk}$$

The equation (38) reduces itself to the following form : (in both cases) :

$$\nabla_j \nabla_k \nu = \frac{1}{6}\left(\overset{1}{\lambda} + 4\overset{\circ}{\lambda}\nu\right)\overset{\circ}{g}_{jk} \tag{43}$$

From the eqs. (43) we can easily see that $\partial_\theta \nu = \partial_\varphi \nu = 0$, so that the only deformations verifying (43) are spherically symmetric.

Then, it is easy to see that for the spherically-symmetric deformations (i.e. $\nu = \nu(r,t)$) the only possible solution is :

$$\nu = \text{Const}, \qquad \nu = -\frac{\overset{1}{\lambda}}{4\overset{\circ}{\lambda}} \tag{44}$$

which corresponds just to the dilatation in $E^5_{1,4}$. Therefore the same will be true <u>for any order</u> . We have thus the following <u>Theorem 1</u>: any <u>Einsteinian</u> deformations leaving $\overset{\circ}{V}_4$ in $E^5_{1,4}$ are in all orders just simple dilatations (conformal mappings).

This extreme rigidity of the spherically-symmetric space times is due of course to the limited class of deformations we have considered up to now. If we admit the deformations going beyond $E^5_{1,4}$, e.g. in the sixth direction in $E^6_{1,4}$, then we obtain the equations of the second order :

$$2\Theta_{jk}(\overset{1}{v}^A, \overset{1}{v}^B) - 2\eta_{AB}\,\overset{\circ}{g}^{im}\,\overset{\circ}{R}_{ijk}{}^{l}\,\nabla_m \overset{1}{v}^A \nabla_l \overset{1}{v}^B = -\overset{\circ}{\lambda}\,\eta_{AB}\nabla_j \overset{1}{v}^A \nabla_k \overset{1}{v}^B - \overset{2}{\lambda}\,\overset{\circ}{g}_{jk} \tag{45}$$

where $\nu^A = \nu\delta^A_6$.

In the case of $\nu = \nu(r,t)$ the system (45) is of Monge-Ampère type and admits plenty of local solutions.

3.2. <u>Deformations of the Minskowskian space-time M_4</u>

We shall consider M_4 as a pseudo-Euclidean plane embedded in some pseudo-Euclidean space $E^N_{p,q}$. The first four coordinates in $E^N_{p,q}$ are identical with the coordinates in M_4, the other ones being orthogonal to the first four in $E^N_{p,q}$. The non-trivial deformations are thus :

$$\nu^A = 0 \quad \text{for} \quad A = 1, 2, 3, 4$$

$$\nu^A \neq 0 \quad \text{for} \quad A = 5, 6, \ldots N$$

The equations to be verified are :

$$2\eta_{AB}\,\overset{\circ}{g}^{im}\left[\nabla_i\nabla_m v^A \nabla_j\nabla_k v^B - \nabla_j\nabla_m v^A \nabla_i\nabla_k v^B\right] = -\overset{2}{\lambda}\,\overset{\circ}{g}_{jk} \tag{46}$$

where A, B, = 5, 6, ... N.

Now, two types of deformations have to be dealt with separately :

a) the deformations preserving the null scalar curvature, i.e. with $\overset{2}{\lambda} = 0$.

b) the deformations giving rise to a non-null constant curvature, with $\overset{2}{\lambda} \neq 0$.

ad a) In the case of the plane symmetry, with $\overset{2}{\lambda} = 0$, one of the most obvious solutions (in $E^5_{1,4}$ or $E^5_{2,3}$) is $\upsilon = \upsilon(k_i\, x^i)$, k_i being a constant vector.

Then
$$g_{jk} = \overset{o}{g}_{jk} \pm \varepsilon^2 \dot{\upsilon}^2 k_j k_k \tag{47}$$
which corresponds to a plane wave propagating with an arbitrary velocity ($k^2 \neq 0$). However, if we want to verify :

$$g_{jk}\, g^{kl} = \delta^l_j \tag{48}$$

exactly, instead of the approximate identity :

$$g_{jk}\, g^{kl} = \delta^l_j + \mathcal{O}(\varepsilon^2) \tag{49}$$

then it does impose the condition $k^2 = 0$.

In E^6, E^7, etc., our solution can be generalized :

$$\upsilon^A = \upsilon^A\left(k^{(A)}_i x^i\right) \tag{50}$$

$k^{(A)}_i$ being linearly independent null vectors. This makes sense up to E^7 ; afterwards any other $k^{(A)}$ can be expressed as a linear combination of the former ones.

ad b) We can not solve the equations (46) in general, but we can do it for the spherically-symmetric deformations in $E^5_{p,q}$. In this case the system (46) reduces itself to the following four equations :

$$\text{I} \qquad \frac{2}{r}\,\partial_r v\,\partial^2_{tr} v = 0$$

$$\text{II} \qquad (\partial^2_{tr} v)^2 - \partial^2_{tt} v\left(\partial^2_{rr} v + \frac{2}{r}\partial_r v\right) = -\overset{2}{\lambda} \tag{51}$$

$$\text{III} \qquad \partial^2_{tt} v\,\partial^2_{rr} v - \frac{2}{r}\partial^2_{rr} v\,\partial_r v - (\partial^2_{tr} v)^2 = \overset{2}{\lambda}$$

$$\text{IV} \qquad \left[\partial^2_{tt} v - \partial^2_{rr} v + \frac{1}{r}\partial_r v\right]\partial_r v = \overset{2}{\lambda}\, r$$

In the case $\overset{2}{\lambda} = 0$, the only possible solution is $v = 0$. For $\overset{2}{\lambda} \neq 0$ we have :

$$\partial_{tr} v = 0 \quad \Rightarrow \quad v(r,t) = T(t) + R(r) \tag{52}$$

and our equations are :

$$\text{II} \qquad \ddot{T}\left(\ddot{R} + \frac{2}{r}\dot{R}\right) = \overset{2}{\lambda}$$

$$\text{III} \qquad \ddot{R}\left(\ddot{T} - \frac{2}{r}\dot{R}\right) = \overset{2}{\lambda} \tag{53}$$

$$\text{IV} \qquad \left(\ddot{T} - \ddot{R} - \frac{1}{r}\dot{R}\right)\dot{R} = \overset{2}{\lambda}\, r$$

From II we obtain now $\ddot{T} = \overset{2}{\lambda}\beta = \text{Const}$ (54)

and $$\frac{1}{r^2}\frac{d}{dr}\left(r^2 \frac{dR}{dr}\right) = \beta$$

so that $$\dot{R} = \frac{\beta r}{3} + \frac{\gamma}{r^2}$$

and $$\ddot{R} = \frac{\beta}{3} - \frac{2\gamma}{r^3} \qquad \gamma = \text{Const} \tag{55}$$

From the consistency with III and IV we obtain now $\gamma \equiv 0$, and

$$\beta = \pm \frac{3\sqrt{\lambda}}{\sqrt{3\lambda - 2}} \tag{56}$$

For β to be real, either $\lambda < 0$, either $\lambda > \frac{2}{3}$; the second condition makes no sense because of the continuity of the deformation.

We have obtained thus a

Theorem 2 :

The only admissible spherically-symmetric deformations of the M_4 embedded in $E^5_{p,q}$ are leading to the solutions with a negative cosmological constant, $\lambda < 0$.

The corresponding metric is then :

$$g_{tt} = 1 \pm \varepsilon^2 \lambda^{2} \beta^2 t^2$$
$$g_{rr} = -1 \mp \varepsilon^2 \frac{\lambda}{3\lambda - 2} r^2 \qquad (57)$$

It can be considered as an approximate solution only for finite t, r, otherwise it becomes unbounded and can not be taken into consideration as a global deformation.

The last Theorem is true only for the embedding in $E^5_{p,q}$, and is not valid for higher dimensions.

References.

1. Marsden J.E., Fischer A.E., Journal of Mathematical Physics, Vol.13, N°4, (1972)
2. Choquet-Bruhat Y., G.R.G - Journal, Vol. 5, N°1 (1974)
3. Fischer A.E., Marsden J.E., Springer Notes in Physics, 14, N.Y. 1972
4. Choquet-Bruhat Y., Commun. Math. Phys., Vol. 21, (1971)
5. Brill D.R., Isolated Solutions in General Relativity, in "Gravitation" Naukova Dumka , Kiev, (1972)
6. Kerner R., Approximate Solutions of Einstein's Equations in "Relativity & Gravitation", Gordon & Breach, N.Y. (1972)
7. Fischer A.E., Marsden J.E., GRG journal, Vol.4, N°4, (1973)
8. Moncrieff V., Taub A., preprint "Second variation and stability of the Relativistic, Nonrotating stars"
9. Brill D.R., Deser S., Annals of Physics, Vol. 50, N°3 (1968)

The Causal Structure of Singularities

by

Hans-Jürgen Seifert

Hochschule der Bundeswehr Hamburg

Extended version of a lecture given at the conference on "Differential Geometrical Methods in Mathematical Physics" (Bonn; 1.-4. July 1975)

based on an unpublished manuscript by H. Müller zum Hagen, H.J.Seifert, P. Yodzis: "When is a Singularity Innocuous?" (Hamburg 1974)
Work partly supported by the Deutsche Forschungsgemeinschaft.

Abstract:

Two definitions of singularities are proposed which include breakdown of continuity and of causality as well as incompleteness. Some hypotheses about the causal relation between singularities in gravitational collapse and the outside region are discussed.

1. Introduction

Perhaps the most thrilling objects in classical (i.e. no quantum effects whatever considered) general relativity in the last 12 years have been the singularities of space-time; the following problems have been investigated:

(i) the *occurrence:* It has turned out that they are a quite general phenomenon.

(ii) the *description:* A highly technical but mathematically elegant description of singular points (b-boundary) has been found.

(iii) the *causal interaction:* The question, where they occur, or, more precisely, which parts of the universe they can influence or be influenced by, is still unclear.

As this last point is the topic of my lecture I can only present *belief, notions, conjectures, arguments,* and *hopes* rather than *knowledge, definitions, theorems, proofs* and *results.*

Everything is done in the framework of classical relativity, hence the fascinating new results of Hawking on quantum fields near black holes [8] are not taken into consideration, despite the fact that they give hints that the classical theory becomes invalid in situations where most people had expected it to be applicable up to a very high accuracy[1]. Nor is it explicitly discussed what the occurrence of a singularity in some solution of relativity means for the physics of nature. At first it displays a shortcoming of the theory[2]. Furthermore it is plausible that at least certain classes of singularities (discussed in § 6) do not only indicate a breakdown of the theory but demonstrate (within the region where the theory can be applied) that physically very uncomfortable situations occur in nature[3]. Even if one possessed a new theory which describes these situations by large but finite values of some quantities, one might prefer classical relativity to detect these situations, as mathematically, in most cases it is easier to detect "infinities" rather than "very large values". I shall concentrate on the question of *collapse*[4] and neglect the other "singular end" of the cosmos: the big bang, which also raises many interesting questions of causal structure (particle horizons, Mixmaster universe, etc.).

2. Singularities

The Einstein field equations are hyperbolic, nonlinear and describe the attractive gravitational interaction, hence one expects singularities to be a quite general phenomenon. But in contrast to the other classical fields which are functions on a given, everywhere regular background, the gravitational field *is* geometry. So, if it becomes too singular, the background is no longer well defined and the place of singularity fades away. Only a more sophisticated concept can bypass this difficulty: singularity is the breakdown of some basic structure of general relativity, or, more cautiously formulated, is the violation of a regularity condition of some basic structure.

---

[1] In fact, they remember us that quantum effects may alter qualitatively predictions of classical theory not only for space-time regions of very high curvature but also for regions with comparatively small curvature if there is enough time (10^{66} years in the case of a black hole of one solar mass). The appearance of the black hole is not affected during most of this period, but the basis for the classical stationary end-state considerations has disappeared.

[2] This "negative" aspect of singularity: breakdown of some basic mathematical structure (notion 1) is discussed in § 2; regularity conditions for these three structures are given in §§ 3, 4, 5.

[3] This "positive" aspect of singularity: occurrence of some physically critical situation (notion 3) is discussed in § 6, conditions for their generality and strength are given in §§ 7, 8.

[4] This application is discussed in the remaining sections; the notion of collapse is given in § 9, and two hypotheses about its causal structure are discussed in §§ 10, 11.

Notion 1: *Singularity*

Given

(a) a *model* (e.g. vacuum field equations; energy inequality),

(b) an *initial situation* (e.g. a compact spacelike surface),

one obtains a *singular development*, if any space-time V which fulfills the condition for the model (a) throughout and in which the situation (b) is imbedded,

either (i) contains strong *discontinuities*
(e.g. violations of the junction conditions)

or (ii) is *acausal*
(e.g. time-periodical world-lines are possible)

or (iii) is *incomplete*
(e.g. the history of some test-particle is not described for any value of proper time)

For any of these basic (case i: *metric*; case ii: *conformal*; case iii: *affine*) structures there is a complicated hierarchy of regularity conditions, of which parts will be presented in the next sections (§§ 3 - 5).

Remarks: This notion is a negative one, as it does not tell us what the singular behaviour really is. Peculiar examples as given in § 6 will display this disadvantage. On the other hand, such an approach seems to be natural in the context of the famous Hawking-Penrose theorems on existence of singularities [9; 256-275]. In the common definitions of singularities case (i) is forgotten (whether discontinuities seem to be too trivial or too prerelativistic as they also occur in hydrodynamics etc. or too useful for describing highly idealized regular situations as surface layers and shock waves, I do not know; but they are normally, without any further discussion, excluded by imposing certain smoothness conditions on space-time), case (ii) is, for good reasons, treated somewhere else, therefore there remains the simple equivalence: singularity is incompleteness. case (i) is a purely local property; case (ii) is essentially global; case (iii) is, by definition, global, but it is left open whether one can localize the incompleteness by looking where the incomplete curves are going. In fact, by a mathematically very elegant construction, Schmidt [16], [9; 276-284] has given an extension of space-times in which any incomplete curve has endpoints.

3. Continuity

The underlying manifold V can be assumed to be C^{∞}, as any C^k-atlas ($k \geq 1$) contains a C^{∞} subatlas and none of the following conditions would be affected by such a transition. But the question which C^{∞}-structure has to be chosen is, in particular cases, neither trivial nor academic: a wrong choice leads to "coordinate singularities", and a lot of work has been done to find coordinates which are as smooth as possible (of particular interest are normal and harmonic coordinates). Any of the following conditions should be read as follows: "There exists a C^{∞}-atlas on V such that the metric g fulfills ...".

What is regarded as a suitable *differentiability class* of g depends on the context. For guaranteeing existence and uniqueness of geodesics, as required for most investigations of existence and structure of singularities, one assumes [4]:

> *Strong curvature differentiability class sc-C^{2-}*:
> The metric tensor g_{ab} is Lipschitz continuous (C^{1-}) and the Riemann tensor $R^a{}_{bcd}$ is bounded (C^{0-}).

In the proof of existence and uniqueness of the solution of the Cauchy initial value problem in [5; 226 ff] the following assumption is used:

> *Sobolev class W^4:* g and its derivatives up to the 4th order exist as distributions and are locally square integrable.

For smooth C^k-dependence on initial data these conditions have to be strengthened to sc-$C^{(2+k)-}$ resp. W^{4+k}.

Other conditions do not reflect what seems to be mathematically necessary but what one thinks to be physically desirable:

> *Junction Conditions:* g is C^1, piecewise C^3 (The points where $g \notin C^3$ form a finite collection of smoothly imbedded hypersurfaces; on each of their sides g possesses C^3 limits).

This is equivalent to the continuity of the normal components of the energy momentum tensor $T^a{}_b f_{,a}$ across the surfaces f = const.. In order to describe surface layers or shock waves one has to weaken these assumptions (see [14; 551-556]).

Remark: Any C^0 function can be approximated by C^∞functions to any required accuracy (e.g. of some measuring instrument). This fact demonstrates the difficulties of physical interpretation of differentiability, but does not make such considerations superfluous: Generally, the smooth approximation of some g cannot be chosen as an exact solution of the nonlinear field equations; in many cases of idealized situations the problem after having smeared out some jumps cannot be solved explicitly any longer; finally, near singularities with poles in some quantities this remark does not apply anyhow.

4. Causality

If the metric g fulfills the condition of

Nondegeneracy: g is of signature (+++-) everywhere,

it is called a Lorentz metric and distinguishes three classes of nonvanishing vectors v:

$$g_{ab}\, v^a\, v^b \begin{cases} < 0 \text{ timelike} & (t\text{-}) \\ = 0 \text{ null} & (n\text{-}) \\ > 0 \text{ spacelike} & (s\text{-}) \end{cases} \quad \text{(t-), (n-): causal (c-)}$$

and if g fulfills:

Time-orientability: The light-cones formed by the n-vectors can be continuously separated into two classes: the future- and the past-directed,

one obtains (smooth future-directed "causal") c-curves x(t) if $g_{ab}\, \dot{x}^a\, \dot{x}^b \leq 0$ and $\dot{x}$ future directed for all t and the causal future $J^+(p)$ for any event $p \in V$ as $\{q \in V \mid \exists$ c-curve $\widehat{pq}$; a c-curve is a ("timelike") t-curve if $g_{ab}\, \dot{x}^a\, \dot{x}^b < 0$ for all t. Similarly, one defines the timelike future $I^+(p)$ and the causal resp. timelike pasts $J^-(p)$, $I^-(p)$.

Global behaviour of causal relations in V can be characterized by properties of $J^\pm$ resp. $I^\pm$ [11]:

(i) *Causality:* No closed world-lines exist; or: J^+ is a partial ordering.
$(\forall\, p \in V\colon J^+(p) \cap J^-(p) = \{p\})$

(ii) *Stable Causality:* Causality is stable under small perturbations of the light-cones
$(\exists \hat{g}$: (a) $\forall\, p \in V\colon J^+(\hat{g};p) \cap J^-(\hat{g};p) = \{p\}$ ($\hat{g}$ is causal)
(b) $\forall\, v^a\colon g_{ab}\, v^a\, v^b \leq 0 \Rightarrow \hat{g}_{ab}\, v^a\, v^b < 0$
(the light-cones of $\hat{g}$ are wider than those of g))

This condition is equivalent to the existence of a global scalar function which increases along any c-curve; for this and other characterisations see [9; 198 ff], [18].

(iii) *Causal Continuity:* V is causal, and the timelike futures and pasts $I^+(p)$, $I^-(p)$ depend continuously on the base event p

$(\forall\, p, q, r \in V: q \in V \setminus \overline{I^+(p)},\ r \in V \setminus \overline{I^-(p)} \implies \exists\, U$ (neighbourhood of p)$: \forall\, s \in U: q \in V \setminus \overline{I^+(s)},\ r \in V \setminus \overline{I^-(s)})$.

For further characterisations see [10].

(iv) *Causal Simplicity:* V ist causal, and the light-cones contain the boundaries of the futures and pasts

$(\forall\, p \in V: \dot{I}^+(p) = \dot{J}^+(p);\ \dot{I}^-(p) = \dot{J}^-(p))$

(v) *Global Hyperbolicity:* V is causal, and the "causal interval" between any two events is compact (it might be empty, of course)

$(\forall\, p,q \in V: J^+(p) \cap J^-(p)$ compact)

This is equivalent to: The set of c-curves $\widehat{pq}$ between any two events p,q is compact with respect to the continuous convergence, and also to: the existence of a subset $S \subset V$ (*"Cauchy surface"*) which is met by any inextensible c-curve exactly once. For this and further characterisations see [6], [18], [9, 206 ff].

Definition: Given a spacelike hypersurface S, the maximal subspace of V for which S is a Cauchy surface is called the *Cauchy development* D(S); its future part $D(S) \cap J^+(S)$ is denoted by $D^+(S)$.

Proposition: (v) $\Rightarrow$ (iv) $\Rightarrow$ (iii) $\Rightarrow$ (ii) $\Rightarrow$ (i)
(i) $\not\Rightarrow$ (ii) $\not\Rightarrow$ (iii) $\not\Rightarrow$ (iv) $\not\Rightarrow$ (v)

The proof of the implications can be found in [6], [10], [18]; only the step (iii) $\Rightarrow$ (ii) is not obvious. It is more instructive to study counterexamples which disprove equivalences (the four examples given below are constructed by identifications or mutilations in Minkowski space M: $ds^2 = -dt^2 + dx^2 + dy^2 + dz^2$)

(i) $\not\Rightarrow$ (ii) Identify (t_1, x_1, y_1, z_1) and (t_2, x_2, y_2, z_2) iff $x_1 - t_1 = x_2 - t_2 + g$; $y_1 = y_2 + gr$; $z_1 = z_2$; $x_1 + t_1 = x_2 + t_2$ for a fixed irrational number r and some arbitrary integer g. Now, along any c-curve in M, x + t does not decrease; hence, by this identification, we would only produce a closed c-curve γ if x + t remains constant. The only curves of this type are the null-geodesics with x + t, y, and z const. As r is irrational, even such a γ is not closed but meets a dense subset of the torus z, x + t const. (see fig.). V is causal, but obviously any widening of the null-cones would produce closed c-curves.

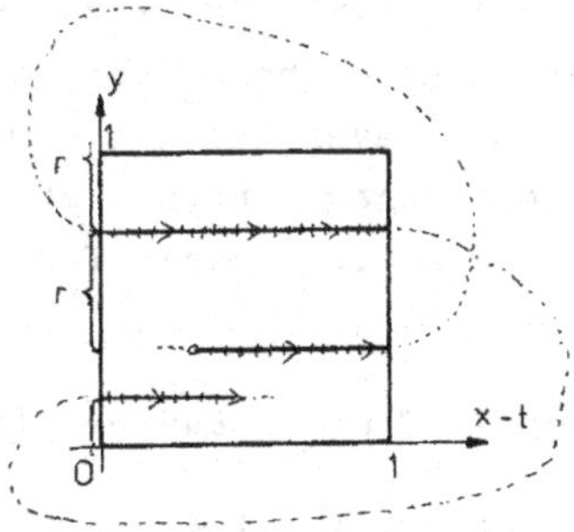

(ii) $\not\Rightarrow$ (iii) Remove the half plane t = 0, x $\leqslant$ 0. The future of p: (-1; -1; 0; 0) does not contain any q with a t-value ≥ 0, but the future of any event r in the past of p "passes the edge x = 0 = t". I^+ is not continuous at p; but V is stably causal as M is.

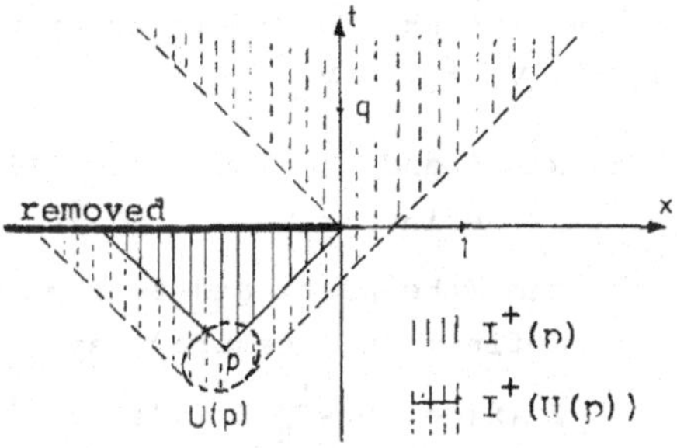

(iii) $\not\Rightarrow$ (iv) Remove the origin (0; 0; 0; 0). Evidently, the sets $I^+(p)$, $I^-(p)$ depend continuously on p, but for p: (-1; -1; 0; 0), q: (1; 1; 0; 0) it holds $q \in \overline{I^+(p)} \setminus J^+(p)$.

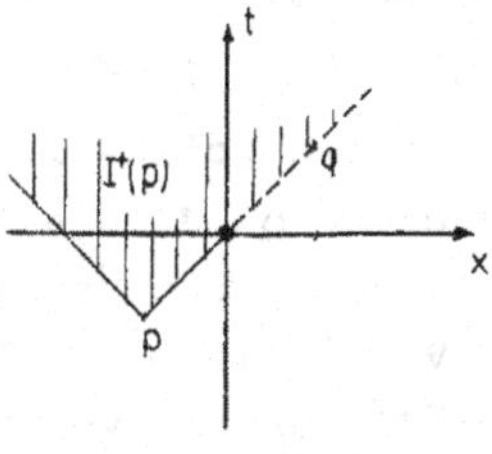

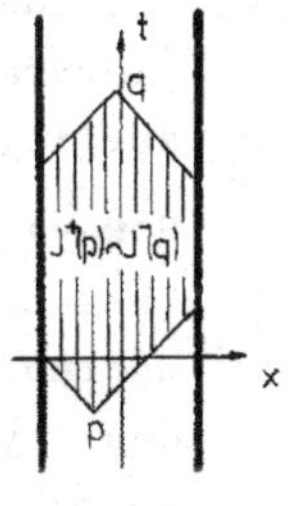

(iv) $\not\Rightarrow$ (v) Remove all points with $|x| \geq 1$. $J^+(p) \cap J^-(q)$ cannot be compact if the time-coordinates of q and p differ by more than 4. ■

While this list of conditions is well suited to classify the *regularity* of causal structure, for an analysis of causal *relations* in some space-time V a different concept is used. Let us consider the subsets A and B of Minkowskispace: A has the property: stably causal, $s \in V$ $\forall p,q$: $J^{+}(p) \cap J^{-}(q) \neq \emptyset$, but in B an observer at q, no matter how he travels later on, never can get some information about what happened at p. A suitable theoretical basis for such pictures is:

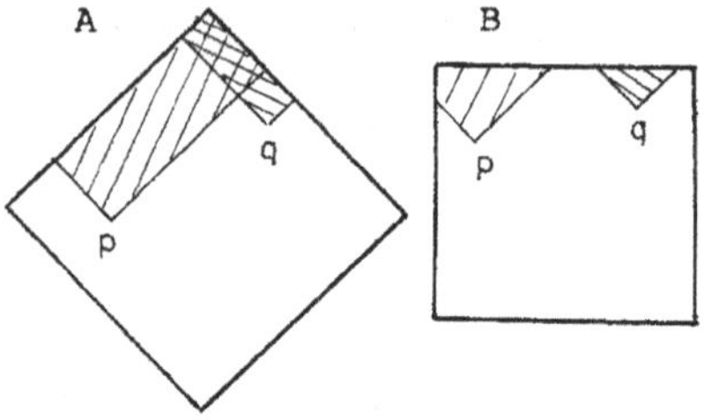

Notion 2: *Causal boundary* [19], [7], [2]

For any inextensible c-curve $x(t)$ $(t \in]-\infty,\infty[)$ the endpoints $x(\infty)$, $x(-\infty)$ are attached to space-time V as *ideal points*.

There are several stages of integrating the set of ideal points (the *causal boundary* $\dot{V}$), in other words, extending the structures of V onto $\bar{V} := V \cup \dot{V}$:

a) $x(-\infty)$ and $x(+\infty)$ are for illustration only, no structure is defined on $\dot{V}$.

b) timelike pasts and futures of the ideal points can be defined in a natural way:

$$I^{-}(x(+\infty)) := \bigcup_t I^{-}(x(t)); \quad I^{+}(x(+\infty)) := \bigcap_t I^{+}(x(t))$$

$$I^{-}(x(-\infty)) := \bigcap_t I^{-}(x(t)); \quad I^{+}(x(-\infty)) := \bigcup_t I^{+}(x(t)).$$

For the purpose of this lecture (to get an intuitive picture for discussing causal properties of V) this is sufficient.

c) However for mathematicians it might be very unsatisfactory that the structure of V is extended to $\bar{V}$ only to a very low degree; all ordering properties of I^{+} are lost, e.g. the duality: $p \in \dot{V}$, $q \in V \Rightarrow q \notin I^{+}(p)$ even if $p \in I^{-}(q)$. The causal future J^{+} cannot even be defined analoguously. In [19], [7], [2], some progress into this direction has been made:

Point set structure: The proposition: $I^{+}(r) = I^{+}(s)$ and $I^{-}(r) = I^{-}(s)$ iff $r = s$

(i) holds for all $r, s \in \dot{V}$ if V is stably causal
(ii) cannot hold for any $r \in \dot{V}$ if V is stably causal, $s \in V$
(iii) can be taken as definition for $r = s$ if $r, s \in \dot{V}$

How one can extend *causal ordering* and *topology* to $\bar{V}$ is quite complicated and cannot be described here. Only the characterizations of causal conditions in terms of $\dot{V}$ will be given here:

(i) One can introduce a partial ordering $\hat{J}^+$ on V which includes the old one (i.e.: $\forall\ p,q \in \bar{V}$: $q \in J^+(p) \Rightarrow q \in \hat{J}^+(p)$) iff V is stably causal.

(ii) The timelike futures $\hat{I}^+$ coincide on V with the original ones: $\hat{I}^+(p) \cap V = I^+(p)\ \forall\ p \in V$ iff V is causally continuous.

(iii) The causal futures $\hat{J}^+$ coincide on V with the original ones: $\hat{J}^+(p) \cap V = J^+(p)\ \forall\ p \in V$ iff V is causally simple.

(iv) Ideal points have either timelike pasts or futures only: $\forall\ p \in \dot{V}$: $\hat{I}^+(p) = \emptyset$ or $\hat{I}^-(p) = \emptyset$ iff V is globally hyperbolic.

Remarks: Generally, $\dot{V}$ is different from the conformal boundary constructed by Schmidt[5] [17].

Causality does not distinguish between ideal points at infinity resp. at a singularity as shown by the following theorem [18] : Let V be stably causal. Then there exist conformally related (i.e. causally equivalent) space-times V', V" where V' contains no incomplete c-geodesic but in V" any c-geodesic is incomplete in both directions (hence their ideal end-points are singular). For definitions and relevance compare the next section.

Let us finish this section by giving pictures of $\bar{V}$ for some important spherically symmetric space-times V ($ds^2 = g_{AB}\, dx^A\, dx^B + R^2(x^A)\cdot(d\theta^2 + \sin^2\theta\cdot d\phi^2)$; $x^A = r,t$); one can describe V as a 2-dim Lorentz-space (r,t-plane) with the additional function R(r,t) (the curvature radius of the spherical symmetry group orbits), furthermore one can choose the coordinates ("double null coordinates") such that the light-rays in the r-t plane have angles 45° or 135°.

[5] The first example in § 6 has a $\dot{V}$ consisting of two points, and a c-boundary of two points and four circles. The Einstein-De Sitter universe has an empty c-boundary but two ideal points in $\dot{V}$.

For the simplest case, the Minkowski space M, the explicite calculation goes as follows:

$$ds^2 = (-\,dt + dr)(dt + dr) + r^2(d\theta^2 + \sin^2\theta\cdot d\phi^2)$$

for $u = \arc\tan(t + r)$, $v = \arc\tan(-\,t + r)$ one gets:

$$0 \leq u + v;\; u,\, v < \frac{\pi}{2}$$

$$ds^2 = \frac{du\,dv}{(\cos u\cdot\cos v)^2} + \frac{1}{4}\cdot(\tan u + \tan v)^2(d\theta^2 + \sin^2\theta\cdot d\phi^2)$$

$$V = I^- \cup \mathscr{I}^- \cup \mathscr{I}^+ \cup I^+; \qquad I^\pm := (u = \pm\frac{\pi}{2},\, v = \mp\frac{\pi}{2});$$

$$\mathscr{I}^+ := \{u = \frac{\pi}{2}\},\; \mathscr{I}^- := \{v = \frac{\pi}{2}\}$$

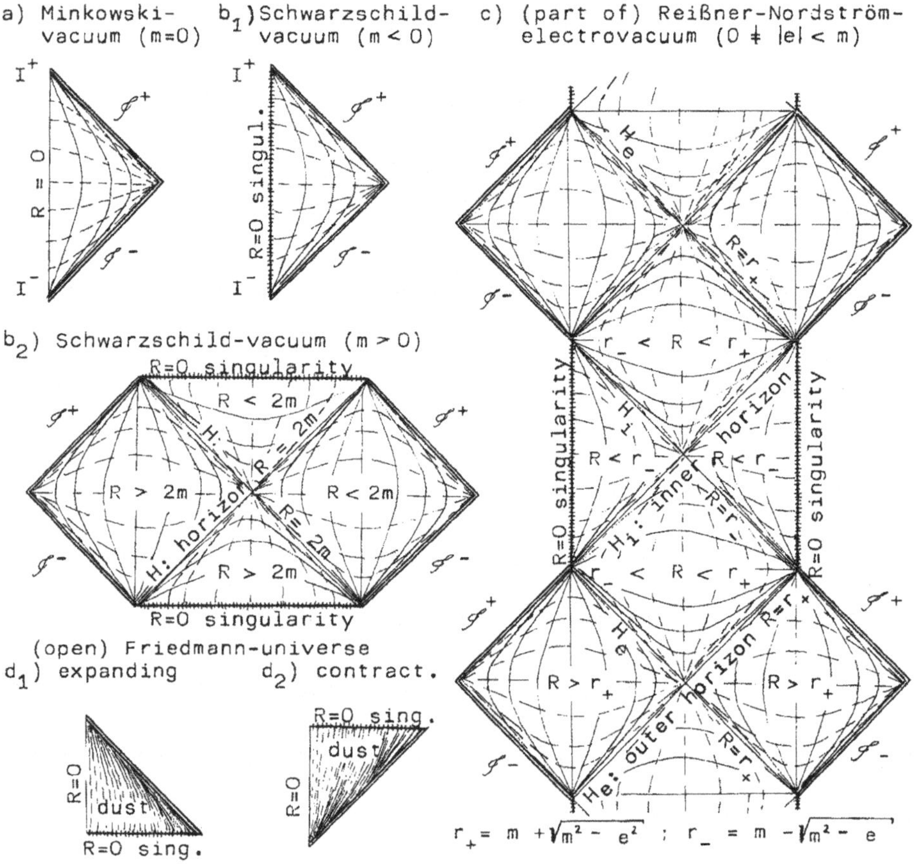

The line R = 0 in fig. a, d_1, d_2 does not consist of points of $\dot{V}$ but of centres of symmetry: one gets the full space V by rotating the picture around these points which then become regular interior points of V. Double lines represent infinitely far boundary points, hatched lines represent singular points of $\dot{V}$. ——— :R=const. - - - :t=const.

5. Completeness

In Riemannian spaces R there is a natural definition of completeness: R is complete

$$\iff \delta(p,q) := \inf_{\widehat{pq}} \int ds \text{ is a complete metric distance function}$$

$\iff$ any geodesic can be extended to arbitrarily large values of its length

$\iff$ any inextensible curve has infinite length.

In Lorentz-spaces V this simple situation drastically changes: The distance function δ does not exist any longer, since there are inextensible curves of arc-length zero (the light-rays). Even in the most regular case, Minkowski space M, one can find inextensible timelike curves of finite length (the curve $x - t = (x + t)^{-2}$ from $(x = 1;\ t = 0)$ to infinity has length $\sqrt{8}$), but, fortunately, this does not happen in the case of t-curves with bounded acceleration (which include the physically interesting worldlines, as particles with restmass nonzero need infinite energy to be accelerated unboundedly). For an analysis of all t-curves the arc-length $\int ds$ should be replaced by the *"affine length"*:

$\mathcal{L} : \int_{x(t)} \| \dot{x}(t) \| \cdot dt$, where $\| \dot{x}(t) \|$ is the Euclidian norm of the tangent vector measured in a parallely propagated frame

which, for geodesics, is proportional to the arc-length and is infinite for all inextensible curves in M. The numerical value but not the finiteness of $\mathcal{L}$ depends on the choice of the frame.

According to the class of (inextensible) curves whose affine length is required to be infinite one gets the following hierarchy for Lorentz spaces of dimension greater two, [12], [6]:

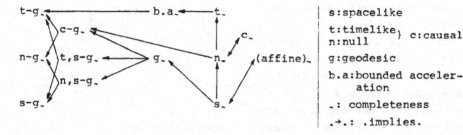

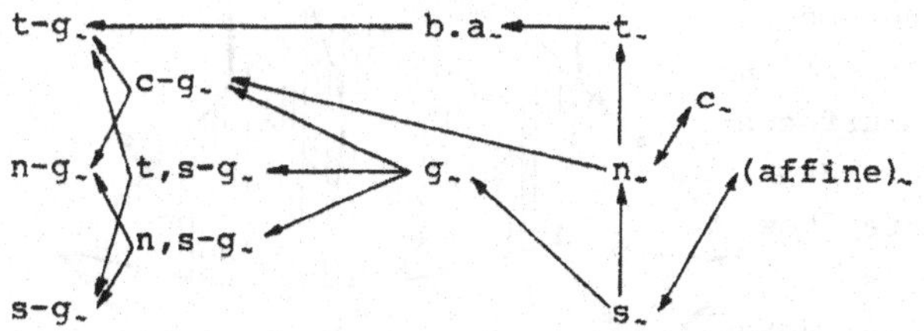

s: spacelike
t: timelike } c: causal
n: null
g: geodesic
b.a: bounded acceleration
.: completeness
→: implies.

All implications in this scheme which are not indicated by an arrow or following by transitivity can be disproved by counterexamples. For example, completeness of all spacelike curves implies completeness of timelike curves; but if all s-geodesics are complete, the t-geodesics might not be complete (this is due to the fact that any p.p. framed t-curve can be approximated by p.p. framed s-curves but not by s-geodesics. Fortunately, in the existence theorems of Hawking & Penrose the occurrence of the very strong c-g *incompleteness* is proven while in the completion procedure of Schmidt [16] only spaces of the strongest (namely affine) *completeness* are not extended, every affinely incomplete curve gets an endpoint.

6. Nocuous Singularities

In order to understand the reason for restricting the concept of singularity, let us consider three examples of singular space times:

(a) take the region $M^+ := \{t>|x|\}$ of Minkowski space M^2 and some Lorentz transformation Λ generating the cyclic group Λ^g ($g \in Z$); identify any two points p,q of M^+ iff there exists a $k \in Z$ such that Λ^k maps p onto q. This new manifold M^+/Λ is flat ($R^a{}_{bcd} = 0$), it cannot be embedded into a complete manifold as in the tangent space of the "endpoint" of γ: $\{x = 0;\ t > 0\}$ all the vectors $\Lambda^g(0,1)$ have to be identified. M^+/Λ is the solution of the vacuum vectors field equations with initial data given on $\{t^2 - x^2 = 1\}/\Lambda$.

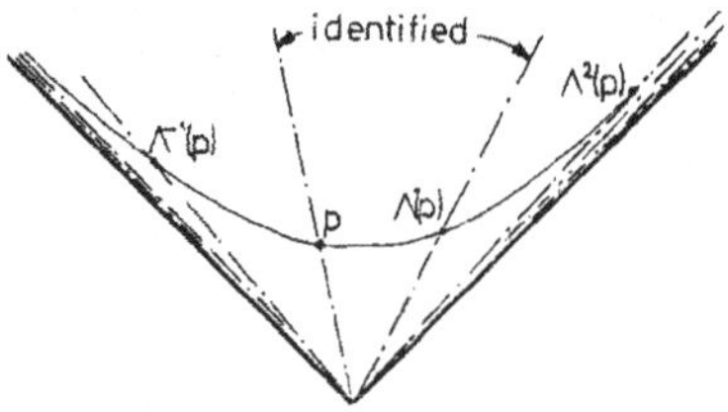

(b) In the Reißner-Nordström solution V (electrovac; initial data may be given on S. $D^+(S)$ contains no incomplete curve which cannot be extended into some regular extension of V. But as trapped surfaces occur in $D^+(S)$, the Penrose-Hawking theorems show that no matter how one extends D^+ beyond H_1 as a spacetime with non-negative energy density, the occurrence of singularities is inevitable.

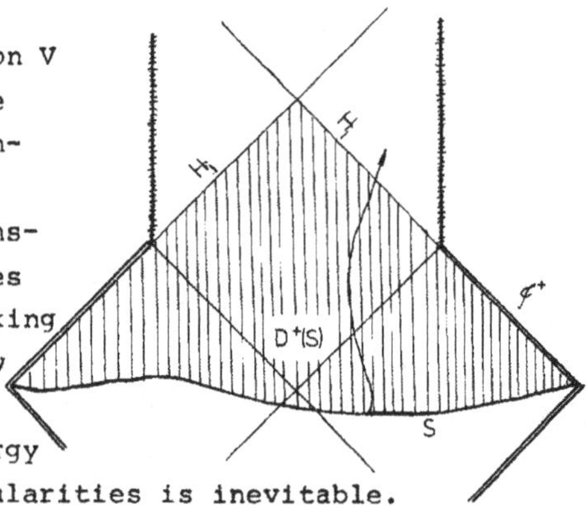

(c) There are spherically symmetric solutions for a perfect fluid ball with vacuum outside [23] for any arbitrarily bounded equation of state ($p(\mu)<p_\infty<\infty$). It seems very likely that even for some classes of unbounded $p(\mu)$ and slight deviations from symmetry the singularities will still exist, but that the viscosity of realistic fluid will prevent them.

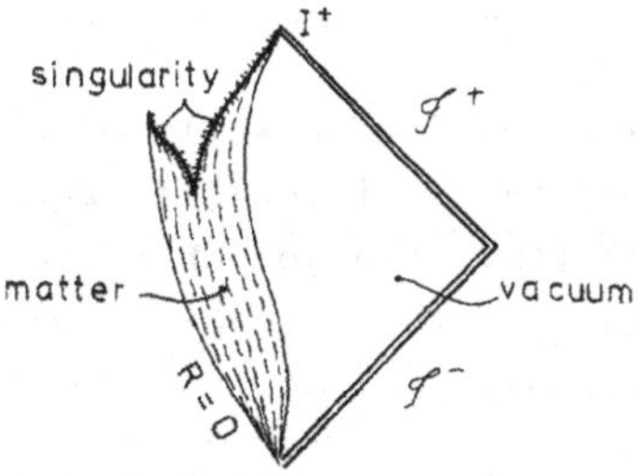

For physical reasons one feels uncomfortable with such examples included in the class of singular space-times: in the first example no local physical quantity goes wrong or gets large values, therefore one cannot say: near a singularity the classical theory has to be replaced by a new (quantum-)theory. In the second example in the region determined by the initial data $D^+(S)$ nothing goes wrong at all, D^+ can be extended beyond the whole boundary H_i. The third example displays a feature which has been discussed in the context of singularities for a long time: The occurrence of singularities might be (at least partially) due to the fact that one only knows too special (idealized, symmetric) solutions, general cases might behave differently. These examples suggest that a singularity should only be taken seriously (from the physical standpoint) if it is a:

Notion 3: *Nocuous Singularity*

Let (a) a *realistic model* and (b) *realistic initial data,* given on a hypersurface S have a singular development in the sense of notion 1.

This *singularity* is called *nocuous* if:

(i) the development $D^+(S)$ contains a point p of discontinuity or an incomplete world line γ which cannot be extended into some regular extension of D^+ *(Cauchy singularity)*,

(ii) in any neighbourhood of p resp. along γ geometrical quantities (curvature) do not tend to a limit *(Curvature singularity)*,

(iii) small changes in (a) and (b) do not affect the occurrence and properties (i,ii) of singularity *(Stable singularity)*. Note that by definition, $D^+(S)$ fulfils the very strong causality condition of being globally hyperbolic, hence a Cauchy singularity cannot be an acausality).

Now, condition (i) is well defined; but depending on the choice of quantities in condition (ii) resp. the choice of topology for condition (iii) one gets hierarchies for the "strength" resp. "stability" of the singularities which will be discussed in the next sections.

What one considers as "realistic" models or data seems to be a matter of taste rather than of definition and, even worse, this requirement seems to exclude all solutions which ever can be explicitly calculated as being too special. But one has to bear in mind that special solution *idealizing* realistic situations are *realistic* with respect to properties which are *stable*. In any case, after having chosen the basic situation (e.g. collapse as in this lecture), one should avoid any further reference to this part of the notion in order to prevent a confusion of formal (mathematical) and physical arguments.

Remark [3]: For incomplete c-curves in condition (ii) the *limit* of curvature is considered, the *extension* strength of singularities is not taken into account. Now, for any incomplete c-curve γ in D(S) one can find a neighbourhood U which can be extended as a C^0 or C^1 Lorentz space as one can choose "Fermi"-coordinates along γ with $g_{ab} \underset{*}{=} \eta_{ab}$, $\Gamma^a_{bc} \underset{*}{=} 0$ on γ, a C^2-extension of U existing iff the C^2-limits along γ (curvature at the endpoint) behave well. Breakdown of higher C^k-structure seems to be irrelevant for the definition of singularity. *Global extensions* beyond the endpoint of γ are impossible even at the C^0 level in some cases in which the singularity seems to be very weak (our example (a)) hence this property seems to have no connection with physical quantities (at least those which have been discussed up to now).

7. Topology of fields

In Riemannian spaces the distance function $\delta(x,y)$ induces a vector norm and a natural uniform structure. The uniform structure for compact spaces or subspaces is independed of the particular choice of the metric tensor g_{ab}; hence, by choosing a positive definite auxiliary metric h_{ab}, also for compact subsets of Lorentz spaces one gets a topology of *uniform convergence*.

For vector or tensor fields over incompact sets (e.g.: initial data on some S^3 or a vector field along some curve) this is lost, but there are two simple topologies which easily can be described by means of tensor bundles. The tensors of type (m,n) over V form a bundle B with base space V, projection $\pi: B \to V$, and the linear space of (m,n)-tensors as fibre; B has a natural topology and C^∞-structure. For $S \subset V$: $B_{|S}$ is by definition $\{\pi^{-1}(x) | x \in S\}$. A tensor field t over S corresponds to a cross-section of $B_{|S}$ (i.e. a subset {t} of $B_{|S}$ such that π is an isomorphism between {t} and S). Any neighbourhood $\hat{U}$ of {t} in $B_{|S}$ corresponds to a set U of cross-sections: $\left\{\tilde{t}\text{: (m,n) tensor field over } S \middle| \{\tilde{t}\} \subset \hat{U}\right\}$. Taking all these sets U as the neighbourhoods of t one defines a topology on the set of fields: <u>the fine topology.</u> If one considers instead all compact subsets C of S and defines via the neighbourhoods $\hat{U}_C$ of $\{t_{|C}\}$ in $B_{|C}$ the sets U_C of cross-sections $\left\{\tilde{t}\text{(m,n) tensor field over } S \middle| \{\tilde{t}_{|C}\} \subset \hat{U}_C\right\}$ as neighbourhoods of t, one gets the <u>compact open topology.</u> By including the first k derivatives in these definitions one similarly obtains the C^k-f resp. the C^k-c.o.-topology.

For most purposes the c.o.-topology is too coarse (e.g.: any c.o.-neighbourhood of the Minkowski metric η_{ab} over R^4 contains acausal metrics) and the f-topology is too fine (e.g.: if f_n is a sequence of functions over R^n with f-lim $f_n =: f_o \equiv 0$, almost all f_n must have a compact support C_n; for the description of asymptotic behaviour this topology is almost useless).

An intermediate *uniform* topology exists only for a few (but important) special cases:

(i) for scalar fields: U_ε (t): $\left\{\tilde{t} \text{ scalar} \middle| |\tilde{t}(x) - t(x)| < \varepsilon, \forall x \in S\right\}$

(ii) for tensor fields along some curve if one considers the components with respect to a parallely propagated frame as scalars. In contrast to Riemannian spaces the restriction to p.p. frames is necessary, as the components of any non-vanishing vector v with respect to (orthonormal) frames are unbounded: by "boosts" (i.e.: proper Lorentz transformations) one can make them arbitrarily large or small. The neighbourhood basis U_ε, but not the topology depends on the choice of the p.p. frame.

(iii) if one imposes suitable restrictions on the asymptotic behaviour (roughly speaking, this is equivalent to compactifying S and hence to getting around the difficulties of incompactness).

These exceptional cases might be helpful for choosing suitable topologies for initial data (which normally are given on a compact S or are underlying some asymptotic conditions) resp. for the model (which is often given in terms of scalars: energy density inequality, equation of state, etc.).

8. Limits of Curvature

If one looks for suitable geometrical quantities for condition (ii) in notion 3 one has to remember the following facts:

(a) There are no invariantly defined non-constant scalars of order zero or one (built up by g_{ab} and $g_{ab,c}$ only) or - more generally - quantities of order zero or one whose limiting behaviour along curves (boundedness etc.) is an invariant property.

(b) The scalar invariants of second order (scalar polynomials in g_{ab}, R_{abcd}, and the totally antisymmetric tensor η_{abcd}) do not characterize the curvature completely (in some plane wave solutions, the scalar polynomials vanish identically while the curvature tensor in parallely propagated frames along some incomplete curves diverges[6]).

(c) The frame components of the Riemann tensor have a physical meaning: they characterize the tidal effects of the gravitational field.

(d) The sectional curvatures (Gauß curvatures of 2-planes in V) at some point x are unbounded unless V is of constant curvature at x (all sectional curvatures are equal). Along any curve on which the curvature in some p.p. frame behaves well (e.g.: a curve with an endpoint in V) there are frames in which the curvature components behave badly. While this is a typical situation in Lorentz spaces (cf. remark about p.p. frames in last section) it might not be surprising that the inverse also can occur: In the plane waves mentioned above the incomplete curves with diverging curvature in a p.p.frame

---

[6] If one wants to describe these singularities by unbounded scalars, one has to consider rational (rather than polynomial) scalars of higher order [13] or the curvature scalar in the bundle of linear frames over the space-time, see [21].

possess frames (with unbounded rotational acceleration) in which the components of $R^a{}_{bcd}$ have nice limits, a simple consequence of the fact that these space-times are homogeneous, hence there is a set of isometries which map a basis point $x(0)$ together with a frame $e_k(0)$ to any point $x(t)$ on the considered curve, such that the values of the components of $R^a{}_{bcd}$ with respect to $e_k(t)$ coincide for all t.

These remarks suggest the following definition:

Curvature Singularity:

An incomplete c-curve γ terminates at a curvature singularity if the Riemann tensor components in a parallely propagated (p.p.) frame are not all bounded along γ.

Remark: In the terminology of [5] *our* curvature singularities include "intermediate" as well as "curvature" singularities, the word "curvature singularity" being reserved for spaces in which, along some curve γ, $R^a{}_{bcd}$ is unbounded in any frame along γ; hence the plane waves mentioned above as well as some cosmological models (a "whimper" instead of a "bang") are not curvature singular in the sense of [5]. But we cannot see any reason to pay frames of reference with unbounded rotational acceleration (with respect to the affine parameter) so much attention. Also we do not think that the incomplete information given by the scalar polynomials of second order should be made the only criterion for singularity ("s.p. singularities" in [9; 260]). In these two references [5], [9] it is conjectured that the occurrence of curvature singularities (in our sense) which have regular 2nd order scalar polynomials is unstable.

While a classification of limiting behaviour of higher order quantities is straightforward but presumably uninteresting (one should not call spaces with regular second order quantities singular at all, see [3]), the following example shows the necessity of distinguishing lower (at least: first) order singularities.

Example: V is the spherically symmetric field of a surface layer of mass m and radius r_o with mass density

$$\mu = \begin{cases} \dfrac{m}{8\pi h r^2}\cdot\log\left|\dfrac{r-r_o}{h}\right| & |r-r_o|<h \\ 0 & |r-r_o|\geq h \end{cases} \qquad 0<h\ll r_o$$

for: $r \leq r_o - h$: Minkowski vacuum, for $r \geq r_o + h$: Schwarzschild vacuum. (The exact expression for μ is not important; only the integrability and the pole at r_o is essential. One also could use the limit for $h \to 0$: $\mu = \frac{m}{4\pi r^2}\,\delta(r-r_o)$ instead, if one treats this distribution as what it really is: an operator on test functions rather than a point-function). After a suitable choice of radial pressure one gets the following properties: V is not C^2 or C^{2^-} but C^{1^-}, the geodesics as stationary values of the operator $\int ds$ can be uniquely extended across $\{r = r_o\}$ and no observer (c-curve with bounded acceleration and p.p. frame of reference) is torn apart by infinite tidal stresses[7]. So these discontinuous spaces occur quite often in the literature as idealisations of perfectly regular situations. On the other hand, if one removes $\{r \leq r_o\}$, the rest is not only incomplete (of course) but has a curvature singularity, hence (according to the general opinion) has a physical singularity. Only the neglect of discontinuities in the general discussion of singular spaces conceals this incoherence.

The study of this and similar examples leads us to the following concepts:

Tidal Stress Singularity: An incomplete c-curve γ terminates on a t.s. singularity if the tidal stresses (on some test body travelling along γ) are unbounded.

Geodesic Singularity: An incomplete c-geodesic $\gamma(s)$, $s \in]0;1]$ (s being an affine distance from the singularity) terminates on a g-singularity if one

either (i) has an infinite sequence of conjugate points for $s \to 0$

or (if case (i) does not apply, one always can find a point $\gamma(s_o)$ which is not separated from the singularity by conjugate points) (ii) the Christoffel symbols $\Gamma^a_{bc}(s_1,s)$ at $\gamma(s)$ calculated in normal coordinates with origin $\gamma(s_1)$, $(s_1 < s_o)$ do not possess the limit: $\lim_{s_1 \to 0}\left(\lim_{s \to 0} \Gamma^a_{bc}(s_1,s)\right) = 0$

[7] For "the fate of an observer who falls into the singularity", i.e. the explicite calculation of tidal stresses by integrating the tidal force producing components of the Riemann tensor over the body of the observer, see [14; p. 860 ff].

Roughly speaking, t.s.- as well as g-singularities are of first order: The tidal stresses are *integrals* over curvature tensor components, the Γ^a_{bc} in normal coordinates are in first order equal to $R^a{}_{(bkc)} \cdot s\, \dot{\gamma}^k$, which, for integrable curvature tensor, fulfills the limit condition. If normal coordinates do not exist (case i), for some components even $R^a{}_{bcd} \cdot s^2$ must be unbounded; whether the unboundedness of $R^a{}_{bcd} \cdot s^k$ for $k > 2$ has some geometrical significance, I do not know.

9. Collapse

The classical model of collapse is the spherically symmetric uniform dust cloud with vacuum outside. Regions taken from (time reflected) Friedmann resp. Schwarzschild space-time are joined together (compare figures b_2, d_2 of section 4 and figure on the left). The causal picture does not display the metric proportions correctly, at $t = -\infty$ the dust cloud is represented by one point while it has infinite diameter, at the final singularity the cloud has collapsed to a point, but is represented by a line in the causal picture (if one adds the suppressed θ, ϕ-coordinates, one obtains a cylinder $R \times S^2$).

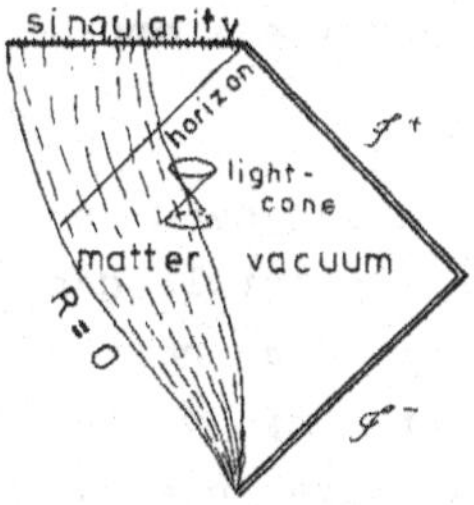

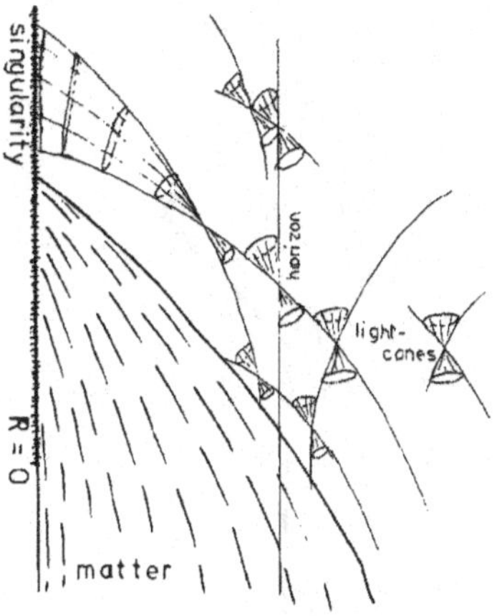

This corresponds to the fact that although the particles come arbitrarily close together they cannot interact with each other at arbitrarily late time; especially the singularity itself cannot be seen by any observer before he falls down to it broken to pieces by the infinite tidal stress.

Notion 4: *Gravitational Collapse*

An *isolated system* (i.e.: there exists a regular spacelike ("initial") surface S on which the energy momentum tensor has a compact support), which at far distances behaves like Minkowski-space is undergoing *gravitational collapse* if the development of the data leads to a singularity in the future of S. If the singularity is nocuous, the collapse is called a *catastrophic gravitational collapse.* (A generalisation to systems with electrovac instead of vacuum outside is straightforward).

As a criterion for V to be "asymptotically like Minkowski space", Penrose and Hawking have invented "weakly asymptotic simplicity and emptiness" [9; 222, 225], a highly technical but very powerful concept which cannot be described here in detail. What we need here is

(a) the causal boundary $\dot{M}$ of Minkowski space and

(b) a condition that the causal boundary $\dot{V}$ of V contains a part ($\mathscr{I}^{\pm}$) which corresponds to $\dot{M}$ and on which the null geodesics running away from the matter ball and traveling through the far distant region have their ideal endpoints.

Definitions: $I^{-}(\mathscr{I}^{+})$ is the region of all events which can send information to the asymptotic region resp. from which a particle can escape the collapsing body to that region (*"domain of outer communication"*). The boundary $\dot{I}^{-}(\mathscr{I}^{+})$ is called the future *event horizon*, the region $V \setminus I^{-}(\mathscr{I}^{+})$ is called the *black hole*. A singular point $p \in \dot{V}$, $p \in J^{+}(S)$ is called *naked* if $p \in I^{-}(\mathscr{I}^{+})$, p otherwise is called *hidden*[8].

Let us finish this section by giving an example of a collapse which is not catastrophic in the sense of our definition:

A ball with mass m and charge e ($|e| < m$) which shrinks down below its event horizon, might stabilize and finally expand again to arbitrarily large volume. In between, a singularity develops *outside* the matter repelling all t-geodesics. An outside observer $x_1(t)$ does not see any great difference from the (catastrophic) collapse of an uncharged dust cloud, but an observer $x_2(t)$ who falls into the black hole can escape the singularity as well as all the matter world-lines do; [9; 360, f].

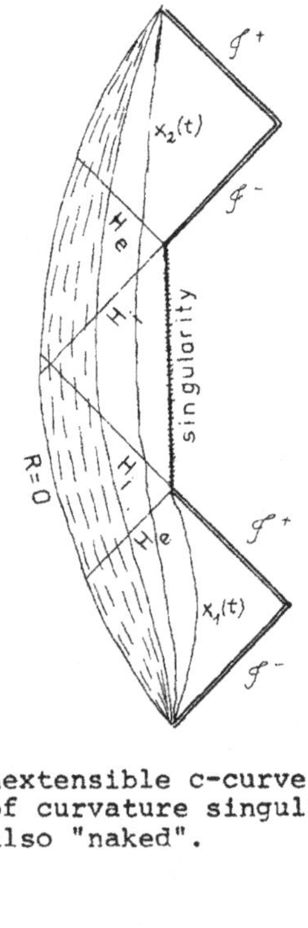

[8] If a singular point lies on the horizon, a nonextensible c-curve lies completely in $\dot{I}^{-}(\mathscr{I}^{+})$; especially in the case of curvature singularities it is sensible to call this singularity also "naked".

10. Cosmic Censorship

A natural generalisation of the features of the collapsing dust cloud is:

Conjecture 1: Cosmic Censorship Hypothesis (CCH)

The singularities in a gravitational collapse are *hidden* behind the event horizon.

In this simple form the CCH is definitely wrong as counterexamples show: The Kerr solution with $|a| > m$, the Reißner-Nordström solution with $|e| > m$. If one excludes these solutions as having too unrealistic values of angular momentum or charge, one still has to argue about some spaces which are very close to the standard examples of Schwarzschild

(i) in [22] a family of static vacuum spaces is given with mass m and quadrupole moment q, which for $q = 0$ gives the Schwarzschild metric, but for arbitrarily small $q > 0$ has a singular event horizon. This example shows that the occurrence of regular event horizons is not stable against small changes in the model (physical parameters). On the other hand, these solutions do not fulfil the requirements for a collapse space time as no interior matter source which collapses from a regular state down to this singular horizon is given; power series calculations of Price make it plausible that such multipole moments will be radiated away before the star collapses. But one still feels uneasy about the instability of horizons hinted at by this example.

(ii) in [23] exact dynamical solutions for spherically symmetric balls of perfect fluid ($p = p(\mu)$) are considered. It is shown that for any bounded p ($p(\mu) < p_\infty < \infty$) one can find solutions with naked "caustic" curvature singularities. It is not clear, whether the nakedness of singularities is stable against nonsymmetric perturbations of initial data or inclusion of viscosity of realistic fluid. The tidal stresses remain finite.

Discussing these solutions [23] with colleagues, we (P. Yodzis, H. Müller zum Hagen, H.J. Seifert) collected the following possible modifications of CCH:

Conjecture 1a:

The singularities in a *realistic* gravitational collapse are hidden. It is hardly possible to discuss this conjecture.

<u>Conjecture 1b:</u>
In *almost all cases* the singularities in a gravitational collapse are hidden.

This requires a measure on the set of all solutions which is unlikely to be found. A similar conjecture which is based on topology only is discussed in the next section.

<u>Conjecture 1c:</u>
The *gravitational* singularities in gravitational collapse are hidden.

The basic idea of this formulation is the following objection against example (ii): The caustics also can occur in special relativity, they are caused by the nonlinearity of classical fluid mechanics.

We will call a singularity *hydrodynamical*, if one can find an analogue in special relativity or prerelativistic classical physics with neglection of gravitational field, otherwise *gravitational*. It is far from clear whether this notion can be given a precise meaning by saying what "analogous" is. On the other hand, via the field equations, the gravitational field (i.e. the geometry) becomes singular (otherwise we would not call space-time singular at all); one only has the vague feeling that in really *gravitational* singularities the feed back of gravitation is the only constitutive element in the formation of singularities, in the other cases gravitation plays essentially a passive role only. Finally, one should mention that collapsing dust clouds as well as the Friedman cosmos, *the* standard models for final, respectively initial, singularities, are hydrodynamical in this sense, and that in the Friedman cosmos, in contrast to the caustics in example (ii) the Weyl tensor $C^{a}{}_{bcd}$ remains regular, only the matter part $R^{a}{}_{b}$ diverges. Hence a distinction between Weyl-singularities (which include the typical examples of gravitational singularities: the vacuum singularities with trivially regular $R^{a}{}_{b}$) and Ricci-singularities with regular $C^{a}{}_{bcd}$ (as hydrodynamical singularities) is not useful.

<u>Conjecture 1d:</u>
The *inevitable* singularities in gravitational collapse are hidden.
What does "inevitable" mean? As the initial data with a specified model determine the development uniquely, at least any Cauchy singularity is inevitable in some sense. But normally this attribute is applied only in cases where the occurrence of singularities is guaranteed under very weak restrictions on the model (e.g. energy inequality). This occurs in the context of the singularity theorems of Penrose & Hawking,

but here via the Raychaudhuri equation and the ingenious application of the theory of second variation inevitability and forming of horizons is equivalent. Hence with this interpretation conjecture 1d does not say anything about physics: it only interprets a beautiful mathematical method.

Conjecture 1e:
Any *nocuous* singularity in gravitational collapse is hidden.

See next section.

11. Strong Cosmic Censorship

In order to avoid the almost fruitless discussion about what is "physically reasonable" or "really gravitational" Penrose [15] anticipated another conjecture which strengthens the CCH and uses mathematical terms only. Hence it is much clearer in which directions investigations have to go; it is, on the other hand, obvious that they will be very awkward and likely fail to produce relevant results.

Conjecture 2: Strong Cosmic Censorship Hypothesis (SCCH)
Generically, solutions of Einstein's field equations do *not* have *timelike causal boundary* points.

The causal boundary $\dot{V}$ is called *timelike* at $p \in \dot{V}$ if $\hat{I}^+(p) \neq \emptyset$ and $\hat{I}^-(p) \neq \emptyset$. According to the characterisation of global hyperbolicity in § 4 (see [19], [2]) we have an equivalent formulation:

Conjecture 2:
Generically, space-times of general relativity are *globally hyperbolic* (possess a Cauchy surface). What "generic" means has not been explained in [15]. As it seems to me to be hopeless to look for a natural measure on the space of space-times which would define "generic" by "except a set of measure zero". But as on many interesting subsets one can define a topology, let us propose the following

Definition: Let R be a topological space. A property "p" is called *generic* if $P := \{x \in R | p(x)\}$ contains an open dense subset. A property "q" is called *stable* at $x_o \in R$ if $\{x \in R | q(x)\}$ is a neighbourhood of x_o.

Remark: Stability and genericity are dual properties:

(i) If "p" is generic, "non-p" is nowhere stable
(Proof by contradiction: Let "non-p" stable at $x_o \in R$, then x_o cannot be an element of $\bar{P}$, but $\bar{P} = R$)

(ii) If "p" is generic, "stably-p" is also generic.
(P_o, the set of all $x \in R$ where p is stable, is the open interior of P, which by definition of "generic" must be dense: $\bar{P}_o = R$, hence represents also a generic property).

Unfortunately, in Euclidean spaces with Lebesgue measure there is only a loose connection: if "p" is generic or if "p" is stable, P cannot be of measure zero, but it can be of arbitrarily small measure.

Example: Let $\{q_n\}$ be a sequence which includes all rational numbers $\mathbb{Q}$, U_n be the neighbourhood of q_n with diameter $\varepsilon \cdot 2^{-n}$ and "p" the property to lie in at least one of the U_n. p is generic but P has measure $< 2 \cdot \varepsilon$. But with respect to reasonable physical properties described by a set of parameters εR^n, I do not know any example where, for generic "p", "non-p" has measure > 0.

In [20] Simpson & Penrose studied electromagnetic test fields on the Reißner-Nordström electrovacuum and found by numerical calculations that they blow up along the inner horizon H_i: (see fig. c in § 4). They conjecture that this is a quite general phenomenon. Now, as they point out several times, their argument is far from being a rigorous investigation even of this special space-time. Test fields are not solutions of the Einstein-Maxwell equations (not even of linearized ones), numerical values cannot prove the occurrence of a singularity, they might give a hint that the assumption (test field) breaks down. Even if one takes singularities at H_i for granted, they seem to be caused by the infinite blue-shift along any c-curve crossing H_i (events behind H_i have the complete asymptotic region for all times in their past) rather than by the timelike singularity, which cannot influence the region below H_i as the c-curves terminating on the "edge" I^+ are affinely complete. Now, infinite blue-shift occurs in space-time if the causal boundary is timelike at some ideal point at infinity, but not necessarily, if the ideal point is at the singularity (counter-examples are (ii) in § 10). But the basic idea of [1] is much more convincing than this special example: If a singularity occurs, the hyperbolic field equations will propagate irregularities along the characteristics (the light cones), if the singularities are not very specially arranged (e.g.: consider the singular focus of incoming radiation and then take the time reflection of it). Such a conjecture could be in principle decided by investigation of nonlinear hyperbolic differential equations. One striking cosmological result one would get if SCCH was settled this way: The big bang is a specially arranged singularity, otherwise the future of it, the whole cosmos, would be interspersed with singularities. Some results of Lifshitz and coll. [1] would support this view: they claim to have found a general solution with a quite special singularity: it can be described by spacelike hypersurface {t = 0} (more precisely: an open set of solutions in the space of spacetimes); but as their solutions even in the regular region have a special geometrical feature: there exists a congruence of curves whose principal directions of shear are Lie-shifted with respect to this congruence, it seems to me doubtful whether their methods (power series expansion; counting of arbitrary functions for showing generality) are really reliable.

Finally, let us consider what validity of conjecture 2: SCCH (at least for some classes space-times) yields for the problems of the last section:

(a) The question for which space-times SCCH is valid, can be asked for a much wider class of space-times, as for stably causal spaces, the property of an ideal point to be time-like is well defined, while the formulation of CCH requires the specially arranged asymptotically flat and empty region $\mathscr{I}^+$ which does not occur for a collapsing body imbedded in a cosmos.

(b) If SCCH is valid for some space V, CCH can be only marginally violated: the event horizon might be singular (see example (i) in § 10).

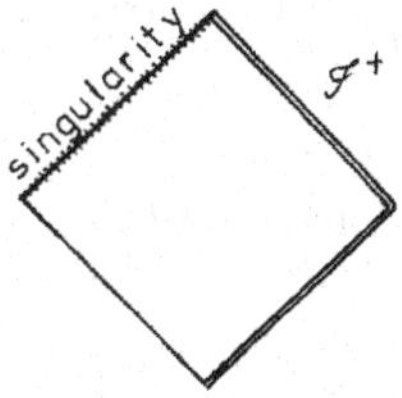

(c) If conjecture 2 holds in the space of space-times, conjecture 1e also must hold there (with the possibility of margial violation as discussed in (b)).

Proof: (compare the three requirements (i, ii, iii) for nocuity in notion 3)

(i) The property of a space-time of being globally hyperbolic is *generic*, hence the property of a singularity of being a non-Cauchy-singularity is nowhere stable, hence innocuous (see statement (i) in the remark in this section).

(ii) According to a theorem of Clarke [4]:

Let p be some singular point of the causal boundary of $D^+(S)$ such that along any c-curve terminating on p the p.p. frame components are bounded, then either D^+ is *special* at p (there exists a group of boosts such that $R^a{}_{bcd}$ is invariant under this group) or space-time can be extended beyond p, hence p is not a singular point,

the property of a Cauchy singularity *not* to be a *curvature* singularity is nowhere stable, hence innocuous.

(iii) If the property of singularities to be curvature- and Cauchy-singularities is generic (result of (i, ii), the stability of this property is also generic (statement (ii) in the remark in this section. ■

Acknowledgements

The main part of this lecture is based on the collaboration of Henning Müller zum Hagen, Peter Yodzis, and myself in Hamburg 1973/4. We also had discussed these matters with many colleagues; especially helpful had been conversations of some of us with Stephen Hawking, Andrew King, Wolfgang Kundt, Roger Penrose, Bernd Schmidt, and Nick Woodhouse.

References:

[1] Belinski, V.A.; Khalatnikov, I.M.; Lifshitz, E.M.: *Adv. in Phys.* 19, 523-573 (1970)

[2] Budic, R.; Sachs, R.K. *J. Math. Phys.* 15, 1302-9 (1974)

[3] Clarke, C.J.S. *GRG Journ.* 6, 35-40 (1975)

[4] Clarke, C.J.S. *Commun. Math. Phys.* 41, 65-78 (1975)

[5] Ellis, G.F.R.; King, A.R.: *Commun. Math. Phys.* 38, 119-156 (1974)

[6] Geroch, R.P.: *"Singularities in the Space-Time of General Relativity"* Ph. D. Thesis (Princeton 1967)

[7] Geroch, R.P.; Kronheimer, E.H.; Penrose, R.: *Proc. Roy. Soc. London A* 327, 545-67 (1972)

[8] Hawking, S.W.: *Commun. Math. Phys.* 43, 199-220 (1975)

[9] Hawking, S.W.; Ellis, G.F.R.:
"The Large Scale Structure of Space-Time"
Cambridge: At the University Press 1973

[10] Hawking, S.W.; Sachs, R.K.:
Commun. Math. Phys. 35, 287-296 (1974)

[11] Kronheimer, E.H.; Penrose, R.
Proc. Camb. Phil. Soc. 63, 481-501 (1967)

[12] Kundt, W.:
Zs. f. Phys. 172, 488/89 (1963)

[13] Kundt, W.:
in: Recent Developments in General Relativity (Infeld-Festschrift), Pergamon Oxford, PWN Warsaw (1962)

[14] Misner, C.W.; Thorne, K.S.; Wheeler, J.A.:
"Gravitation"
Freeman San Francisco (1973)

[15] Penrose, R.:
"Gravitational Collapse"
in De Witt-Morette, C. (Ed.): Gravitational Radiation and Gravitational Collapse, Proc. I.A.U. meetings No. 64, Warsaw 1973

[16] Schmidt, B.G.
GRG Journ. 1, 269-80 (1971)

[17] Schmidt, B.G.
Commun. Math. Phys. 36, 73-90 (1974)

[18] Seifert, H.J.:
"Kausale Lorentzräume"
Ph. D. Thesis (Hamburg 1968)

[19] Seifert, H.J.:
GRG Journ. 1, 247-59 (1971)

[20] Simpson, M.; Penrose, R.:
Int. Journ. Theor. Phys. 7, 183-97 (1973)

[21] Störmer, O.:
"Die Struktur der Bündelmetrik der ebenen Gravitationswellen"
Ph. D. Thesis (Hamburg 1971)

[22] Winicour, J.; Janis, A.I.; Newman, E.T.:
Phys. Rev. 176, 1507-13 (1968)

[23] Yodzis, P.; Seifert, H.J.; Müller zum Hagen, H.:
Commun. Math. Phys. 34, 135-148 (1973); 37, 29-40 (1974)

TOWARDS QUANTUM GRAVITY

Cecile DeWitt-Morette

Department of Astronomy and Center for Relativity
University of Texas, Austin, Texas 78712

Gravitation is the most difficult physical system to quantize. Methods which have been used to quantize other physical systems have to be re-expressed from a more fundamental point of view before they can be generalized to quantize Einstein's theory of relativity. For this reason I am very grateful to Professor Bleuler for having included quantum gravity as one of the topics of this conference. Indeed many of the contributions presented here have a bearing on the problem I shall discuss.

The quantum theory of gravity has been approached from two directions which have become known as "canonical quantization" and "covariant quantization" respectively. A significant progress in canonical quantization was made by Dirac [1] who generalized hamiltonian dynamics to include fields with constraints. Dirac's generalized hamiltonian dynamics was used to quantize the gravitational field first by Schild and Pirani [2]. The canonical quantization has been further developed by Arnowitt, Deser and Misner [3]. Covariant quantization is a formal offspring of the Feynman functional integral formalism; it has been investigated primarily by Feynman, DeWitt and Faddeev. References can be found in three articles by DeWitt [4]. To this day, it is not known whether the two types of quantization lead to the same theory.

1°) Can we develop the Feynman formalism beyond a formal procedure?

2°) Can we use the Feynman formalism to obtain the constraints of the covariant formalism? These questions have not been answered even for models with a finite number of degrees of freedom which have the same type of stumbling blocks one encounters when attempting to quantize the gravitation field.

All properties of a quantized physical system - expectation values, energy levels, S-matrix, decay rates - can be inferred from the knowledge

of a suitable set of transition amplitudes for the system. The most challenging expression that has ever been written down to express a transition amplitude is that of the Feynman path integral. Let C^n be the configuration space of a system S with n degrees of freedom; let $a, b \in C^n$, we want to find the probability for the transition $A \to B$ where $A = (a,t_a)$ and $B = (b,t_b)$.
Analysing the physical characteristics of quantum effects, Feynman came to the conclusion that the probability for the transition $A \to B$ can be expressed by

$$|K(B;A)|^2 = |\text{"}\textstyle\sum\text{"} \exp \frac{i}{\hbar} S(f) \;|^2 \quad \text{with } f \in F$$

where S is the action of the system, F is the space of all continuous paths mapping the time interval $T = (t_a,t_b)$ into C^n such that $f(t_a) = a$ and $f(t_b) = b$, and "$\sum$" is the "sum over all paths". What is the mathematical meaning of this "sum over all paths." How can it be computed? To answer this question I need to bring together so much analysis, geometry and physics that I shall use the following diagram to discuss each aspect of the question in terms which are, I hope, mathematically simple and physically meaningful.

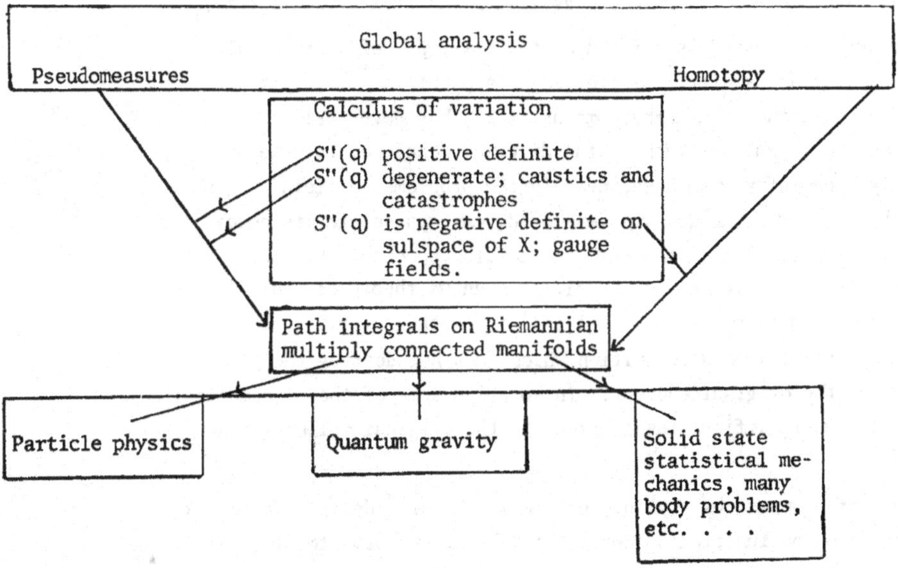

First of all, we note that path integrals are more than solutions of partial differential equations, they state the problem in different terms: In the Feynman formalism of quantum physics, the data necessary to analyze a physical system are:

- the space F of all possible, non equivalent paths (histories) f of the system between some fixed end points $a,b \in C^n$

- a function S on F called the action of the system

In this formalism, the global properties of the configuration space take their right place; they , indeed, contain much information on the generic properties of the system; nevertheless they have received little attention. Why? Possibly because global problems are difficult, but also because dynamical laws have, since Newton, been stated as differential equations and until recently been investigated locally.

The Feynman formalism is the only global formalism of physics. This is indeed its essential feature. For this reason, it is the formalism best suited to quantize gravitation which is basically a global theory. Although much progress can be done with minor modifications of quantum field theories designed for flat space time, some fundamental problems can only be solved in the framework of a global formalism of quantum physics.

Unfortunately, the original definition of path integrals blurs the global aspect of the Feynman formalism. There the time interval is divided in p+1 small intervals, $t_a<t_1<\dots<t_p<t_b$, the path f is replaced by a broken classical path, consisting of the classical paths from $f(t_k)$ to $f(t_{k+1})$. The "sum over all paths" is defined as the limit $p\to\infty$ of an integral over the np variables $\{f(t_i)\}$, treated as a point in R^{pn}. This definition, undefined and cumbersome as it is, has nevertheless yielded some tremendous results - such as the diagram technique, to name only one. Only those of us who have investigated problems in field theory before the diagram technique was developed, can appreciate its power. We may also regret that in the newly discovered power of this technique, some investigations, for example, Heitler quantum theory of radiation damping have been swept away - But this is another story.
I shall now propose a new definition which does not destroy the global character of the original concept. In so doing some of the mathematical and computational difficulties inherent to the original definition are bypassed.

The theory of promeasures (cylindrical measures) provides the framework for integration on function spaces. Let X be a Hausdorff topological vector space locally convex. A promeasure is a family of bounded measures defined on a family of finite dimensional spaces suitably related to X, satisfying some coherence conditions. The restriction to bounded measures makes it impossible to use the theory of promeasures for Feynman integration. However there is a one to one correspondence between the set of promeasures on X and their Fourier transforms on its dual X'. One can

thus define a promeasure by its Fourier transform and states the coherence conditions as conditions satisfied by the Fourier transforms.
At this point it is possible to remove the restriction to bounded measures and to generalize the concept of promeasures: Indeed the Fourier transforms of measures, considered as distributions of order zero, are defined for all measures, bounded or not. This new concept, given for convenience a name "pseudomeasure"+) and a symbol w enters our work only by its Fourier transform.

Which pseudomeasures are needed to express Feynman path integrals in a form convenient for their computation? The second variation of the action dominates quantum physics, very much like the first variation dominates classical physics. Hence, we shall expand

$$S(f) = S(q) + \frac{1}{2} S''(q)xx + \Sigma(q,x) \qquad\qquad S'(q) = o$$

where q is a classical path from a to b. Systems of physical interests: harmonic oscillator , a particle with spin, systems of indistinguishable particles, prototypes for the gravitational field etc. are systems above configuration space is a Riemannian manifold, usually multiply connected. Hence we do not have $f = q + x$. Instead we must consider a one parameter variation $\bar{\alpha}(u)$ of q such that $\bar{\alpha}(0) = q$, $\bar{\alpha}(1) = f$. Set $\bar{\alpha}(u)(t) = \alpha(u,t)$ then $\partial\alpha(u=0,t)/\partial u = x(t)$.
We shall first consider the case where $S''(q)$ is positive definite, and later on remove this restriction.

1°) $S''(q)$ is positive defininte; $f: T \to N \subset C^n$

f maps T into a geodesically convex neighborhood of C^n:$a,b \in N$ implies that there is a unique classical path from a to b which lies entirely in N.
A natural pseudomeasure for path integrals for a system S is w defined by

$$Fw = w(X) \exp(-iW/2)$$

i.e. w is a gaussian complex pseudomeasure. W is a quadratic form on the dual X' of X

$$\langle \mu,x\rangle = \int_T x^\alpha(t)\, d\mu_\alpha(t) \qquad\qquad \mu\varepsilon X',\ x\varepsilon X$$

$$W(\mu) = \underline{W}(\mu,\mu) = \int_T d\mu_\alpha(r) \int_T d\nu_\beta(s)\, G^{\alpha\beta}(r,s)$$

where G(r,s) is the Feynman Green function, i.e. the green function of the small disturbance operator that vanishes on the boundary: r or s equal to t_a or t_b.

+) Note added in proof: P. Kree uses the name "prodistribution".

$$w(X) = \left(\det M(t_a,t_b)\right)^{\frac{1}{2}} / (2\pi i)^{n/2}$$

where $M(t_a,t_b)$ is the Van Vleck matrix, the off diagonal block of the hessian of the action function $\tilde{S}(a,b) = S(q)$:

$$M(t_a,t_b) = -\partial^2\tilde{S}(a,b) / \partial a \partial b$$

With this pseudomeasure the probability amplitude is

$$K(B;A) = \exp \frac{i}{\hbar} S(q) \int_X \exp \frac{i}{\hbar} \Sigma(q,x)\, dw(x)$$

If f is replaced by a broken-classical path, $\Sigma(q,x)$ becomes a cylindrical function; i.e. a function of the (np)-tuple $\left\{x^\alpha(t_i)\right\}$. The integral over X can then be reexpressed as an integral over R^{pn}. If the configuration space is R^n, the limit of this integral when $p \to \infty$ is Feynman's original definition of a path integral.

w was called a "natural" pseudomeasure because it can be obtained from the canonical gaussian pseudomeasure on the Hilbert space of square integral functions:

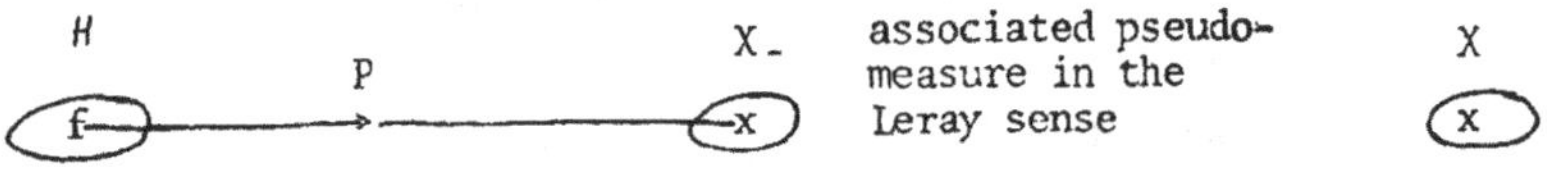

$P: H \to X_-$ by $x(t) = K(t,t_b) \int_{t_a}^{t} N(t_b,s)\, f(s)\, ds$

The Fourier transform of the canonical gaussian on H is $\exp(-i||f||^2/2)$ its image on X_- induced by P is $\mathcal{F}w_- = w_-(X) \exp(-iW_-/2)$ where

$$w_-(X) = \left(\det K(t_b,t_b) / \det K(t_a,t_b)\right)^{\frac{1}{2}}$$

$$G_-(r,s) = Y(s-r)\, J(r,t_a)\, \tilde{N}(t_a,t_b)\, \tilde{K}(t_b,s) + r \leftrightarrow s$$

Its associated pseudomeasure on $X \subset X_-$, in the Leray sense, is w.

2°) $S''(q)$ is degenerate; a is conjugate to b along the geodesic q.

The det $M(t_a,t_b)$ is infinite. The previous scheme breaks down because $K(B;A)$ is the solution to a Dirichlet problem; indeed when a and b are conjugate, Dirichlet boundary conditions do not characterize Jacobi fields. If we go back to the space X_-, we can compute $K(B';A)$ where $B' = (b',t_b)$ such that b' is an arbitrary point, while keeping the reference path q fixed; the limit $K(B';A)$ when $b' \to b$ gives the value of $K(B;A)$ on the caustic.

3°) $S''(q)$ is negative definite on a subspace of X; $f: T \to M$ where M is multiply corrected.

Its fundamental group Π is given by Morse theory.

The propagator

$$|K(B;A)| = |\sum_{\alpha\in\Pi} X(\alpha)\ K_\alpha\ (B;A)\ |$$

where $\{X(\alpha);\ \alpha\in\Pi\}$ is a character of the fundamental group. One can obtain in particular by this method the propagator for a particle with spin, for a system of indistinguishable particles [5], etc...

This scheme for defining and computing Feynman path integrals works out beautifully in many problems and has been used to obtain analytical non singular answers in a number of cases where this was thought to be impossible - the anharmonic oscillator,[6] an electron gas in a random potential,[7] etc. . .

But, it remains to apply the acid test of quantization, namely the quantization of a free particle on a curved space:

$$S(f) = \frac{1}{2} \int_T g_{\alpha\beta} \left(f(r)\right) \dot{f}^\alpha\ (t)\ \dot{f}^\beta\ (t)\ dt$$

$K(B;A)$ can be obtained by expanding $\exp \frac{i}{\hbar}\ \Sigma(q,x)$. This calculation considered as one of the very difficult ones, (called, among specialists, "the two-loop contributions") has not yet been completed unambiguously. It would be straightforward in the formalism presented here if it were not for an unwarranted and ambiguous change of order of integration over X and over **T**.

Indeed, $\sum(q,x)$ includes terms such as

$$I = \int_X dw(x) \int_T R_{\alpha\beta\gamma\delta}\ q(t)\ \langle \delta_t^\alpha, x\rangle \langle \delta_t^\beta, x\rangle \langle \dot{\delta}_t^\gamma, x\rangle \langle\ \delta_t^\delta, x\rangle\ dt$$

The Fourier transform Fw of w is defined on X' which is the space of distributions of order zero and not the space of distributions of higher order such as $\dot{\delta}$, hence we cannot integrate $\dot{x}(t)$ over X, we have to integrate it over T first. Such terms would not appear if we had integrated over the space of paths in phase space, paths mapping the time interval T into the cotangent bundle of the configuration space C^n. Preliminary calculations indicate that the pseudomeasures which can be defined for Feynman integrals over the space of paths in phase space are expressible as product of pseudomeasures one over momentum space, one over configuration space only in the simple cases where path quantization on configuration space is adequate. In such cases, integration over momentum space contributes the normalisation factor $w(X)$ and integration over configuration space gives the remaining terms. The case of a free particle on a curved space is precisely a case where the pseudomeasure is <u>not</u> expressible as a product of pseudomeasures.

The mathematical motivation for setting up the problem in phase space is paralled by a physical motivation: Working on the configuration space rather than on its cotangent bundle prevents us from incorporating fully the uncertainly principle in the formalism. We speak of probability transition from (a,t_a) to (b,t_b) where a and b are points in C^n. This is possible only if, simultaneously, we say that at t_a and t_b the momentum is completely undetermined. Between t_a and t_b, the wave packet spreads and then contracts; in a loose sense, the precision with which the momentum can be known increases as the wave packet spreads and decreases as it contracts. In some cases this correlation cannot be ignored; they are the cases where the pseudomeasure on phase space does not decouple into a pseudomeasure on configuration space and a pseudomeasure on momentum space.

References

[1] Dirac P.A.M. Can. J. Math. 2 129 (1950)

[2] Pirani F.A.E. and Schild A. Phys. Rev. 79 986 (1950)

[3] Arnowitt R., Deser S., and Misner C.W. "The Dynamics of General Relativity" in Gravitation Ed. by Witten L.John Wiley and Sons 1962

[4] DeWitt B.S. Phys. Rev. 160 1113-1148; 162 1195-1239; 162 1239-1256 (1967)

[5] Laidlaw, M.G.G., DeWitt-Morette, C.: Phys. Rev. D3, 1375-1378 (1971)

[6] Mizrahi, M.M. Ph. D Thesis, University of Texas at Austin 1975

[7] Maheshwari, A. XIIth Winter School of Theoretical Physics in Karpacz. Acta Universitatis Wratislaviensis 1975.

For more details on the matters presented in this talk see

DeWitt-Morette, C. Commun. Math. Phys. 28 47-67 (1972) and 37 63-81 (1974)

DeWitt-Morette, C. The Semi-Classical Expansion, Preprint.

REMARKS ABOUT DIRAC'S IDEA OF COSMOLOGICAL VARIATION OF SO CALLED 'CONSTANTS OF NATURE'

by

Pascual Jordan

§ 1. It is well known that since 1937 Dirac put forward the idea that several "Constants" of Nature, as we usually call them, might be in reality slowly varying functions of the age of the universe. This daring idea seems to a certain degree natural in the frame of a Friedmann cosmos where we have a naturally defined time coordinate together with a homogeneous space. For instance the quotient of the electrical and the gravitational attraction of the two parts of the hydrogen atom may be approximately (though only in a poor approximation) proportinal with the age A of the universe, making Newton's gravitational constant G roughly inversely proportional with this age:

$$G \sim A^{-1} .$$

And the number of nucleons in the universe may, according to Dirac, be approximately proportional with the square of the age: $N \sim A^2$. Occasionally Dirac remarked that also the famous fine structure constant α might be variable, being perhaps inversely proportional to the logarithm of the age:

$$\alpha \sim (\log A)^{-1}$$

My first remark is that probably there is a possibility to make an empirical test about this point. The frequency differences of optical spectroscopic multiplett terms are theoretically proportional with a power of the fine structure constant; therefore any variability of this fine structure constant in the course of very long time may be detectable in the spectra of remote, extremely distant galaxies.

When several years ago I mentioned for the first time this point, I had the impression that from these spectra a real, absolute constancy of the fine structure constant might be concluded. Just a few weeks ago I was not convinced that that first impression was justified and I thought, that perhaps in the contrary a confirmation of Dirac's idea might be derived by close and careful inspection of the spectra of remote quasars.

And in the meantime Abdus Salam found a theoretical basis given by the theory of renormalisation for a relation between the gravitational constant G and the fine structure constant $\alpha \approx 1/137$.

His result is

$$\alpha \cdot \log \frac{e^2}{Gm_e^2} = \frac{105}{137}$$

with m_e = mass of the electron. Deriving this result, Salam made use of Einsteins general theory of relativity and of the Schwarzschild solution of Einsteins field equations. Now, if Dirac's gravitational hypothesis is correct, we have to replace Einstein's "tensorial" theory of gravitation by a modified "tensorial-scalar" theory, formulated by Y. Thiry and Bergmann-Einstein and myself and Brans and Dicke. The Schwarzschild solution then is to be replaced by what I called the Heckmann solution. Therefore the derivation of Salam's result is to be modified a little, but perhaps not principally. The result of Salam therefore remains perhaps a strong argument of a connection between α and G, making with G also α a function of the age of the universe.

But such inspection has to a certain extent already been done by Bahcall and Schmidt, in agreement with my mentioned old impression they concluded from the measurement of a dublett line in the spectrum of O^{++} that a variability of α with time is not to be found. I learned this from an article of F.J. Dyson in the book "Aspects of Quantum Theory", edited by A. Salam and E. Wigner (Cambridge 1972). This article brings forward also other arguments which are unfavorable for Dirac's idea. This is the first time since 1937 that I myself feel a little unsure about Dirac's hypothesis.

§ 2. Dirac's hypothesis

$$\dot{G}/G < 0$$

is not yet tested empirically. Shapiro from his admirable radio echos from planets concluded that the variation of G, if real, must be surely slow, so that

$$-\dot{G}/G < 4.10^{-10} \text{ per year.}$$

Extremely exact measurements of the motion of the Moon may allow a decision about the variability or constancy of G.

But it has been pronounced that the method of determination of the motion of the Moon from measuring with atom clocks the occultations of distant stars by the Moon cannot allow success because the Moon is too strongly deviating from a mathematically exact sphere. The best hope is probably given by the fact, that by rockets and astronauts three laser reflectors have been put on the Moon. Their distances from terrestrial laboratories can be controlled with a precision making errors of one meter already impossible. From this and other measurements being now in progress Dirac hopes that a decision about his gravitational hypothesis might be expected already in one of the next future years.

If G really is decreasing, then our Earth must be expanding slowly - and I am convinced that that is really to be inferred from modern results of oceanography. I discussed this matter in my contribution to the Dirac-Symposium at Trieste.

§ 3. Concerning the increase of the number of nucleons, Dirac discussed recently two different more special hypotheses. The first of these is similar to what the now discarded cosmological steady state theory stated hypothetically: The vacuum itself may have a small amount of radioactivity giving rise to a small production probably of hydrogen in the intergalactic space. According to steady state theory this production ought to be great enough in order to make the mean mass density of the universe constant in time; but according to Dirac this production might be considerably smaller, the mean mass density in the universe being according to Dirac inversely proportional to the age of the universe. (Then $\rho \sim A^{-1}$).

The second hypothesis discussed by Dirac says, that production of new matter in the universe may occur there, where matter is already present. In this case the single stars must have themselves increasing mass; and the increasing masses of the Moon and the Earth would cause that the distance between these must decrease - though with constant masses of these two bodies a decreasing gravitational constant would give an increasing orbital radius of the Moon around the Earth. Also this question would be to be answered by an extremely precise measurement of the motion of the Moon.

§ 4. It is well known that Wheeler in a very suggestive and inspiring manner put forward the idea that the real manifold of space and time might show also topological deviations from the classical theory: There may exist "wormholes" of space. I do not believe in Wheeler's idea that such wormholes may exist in the frame of microphysics. For long ago my late friend Pauli criticised attempts to formulate a continuum theory of the electron, talking about an electrical field also in the interior of the electron. Pauli said that such a field would be purely fictitious because there would be no physical possibility to measure it. In the same manner the metric in the interior of wormholes would be fictitious if the dimensions of the wormholes would be microphysical. Wormholes, if they exist, must be of at least macrophysical or of astrophysical dimensions. Therefore I should prefer to believe that only one quasi-empirical fact can be discussed in relation to wormholes: I think that Ambarzumian explosions - perhaps really occuring in empirical astrophysics - are to be interpreted in this manner. And probably the space in the interior of wormholes is identical with the zones where production of new matter is going on - what Ambarzumian calls "prestellar matter" may be in reality new matter still inclosed in wormholes.

In this manner the concept of Ambarzumian explosions may win a simple connection with Dirac's hypothesis of cosmological increase of matter. You may say that this is a connection between two extremely hypothetical ideas - but perhaps in spite of all doubts, partially well founded, this connection may contain a piece of truth.

Vol. 399: Functional Analysis and its Applications. Proceedings 1973. Edited by H. G. Garnir, K. R. Unni and J. H. Williamson. II, 584 pages. 1974.

Vol. 400: A Crash Course on Kleinian Groups. Proceedings 1974. Edited by L. Bers and I. Kra. VII, 130 pages. 1974.

Vol. 401: M. F. Atiyah, Elliptic Operators and Compact Groups. V, 93 pages. 1974.

Vol. 402: M. Waldschmidt, Nombres Transcendants. VIII, 277 pages. 1974.

Vol. 403: Combinatorial Mathematics. Proceedings 1972. Edited by D. A. Holton. VIII, 148 pages. 1974.

Vol. 404: Théorie du Potentiel et Analyse Harmonique. Edité par J. Faraut. V, 245 pages. 1974.

Vol. 405: K. J. Devlin and H. Johnsbråten, The Souslin Problem. VIII, 132 pages. 1974.

Vol. 406: Graphs and Combinatorics. Proceedings 1973. Edited by R. A. Bari and F. Harary. VIII, 355 pages. 1974.

Vol. 407: P. Berthelot, Cohomologie Cristalline des Schémas de Caracteristique p > o. II, 604 pages. 1974.

Vol. 408: J. Wermer, Potential Theory. VIII, 146 pages. 1974.

Vol. 409: Fonctions de Plusieurs Variables Complexes, Séminaire François Norguet 1970–1973. XIII, 612 pages. 1974.

Vol. 410: Séminaire Pierre Lelong (Analyse) Année 1972–1973. VI, 181 pages. 1974.

Vol. 411: Hypergraph Seminar. Ohio State University, 1972. Edited by C. Berge and D. Ray-Chaudhuri. IX, 287 pages. 1974.

Vol. 412: Classification of Algebraic Varieties and Compact Complex Manifolds. Proceedings 1974. Edited by H. Popp. V, 333 pages. 1974.

Vol. 413: M. Bruneau, Variation Totale d'une Fonction. XIV, 332 pages. 1974.

Vol. 414: T. Kambayashi, M. Miyanishi and M. Takeuchi, Unipotent Algebraic Groups. VI, 165 pages. 1974.

Vol. 415: Ordinary and Partial Differential Equations. Proceedings 1974. XVII, 447 pages. 1974.

Vol. 416: M. E. Taylor, Pseudo Differential Operators. IV, 155 pages. 1974.

Vol. 417: H. H. Keller, Differential Calculus in Locally Convex Spaces. XVI, 131 pages. 1974.

Vol. 418: Localization in Group Theory and Homotopy Theory and Related Topics. Battelle Seattle 1974 Seminar. Edited by P. J. Hilton. VI, 172 pages 1974.

Vol. 419: Topics in Analysis. Proceedings 1970. Edited by O. E. Lehto, I. S. Louhivaara, and R. H. Nevanlinna. XIII, 392 pages. 1974.

Vol. 420: Category Seminar. Proceedings 1972/73. Edited by G. M. Kelly. VI, 375 pages. 1974.

Vol. 421: V. Poénaru, Groupes Discrets. VI, 216 pages. 1974.

Vol. 422: J.-M. Lemaire, Algèbres Connexes et Homologie des Espaces de Lacets. XIV, 133 pages. 1974.

Vol. 423: S. S. Abhyankar and A. M. Sathaye, Geometric Theory of Algebraic Space Curves. XIV, 302 pages. 1974.

Vol. 424: L. Weiss and J. Wolfowitz, Maximum Probability Estimators and Related Topics. V, 106 pages. 1974.

Vol. 425: P. R. Chernoff and J. E. Marsden, Properties of Infinite Dimensional Hamiltonian Systems. IV, 160 pages. 1974.

Vol. 426: M. L. Silverstein, Symmetric Markov Processes. X, 287 pages. 1974.

Vol. 427: H. Omori, Infinite Dimensional Lie Transformation Groups. XII, 149 pages. 1974.

Vol. 428: Algebraic and Geometrical Methods in Topology, Proceedings 1973. Edited by L. F. McAuley. XI, 280 pages. 1974.

Vol. 429: L. Cohn, Analytic Theory of the Harish-Chandra C-Function. III, 154 pages. 1974.

Vol. 430: Constructive and Computational Methods for Differential and Integral Equations. Proceedings 1974. Edited by D. L. Colton and R. P. Gilbert. VII, 476 pages. 1974.

Vol. 431: Séminaire Bourbaki – vol. 1973/74. Exposés 436–452. IV, 347 pages. 1975.

Vol. 432: R. P. Pflug, Holomorphiegebiete, pseudokonvexe Gebiete und das Levi-Problem. VI, 210 Seiten. 1975.

Vol. 433: W. G. Faris, Self-Adjoint Operators. VII, 115 pages. 1975.

Vol. 434: P. Brenner, V. Thomée, and L. B. Wahlbin, Besov Spaces and Applications to Difference Methods for Initial Value Problems. II, 154 pages. 1975.

Vol. 435: C. F. Dunkl and D. E. Ramirez, Representations of Commutative Semitopological Semigroups. VI, 181 pages. 1975.

Vol. 436: L. Auslander and R. Tolimieri, Abelian Harmonic Analysis, Theta Functions and Function Algebras on a Nilmanifold. V, 99 pages. 1975.

Vol. 437: D. W. Masser, Elliptic Functions and Transcendence. XIV, 143 pages. 1975.

Vol. 438: Geometric Topology. Proceedings 1974. Edited by L. C. Glaser and T. B. Rushing. X, 459 pages. 1975.

Vol. 439: K. Ueno, Classification Theory of Algebraic Varieties and Compact Complex Spaces. XIX, 278 pages. 1975

Vol. 440: R. K. Getoor, Markov Processes: Ray Processes and Right Processes. V, 118 pages. 1975.

Vol. 441: N. Jacobson, PI-Algebras. An Introduction. V, 115 pages. 1975.

Vol. 442: C. H. Wilcox, Scattering Theory for the d'Alembert Equation in Exterior Domains. III, 184 pages. 1975.

Vol. 443: M. Lazard, Commutative Formal Groups. II, 236 pages. 1975.

Vol. 444: F. van Oystaeyen, Prime Spectra in Non-Commutative Algebra. V, 128 pages. 1975.

Vol. 445: Model Theory and Topoi. Edited by F. W. Lawvere, C. Maurer, and G. C. Wraith. III, 354 pages. 1975.

Vol. 446: Partial Differential Equations and Related Topics. Proceedings 1974. Edited by J. A. Goldstein. IV, 389 pages. 1975.

Vol. 447: S. Toledo, Tableau Systems for First Order Number Theory and Certain Higher Order Theories. III, 339 pages. 1975.

Vol. 448: Spectral Theory and Differential Equations. Proceedings 1974. Edited by W. N. Everitt. XII, 321 pages. 1975.

Vol. 449: Hyperfunctions and Theoretical Physics. Proceedings 1973. Edited by F. Pham. IV, 218 pages. 1975.

Vol. 450: Algebra and Logic. Proceedings 1974. Edited by J. N. Crossley. VIII, 307 pages. 1975.

Vol. 451: Probabilistic Methods in Differential Equations. Proceedings 1974. Edited by M. A. Pinsky. VII, 162 pages. 1975.

Vol. 452: Combinatorial Mathematics III. Proceedings 1974. Edited by Anne Penfold Street and W. D. Wallis. IX, 233 pages. 1975.

Vol. 453: Logic Colloquium. Symposium on Logic Held at Boston, 1972–73. Edited by R. Parikh. IV, 251 pages. 1975.

Vol. 454: J. Hirschfeld and W. H. Wheeler, Forcing, Arithmetic, Division Rings. VII, 266 pages. 1975.

Vol. 455: H. Kraft, Kommutative algebraische Gruppen und Ringe. III, 163 Seiten. 1975.

Vol. 456: R. M. Fossum, P. A. Griffith, and I. Reiten, Trivial Extensions of Abelian Categories. Homological Algebra of Trivial Extensions of Abelian Categories with Applications to Ring Theory. XI, 122 pages. 1975.

Vol. 457: Fractional Calculus and Its Applications. Proceedings 1974. Edited by B. Ross. VI, 381 pages. 1975.

Vol. 458: P. Walters, Ergodic Theory – Introductory Lectures. VI, 198 pages. 1975.

Vol. 459: Fourier Integral Operators and Partial Differential Equations. Proceedings 1974. Edited by J. Chazarain. VI, 372 pages. 1975.

Vol. 460: O. Loos, Jordan Pairs. XVI, 218 pages. 1975.

Vol. 461: Computational Mechanics. Proceedings 1974. Edited by J. T. Oden. VII, 328 pages. 1975.

Vol. 462: P. Gérardin, Construction de Séries Discrètes p-adiques. »Sur les séries discrètes non ramifiées des groupes réductifs déployés p-adiques«. III, 180 pages. 1975.

Vol. 463: H.-H. Kuo, Gaussian Measures in Banach Spaces. VI, 224 pages. 1975.

Vol. 464: C. Rockland, Hypoellipticity and Eigenvalue Asymptotics. III, 171 pages. 1975.

Vol. 465: Séminaire de Probabilités IX. Proceedings 1973/74. Edité par P. A. Meyer. IV, 589 pages. 1975.

Vol. 466: Non-Commutative Harmonic Analysis. Proceedings 1974. Edited by J. Carmona, J. Dixmier and M. Vergne. VI, 231 pages. 1975.

Vol. 467: M. R. Essén, The Cos $\pi\lambda$ Theorem. With a paper by Christer Borell. VII, 112 pages. 1975.

Vol. 468: Dynamical Systems – Warwick 1974. Proceedings 1973/74. Edited by A. Manning. X, 405 pages. 1975.

Vol. 469: E. Binz, Continuous Convergence on C(X). IX, 140 pages. 1975.

Vol. 470: R. Bowen, Equilibrium States and the Ergodic Theory of Anosov Diffeomorphisms. III, 108 pages. 1975.

Vol. 471: R. S. Hamilton, Harmonic Maps of Manifolds with Boundary. III, 168 pages. 1975.

Vol. 472: Probability-Winter School. Proceedings 1975. Edited by Z. Ciesielski, K. Urbanik, and W. A. Woyczyński. VI, 283 pages. 1975.

Vol. 473: D. Burghelea, R. Lashof, and M. Rothenberg, Groups of Automorphisms of Manifolds. (with an appendix by E. Pedersen) VII, 156 pages. 1975.

Vol. 474: Séminaire Pierre Lelong (Analyse) Année 1973/74. Edité par P. Lelong. VI, 182 pages. 1975.

Vol. 475: Répartition Modulo 1. Actes du Colloque de Marseille-Luminy, 4 au 7 Juin 1974. Edité par G. Rauzy. V, 258 pages. 1975. 1975.

Vol. 476: Modular Functions of One Variable IV. Proceedings 1972. Edited by B. J. Birch and W. Kuyk. V, 151 pages. 1975.

Vol. 477: Optimization and Optimal Control. Proceedings 1974. Edited by R. Bulirsch, W. Oettli, and J. Stoer. VII, 294 pages. 1975.

Vol. 478: G. Schober, Univalent Functions – Selected Topics. V, 200 pages. 1975.

Vol. 479: S. D. Fisher and J. W. Jerome, Minimum Norm Extremals in Function Spaces. With Applications to Classical and Modern Analysis. VIII, 209 pages. 1975.

Vol. 480: X. M. Fernique, J. P. Conze et J. Gani, Ecole d'Eté de Probabilités de Saint-Flour IV–1974. Edité par P.-L. Hennequin. XI, 293 pages. 1975.

Vol. 481: M. de Guzmán, Differentiation of Integrals in R^n. XII, 226 pages. 1975.

Vol. 482: Fonctions de Plusieurs Variables Complexes II. Séminaire François Norguet 1974–1975. IX, 367 pages. 1975.

Vol. 483: R. D. M. Accola, Riemann Surfaces, Theta Functions, and Abelian Automorphisms Groups. III, 105 pages. 1975.

Vol. 484: Differential Topology and Geometry. Proceedings 1974. Edited by G. P. Joubert, R. P. Moussu, and R. H. Roussarie. IX, 287 pages. 1975.

Vol. 485: J. Diestel, Geometry of Banach Spaces – Selected Topics. XI, 282 pages. 1975.

Vol. 486: S. Stratila and D. Voiculescu, Representations of AF-Algebras and of the Group U (∞). IX, 169 pages. 1975.

Vol. 487: H. M. Reimann und T. Rychener, Funktionen beschränkter mittlerer Oszillation. VI, 141 Seiten. 1975.

Vol. 488: Representations of Algebras, Ottawa 1974. Proceedings 1974. Edited by V. Dlab and P. Gabriel. XII, 378 pages. 1975.

Vol. 489: J. Bair and R. Fourneau, Etude Géométrique des Espaces Vectoriels. Une Introduction. VII, 185 pages. 1975.

Vol. 490: The Geometry of Metric and Linear Spaces. Proceedings 1974. Edited by L. M. Kelly. X, 244 pages. 1975.

Vol. 491: K. A. Broughan, Invariants for Real-Generated Uniform Topological and Algebraic Categories. X, 197 pages. 1975.

Vol. 492: Infinitary Logic: In Memoriam Carol Karp. Edited by D. W. Kueker. VI, 206 pages. 1975.

Vol. 493: F. W. Kamber and P. Tondeur, Foliated Bundles and Characteristic Classes. XIII, 208 pages. 1975.

Vol. 494: A Cornea and G. Licea. Order and Potential Resolvent Families of Kernels. IV, 154 pages. 1975.

Vol. 495: A. Kerber, Representations of Permutation Groups II. V, 175 pages. 1975.

Vol. 496: L. H. Hodgkin and V. P. Snaith, Topics in K-Theory. Two Independent Contributions. III, 294 pages. 1975.

Vol. 497: Analyse Harmonique sur les Groupes de Lie. Proceedings 1973–75. Edité par P. Eymard et al. VI, 710 pages. 1975.

Vol. 498: Model Theory and Algebra. A Memorial Tribute to Abraham Robinson. Edited by D. H. Saracino and V. B. Weispfenning. X, 463 pages. 1975.

Vol. 499: Logic Conference, Kiel 1974. Proceedings. Edited by G. H. Müller, A. Oberschelp, and K. Potthoff. V, 651 pages 1975.

Vol. 500: Proof Theory Symposion, Kiel 1974. Proceedings. Edited by J. Diller and G. H. Müller. VIII, 383 pages. 1975.

Vol. 501: Spline Functions, Karlsruhe 1975. Proceedings. Edited by K. Böhmer, G. Meinardus, and W. Schempp. VI, 421 pages. 1976.

Vol. 502: János Galambos, Representations of Real Numbers by Infinite Series. VI, 146 pages. 1976.

Vol. 503: Applications of Methods of Functional Analysis to Problems in Mechanics. Proceedings 1975. Edited by P. Germain and B. Nayroles. XIX, 531 pages. 1976.

Vol. 504: S. Lang and H. F. Trotter, Frobenius Distributions in GL_2-Extensions. III, 274 pages. 1976.

Vol. 505: Advances in Complex Function Theory. Proceedings 1973/74. Edited by W. E. Kirwan and L. Zalcman. VIII, 203 pages. 1976.

Vol. 506: Numerical Analysis, Dundee 1975. Proceedings. Edited by G. A. Watson. X, 201 pages. 1976.

Vol. 507: M. C. Reed, Abstract Non-Linear Wave Equations. VI, 128 pages. 1976.

Vol. 508: E. Seneta, Regularly Varying Functions. V, 112 pages. 1976.

Vol. 509: D. E. Blair, Contact Manifolds in Riemannian Geometry. VI, 146 pages. 1976.

Vol. 510: V. Poènaru, Singularités C^∞ en Présence de Symétrie. V, 174 pages. 1976.

Vol. 511: Séminaire de Probabilités X. Proceedings 1974/75. Edité par P. A. Meyer. VI, 593 pages. 1976.

Vol. 512: Spaces of Analytic Functions, Kristiansand, Norway 1975. Proceedings. Edited by O. B. Bekken, B. K. Øksendal, and A. Stray. VIII, 204 pages. 1976.

Vol. 513: R. B. Warfield, Jr. Nilpotent Groups. VIII, 115 pages. 1976.

Vol. 514: Séminaire Bourbaki vol. 1974/75. Exposés 453 – 470. IV, 276 pages. 1976.

Vol. 515: Bäcklund Transformations. Nashville, Tennessee 1974. Proceedings. Edited by R. M. Miura. VIII, 295 pages. 1976.